PRESBYOPIA RESEARCH

From Molecular Biology to
Visual Adaptation

PERSPECTIVES IN VISION RESEARCH

Series Editor: Colin Blakemore
University of Oxford
Oxford, England

Biochemistry of the Eye
Elaine R. Berman

Development of the Vertebrate Retina
Edited by Barbara L. Finlay and Dale R. Sengelaub

Parallel Processing in the Visual System
THE CLASSIFICATION OF RETINAL GANGLION CELLS AND ITS IMPACT ON THE NEUROBIOLOGY OF VISION
Jonathan Stone

Presbyopia Research
FROM MOLECULAR BIOLOGY TO VISUAL ADAPTATION
Edited by Gérard Obrecht and Lawrence W. Stark

A Continuation Order Plan is available for this series. A continuation order will bring delivery of each new volume immediately upon publication. Volumes are billed only upon actual shipment. For further information please contact the publisher.

PRESBYOPIA RESEARCH

From Molecular Biology to Visual Adaptation

Edited by

Gérard Obrecht

Essilor
Creteil, France

and

Lawrence W. Stark

University of California, Berkeley
Berkeley, California

PLENUM PRESS • NEW YORK AND LONDON

Library of Congress Cataloging in Publication Data

International Symposium on Presbyopia (4th: 1989: Marrakech, Morocco)
Presbyopia research: from molecular biology to visual adaptation / edited by Gérard Obrecht and Lawrence W. Stark.
p. cm. —(Perspectives in vision research)
"Based on the proceedings of the Fourth International Symposium on Presbyopia, held in Marrakesh, Morocco, on June 5–10, 1989"—T.p. verso.
Includes bibliographical references and index.
ISBN 0-306-43659-0
1. Presbyopia—Molecular aspects—Congresses. 2. Eye—Adaptation—Congresses. I. Obrecht, Gérard. II. Stark, Lawrence. III. Title. IV. Series.
[DNLM: 1. Presbyopia—congresses. WW 300 I61p 1989]
RE938.5.I57 1989
617.7′55—dc20
DNLM/DLC 91-21100
for Library of Congress CIP

Based on the proceedings of the Fourth International Symposium on Presbyopia, held in Marrakesh, Morocco, on June 5–10, 1989

ISBN 0-306-43659-0 

A Division of Plenum Publishing Corporation
233 Spring Street, New York, N.Y. 10013

Printed in the United States of America

Contributors

C. Arruti • Gerontology Research Unit, INSERM U118, 75016 Paris, France

O. A. Bateman • Laboratory of Molecular Biology, Department of Crystallography, Birkbeck College, London WC1E 7HX, Great Britain

B. Bax • Laboratory of Molecular Biology,Department of Crystallography, Birkbeck College, London WC1E 7HX, Great Britain

Ennio Lucio Benedetti • Institute Jacques Monod, CNRS, University of Paris VII, 75251 Paris, France

P. R. Blanquet • Gerontology Research Unit, INSERM U118, 75016 Paris, France

T. L. Blundell • Laboratory of Molecular Biology, Department of Crystallography, Birkbeck College, London WC1E 7HX, Great Britain

Bruce Bridgeman • Program in Experimental Psychology, University of California, Santa Cruz, California 95064

Ana B. Chepelinsky • Laboratory of Molecular and Developmental Biology, National Eye Institute, National Institutes of Health, Bethesda, Maryland 20892

Kenneth Ciuffreda • Department of Vision Sciences, State University of New York, State College of Optometry, New York, NY 10010

Valérie Cornilleau-Peres • Essilor International, Laboratory of Physiological Optics, 94000 Creteil, France, and Laboratory of Neurosensory Physiology, CNRS, 75270 Paris Cedex 06, France

Yves Courtois • Gerontology Research Unit, INSERM U118, 75016 Paris, France

Heiner Deubel • Max-Planck-Institute for Behavioral Physiology, D-8130 Seewiesen, Germany

H. Driessen • Laboratory of Molecular Biology, Department of Crystallography, Birkbeck College, London WC1E 7HX, Great Britain

Jacques Droulez • Laboratory of Neurosensory Physiology, CNRS, 75270 Paris Cedex 06, France

Robert A. Dubin • Laboratory of Molecular and Developmental Biology, National Eye Institute, National Institutes of Health, Bethesda, Maryland 20892

George Duncan • School of Biological Sciences, University of East Anglia, Norwich NR4 7TJ, Great Britain

Irène Dunia • Institute Jacques Monod, CNRS, University of Paris VII, 75251 Paris, France

Paul Erickson • School of Optometry, University of California, Berkeley, California 94720

John Findlay • Department of Psychology, University of Durham, Durham DH1 3LE, Great Britain

G. M. Gauthier • Sensorimotor Control Laboratory, University of Provence, 13397 Marseilles Cedex 13, France

Richard Haines • Research Institute for Advanced Computer Science, Ames Research Center–NASA, Moffett Field, California 94035

John J. Harding • Nuffield Laboratory of Ophthalmology, Oxford University, Walton Street, Oxford OX2 6AW, Great Britain

David Henson • Department of Optometry, University of Wales, Cardiff CF1 3XF, Great Britain

Ian P. Howard • Human Performance in Space Laboratory, York University, North York, Ontario, M3J 1P3, Canada

Michel Imbert • Département des Neurosciences de la Vision, Université Pierre et Marie Curie, 75005 Paris, France

Cynthia J. Jaworski • Laboratory of Molecular and Developmental Biology, National Eye Institute, National Institutes of Health, Bethesda, Maryland 20892

J. C. Jeanny • Gerontology Research Unit, INSERM U118, 75016 Paris, France

Stuart Judge • University Laboratory of Physiology, University of Oxford, Oxford OX1 3PT, Great Britain

R. Lapatto • Laboratory of Molecular Biology, Department of Crystallography, Birkbeck College, London WC1E 7HX, Great Britain

P. F. Lindley • Laboratory of Molecular Biology, Department of Crystallography, Birkbeck College, London WC1E 7HX, Great Britain

Julia M. Marcantonio • School of Biological Sciences, University of East Anglia, Norwich NR4 7TJ, Great Britain

F. Mascarelli • Gerontology Research Unit, INSERM U118, 75016 Paris, France

Joan B. McDermott • Laboratory of Molecular and Developmental Biology, National Eye Institute, National Institutes of Health, Bethesda, Maryland 20892

Susan A. Menditto • Department of Biomedical Engineering, Rutgers University, Piscataway, New Jersey 08855

Reuben S. Mezrich • Department of Biomedical Engineering, Rutgers University, and Laurie Imaging Center, New Brunswick Allied Hospitals, New Brunswick, New Jersey 08901

Frederick Miles • Laboratory of Sensorimotor Research, National Eye Institute, National Institutes of Health, Bethesda, Maryland 20892

D. S. Moss • Laboratory of Molecular Biology, Department of Crystallography, Birkbeck College, London WC1E 7HX, Great Britain

S. Najmudin • Laboratory of Molecular Biology, Department of Crystallography, Birkbeck College, London WC1E 7HX, Great Britain

An Nguyen • School of Optometry, University of California, Berkeley, California 94720

G. Obrecht • Essilor International, Laboratory of Physiological Optics, 94000 Creteil, France

Kevin O'Regan • Laboratory of Experimental Psychology, CNRS, Université René Descartes Paris V, 75006 Paris, France

C. Pedrono • Essilor International, Laboratory of Physiological Optics, 94000 Creteil, France

Joram Piatigorsky • Laboratory of Molecular and Developmental Biology, National Eye Institute, National Institutes of Health, Bethesda, Maryland 20892

Mark Rosenfield • Department of Vision Sciences, SUNY State College of Optometry, New York, New York 10010

Clifton Schor • School of Optometry, University of California, Berkeley, California 94720

John Semmlow • Department of Biomedical Engineering, Rutgers University, and Department of Surgery (Bioengineering), UMDNJ, Robert Wood Johnson Medical School, Piscataway, New Jersey 08855

Christine Slingsby • Laboratory of Molecular Biology, Department of Crystallography, Birkbeck College, London WC1E 7HX, Great Britain

Lawrence Stark • School of Optometry, University of California, Berkeley, California 94720

Annette Tardieu • Centre de Génétique Moléculaire, CNRS, 91198 Gif sur Yvette Cedex, France

Julie Tomlinson • School of Biological Sciences, University of East Anglia, Norwich NR4 7TJ, Great Britain

J. Treton • Gerontology Research Unit, INSERM U118, 75016 Paris, France

Christopher W. Tyler • Smith–Kettlewell Eye Research Institute, San Francisco, California 94115

Corina Vandepol • School of Optometry, University of California, Berkeley, California 94720

J. L. Vercher • Sensorimotor Control Laboratory, University of Provence, 13397 Marseilles Cedex 13, France

Francoise Vérétout • Centre de Génétique Moléculaire, CNRS, 91198 Gif sur Yvette Cedex, France

Eric F. Wawrousek • SmithKline and French Laboratories, Research and Development, King of Prussia, Pennsylvania 19406

Robert Weale • Age Concern Institute of Gerontology, King's College London, University of London, London SE1 8TX, Great Britain

Jeremy Wolfe • Department of Brain and Cognitive Sciences, Massachusetts Institute of Technology, Cambridge, Massachusetts 02139

Foreword

Seeing is life. Seeing is transforming luminous colored stimulations and shapes into a mental representation, structured in space and in time. But seeing is also opening onto the world that surrounds us: it is thus a means for communicating and learning.

Jean-Jacques Rousseau, a philosopher worth quoting during the bicentennial of the French Revolution of which he was an instigator, stated, "of all the senses, vision is that which can be the least readily separated from judgments of the mind."

Sight is increasingly called on in our modern world. Maturity is affected at about 40–45 years by the onset of presbyopia. At that age, which demands all our intellectual and physical means, our sight should be irreproachable. Our efficiency must not be diminished. In the world, one person out of four is a presbyope; nobody can escape from this state after entering the 40-year age group.

It is an enormous challenge to provide all presbyopes with natural, faithful vision. It is this challenge that has drawn us together in the marvelous imperial town of Marrakesh.

Since the days of the Arab scientist Alhazen, who discovered the role of the crystalline lens in approximately the year 1000, up to our time, at the dawn of the year 2000, attempts to improve sight have engendered a colossal amount of research, experiments, errors, and successes.

The first two symposia in 1977 and in 1981 were essentially professional; the third was predominantly scientific. This year, we sought to create a "Janus" symposium with a scientific congress on the one hand and professional workshops on the other. Indeed, we believe that Essilor's role lies at the interface between scientists and professionals. We wish to extend our academic and theoretical knowledge and also to complete and exchange our technical and professional experience to prepare corrective means for the future.

Numerous questions have yet to be answered, such as:

- Will it one day be possible to defer or stop the aging of the accommodative apparatus?
- Is further improvement of the current corrective means possible, whether spectacles or contact lenses?
- How are behavioral and psychological presbyope typologies to be integrated in the course of examination, prescription, and fitting with corrective aids?
- How can the various professional partners involved in vision correction render the best possible service when fitting patients and customers?

This Fourth symposium drew together participants from 22 nations and many varied professions. The numerous exchanges it allowed have certainly enabled us to further our knowledge and to come up with more satisfactory answers to these questions.

This publication, *Presbyopia Research: From Molecular Biology to Visual Adaptation*, presents the lectures given at the Scientific Congress of the Fourth Symposium in their entirety. We hope that this work will contribute toward increased knowledge about presbyopia.

Bernard Maitenaz
Essilor Managing Director

Creteil, France

Preface

Presbyopia is a challenge to the visual science community and an opportunity! The scientific committee considered the various research developments that are progressing worldwide related to presbyopia and decided on three themes.

Molecular biology is in a revolutionary state. Advances abound in genetics, biochemistry, molecular clocks in evolution, diagnostic and therapeutic genetic engineering applications, and especially deeper understanding of life processes. The lens has been the focus of intense studies, and the contributors to our conference were able to present a wide sampling of recent approaches to lens development and aging.

The ability of the visual and visual–motor systems to adapt enables the utilization of spectacles, contact lenses, and implanted lenses. The interaction of vision, accommodation, and vergence eye movements influences the adaptive processes. These areas formed exciting sessions with speakers providing both overviews of the current knowledge and descriptions of ingenious recent experimental findings.

If bifocal implanted lenses were to perform and supply an elegant, if invasive, solution for presbyopia, still we would be faced with a problem. Two images, from the near and from the far prescription, would be projected onto the retina. How are the retina, visual cortex, and other brain structures to separate these superimposed images? Two levels of processes could contribute. There are fixed, preattentive candidate mechanisms to make such distinctions. Presentations were made describing fundamental analyses on a variety of these visual functions. However, if the two images are both in focus and not otherwise distinguishable, then higher-level top-down cognitive–perceptual processes must be called on. Even these may fail, and the distraction in heads-up displays suggests a residual problem.

Accommodation could participate to separate simultaneously viewed images with different clear vision distances. It also plays a crucial visual–motor role in presbyopia. Our sessions concluded with recent researches concerning this final theme.

The scientific participants were very stimulated by their interaction with their own disciplinary colleagues and perhaps even more by the education they received from presentations of workers in unfamiliar areas. Here they were in the same boat as the general conference attendees who participated in clinical workshops but also were welcomed in and appreciated the visual science material incorporated into this volume. Indeed, the existence of this book is a tribute to the vision care professions and to their support by Essilor.

G. Obrecht
L. Stark

Creteil, France
Berkeley, California

Acknowledgment

We would like to thank all those persons who have contributed toward this publication: first and foremost the chairmen of the sessions, Y. Courtois, K. O'Regan, J. Findlay, L. Stark, and R. Weale, who grouped the lecturers to ensure coverage of the various scientific subjects. We would like to thank the authors of the chapters that constitute the basis of this book, our assistant Martine Dupont, and the translator Sally Jane Norman.

Our thanks also go to Essilor International, which bore the costs of this conference and allowed numerous researchers from various fields to meet and to confront their ideas on presbyopia.

G. Obrecht
L. Stark

Contents

Part I

MOLECULAR BIOLOGY OF THE LENS: NEW APPROACHES IN LENS DEVELOPMENT AND AGING

Part I

Introduction

YVES COURTOIS

THE LENS

Because it is an apparently very simple, easily isolated organ that plays an important role in vision, the lens has been the object of numerous studies. Nevertheless, research over the past few years concerning its structure, development, and composition has provided data of a more spectacular and important nature, mainly thanks to the new techniques of molecular biology or modern physics.

THE CRYSTALLINS: FOSSIL ENZYMES

The crystallins, the principal lens proteins, are distributed in man into three main families: the α-, β-, and γ-crystallins. Synthesis of these proteins varies during development and as a function of time, allowing their association in variably sequenced aggregates and the production of an organ with remarkable optical and elastic properties.

Thanks to progress in molecular biology, particularly that brought about by the team of Dr. Piatigorsky at the National Institutes of Health in Bethesda, most crystallins of the various species have been sequenced, and genetic analysis is currently underway. The first major surprise has been the fact that data bank comparison of these protein sequences has shown large regions that are also found in enzymes such as enolase or argininosuccinatelyase. This upsets our conceptions concerning structural proteins and evolution, especially in that crystallins have now been found in other organs such as the heart.

The second big surprise comes from the molecular biologists' capacity to "manufacture" transgenic mice (i.e., to induce integration into the murine genome of coding sequences for new proteins). By using the promotor part of the crystallin genes, Piatigorsky and his team have induced the expression of cancer genes within the lens. More than a mere curiosity, this finding demonstrates that this fantastic technique allows analysis of the basic mechanisms of genetic expression, with possible therapeutic applications.

Yves Courtois • Gerontology Research Unit, INSERM U118, 75016 Paris, France.

THE CRYSTALLINE LENS: AN ORGAN DEPENDENT ON ITS ENVIRONMENT

One cannot understand the perfect harmony characterizing crystalline lens development without posing questions about the messages that control this development. A number of these messages, derived from the retina, have recently been identified. These are the growth factors. Our laboratory has demonstrated the mitogen and stimulator role of protein synthesis by factors called "fibroblast growth factors." These are not the only signals exchanged by the intraocular organs. In 1987–1988, investigation in this context was begun for at least three new growth factors.

THE LENS: A WELL-STRUCTURED ORGAN

The extraordinary fiber stack constituting the heart of the lens is built on an architecture of microfibrillae that form a cytoskeleton, the components of which are associated with the membranes according to highly specific interactions. Although they do not assume a muscular-type contractile role, it is reasonable to attribute to these microfibrillae a role in lens elasticity or resistance to deformation in addition to their role of compartmentalizing nonmembrane constituents.

Lenses are aggregates with superstructures that physicists using diffraction techniques with various methods are finally able to analyze to explain the transparency of this organ. The three-dimensional structure of crystalline lens proteins can even be described, thanks to collaboration between molecular biologists and crystallography physicists using powerful computer means to obtain three-dimensional structural models of these molecules.

THE LENS: AN ORGAN THAT EVOLVES WITH AGE

Among the various biochemical causes leading to changes in transparency, one very important factor is the accumulation of calcium in the lens. Nonetheless, calcium, like the other ions, participates in the lens physiology and intercellular communication. Finally, a large number of posttranslational changes arise as a function of age and nutrition. A number of these changes, such as carbamylation, have been discovered only recently and seem to account for the higher incidence of cataracts in certain populations. These biochemical modifications observed as a function of age are manifest as increasing absorption of visible light and UV. The influence of this phenomenon on the lens itself and on light attaining the retina deserves close attention.

In conclusion, Part I of this volume will provide new, highly documented information on the formation, structure, and composition of the lens and its reaction to the environment. This will allow better understanding of the nature of the multiple mechanisms leading to aging of the lens, to presbyopia, and to cataracts.

1

Transcriptional Control of the α-Crystallin Gene Family

ANA B. CHEPELINSKY, ERIC F. WAWROUSEK, ROBERT A. DUBIN, CYNTHIA J. JAWORSKI, JOAN B. McDERMOTT, AND JORAM PIATIGORSKY

INTRODUCTION

Crystallins constitute 80–90% of the soluble protein of the ocular lens. There are a surprisingly large number of crystallin gene families; some (α- and βγ-crystallins) are present in all vertebrate lenses, whereas others (the enzyme–crystallins) are found only in the lenses of certain species. The enzyme-crystallins are not lens-specific and appear to have a double function, i.e., as structural proteins in the lens and as metabolic enzymes in other tissues (see Wistow and Piatigorsky, 1988; Piatigorsky and Wistow, 1989). In addition to their lens-specific or lens-preferred expression, the synthesis of the crystallin polypeptides is developmentally regulated in a temporal and spatial manner in the lens (see McAvoy, 1981; Piatigorsky, 1981).

The α-crystallin family consists of two similar, highly conserved genes (αA and αB) that are situated on different chromosomes (Kaye *et al.*, 1985; Quax-Jeuken *et al.*, 1985; Skow and Donner, 1985; Hawkins *et al.*, 1987; Ngo *et al.*, 1989). α-Crystallin is the first crystallin to appear during lens development of the mouse and is expressed in both the epithelia and fibers (Zwaan, 1983).

ANA B. CHEPELINSKY, ERIC F. WAWROUSEK, ROBERT A. DUBIN, CYNTHIA J. JAWORSKI, JOAN B. MCDERMOTT, AND JORAM PIATIGORSKY • Laboratory of Molecular and Developmental Biology, National Eye Institute, National Institutes of Health, Bethesda, Maryland 20892. Present address of E. F. W.: SmithKline and French Laboratories, Research and Development, King of Prussia, Pennsylvania 19406.

In order to understand the mechanisms involved in the regulation of expression of these two genes, their 5′ flanking and some noncoding sequences were analyzed in functional assays.

STRUCTURAL FEATURES

αA-Crystallin and αB-crystallin are evolutionarily related, exhibiting 56% homology at the amino acid level in the bovine lens (van der Ouderaa *et al.*, 1974); in addition, both crystallins are partially homologous to small heat shock proteins and to an egg antigen (p40) of the blood fluke *Schistosoma mansoni* (Ingolia and Craig, 1982; Wistow, 1985; de Jong and Hendriks, 1986; Nene *et al.*, 1986; de Jong *et al.*, 1988). The αA and αB polypeptides undergo posttranslational modifications (Spector *et al.*, 1985; Chiesa *et al.*, 1987; Voorter *et al.*, 1986) that might modulate the aggregation of α-crystallin subunits (Delcour and Papaconstantinou, 1974; Vermorken *et al.*, 1978) and their interaction with lens plasma membranes (Mulders *et al.*, 1985; Rameakers *et al.*, 1980).

The αA-crystallin gene has been cloned from mouse (King and Piatigorsky, 1983), hamster (van den Heuvel *et al.*, 1985), chicken (Thompson *et al.*, 1987), mole rat (Hendriks *et al.*, 1987), and human (McDevitt *et al.*, 1986; Jaworski and Piatigorsky, 1989); the αB-crystallin gene has been cloned from hamster (Quax-Jeuken *et al.*, 1985), mouse (Dubin *et al.*, 1989), and human (Dubin *et al.*, 1990). They show a highly conserved structure; for αA, exon 1 codes for amino acids 1–63 (1–67 in αB), exon 2 for amino acids 64–104 (68–108 in αB), and exon 3 for amino acids 105–173 (109–175 in αB). The αA-crystallin gene of several mammals codes for two polypeptides produced by alternative splicing, αA and αA^{ins}. αA^{ins} contains a 23-amino acid insert at amino acid 63 and has been detected in mice, rats, and related rodents (Cohen *et al.*, 1978a,b; King and Piatigorsky, 1983, 1984; van den Heuvel *et al.*, 1985) as well as in pika, bats, and hedgehogs (Hendriks *et al.*, 1988). The αA^{ins} protein has been detected in neither chicken nor humans. However, a comparison of human and mouse sequences reveals the presence of a human pseudoexon in this region, corresponding to the alternatively spliced insert exon (Jaworski and Piatigorsky, 1989). A single base deletion in the pseudoexon results in a frameshift mutation creating two stop codons in exon number 2 (see Fig. 1-1).

GENE EXPRESSION

When the levels of αA and αB mRNA were analyzed in different mouse tissues by Northern blot hybridization, αA mRNA was found only in the lens. While αB mRNA was also abundantly expressed in the lens, it was detected at lower levels in some other tissues. An approximately 900 base mRNA was observed in lens, heart, skeletal muscle, and kidney. A larger transcript was observed in lung and brain (see Fig. 1-2; Dubin *et al.*, 1989); further work demonstrated that the 900 base mRNA is also present in lung at very low levels (Dubin and Piatigorsky, unpublished). Although these two similar genes are closely related in structure, they clearly have very different patterns of expression.

CIS REGULATORY SEQUENCES OF THE MURINE αA-CRYSTALLIN GENE

To determine the location of the regulatory elements of the murine αA-crystallin promoter, 5′ flanking sequences of the murine αA-crystallin gene, containing 46 bp of exon 1, were introduced into the pSV0-CAT expression vector (Gorman *et al.*, 1982b) and tested for their ability to activate CAT (chloramphenicol acetyltransferase) gene expression in transient assays and in transgenic mice (Fig. 1-3).

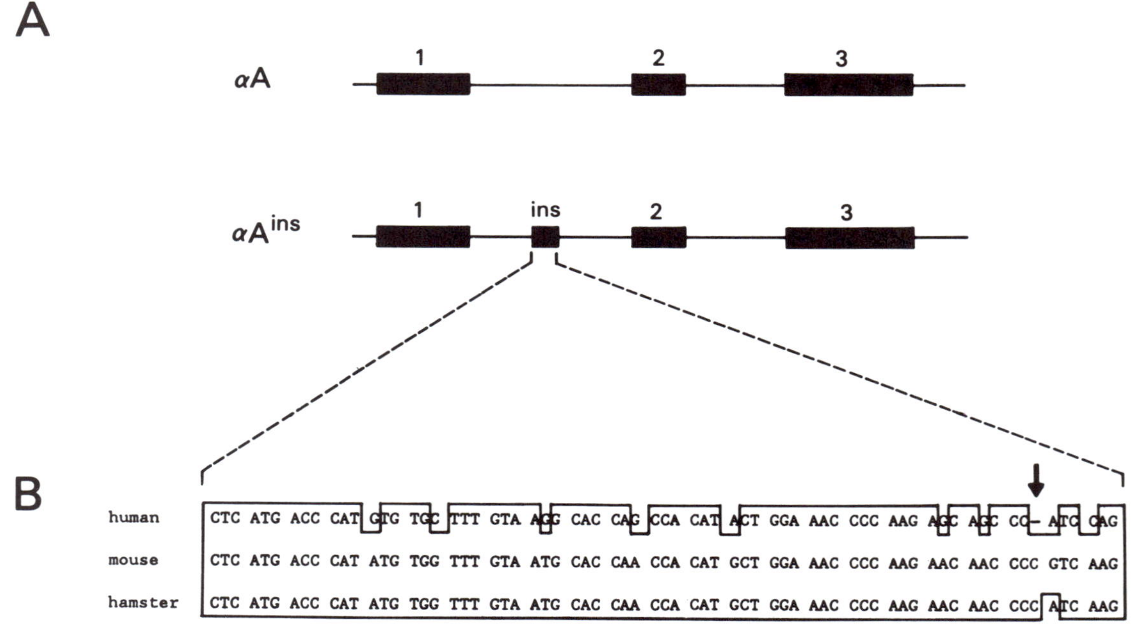

FIGURE 1-1. Structure of the αA-crystallin gene. Solid bars, exons; lines, introns or flanking sequences. A: Exons used in αA and αA^{ins}. B: Comparison of αA^{ins} sequence from mouse (King and Piatigorsky, 1983), hamster (van den Heuvel *et al.*, 1985), and human (Jaworski and Piatigorsky, 1989); arrow indicates base deletion that creates stop codons in exon 2.

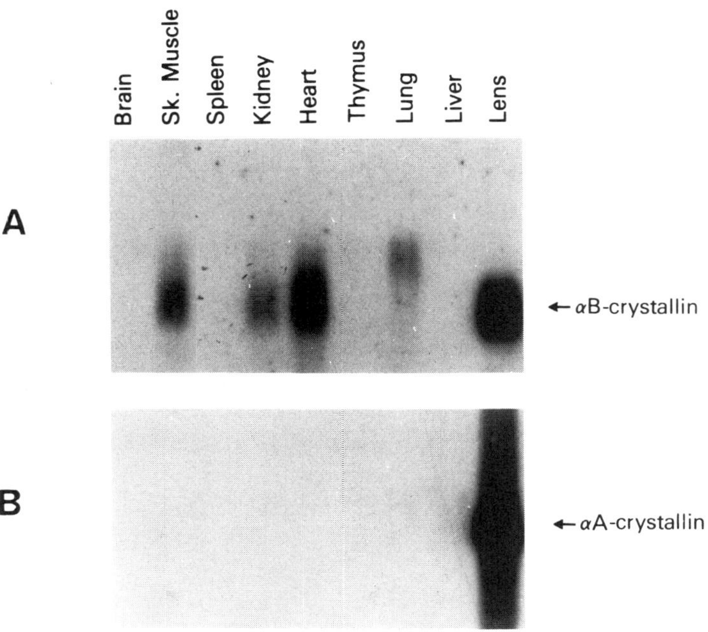

FIGURE 1-2. αA-Crystallin and αB-crystallin mRNA synthesis in the mouse. Northern blot hybridizations of mRNA isolated from different tissues of 2-week-old mice. A: Hybridization to mouse αB-crystallin probe. B: Hybridization to mouse αA-crystallin probe (Dubin *et al.*, 1989).

Transient Assays

The analysis of these hybrid genes in explanted chicken lens epithelia showed that sequences −111 to −60 were essential for promoter activity; −111/+46 contained an active promoter, and deletion of the sequence −88/−60 abolished promoter activity (Fig. 1-4). Sequence −111/−84 contained an element able to activate, in either orientation, the promoter element present in −88/+46 (Fig. 1-4B) (Chepelinsky *et al.*, 1985, 1987). Regulatory elements present in both the proximal (−88 to −55) and distal (−111 to −84) domains interact with chicken lens nuclear factors (Sommer *et al.*, 1988).

The plasmid constructs containing the −111/+46 and −88/+46 promoter fragments were active in a rabbit lens cell line (Reddan *et al.*, 1986) (Fig. 1-5). These plasmid constructs did not express the

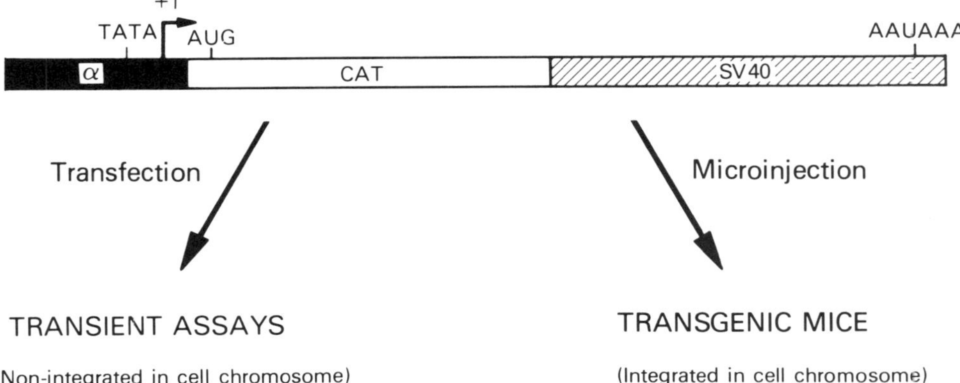

FIGURE 1-3. αA-Crystallin–CAT hybrid gene tested in transient assays and in transgenic mice. Solid bar, murine αA-crystallin sequence containing initiation site of transcription (+1) (Chepelinsky *et al.*, 1987); empty bar, CAT gene coding sequence containing translation initiation site (AUG); hatched bar, SV40 sequence containing RNA-splicing signals and polyadenylation signals (AAUAAA) (Gorman *et al.*, 1982b).

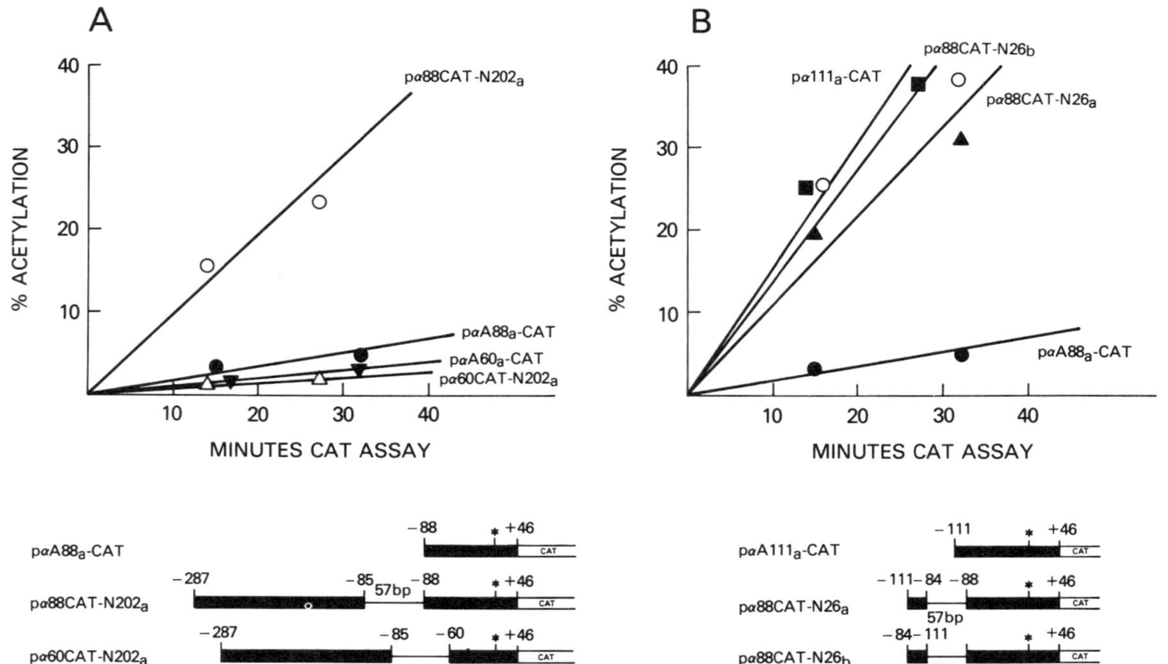

FIGURE 1-4. Dissection of the murine αA-crystallin promoter in transient assays in explanted embryonic chicken lens epithelia. A: Effect of deletion of sequence −88/−60 on CAT activity. B: Activator properties of sequence −111/−84. Solid bars: αA-Crystallin gene sequences; sequences 5′ or 3′ to the initiation site of transcription (*) are preceded by − or +, respectively. Open bar, CAT gene sequence; line, pBR322 sequence. CAT activity, as percentage of chloramphenicol acetylated products, is indicated for each construction (Chepelinsky *et al.*, 1987).

CAT gene in nonlens cells (Fig. 1-5; Chepelinsky *et al.*, 1985; Reddan *et al.*, 1986).

Transgenic Mice

When a hybrid gene containing −366/+46 bp of the αA-crystallin gene fused to the CAT gene was microinjected into the pronuclei of fertilized mouse eggs, transgenic mice that expressed the CAT gene specifically in the lens were obtained (Overbeek *et al.*, 1985). When the same αA-crystallin sequence (−366/+46) was fused to the SV40 coding early region, differentiation of the lens was disrupted, and a lens tumor appeared as a consequence of SV40 large T antigen expression in the lens (Mahon *et al.*, 1987). Recently, other groups have also altered lens phenotype by directing the expression of ricin (Landel *et al.*, 1988) or diphtheria toxin (Kaur *et al.*, 1989) or polyoma large T antigen (Griep *et al.*, 1989) to the lens of transgenic mice with this αA-crystallin promoter fragment.

To map more precisely the sequence responsible for the lens-specific expression of the αA-crystallin gene, we obtained transgenic mice containing further 5′ deletions of the αA-crystallin promoter fused to the CAT gene. Mice containing the −111/+46 CAT or the −88/+46 CAT hybrid transgene expressed CAT exclusively in the lens; no CAT activity was observed in 14 other tissues analyzed. The transcription of the CAT gene initiated at the same site as the endogenous αA-crystallin gene (Fig. 1-6), indicating that the CAT gene expression observed in the lens was directed properly by the αA-crystallin promoter. In contrast, the −34/+46 CAT hybrid transgene, containing no sequences upstream of the TATA box, was inactive in transgenic mice (Fig. 1-7). Expression of the CAT gene in −88/+46 CAT transgenic mice appears during embryonic development at approximately the same time as the endogenous gene (Wawrousek *et al.*, 1990). These results indicate that the mouse αA-crystallin sequence −88 to +46 contains information sufficient to direct the expression of the αA-crystallin gene specifically to the lens and for its proper initiation of expression during development.

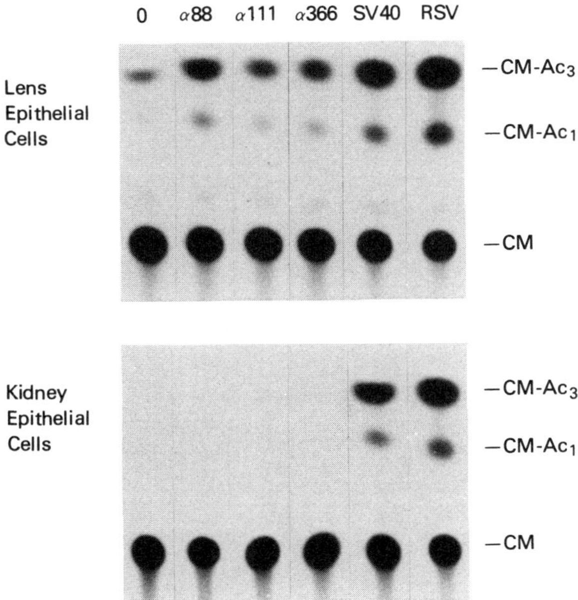

FIGURE 1-5. Lens specificity of the murine αA-crystallin promoter: CAT gene expression after transfection of a rabbit lens epithelial cell line (Reddan *et al.*, 1986) and in a kidney epithelial cell line. 0, pSV0-CAT vector; SV40, pSV2-CAT; RSV, pRSV-CAT (Gorman *et al.*, 1982a,b); α88, pαA88$_a$-CAT; α111, pαA111$_a$-CAT; α366, pαA366$_a$-CAT (Chepelinsky *et al.*, 1985, 1987); CM, chloramphenicol; CM-Ac$_1$ and CM-Ac$_3$, chloramphenicol monoacetylated products.

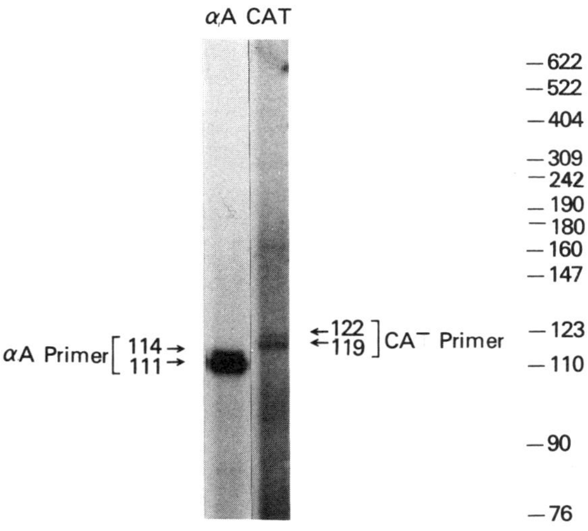

FIGURE 1-6. Lens RNA analysis by primer extension of an F1 αA88-CAT transgenic mouse. Extended products with a CAT primer (CAT lane) or an αA-crystallin primer (αA lane) are indicated (Wawrousek *et al.*, 1990).

CIS REGULATORY SEQUENCES OF THE MURINE αB-CRYSTALLIN GENE

As mentioned above, the αB-crystallin gene is expressed in lens and nonlens tissues (Bhat and Nagineni, 1989; Dubin *et al.*, 1989; Duguid *et al.*, 1988; Iwaki *et al.*, 1989). Primer extension studies indicated that the initiation site of transcription was the same in lens, heart, skeletal muscle, and kidney (Dubin *et al.*, 1989).

To determine the *cis* regulatory sequences involved in the expression of this gene, an αB minigene containing 666 bp of 5′ flanking sequence, 75 bp of exon 1, 300 bp of exon 3, and 2400 bp of 3′ flanking sequence (Fig. 1-8; Dubin *et al.*, 1989) was injected into the pronuclei of fertilized mouse eggs, and its expression was analyzed in the resulting transgenic mice. Transgene mRNA was found in the lens and in some of the nonlens tissues where the endogenous αB-crystallin gene is expressed (Fig. 1-8; Dubin *et al.*, 1989). Primer extension analysis showed that the initiation site of transcription of the transgene was the same in lens, heart, skeletal muscle, kidney, and lung. In heart there appears to be an

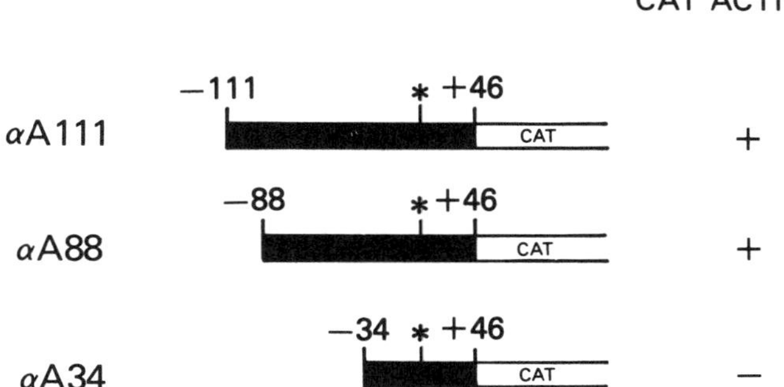

FIGURE 1-7. CAT gene expression directed by the murine αA-crystallin promoter in transgenic mice. CAT activity is observed in αA111-CAT and αA88-CAT but not in αA34-CAT transgenic mice (Wawrousek *et al.*, 1990).

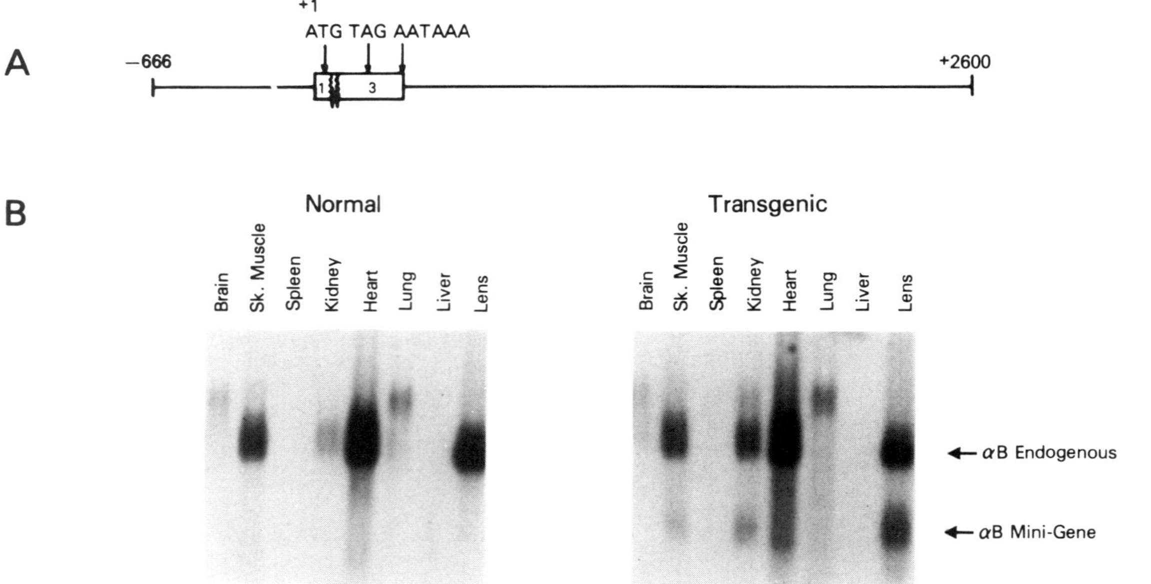

FIGURE 1-8. αB-Crystallin minigene expression in 2-month-old transgenic mice. A: Diagram of minigene. B: Northern analysis of RNA obtained from different tissues of normal and αB-crystallin minigene transgenic mice (Dubin *et al.*, 1989).

additional poorly utilized initiation site(s) approximately 40 bp upstream of the main initiation site for both the endogenous gene and the transgene. The transcript observed in the lung for both the endogenous gene and the transgene is larger than that found in other tissues.

When a hybrid gene containing αB-crystallin sequence −661/+44 fused to the CAT gene was microinjected into mouse embryos, the resulting transgenic mice expressed the CAT gene preferentially in the lens and to a lesser extent in skeletal muscle (R. A. Dubin, E. F. Wawrousek, and J. Piatigorsky, unpublished data). Therefore, the sequences sufficient to direct expression of the αB-crystallin gene in lens and skeletal muscle are located between −661 and +44. Some of the sequences responsible for expression of this gene in other tissues may also be located in this region; it is also possible that other sequences not present in the minigene or the CAT constructs may be necessary for proper regulation of the αB-crystallin gene.

CONCLUSIONS

1. Even though the αA- and αB-crystallin genes arose by duplication and code for related polypeptides that form aggregates in the lens, their genes are differentially regulated.
2. The αA-crystallin gene is expressed specifically in the lens. The murine αA-crystallin sequence −88 to +46 contains *cis* regulatory elements required for its lens specificity and for the correct developmental regulation of this gene.
3. The αB-crystallin gene is not lens-specific and it is also expressed at lower levels in other tissues. The murine αB-crystallin sequence −661 to +44 contains *cis* regulatory elements required for its preferential expression in the lens and in skeletal muscle.
4. The nonlens expression of the αB-crystallin gene suggests that it has a noncrystallin function in nonlenticular tissues and possibly also in the lens. In this way, αB-crystallin may share similarities with the taxon-specific enzyme-crystallins.

SUMMARY

αA- and αB-crystallins are closely related structural proteins in the vertebrate ocular lens that are encoded by separate genes located on different chromosomes. Although mouse αA-crystallin mRNA is found exclusively in the lens, mouse αB-crystallin mRNA is also found in heart, skeletal muscle, kidney, lung, and brain. αA-Crystallin promoter–CAT hybrid genes or an αB-crystallin minigene was introduced into the mouse genome. Several lines of transgenic mice containing αA promoter sequences −111/+46 or −88/+46 expressed the CAT (chloramphenicol acetyltransferase) gene exclusively in the lens. The αA-CAT transgenes were activated in lens during embryonic development at approximately the same time as the endogenous αA-crystallin gene, and the transcription initiation site was the same for the transgenes and the endogenous gene. Transgenic mice containing the αA-crystallin −34/+46–CAT fusion gene did not express the CAT gene, suggesting that the sequence between −88 and −34 is essential for αA-crystallin promoter function in the mouse lens. Transgenic mice containing an αB-crystallin minigene construct (consisting of 666 bp of 5′-flanking sequence and 75 bp of exon 1 fused to 300 bp of exon 3 and 2400 bp of 3′ flanking sequence) synthesized minigene mRNA in other tissues besides the lens. These experiments show that the αA- and αB-crystallin genes differ greatly in their tissue specificity.

REFERENCES

Bhat, S. P., and Nagineni, C. N., 1989, αB subunit of lens-specific protein α-crystallin is present in other ocular and non-ocular tissues, *Biochem. Biophys. Res. Commun.* **158:**319–325.

Chepelinsky, A. B., King, C. R., Zelenka, P. S., and Piatigorsky, J., 1985, Lens-specific expression of the chloramphenicol acetyltransferase gene promoted by 5′ flanking sequences of the murine αA-crystallin gene in explanted chicken lens epithelia, *Proc. Natl. Acad. Sci. U.S.A.* **82:**2334–2338.

Chepelinsky, A. B., Sommer, B., and Piatigorsky, J., 1987, Interaction between two different regulatory elements activates the murine αA-crystallin gene promoter in explanted lens epithelia, *Mol. Cell. Biol.* **7:**1807–1814.

Chiesa, R., Gawinowicz-Kolks, M. A., and Spector, A., 1987, The phosphorylation of the primary gene products of α-crystallin, *J. Biol. Chem.* **262:**1438–1441.

Cohen, L. H., Westerhuis, L. W., Smits, D. P., and Bloemendal, H., 1978a, Two structurally closely related polypeptides encoded by 14S mRNA isolated from rat lens, *Eur. J. Biochem.* **89:**251–258.

Cohen, L. H., Westerhuis, L. W., de Jong, W. W., and Bloemendal, H., 1978b, Rat α-crystallin A chain with insertion of 22 residues, *Eur. J. Biochem.* **89:**259–266.

de Jong, W. W., and Hendriks, W., 1986, The eye lens crystallins: Ambiguity as evolutionary strategy, *J. Mol. Evol.* **24:**121–129.

de Jong, W. W., Leunissen, J. A. M., Leenen, P. J. M., Zweers, A., and Versteeg, M., 1988, Dogfish α-crystallin sequences: Comparison with small heat shock proteins and *Schistosoma* egg antigen, *J. Biol. Chem.* **263:**5141–5149.

Delcour, J., and Papaconstantinou, J., 1974, A change in the stoichiometry of assembly of bovine lens α-crystallin subunits in relation to cellular differentiation, *Biochem. Biophys. Res. Commun.* **57:**134–141.

Dubin, R. A., Wawrousek, E. F., and Piatigorsky, J., 1989, Expression of the murine αB-crystallin gene is not restricted to the lens, *Mol. Cell. Biol.* **9:**1083–1091.

Dubin, R. A., Ally, A. H., Chung, S., and Piatigorsky, J., 1990, Human αB-crystallin gene and preferential promoter function in lens, *Genomics* **7:**594–601.

Duguid, J. R., Rohwer, R. G., and Seed, B., 1988, Isolation of cDNAs of scrapie-modulated RNAs by subtractive hybridization of a cDNA library, *Proc. Natl. Acad. Sci. U.S.A.* **85:**5738–5742.

Gorman, C. M., Merlino, G. T., Willingham, M. C., Pastan, I., and Howard, B. H., 1982a, The Rous sarcoma virus long terminal repeat is a strong promoter when introduced into a variety of eukaryotic cells by DNA-mediated transfection, *Proc. Natl. Acad. Sci. U.S.A.* **79:**6777–6781.

Gorman, C. M., Moffat, L. F., and Howard, B. H., 1982b, Recombinant genomes which express chloramphenicol acetyltransferase in mammalian cells, *Mol. Cell. Biol.* **2:**1044–1051.

Griep, A. E., Kuwabara, T., Lee, E. J., and Westphal, H., 1989, Perturbed development of the mouse lens by polyomavirus large T antigen does not lead to tumor formation, *Genes and Development* **3:**1075–1085.

Hawkins, J. W., Van Keuren, M. L., Piatigorsky, J., Law, M. L., Patterson, D., and Kao, F.-T., 1987, Confirmation of assignment of the human α1-crystallin gene (CRYA1) to chromosome 21 with regional localization to q22.3, *Hum. Genet.* **76:**375–380.

Hendriks, W., Leunissen, J., Nevo, E., Bloemendal, H., and de Jong, W. W., 1987, The lens protein αA-crystallin of the blind mole rat, *Spalax ehrenbergi:* Evolutionary change and functional constraints, *Proc. Natl. Acad. Sci. U.S.A.* **84:**5320–5324.

Hendriks, W., Sanders, J., de Leij, L., Ramaekers, F., Bloemendal, H., and de Jong, W. W., 1988, Monoclonal antibodies reveal evolutionary conservation of alternative splicing of the αA-crystallin primary transcript, *Eur. J. Biochem.* **174:**133–137.

Ingolia, T. D., and Craig, E. A., 1982, Four small *Drosophila* heat shock proteins are related to each other and to mammalian α-crystallin, *Proc. Natl. Acad. Sci. U.S.A.* **79:**2360–2364.

Iwaki, T., Kume-Iwaki, A., Liem, R. K. H., and Goldman, J. E., 1989, αB-Crystallin is expressed in non-lenticular tissues and accumulates in Alexander's disease brain, *Cell* **57:**71–78.

Jaworski, C. J., and Piatigorsky, J., 1989, A pseudo-exon in the functional human αA-crystallin gene, *Nature* **337:**752–754.

Kaur, S., Key, B., Stock, J., McNeish, J. D., Akeson, R., and Potter, S. S., 1989, Targeted ablation of α-crystallin-synthesizing cells produces lens-deficient eyes in transgenic mice, *Development* **105:**613–619.

Kaye, N. W., Church, R. L., Piatigorsky, J., Petrash, J. M., and Lalley, P. A., 1985, Assignment of the mouse alpha A-crystallin structural gene to chromosome 17, *Curr. Eye Res.* **4:**1263–1268.

King, C. R., and Piatigorsky, J., 1983, Alternative RNA splicing of the murine αA-crystallin gene: Protein-coding information within an intron, *Cell* **32:**707–712.

King, C. R., and Piatigorsky, J., 1984, Alternative splicing of αA-crystallin RNA: Structural and quantitative analyses of the mRNAs for the αA_2- and αA^{ins}-crystallin, *J. Biol. Chem.* **259:**1822–1826.

Landel, C. P., Zhao, J., Bock, D., and Evans, G. A., 1988, Lens-specific expression of recombinant ricin induces developmental defects in the eyes of transgenic mice, *Genes Dev.* **2:**1168–1178.

Mahon, K. A., Chepelinsky, A. B., Khillan, J. S., Overbeek, P. A.,

Piatigorsky, J., and Westphal, H., 1987, Oncogenesis of the lens in transgenic micè, *Science* **235:**1622–1628.

McAvoy, J. W., 1981, Developmental biology of the lens, in: *Mechanisms of Cataract Formation in the Human Lens* (G. Duncan, ed.), Academic Press, New York, pp. 7–46.

McDevitt, D. S., Hawkins, J. W., Jaworski, C. J., and Piatigorsky, J., 1986, Isolation and partial characterization of the human αA-crystallin gene, *Exp. Eye Res.* **43:**285–291.

Mulders, J. W. M., Stokkermans, J., Leunissen, J. A. M., Benedetti, E. L., Bloemendal, H., and de Jong, W. W., 1985, Interaction of α-crystallin with lens plasma membranes, *Eur. J. Biochem.* **152:**721–728.

Nene, V., Dunne, D. W., Johnson, K. S., Taylor, D. W., and Cordingley, J. S., 1986, Sequence and expression of a major egg antigen from *Schistosoma mansoni.* Homologies to heat shock proteins and alpha-crystallins, *Mol. Biochem. Parasitol.* **21:**179–188.

Ngo, J. T., Klisak, I., Dubin, R. A., Piatigorsky, J., Mohandas, T., Sparkes, R. S., and Bateman, J. B., 1989, Assignment of the αB-crystallin gene to human chromosome 11, *Genomics* **5:**665–669.

Overbeek, P. A., Chepelinsky, A. B., Khillan, J. S., Piatigorsky, J., and Westphal, H., 1985, Lens-specific expression and developmental regulation of the bacterial chloramphenicol acetyltransferase gene driven by the murine αA-crystallin promoter in transgenic mice, *Proc. Natl. Acad. Sci. U.S.A.* **82:**7815–7819.

Piatigorsky, J., 1981, Lens differentiation in vertebrates: A review of cellular and molecular features, *Differentiation* **19:**134–153.

Piatigorsky, J., and Wistow, G. J., 1989, Enzyme/crystallins: Gene sharing as an evolutionary strategy, *Cell* **57:**197–199.

Quax-Jeuken, Y., Quax, W., van Rens, G., Khan, P. M., and Bloemendal, H., 1985, Complete structure of the αB-crystallin gene: Conservation of the exon–intron distribution in the two non-linked α-crystallin genes, *Proc. Natl. Acad. Sci. U.S.A.* **82:**5819–5823.

Rameakers, F. C. S., Selten-Versteegen, A. M. E., and Bloemendal, H., 1980, Interaction of newly synthesized α-crystallin with isolated lens plasma membranes, *Biochim. Biophys. Acta* **596:**57–63.

Reddan, J. R., Chepelinsky, A. B., Dziedzic, D. C., Piatigorsky, J., and Goldenberg, E. M., 1986, Retention of lens specificity in long-term cultures of diploid rabbit lens epithelial cells, *Differentiation* **33:**168–174.

Skow, L. C., and Donner, M. E., 1985, The locus encoding αA-crystallin is closely linked to H-2K on mouse chromosome 17, *Genetics* **110:**723–732.

Sommer, B., Chepelinsky, A. B., and Piatigorsky, J., 1988, Binding of nuclear proteins to promoter elements of the mouse αA-crystallin gene, *J. Biol. Chem.* **263:**15666–15672.

Spector, A., Chiesa, R., Sredy, J., and Garner, W., 1985, cAMP-dependent phosphorylation of bovine lens α-crystallin, *Proc. Natl. Acad. Sci. U.S.A.* **82:**4712–4716.

Thompson, M. A., Hawkins, J. W., and Piatigorsky, J., 1987, Complete nucleotide sequence of the chicken αA-crystallin gene and its 5′ flanking region, *Gene* **56:**173–184.

van den Heuvel, R., Hendriks, W., Quax, W., and Bloemendal, H., 1985, Complete structure of the hamster αA crystallin gene: Reflection of an evolutionary history by means of exon shuffling, *J. Mol. Biol.* **185:**273–284.

van der Ouderaa, J., de Jong, W. W., Hilderink, A., and Bloemendal, H., 1974, The amino-acid sequence of the αB2 chain of bovine α-crystallin, *Eur. J. Biochem.* **49:**157–168.

Vermorken, A. J. M., Hilderink, J. M. H. C., van de Ven, W. J. M., and Bloemendal, H., 1978, Lens differentiation: Crystallin synthesis in isolated epithelia from calf lenses, *J. Cell Biol.* **76:**175–183.

Voorter, C. E. M., Mulders, J. W. M., Bloemendal, H., and de Jong, W. W., 1986, Some aspects of the phosphorylation of α-crystallin A, *Eur. J. Biochem.* **160:**203–210.

Wawrousek, E. F., Chepelinsky, A. B., McDermott, J. B., and Piatigorsky, J., 1990, Regulation of the murine αA-crystallin promoter in transgenic mice, *Dev. Biol.* **137:**68–76.

Wistow, G., 1985, Domain structure and evolution in α-crystallins and small heat-shock proteins, *FEBS Lett.* **181:**1–6.

Wistow, G. J., and Piatigorsky, J., 1988, Lens crystallins: The evolution and expression of proteins for a highly specialized tissue, *Annu. Rev. Biochem.* **57:**479–504.

Zwaan, J., 1983, The appearance of α-crystallin in relation to cell cycle phase in the embryonic mouse lens, *Dev. Biol.* **96:**173–181.

2

Molecular and Cellular Mechanisms of Lens Growth Control during Development and Aging

Y. COURTOIS, C. ARRUTI, P. R. BLANQUET, J. C. JEANNY, F. MASCARELLI, and J. TRETON

INTRODUCTION

Studies on the regulation of lens growth during development and aging have led to the identification of external signals that may be provided during the whole life span by surrounding tissues, retina, vitreous, or iris. These signals may be positive—i.e., they may be involved in triggering cell division and cell elongation—or negative—i.e., they can prevent cell division in order to keep the growth of the lens under control. In addition, they should be aimed at some precise area of the lens epithelium or fibers, since lens growth depends on stimulation of cells at the lens periphery. They should also be able to cross the lens capsule, which completely surrounds the lens. It has been proposed that polypeptide growth factors can fulfill this function in the eyes. The purpose of this chapter is to show, on the basis data obtained in our laboratory as well as in other laboratories, that a family of growth factors—fibroblast growth factors (FGF, formerly called EDGF)—can control lens growth.

The need for a universal mechanism of control of lens growth is discussed first with reference to an analysis of the different growth curves of various species (Treton and Courtois, 1989). Then *in vitro* experiments on the effect of growth factors on cell proliferation and differentiation in the lens are described (Cirillo *et al.*, 1985; Mascarelli *et al.*, 1989). The mechanism of action of growth factors is a function of specific receptors found at the cell surface. Such receptors have been identified and purified from freshly extracted bovine lens epithelium (Blanquet *et al.*, 1989b). Low-affinity receptors are also present in the lens capsule (Jeanny *et al.*, 1987). Finally, the expression of FGFs themselves within the lens is demonstrated.

EVIDENCE FOR A RELATIONSHIP BETWEEN LONGEVITY OF MAMMALIAN SPECIES AND LENS GROWTH

This study was performed by J. Treton.

Numerous mammalian lens growth curves are available. Since Lord (1959) described the use of the dry weight of the eye lens for aging cottontail rabbits, this technique has been applied to a number of other mammals and birds. The general shape of the growth curves is biphasic, with a steep slope during the first part of the life-span, followed by a shallow slope (Fig. 2-1). It allows the determination of a theorical parameter that is defined as the crossing point of the two straight lines that best fit each curve. This break point defines a period that corresponds to the lens development stage (LDS). This parameter

Y. Courtois, C. Arruti, P. R. Blanquet, J. C. Jeanny, F. Mascarelli, and J. Treton • Gerontology Research Unit, INSERM U118, 75016 Paris, France.

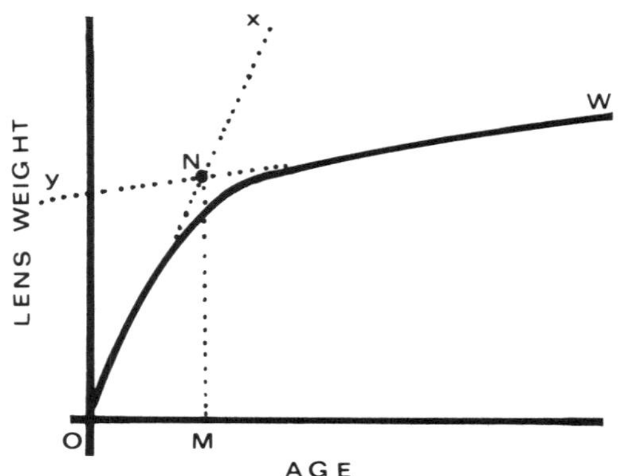

FIGURE 2-1. Typical growth curve of a mammalian lens as a function of age. OM represents the lens development stage (LDS) (from Treton and Courtois, 1989).

has been compared in various animals with different life spans. The maximum life span potential (MLP) of an animal is defined as the maximum life span that members of its species can potentially reach if death by predators or accident is avoided. Most of the life-span data are from Altman and Dihner (1972); compilation of these data is described by Treton and Courtois (1989) for 16 different species. There is a good correlation between LDS and MLP, as shown in Fig. 2-2. However, the MLP and LDS values given here should not be considered highly accurate because of the limited literature data and number of mammals available. The LDS would be 42 months in humans. If brain weight it taken instead of MLP, a better correlation is found.

This similitary of the timing of lens growth in different mammalian species implies that the length of the development stage (LDS) is controlled by a common set of genes conserved during the evolution process. It was previously shown that a growth factor called EDGF (eye-derived growth factor) is present in retina, iris, and vitreous (Arruti *et al.*, 1981; Raulais *et al.*, 1987). The growth of the lens may be directly controlled by the retina to retain harmony in their relationship. Any distortion of that growth would result in some loss in visual acuity. Such regulation must be assumed. Indeed, control of these lens growth factors could be a key in the slowing and eventual cessation of mitotic activity in the

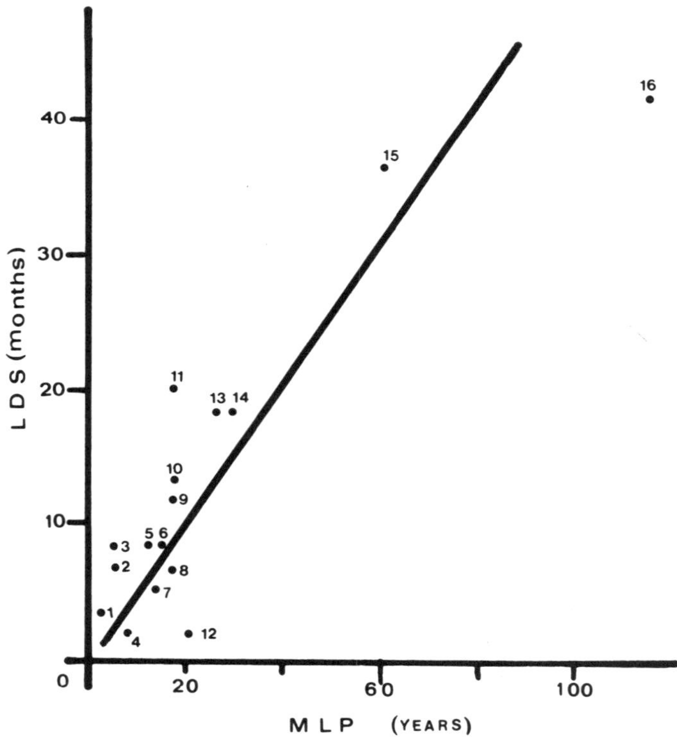

FIGURE 2-2. Lens development stage (LDS) as a function of the maximum life-span potential (MLP). 1, *Mus musculus;* 2, *Rattus norvegicus;* 3, *Syloagus floridanus;* 4, *Cavia porcellus;* 5, *Procyon lotor;* 6, *Wrocyon;* 7, *Squirus carolinens;* 8, *Antilocapra americana;* 9, *Oryctogalus cuniculus;* 10, *Odocoileus hemionus columbianus;* 11, *Odocoileus hemionus hemionus;* 12, *Canis familiaris;* 13, *Calhorinus ursinus;* 14, *Bos taurus;* 15, *Loxondota africana;* 16, *Homo sapiens* (human) (from Treton and Courtois, 1989).

equatorial part of the lens. If such a growth factor is involved, its regulation may be expected at various levels. The growth factor concentration may decrease as a function of age, or its action may be hindered by down-regulation of its cell receptors. Alternatively, its transduction in the cells may be impaired, or growth inhibitors may be produced. These different possibilities are not mutually exclusive. The following data result from our attempts to describe the mechanism of action of EDGF on lens growth and differentiation.

THE EFFECT OF EYE-DERIVED GROWTH FACTOR ON CELL PROLIFERATION AND PROTEIN SYNTHESIS IN ORGAN CULTURE

This work was performed by C. Arruti and F. Mascarelli.

The adult lens is composed of several different cell types. The normal continuous growth of the lens depends on the ordered formation of differentiating lens fibers (Fig. 2-3). The increase in number of differentiating cells is provided by the multiplicative activity of an epithelial subpopulation located in a narrow preequatorial ring. A nondividing group of epithelial cells covers the central region of the lens at the level of pupillary space. Fiber formation results from cell elongation, specific protein synthesis, loss of most cytoplasmic organelles, and nuclear modifications culminating in degeneration (reviewed by Piatigorsky, 1981). Some years ago, we showed that the adult bovine retina contains growth factors (Arruti and Courtois, 1978), which are now identified as acidic and basic fibroblast growth factors (a- and bFGF; Chevallier *et al.*, 1985; Plouet *et al.*, 1989). We are interested in elucidating the role of these external signals in the control of the growth and differentiation of cells in the lens.

Lens organ culture provides a very reliable and accurate model for experimental analysis resembling the *in vivo* situation. Within the adult lens, FGFs stimulate epithelial cell division following an ordered pattern (Arruti *et al.*, 1985). In the absence of serum in the culture medium, EDGFs stimulate cell proliferation in the central epithelium but not in the periphery. Protein synthesis is also stimulated, as described in the following experiment. Whole lenses were incubated with or without FGFs in the presence of [^{35}S]methionine in a methionine-free medium. After different periods of incubation, the lenses were dissected, and three different populations were selected: central epithelial cells, cells of the germinative region, and superficial cortical fibers. The proteins were extracted from these different cell populations and analyzed by chromatography on SDS polyacrylamide gels.

By autoradiography (Fig. 2-4), it can be seen that each pattern of synthesis is specific to one population—central, germinative, or superficial cortical fibers. More proteins are labeled as a function of incubation time. If the same amount of radioactivity is applied on the gel for FGF-treated (+) or untreated (−) samples, there is no apparent difference in the labeling pattern. However, aFGF or bFGF increases the specific activity of each protein fraction by a factor of 1.8 in the central epithelium, 2.2 in the germinative epithelium, and 2.4 in the superficial cortex fibers (Mascarelli *et al.*, 1989). These data indicate that in this system, FGFs have potent effects on cell proliferation and lens growth.

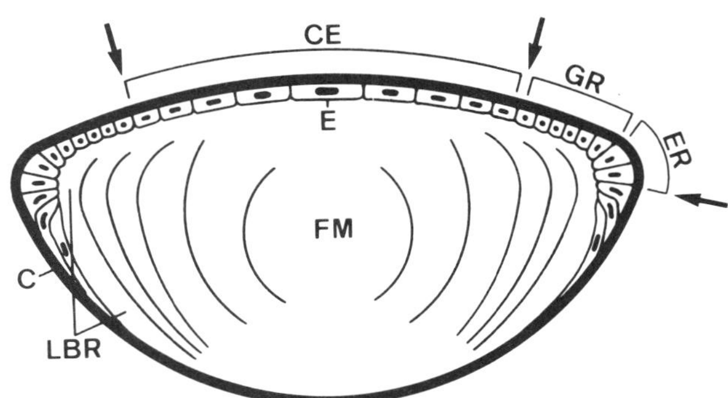

FIGURE 2-3. Diagrammatic cross section of vertebrate lens: C, capsule; E, epithelium; CE, central epithelium; PE, peripheral epithelium; LBR, lens bow region; FM, fiber mass. Arrows show the two regions of dissection used in this study.

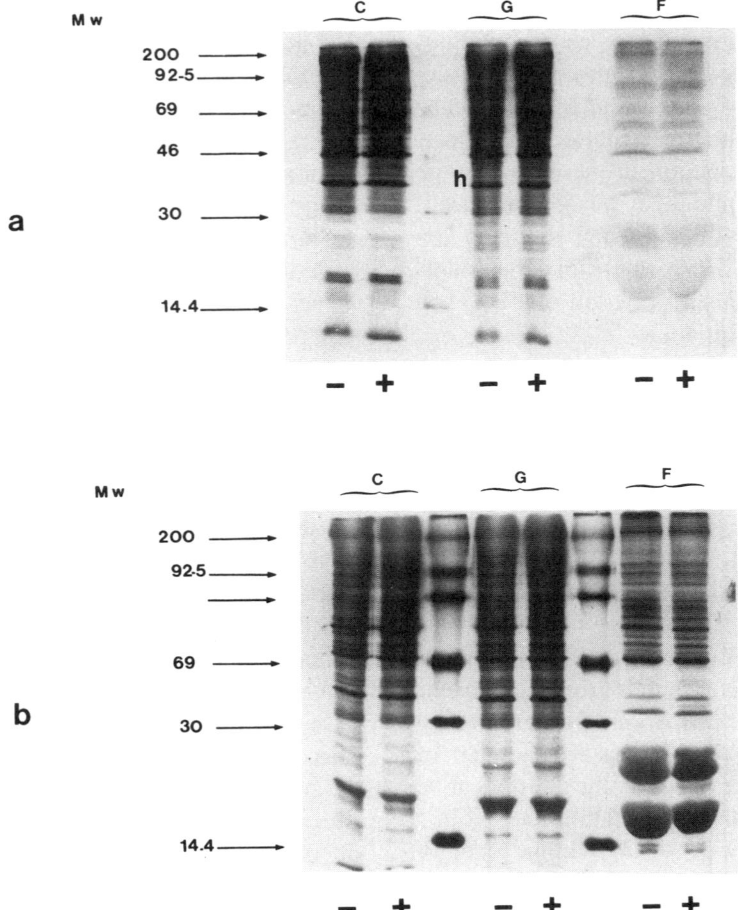

FIGURE 2-4. Fluorography showing newly synthesized polypeptides in different cell populations from cultured lenses. The lens was cultured for 24 or 48 hr (top and bottom, respectively). Lanes C, central epithelial cells; lanes G, germinative epithelium; lanes F, superficial cortex fibers; −, untreated; +, treated with EDGF. Label of 105 cpm was loaded in lanes C and G, and 5104 cpm in lane F (from Mascarelli *et al.*, 1989).

Interestingly, when these experiments were performed on lens containing only preequatorial fibers, with the other cell populations removed, an enhancing effect of FGF on protein synthesis was also observed. This means that these young fibers still contain receptors and an efficient mechanism of transduction. Recently these results were confirmed by Chamberlain and McAvoy (1987), who reported that bFGF can induce crystallin synthesis in newborn rat lens in organ culture. However, our data do not explain the different susceptibilities of the different cell populations. This is a very important feature of the lens, since it determines topologically where growth should and should not take place (Uhlrich *et al.*, 1986). To analyze this phenomenon further, we studied the organization of the basal membrane and cytoskeleton in different parts of the lens epithelium.

EVIDENCE OF STRUCTURAL AND FUNCTIONAL HETEROGENEITY OF THE BASAL MEMBRANE–CYTOSKELETON COMPLEX

This work was performed by P. R. Blanquet.

In our first attempt to identify the early events that could control the transition of adult epithelial stem cells to a terminal state of differentiation, we investigated the membrane–cytoskeleton complex that is involved directly in interactions between the epithelial cells and the capsule of bovine eye lens (Blanquet and Courtois, 1989). In both the absence and presence of the growth factor, the organization of the molecular complex has been determined by cell extraction, immunoprecipitation, and immunoblotting experiments. We have shown that marked differences exist in the organization of this complex.

These data are summarized in Fig. 2-5. They show the following: (1) the organization of several major membrane components in the peripheral epithelium differs from that of the central epithelium; (2) microfilaments and vimentin filaments exist as independent structures in the peripheral epithelium; (3) two surface glycoproteins of 46 and 220 kDa are firmly bound to vimentin filaments in the peripheral region, whereas the intermediate filaments are solely found in close association with the 46- and 220-kDa glycoproteins in the central zone; (4) microfilaments and vimentin filaments mediate the attachment of nuclei to the peripheral capsule, whereas only microfilaments are involved in the anchoring of nuclei to the central capsule. These data make it conceivable that the organization of the capsule-adherent complex in the peripheral epithelium is necessary for a signal across the surface membrane to be propagated through the cytoskeleton network. These data are the basis for the following experiments devised to analyze the mechanism of transduction of FGFs on these cells, including both the presence of FGF receptors and the activation of a metabolic cascade.

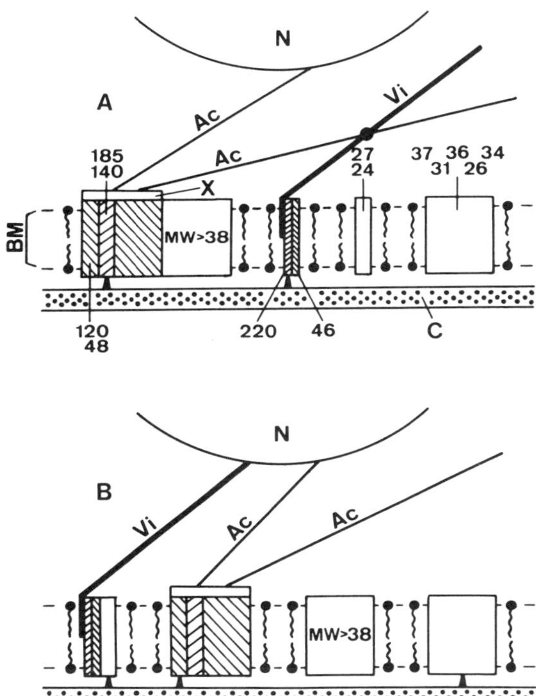

FIGURE 2-5. Highly schematic representations of the molecular organization and protein–protein interactions involved at capsule-attached areas of membrane–cytoskeleton complex in the central (A) and peripheral (B) parts of the lens epithelium. Ac, actin-based filaments; Vi, vimentin-based filaments; BM, the bilayer matrix in which proteins (open) and glycoproteins (shaded) penetrate; X, unknown factor(s) thought to play a role in the association of microfilaments with the cytoplasmic surface of the plasma membrane; N, nuclei; C, the lens capsule. Solid triangles represent the protein components that are tightly associated with the basal membrane. Molecular weights of major proteins and glycoproteins are shown in thousands. It is proposed that in addition to the spatial relationship of microfilaments and groups of components, the protein–protein interactions between some elements of the membrane and the capsule vary from the central to the peripheral epithelium.

THE MECHANISM OF FGF STIMULATION OF LENS EPITHELIUM ANALYZED IN THE BASAL MEMBRANE–CYTOSKELETON COMPLEX

Identification of FGF Receptors

This work was performed by P. Blanquet.

Among the various mechanisms involved in the transduction of growth factors, autophosphorylation of their receptors is the first biochemical event that has been extensively studied, especially for EGF. But the phosphorylation of FGF receptors had not been proven; it was suggested by the observation that the molecular weight of an acidic FGF-stimulated phosphoprotein appears to be identical with that of an acidic FGF-binding component at the surface of a 3T3 cell (Huang and Huang, 1986). The basal membrane–cytoskeleton complex of the lens epithelium was isolated from epithelium pretreated for various times with FGF in the presence of [γ-^{32}P]ATP. The phosphorylated proteins were analyzed on SDS gel, and two of them migrate at apparent molecular weights identical to those obtained by cross-linking [^{125}I]bFGF to cell membranes. The bFGF-stimulated phosphorylation of these proteins was rapid, suggesting an autophosphorylation process. We demonstrated that the phosphorylated proteins can be purified to homogeneity and still be able to bind specifically FGF. We conclude that bFGF stimulates the phosphorylation of two receptors of 130 and 160 kDa in bovine epithelial lens cells (Blanquet *et al.*, 1989b).

Since the central and peripheral epithelia respond differently to FGF stimulation, it is of interest to determine if the phosphorylation of their receptors is identical in the two regions. After incubation of lenses for 15 hr with and without bFGF, cells adhering to the lens capsule were labeled with [^{32}P]ATP for 5, 10, or 30 min (Fig. 2-6). In the central epithelium two polypeptides of about 130

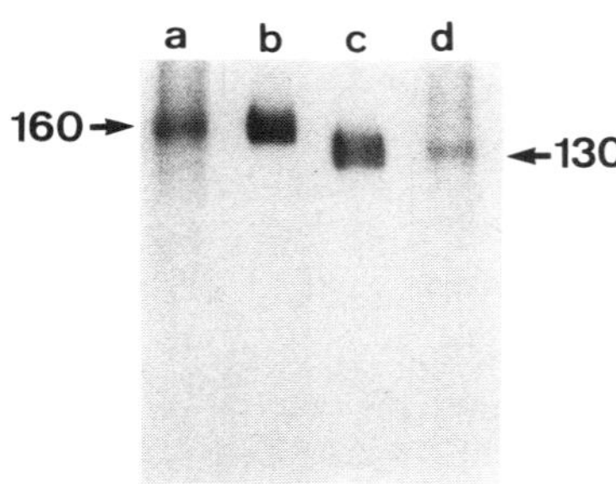

FIGURE 2-6. Identification of bFGF receptors from preparative gels. Following incubation of lenses with bFGF for 15 hr, basal membrane ghosts were prepared from phosphorylated central cells and subjected to preparative electrophoresis. The proteins running at M_r 130 and 160 kDa were then eluted from gels, dialyzed, lyophilized, and analyzed by gel electrophoresis and autoradiography (from Blanquet *et al.*, 1989b). Coomassie blue stain (a, d) and autoradiograms (b, c).

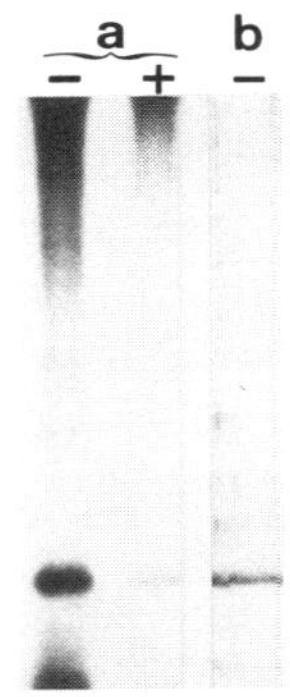

FIGURE 2-7. Purification of p34 and phosphorylation with [^{32}p]ATP of purified p34 with or without bFGF pretreatment. Lenses were incubated for 15 hr with (+) or without (−) bFGF. In lane b, basal membrane extract prepared from peripheral epithelia were treated with EGTA; eluates were electrophoresed and examined with Coomassie blue stain. Peripheral epithelia (a) were phosphorylated for 30 min at 0°C, and the EGTA eluates obtained from basal membrane extracts were electrophoresed and autoradiographed.

and 160 kDa are highly phosphorylated. In the peripheral epithelium, these two proteins are hardly phospholabeled. It is possible that the 130- and 160-kDa proteins and/or the relevant protein kinase do not exist or are present in very low quantities in the peripheral region. An alternative explanation could be that a stimulated phosphatase activity leads to the net dephosphorylation of the receptors.

Phosphorylation of Lipocortin Induced by bFGF in Lens Epithelium

This work was performed by P. Blanquet.

The actions of some growth factors might be intimately linked to those of lipocortins (Brugge, 1986). We therefore hypothesized that the 34-kDa protein (p34) that we described among the basal membrane–cytoskeleton proteins is the second messenger phosphorylated under bFGF stimulation (Fig. 2-7). When assayed *in vitro* for modulation of exogeneous phospholipase A_2 (PLA_2), p34 prepared from unstimulated peripheral epithelium was found to possess anti-PLA_2 activity like lipocortin. When p34 is prepared from bFGF-treated peripheral epithelium, the lipocortin activity of this protein is markedly enhanced, whereas its phosphorylation by [^{32}P]ATP is considerably decreased. These results strongly suggest that p34, a protein with lipocortin activity, may be one of the second messengers involved in FGF stimulation of lens epithelial cell proliferation (Blanquet *et al.*, 1989a). However, so far we have not been able to demonstrate that p34 is the main target of the kinase activity of FGF receptors. It may be only one of multiple steps involved in the cascade of reactions involved. Nevertheless, this is the first time that arachidonic acid metabolism has been shown to be activated by FGF.

LOCALIZATION OF FGF BINDING SITES ON THE LENS CAPSULE

This work has been performed by J. C. Jeanny.

The FGFs have been purified by affinity chromatography on heparin sepharose. Heparin can also modulate the mitogenic activity of aFGF on bovine lens epithelium (BLE) cells (Uhlrich *et al.*, 1986). Thus, it was of interest to investigate whether some of the lens capsule components have the ability to bind FGF. [^{125}I]bFGF or [^{125}I]aFGF was added to frozen lens sections, and the location of its binding sites was revealed by autoradiography. Using this technique we demonstrated that FGF binds to both the lens capsule in the mouse embryo (Jeanny *et al.*, 1987) and the adult bovine lens capsule (Fig. 2-8). The nature and specificity of this binding were demonstrated by incubating the sections with excess unlabeled growth factor or with an excess of unrelated basic proteins. When the sections are treated by heparitinase (5 μg/ml), the binding site disappears. These results confirm that FGFs have a strong affinity for the proteoheparan sulfate that is contained in most basement membrane.

Although we have not clearly demonstrated that the lens capsule contains FGF *in vivo*, it has been reported that other basement membranes in the

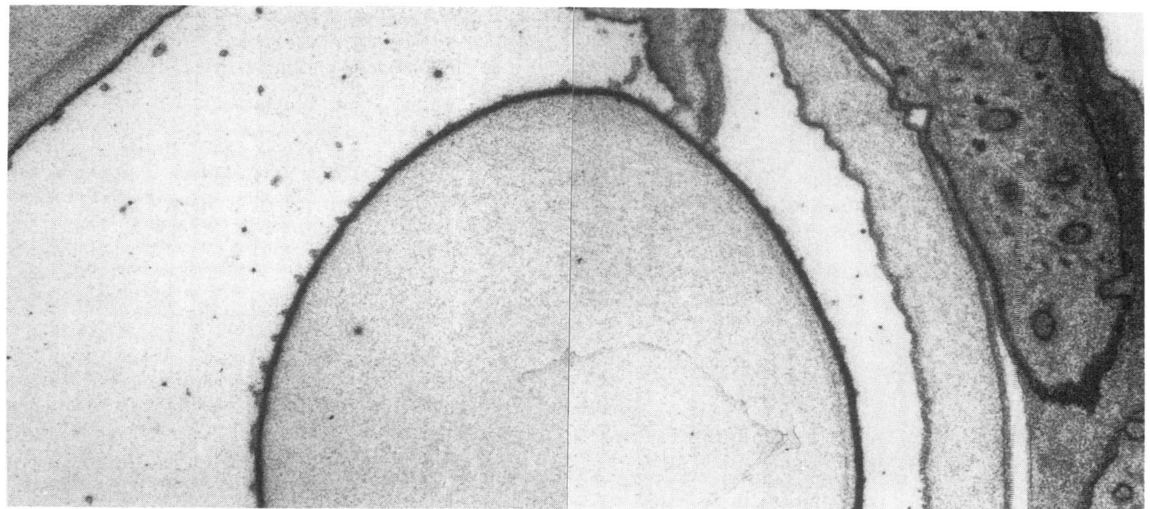

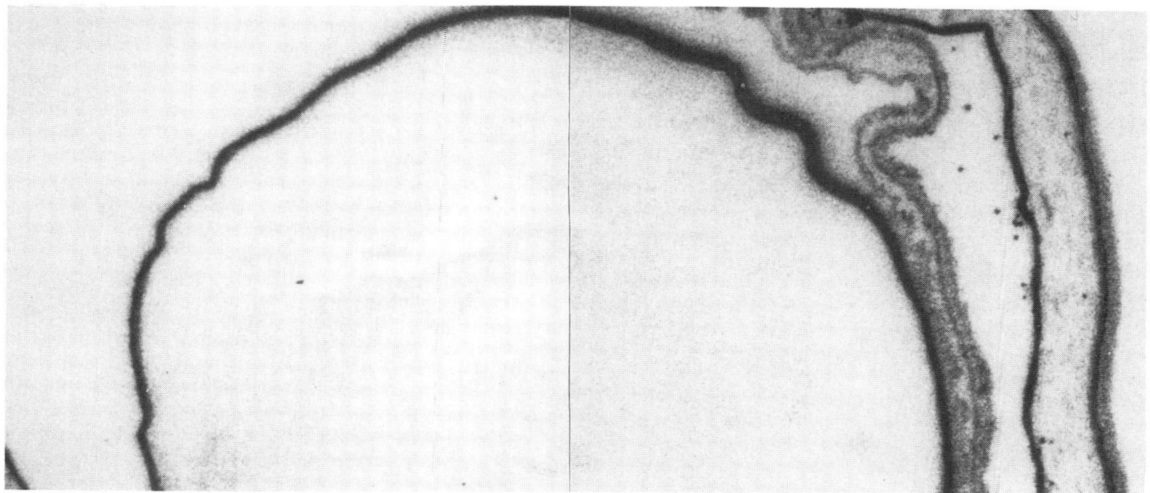

FIGURE 2-8. Binding of [^{125}I]bFGF on lens frozen sections. Frozen sections of adult mouse lenses were incubated with 80 ng/ml of [^{125}I]bFGF. The slides were covered with a photographic emulsion and developed several days later. Top, 1-day-old lens; bottom, 1-month-old lens.

cornea store FGF (Vlodavsky *et al.*, 1987). If FGF is also stored in the lens capsule, does it control the accessibility of FGF cell surface receptors on epithelial cells that are attached to it? Is there a heterogeneity of lens capsule composition in the periphery and in the central part?

CONCLUSION

These data raise further questions about the mechanism of action of FGF in the eye in controlling lens growth. Organ culture experiments with lens incubation have allowed us to produce some interesting information. (1) Lens growth and protein synthesis are dependent the addition of a- or bFGF. (2) The growth factors can bind to lens capsules, and they can stimulate the cells on the inner part of the lens capsules. Thus, basement membranes may trap the growth factors, but they still have the ability to migrate across them to stimulate lens epithelial cell division and/or differentiation. (3) The response of the cells to FGFs in the central part of the lens is different from that at the periphery. This may reflect the differential organization of the basal membrane–cytoskeleton in these regions. The mechanism of FGF transduction is probably dependent on FGF receptor phosphorylation, which in turn triggers the activation of a lipocortin protein. However, a direct link between FGF receptors and lipocortin phosphorylation has not yet been demonstrated.

These data suggest control of the lens physiology by growth factors, but there are still many unanswered questions. For instance, we do not yet know if, during development and aging, there is a modification in the amount of growth factor available or in the affinity of its receptors. There is still very little information on the effect of aging on the transduction mechanism. Finally, recent experiments in our laboratory have demonstrated that a FGF is synthesized *in vitro* by BEL cells (Halley *et al.*, 1988). Thus, an autocrine mechanism may also participate in the regulation of lens growth by growth factors.

ACKNOWLEDGMENTS. This work was performed at various times with the collaboration of N. Fayen, L. Jonet, M. Laurent, S. Paillard, C. Patte, D. Raulais, and M. Vigny. We thank N. Breugnot and H. Coet for their assistance with this manuscript. Our research was supported by grants from the Fondation de la Recherche Medicale and the Fondation de France.

REFERENCES

Altman, P. L., and Dihner, D. S., eds., 1972, *Biology Data Book*, Federation of American Societies for Experimental Biology, Bethesda.

Arruti, C., and Courtois, Y., 1978, Morphological changes and growth stimulation of bovine epithelial lens cells by a retinal extract *in vitro*, *Exp. Cell Res.* **177:**283–292.

Arruti, C., Barritault, D., and Courtois, Y., 1981, Is there a ubiquitous growth factor in the eye? *Differentiation* **18:**29–42.

Arruti, C., Cirillo, A., and Courtois, Y., 1985, An eye-derived growth factor regulates epithelial cell proliferation in cultured lens, *Differentiation* **28:**286–290.

Arruti, C., Mascarelli, F., and Courtois, Y., 1989, The effect of eye-derived growth factor (EDGFs) on methionine incorporation in the different cell populations in bovine adult lens in organ culture, *Exp. Eye Res.* **48:**177–186.

Blanquet, P., and Courtois, Y., 1989, Differential assemblage of the basal membrane cytoskeleton, *Exp. Eye Res.* **48:**187–207.

Blanquet, P., Paillard, S., and Courtois, Y., 1989a, Influence of fibroblast growth factor on phosphorylation and activity of a 34 kDa lipocortin-like protein in bovine epithelial lens cells, *FEBS Lett.* **229:**183–187.

Blanquet, P., Patte, C., Fayein, N., and Courtois, Y., 1989b, Identification and isolation from bovine epithelial lens of two basic fibroblast growth factor receptors that possess bFGF-enhanced phosphorylation activities, *Biochem. Biophys. Res. Commun.* **160:**1124–1131.

Brugge, J. S., 1986, The p35/p36 substrates of protein-tyrosine kinase as inhibitors of phospholipase A_2, *Cell* **46:**149–150.

Chamberlain, C. G., and McAvoy, J. W., 1987, Evidence that fibroblast growth factor promotes lens fibre differentiation, *Curr. Eye Res.* **6:**1165–1168.

Chevallier, B., Loret, C., Barritault, D., Courty, J., Moenner, M., Lagente, O., and Courtois, Y., 1985, Bovine retina contains growth factor activities with different affinity to heparin: Eye-derived growth factor I, II, II, *Biochimie* **67:**265–269.

Cirillo, A., Arruti, C., and Courtois, Y., 1985, A retina derived growth factor regulates epithelial cell proliferation in cultured lens, *Differentiation* **28:**286–290.

Halley, C., Courtois, Y., and Laurent, M., 1988, "Nucleotide sequence of acidic fibroblast growth factor CDNA, *Nucl. Ac. Res.* **16:**10913."

Huang, S. S., and Huang, Y. S., 1986, Association of bovine brain-derived growth factor receptor with protein tyrosine kinase activity, *Biochemistry* **261:**9568–9571.

Jeanny, J. C., Fayein, N., Moenner, M., Chevallier, B., Barritault, D., and Courtois, Y., 1987, Specific fixation of bovine brain and retinal acidic and basic fibroblast growth factor to mouse embryonic eye basement membranes, *Exp. Cell Res.* **171:**63–75.

Lord, R. D., 1959, The lens as an indication of age in cotton tail rabbits, *Wildl. Mgmt.* **23:**338–360.

Mascarelli, F., Raulais, D., and Courtois, Y., 1989, Fibroblast growth factor—rod outer segment interaction: High and low affinity binding sites and release of phosphorylated acidic FUF by light kinase C dependent phosphorylation, *EMBO J.* **8:**2265–2273.

Piatigorsky, J., 1981, Lens differentiation in vertebrates. A review of cellular and molecular features, *Differentiation* **19:**134–153.

Plouet, J., Mascarelli, F., Loret, M. D., Faure, J. P., and Courtois, Y., 1988, Regulation of eye derived growth factor binding to membranes by light ATP, GTP in photoreceptors outer segments, *EMBO J.* **7:**373–376.

Raulais, D., Mascarelli, F., Counis, M. F., and Courtois, Y., 1987, Characterization of acidic and basic fibroblast growth factors in brain, retina and vitreous chick embryo, *Biochem. Biophys. Res. Commun.* **146:**478–486.

Treton, J., and Courtois, Y., 1989, Evidence for a relationship between longevity of mammalian species and a lens growth parameter, *Gerontology* (in press).

Uhlrich, S., Lagente, O., Lenfant, M., and Courtois, Y., 1986, Effects of Heparin on the stimulation of nonvascular cells by human acidic and basic FGF, *Biochem. Biophys. Res. Commun.* **137:**1205–1213.

Vlodavsky, I., Folkman, J., Sullivan, R., Fridman, R., Ishai-Michaeli, R., Sasse, J., and Klagsbrun, M., 1987, Endothelial cell derived basic fibroblast growth factor: Synthesis and deposition into subendothelial extracellular matrix, *Proc. Natl. Acad. Sci. U.S.A.* **84:**2292–2296.

3

Differentiation Patterns in Eye Lens Fibers

ENNIO LUCIO BENEDETTI and IRÈNE DUNIA

INTRODUCTION

The morphology and molecular interactions of lens fiber plasma membrane and cytoskeletal constituents are of particular interest with respect to the formation of specialized membrane domains, to the dynamics of lens accommodation, to aging, and to the cataractous process. We attempt to explore the chemical and ultrastructural features of the interfiber junctions and the existence of physical and/or biochemical relationships between lens fiber plasma membranes and those proteins that in a rather broad terminology are classified as cell motility or cytoskeletal constituents (Burridge *et al.*, 1988; Geiger, 1985; Niggli and Burger, 1987).

It is well documented that during aging and senile cataract, alterations of cortical and nuclear fiber shape and cell surface are present. In extensive reviews Harding *et al.* (1985) and Kistler and Bullivant (1989) summarize the biochemical and structural alterations of various types of human and animal cataracts. In all these cases abnormal cell surface projections and zeiosis have been described, and these types of alterations are commonly found in other cellular systems when the interactions between plasma membrane and cytoskeletal constituents are somehow modified (Godman *et al.*, 1975). Furthermore, changes in the electrochemical gradient and peroxidation throughout the lenticular mass during cataract formation expose the membrane cytoskeletal complex to osmotic swelling, processing of membrane proteins, and new types of interaction between the extrinsic cytoplasmic proteins and the membrane polypeptides. In particular, Kistler and Bullivant (1989) stress the presence of fiber swelling, degeneration and loss of the ball-and-socket structure in a sheep inherited cataract. Junctional alterations have been also described in mouse congenital cataracts (cf. Kistler and Bullivant, 1989).

It is widely believed that the association between cell membrane organelles, including the nuclear envelope and the cytoskeletal scaffold, develops intracellular tension and generates mechanochemical forces that control cell shape, surface architecture, and cell motion.

The erythrocyte has provided a simple, accessible experimental system for the study of plasma membrane–cytoskeleton interactions at both the morphological and the biochemical level (Bennett, 1985; Chasis and Mohandas, 1986; Hall and Bennett, 1987; Lazarides and Moon, 1984; Moon and Lazarides, 1984; Palek, 1987).

Advances in understanding the cell membrane–cytoskeleton complex have provided important explanations of the biochemical and structural basis for inheredited hemolytic anemias in humans and mice (Bennett, 1985; Palek, 1987). Peptide alterations and/or deletion of the cytoskeletal spectrin has been shown to be associated with hereditary spherocytosis or elliptocytosis (Bennett, 1985; Palek, 1987). The impairment of spectrin tetramer assembly described in hereditary elliptocytosis is also in some cases associated with a deficiency in the band 4-1 protein of human red blood cells (Anderson and Lovrien, 1984). Ankyrin–band 3 complex is likely affected in the elliptocytosis type of anemia (Bennett, 1985). Plasma membrane–cytoskeletal instability caused by deletion of the dystrophin polypeptide has recently been described in Duchenne muscular dystrophy (Partridge *et al.*, 1989).

Ennio Lucio Benedetti and Irène Dunia • Institute Jacques Monod, CNRS, University of Paris VII, 75251 Paris, France.

THE MAJOR LENS FIBER PLASMA MEMBRANE INTRINSIC POLYPEPTIDE (MP26)

In the eye lens, emerging molecular and structural patterns characterize the terminal differentiation of the epithelial monolayer, into fibers. In addition to the biosynthesis of specific water-soluble crystallin proteins, terminal differentiation is characterized by the expression of a novel major intrinsic membrane protein, MP26 (Benedetti *et al.*, 1981; Dunia *et al.*, 1985, 1987).

The biosynthesis of this protein parallels the tremendous increase in the total surface of the plasma membrane surrounding each individual fiber. New types of fiber-to-fiber contact and junctions are generated, insuring stable physical association, ionic coupling, and metabolic cooperation throughout the entire lenticular mass. Amino acid analysis (Lien *et al.*, 1985) and c-DNA cloning (Revel *et al.*, 1987) have indicated that MP26 is a polypeptide of 247 amino acid residues with a prevailing degree of hydrophobicity. The predicted molecular model for MP26 is that the putative α-helical domains of the polypeptide transverse the lipid bilayer six times with both amino and carboxyl termini exposed on the cytoplasmic side. Natural proteolytic cleavage of MP26 yields MP22, which is mainly found in aging lenticular nuclear fiber plasma membranes.

Phosphorylation of MP26 mediated by cAMP protein kinase has recently been shown (Johnson *et al.*, 1986). A number of studies demonstrated that MP26 solubilized and reconstituted into liposomes is a channel-forming protein and that the MP26 channels present in one single lipid vesicle are regulated in a calmodulin-dependent way by Ca^{2+} concentration (Girsh and Peracchia, 1985). From our experiments on *in vitro* translation of different classes of lens polyribosomes in a reticulocyte lysate, it appears that the messenger RNA encoding MP26 was predominantly localized on the class of polyribosomes associated with the plasma membrane–cytoskeleton complex, namely, with the actin filaments (Benedetti *et al.*, 1981). During the process of terminal differentiation in the fiber plasma membrane, the junctional assembly is formed simultaneously to the recruitment of the intramembranous particles visualized on cryofractured membrane faces and to biochemical and immunocytochemical evidence of the accumulation of the newly synthetized MP26 in the lipid bilayer (Dunia *et al.*, 1985).

Several lines of experimental evidence lead to the conclusion that MP26 is a general plasma membrane constituent that is concentrated in the junctional domains where it most likely forms the transmembrane communicating pathway. Experiments carried out both on lens fiber membranes and on reconstituted MP26 proteoliposomes showed interesting polymorphic features of the long- and short-range distribution of the MP26 copies within the lipid matrix. The MP26 oligomers are likely formed by a ring of four subunits. The MP26 oligomers are clustered in the junctional domain without any geometric order. Conversely, they might accommodate in orthorhombic arrangement reflecting the existence of preponderant protein-to-protein interaction when a high concentration of the polypeptide or of its proteolytic derivative (MP22) is reached within the natural or artificial lipid bilayer. The square array of particulate entities comprising MP26 and/or MP22 is not necessarily associated with junctional domains and may be the site, especially in the lens nucleus, of preferential and controlled permeability between the intracellular milieu and the narrow extracellular domain (Dunia *et al.*, 1985, 1987; Manenti *et al.*, 1988).

THE PLASMA MEMBRANE CYTOSKELETON COMPLEX IN EYE LENS FIBERS

Electron microscopic observations on thin sections of isolated fiber plasma membrane–cytoskeleton complex (fiber ghost) show that actin filaments and vimentin intermediate filaments (IF) are present with end-on attachment to the inner cytoplasmic surface of the plasma membrane. These two types of filaments are also tightly associated with membrane profiles that are characteristic of the internalization process of plasma membrane domains into the interior of the fibers, connotating the slow but constant assembly and removal of membrane constituents throughout the life of the lens (Benedetti *et al.*, 1981).

It is remarkable that some fiber domains comprise almost exclusively thin filaments that are decorated by crystallin particulate entities and/or ribosomes. Other fiber areas preferentially contain intermediate vimentin filaments with surfaces devoid of beaded structures (Benedetti *et al.*, 1981).

The lens fiber possesses two other molecular forms of actin. One type corresponds to a water-

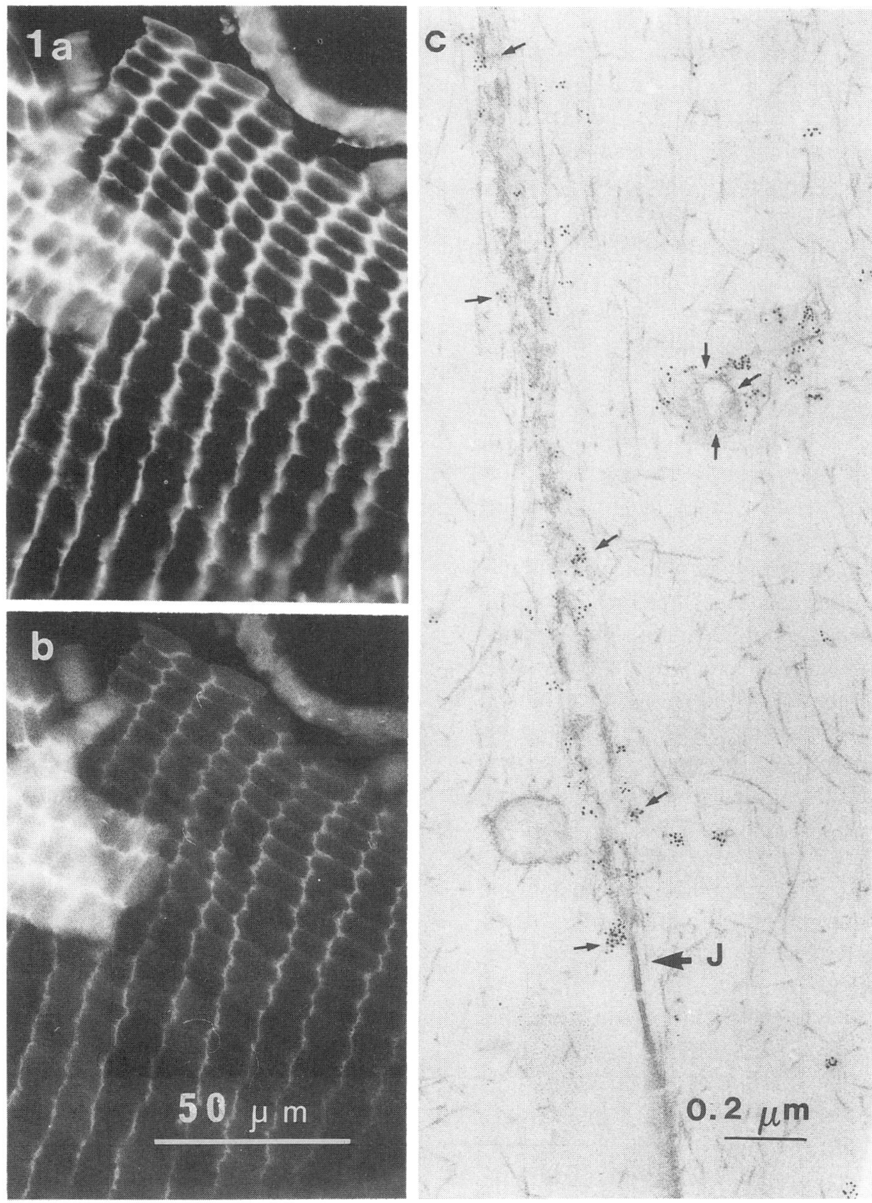

FIGURE 3-1. Cryostat section of cortical fibers. Immunofluorescence with (a) rhodamine-labeled phalloidin and (b) with fluorescein labeled antispectrin. Note the codistribution of actin and spectrin. (c) Isolated fiber ghost incubated with monospecific antibody against spectrin. The immunogold labeling reveals that spectrin is mostly associated with the cytoplasmic surface of the plasma membrane with the exclusion of the junctional domain (arrow). Vesicular profiles with end-on attachment of filaments are also decorated by the gold particles (arrow).

soluble G-actin, which binds specifically to DNAse-I (Benedetti *et al.*, 1981). Monospecific antibody and rhodamine-labeled phalloidin show that in close association with the plasma membrane another macromolecular form of actin is present. This macromolecular form of actin is topographically distributed in close association with lens fiber spectrin, as can be illustrated by double-labeling fluorescence experiments (Fig. 3-1).

Spectrin (fodrin) and actin have also been characterized by immunoblotting experiments.

It is well known that in red blood cells and in other types of nonerythroid cells the association of actin and spectrin forms a subcortical cytoskeletal meshwork that plays an important role in the stability of the plasma membrane, restricting the mobility of transmembrane proteins, and favors and assists the assembly of receptor polypeptides essential for the transmission, transduction, and amplification of regulatory signals (Burridge *et al.*, 1982; Bennett, 1985).

In erythroid cells and likely in other types of cells the association of the spectrin and actin meshwork with the inner plasma membrane surface is mediated by another protein, ankyrin, which binds specifically to β-spectrin (Bennett, 1985; Hall and Bennett, 1987; Niggli and Burger, 1987; Moon and Lazarides, 1984; Lazarides and Nelson, 1982). The firm attachment of ankyrin to the plasma membrane is known to be mediated by two types of molecular interactions. The amphitropic ankyrin subunit (Bum, 1988) interacts with the cytoplasmic domain of band 3, the anion carrier of the red blood cell plasma membrane. The same situation likely occurs in lens fiber plasma membrane, where the presence of band 3 can be demonstrated by immunofluorescence using a monospecific antibody raised against the cytoplasmic domain of this transmembrane protein (Fig. 3-2). Furthermore, in the red blood cell ankyrin is also acylated with long-chain fatty acids. The acylation of this protein is characterized by a rapid turnover, suggesting that this posttranslational modification of the amphitropic ankyrin is of regulatory significance (cf. Niggli and Burger, 1987; Geiger, 1985; Drenckhanh and Bennett, 1987).

The reversibility of the acylation could regulate the transient interaction of ankyrin with the lipid bilayer aside from the binding of this polypeptide to the anion transporter band 3.

So far very little is known about the intimate relationship between amphitropic cytoskeletal proteins and plasma membrane proteins in lens fibers. Recent work in our laboratory has shown that ankyrin is associated with lens fiber plasma membrane, as can be revealed by immunocytochemical probes and immunoblot using a specific antibody against ankyrin (Fig. 3-3). We anticipate that the intermediate vimentin filaments are in close association with the inner cytoplasmic surface of the lens fiber plasma membrane. Freeze-fracture experiments make it possible to visualize the terminal segment of IF apparently exposed at the hydrophobic fracture face of the plasma membrane. Furthermore, newly synthesized vimentin becomes tightly associated with urea-extracted lens fiber plasma membrane added to a reticulocyte lysate programmed with lens polyribosomes (Ramaekers *et al.*, 1982). Once associated with the plasma membrane, vimentin resists 6 M urea extraction, suggesting that this intermediate filament constituent becomes amphitropic in nature and tightly associated to the membrane (Ramaekers *et al.*, 1982). The membrane association of vimentin occurs even when the translation is completed.

New interesting experimental data concern the interaction of vimentin with ankyrin. This latter amphithropic polypeptide constitutes the major attachment site of the vimentin to the inner surface of the plasma membrane in nucleated erythrocytes and possibly elsewhere (Georgatos and Blobel, 1987a; Georgatos *et al.*, 1985; Georgatos and Marchesi, 1985). The ankyrin–vimentin association in a noncooperative manner prevents deploymerization of IF. This interaction likely blocks the NH_2 terminal of vimentin, which is known to be implicated in the polymerization process of IF. In other words, vimentin is "capped" by ankyrin at its NH_2 terminal and thus anchored to the plasma membrane. Conversely, polymerization of IF is triggered by the interaction of vimentin with lamin B, which is a con-

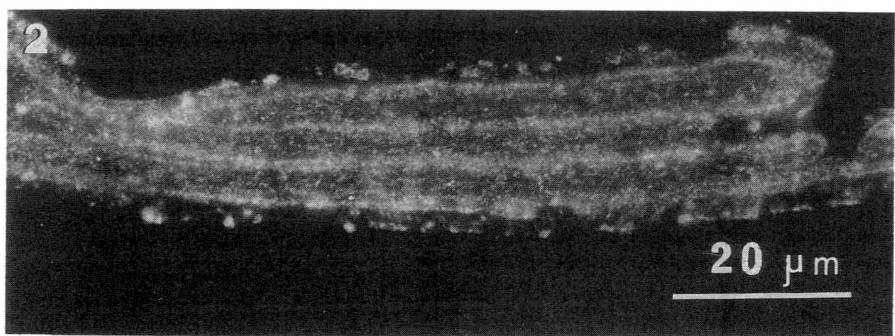

FIGURE 3-2. Immunofluorescence of isolated fiber ghost incubated with antibody raised against band 3 cytoplasmic domain. The band 3 seems to be rather uniformly distributed.

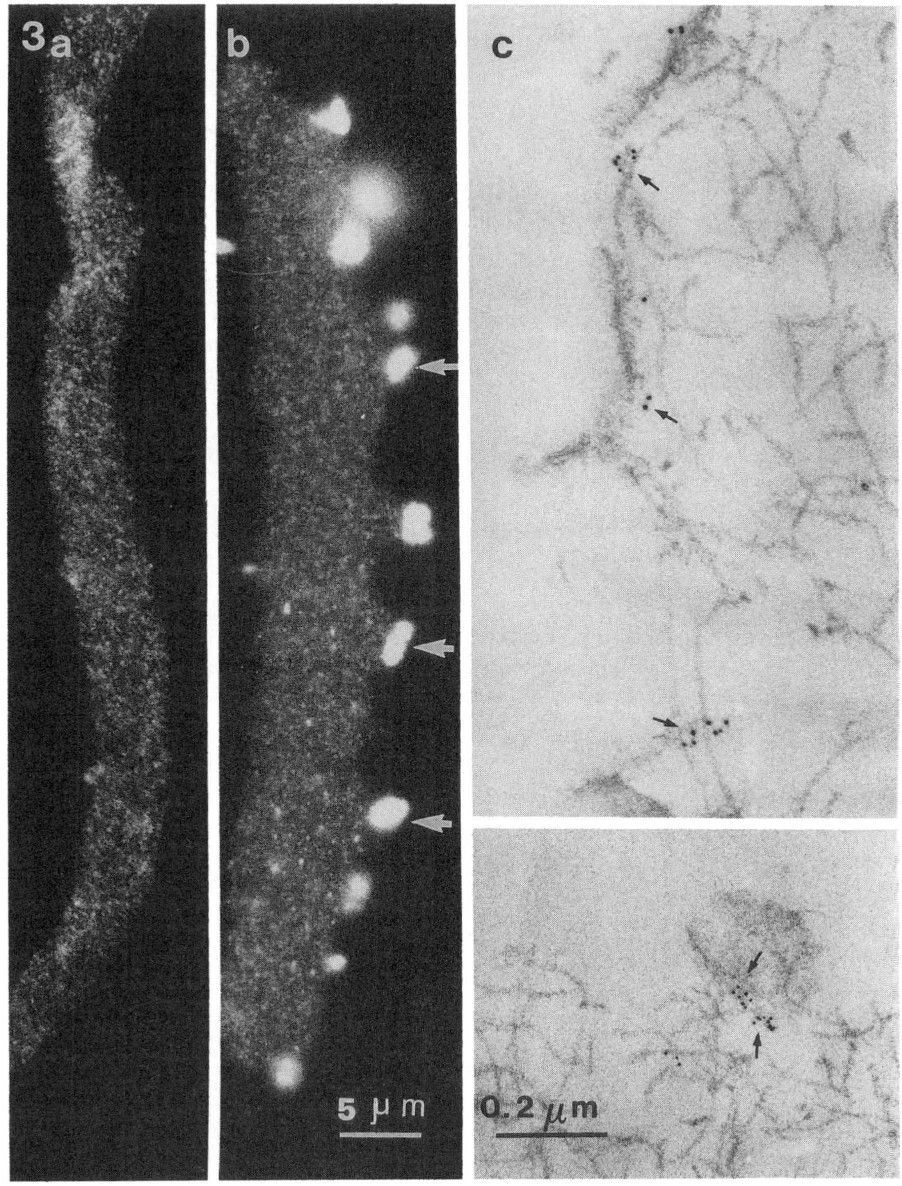

FIGURE 3-3. Isolated fiber ghost immunolabeled with monospecific antibody against ankyrin. (a, b) Note that the fluorescent ankyrin is uniformly distributed in the ghost and also heavily accumulated in the ball and sockets interlocking the fibers (arrow). (c) Immunogold labeling of an isolated ghost showing that the gold particles are distributed along the plasma membrane where IF are attached (arrow).

stituent of the nuclear envelope (Krohne and Benavente, 1986; Georgatos and Blobel, 1987b). These elegant experiments showed that IF subunits may be directly and cooperatively anchored to the nuclear lamina via the carboxyl-terminal segment of vimentin, providing a continuous network connecting the plasma membrane skeleton with the karyoskeleton.

Recent observations by Cartaud *et al.* (1989) demonstrate that *Torpedo marmorata* electrocytes are characterized by two functionally distinct and opposite plasma membrane domains. One plasma membrane domain (postsynaptic membrane) is rich in acetylcholine receptors stabilized by a 43-kDa protein. The other, noninnervated membrane is loaded with Na^+,K^+-ATPase. Intermediate desmin filaments run from the postsynaptic membrane to the opposite, noninnervated one, forming a dense transverse network. Ankyrin is associated with the noninnervated plasma membrane, whereas the

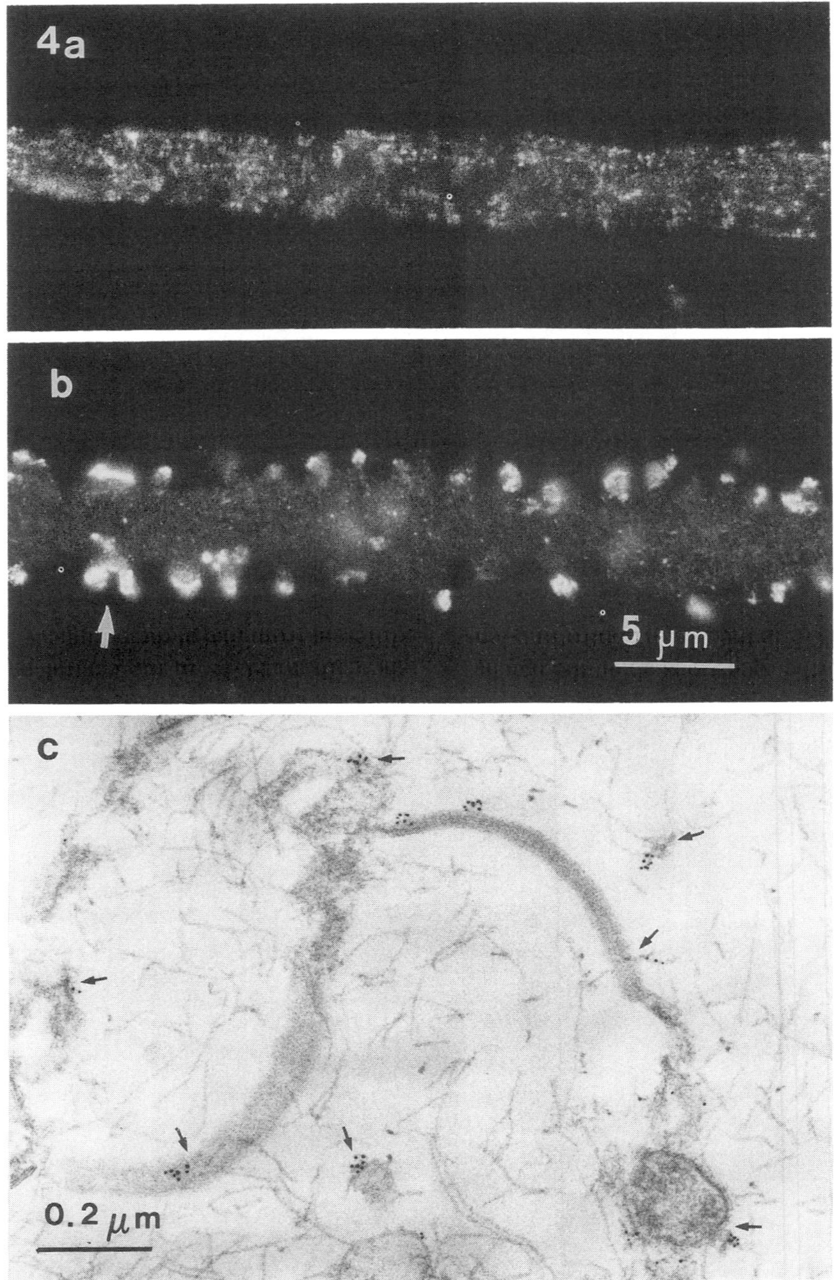

FIGURE 3-4. Isolated fiber ghost immunolabeled with antibody directed against lamin B. (a, b) Immunofluorescence showing that the polypeptide revealed by this antibody has a rather uniform distribution or loads the ball and sockets (arrow). (c) Immunogold labeling of an isolated ghost. Note that small gold particle clusters are associated with membrane profiles where IF have end-on attachment (arrows).

acetylcholine-receptor-rich postsynaptic membrane comprises a protein of about 54 kDa that is recognized by both immunofluorescent labeling experiments and immunoblot with a monospecific antibody for nuclear lamin B.

These interesting results may indicate that intermediate filaments are vectorially inserted into opposite plasma membrane domains in an asymmetric cell type by, respectively, ankyrin and a polypeptide that shares with nuclear lamin B some common epitopes and the capacity of linking intermediate filament subunits.

Our experimental evidence indicates that a polypeptide that is recognized by the same mono-

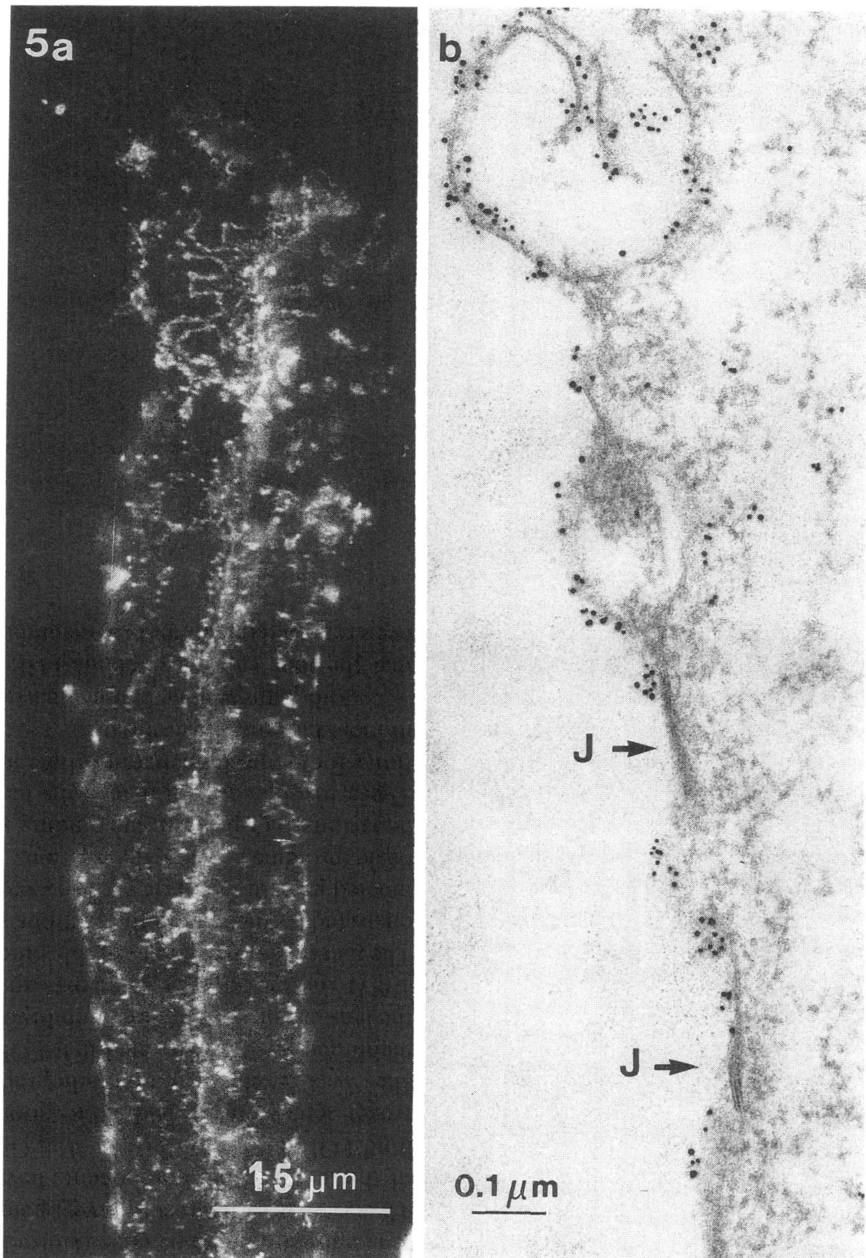

FIGURE 3-5. Immunolabeling of isolated fiber ghost incubated with monospecific antibody against vinculin. (a) The immunofluorescence show that vinculin distribution is not uniform. Spots and linear profiles are labeled. (b) The immunogold labeling technique shows that vinculin is distributed along the general plasma membrane profiles and is excluded from the junctional domains (arrows).

specific antibody against lamin B is also in close association with the inner surface of the lens fiber plasma membrane and with other membrane profiles within the fibers. We may then assume that in lens fiber individual intermediate filaments are alternately polarized, forming a continuous network connecting opposite sides of the lens fiber plasma membranes where ankyrin and lamin-B-like intermediate filaments receptors are codistributed (Figs. 3-3 and 3-4).

Experiments are in progress in our laboratory to better characterize, by immunoblot and two-

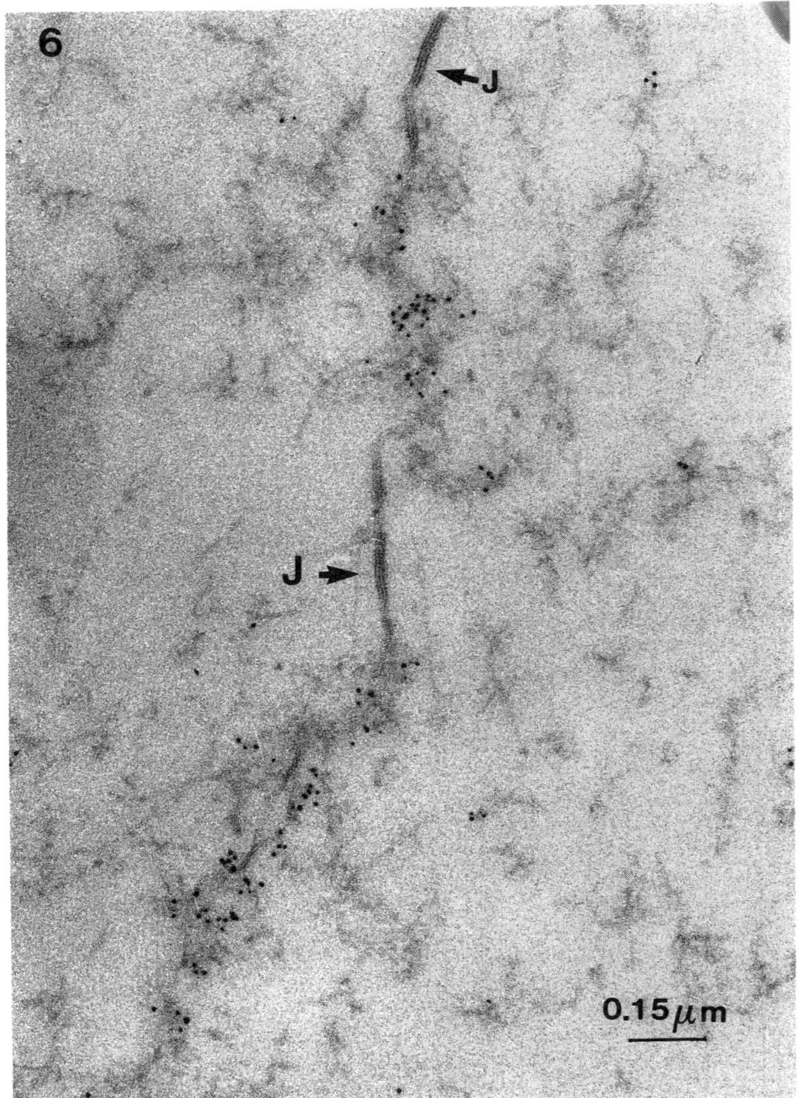

FIGURE 3-6. Immunogold labeling of isolated fiber ghost with antibody directed against α-actinin. The immunogold particles are associated with the plasma membrane where filaments are attached. The junctional domain is not labeled (arrow).

dimensional gel electrophoresis, the biochemical nature of the lens lamin-B-like intermediate filaments receptor protein(s).

Another amphitropic protein, vinculin, has also been revealed by fluorescence and immunogold labeling techniques both in intact lens cryostat sections and in lens fiber ghosts (Fig. 3-5). Vinculin is a good candidate for anchoring the actin filaments to the cytoplasmic inner surface of the plasma membrane. Several observations lead to the assumption that vinculin, via its hydrophobic domain, could be directly inserted into the bilayer (Niggli and Burger, 1987; Geiger, 1985). Another way of anchoring vinculin to the membrane lipid leaflet could be by its acylation. Whatever the molecular nature of the link between vinculin and the plasma membrane will be, it is certainly true that this amphitropic polypeptide represents a nucleation site for actin polymerization that is a unidirectional and polarized assembly process (Buck and Horwitz, 1987; Burridge *et al.*, 1987). Furthermore, vinculin, in addition to its role of anchoring the transmembrane receptor polypeptide that generates cell adhesion (fibronectin, integrin, CAM), plays an important role in the assembly of junctional domains, in particular cell adhesion plaques (Geiger, 1985; Burridge *et al.*,

1988). Recently, interesting observations by Franke *et al.* (1987) have revealed that plakoglobin is a component of the filamentous subplasmalemma coat of lens cells. This protein characterizes the intermembrane adhering junctions in close connection with other cytoskeletal proteins such as vinculin, spectrin, and actin, which are the object of our present investigations.

It is noteworthy that in our immunofluorescence and immunogold labeling experiments vinculin appears not to be directly associated with the inner cytoplasmic surface of the junctional domain. This protein is concentrated in the general plasma membrane inner cytoplasmic surface flanking the gap junctions (Fig. 3-5). Hence, vinculin could generate, in the plane of the lipid bilayer, a tangential segregation of protein oligomers specifically involved in cell-to-cell interaction and communication.

In this view vinculin would not stabilize the junctional transmembrane oligomers directly but indirectly by the formation of a tightly bound filamentous network surrounding the junctional areas. It is widely accepted that vinculin is a linking protein for actin; nevertheless, increasing number of examples suggest that also other actin-binding proteins may be involved in the assembly of the plasma membrane–cytoskeleton complex.

α-Actinin has been localized in actin stress fibers and in the microfilament network associated with the inner cytoplasmic side of the plasma membrane (Langanger *et al.*, 1984). α-Actinin is indeed present as a constituent of the filamentous network found in close association with the cytoplasmic surface of the lens fiber plasma membrane. This polypeptide is codistributed with vinculin and spectrin in the general plasma membrane domain where actin filaments have end-on attachment to the membrane leaflet. It is also remarkable that, like vinculin, α-actinin flanks and segregates the junctional domains (Fig. 3-6).

CONCLUDING REMARKS

Our data imply that MP26 biosynthesis is one of the key steps of cell surface domain formation during the terminal differentiation of the lens fibers. However, the mechanisms controlling the assembly of the various patterns that MP26 may assume in the lipid bilayer and eventually the transition of one type of assembly into another are still not fully understood.

We have anticipated that the formation of various membrane domains could be depicted as a self-assembly of repeating identical or quasiequivalent related protein oligomers that appears to be dependent on the interaction of the transmembrane protein with other membrane and cytoskeletal constituents. A variety of cytoskeletal proteins amphitropic in nature have been identified by immunocytochemical techniques. These constituents, including spectrin, actin, vinculin, and α-actinin, form a filamentous network associated with the inner cytoplasmic surface of the plasma membrane. These proteins, linked to the lipid bilayer in a variety of molecular interactions, are involved in the vectorial assembly of actin filaments. In addition, the amphitropic protein ankyrin may establish a link among the cytoskeleton, transmembrane proteins such as the band 3 anion carrier, and probably the lipid bilayer by acylation.

Our recent observations, still in progress, show that glycoconnectin β is present in the lens fiber plasma membrane, most probably linking the bilayer with the cytoskeletal network (cf. Anderson and Lovrien, 1984).

An original type of membrane intermediate filament interaction is proposed wherein on one side a lamin-B-like receptor of vimentin generates the filament assembly, and ankyrin on the opposite site of the cytoplasmic membrane may "cap" and firmly anchor the intermediate filaments.

The role of vinculin and α-actinin in the junctional assembly appears to be indirect and involves the stabilization of these membrane domains not with a direct chemical and/or physical interaction with the junctional transmembrane oligomers but rather by the segregation of these elements in the plane of the membrane lipid bilayer.

Further investigations will include the biochemical and immunochemical characterization of the amphitropic proteins involved in the architecture and formation of the plasma membrane–cytoskeleton complex during aging of the lens fibers.

Acknowledgments. We are grateful to Prof. H. Bloemendal (University of Nijmegen, Holland) for helpful discussions, and to Dr. D. Louvard (Institut Pasteur, Paris) and Dr. L. A. Pradel (Institut de Biologie, Physico-Chimie, Paris) for providing us with specific monoclonal and/or polyclonal antibodies directed against α-actinin, vinculin, ankyrin, spectrin, and band 3. We thank Dr. L. Luzzatto (Royal College of Medicine, London) for giving us monoclonal antibody against glycoconnectin β. This work was supported by Institut National de la Santé et de la Recherche Médicale (INSERM), France, grant CRE 880002, and by the ALCON Research Institute award (granted to the senior author).

REFERENCES

Anderson, R. A., and Lovrien, R. E., 1984, Glycophorin is linked by band 4.1 protein to the human erythrocyte membrane skeleton, *Nature* **307:**655–658.

Benedetti, E. L., Dunia, I., Ramaekers, F. C. S., and Kibbelaar, M. A., 1981, Lenticular plasma membranes and cytoskeleton, in: *Molecular and Cellular Biology of the Eye Lens* (H. Bloemendal, ed.), John Wiley and Sons, New York, pp. 137–184.

Bennett, V., 1985, The membrane skeleton of human erythrocytes and its implications for more complex cells, *Annu. Rev. Biochem.* **54:**273–304.

Buck, C. A., and Horwitz, A. F., 1987, Cell surface receptors for extracellular matrix molecules, *Annu. Rev. Cell Biol.* **3:**179–205.

Bum, P., 1988, Amphitropic proteins: A new class of membrane proteins, *Trends Biochem. Sci.* **13:**79–83.

Burridge, K., Kelly, T., and Mangeat, P., 1982, Nonerythrocyte spectrins: Actin–membrane attachment proteins occurring in many cell types, *J. Cell Biol.* **95:**478–486.

Burridge, K., Beckerle, M., Croall, D., and Horwitz, A., 1987, A transmembrane link between the extracellular matrix and the cytoskeleton, in: *Molecular Mechanisms in the Regulation of Cell Behavior, Modern Cell Biology,* Vol. 5 (B. H. Satir, ed.), Alan, R. Liss, New York, pp. 147–149.

Burridge, K., Fath, K., Kelly, T., Nuckolls, G., and Turner, C., 1988, Focal adhesions: Transmembrane junctions between the extracellular matrix and the cytoskeleton, *Annu. Rev. Cell Biol.* **4:**487–525.

Cartaud, A., Courvalin, J. C., Ludosky, M. A., and Cartaud, J., 1989, Presence of an immunologically-related form of lamin B in the postsynaptic membrane of *Torpedo marmorata* electrocyte, *J. Cell Biol.* **109:**1745–1752.

Chasis, J. A., and Mohandas, N., 1986, Erythrocyte membrane deformability and stability: Two distinct membrane properties that are independently regulated by skeletal proteins associations, *J. Cell Biol.* **103:**343–350.

Drenckhahn, D., and Bennett, V., 1987, Polarized distribution of M. 210,000 and 190,000 analogs of erythrocyte ankyrin along the plasma membrane transporting epithelia, neurons and photoreceptors, *Eur. J. Cell Biol.* **43:**479–486.

Dunia, I., Lien, D. N., Manenti, S., and Benedetti, E. L., 1985, Dilemmas of the structural and biochemical organization of lens membranes during differentiation and aging. *Curr. Eye Res.* **4**(11):1219–1234.

Dunia, I., Manenti, S., Rousselet, A., and Benedetti, E. L., 1987, Electron microscopic observations of reconstituted proteoliposomes with the purified major intrinsic membrane protein of eye lens fibers, *J. Cell Biol.* **105:**1679–1689.

Franke, W. W., Kapprell, H. P., and Cowin, P., 1987, Plakoglobin is a component of the filamentous subplasmalemmal coat of lens cells, *J. Cell Biol.* **43:**301–315.

Geiger, B., 1985, Microfilament–membrane interaction, *Trends Biochem. Sci.* **10:**456–461.

Georgatos, S. D., and Blobel, G., 1987a, Two distinct attachment sites for vimentin along the plasma membrane and the nuclear envelope in avian erythrocytes: A basis for a vectorial assembly of intermediate filaments, *J. Cell Biol.* **105:**105–115.

Georgatos, S. D., and Blobel, G., 1987b, Lamin B constitutes an intermediate filament attachment site at the nuclear envelope, *J. Cell Biol.* **105:**117–125.

Georgatos, S. D., and Marchesi, V. T., 1985, The binding of vimentin to human erythrocyte membranes: A model system for the study of intermediate filament–membrane interactions, *J. Cell Biol.* **100:**1955–1961.

Georgatos, S. D., Weaver, D., and Marchesi, V. T., 1985, Site specificity in vimentin–membrane interactions: Intermediate filament subunits associated with the plasma membrane via their head domains, *J. Cell Biol.* **100:**1962–1967.

Girsh, S. J., and Peracchia, C., 1985, Lens cell-to-cell channel protein. I. Self assembly into liposomes and permeability regulation by calmodulin, *J. Membr. Biol.* **83:**217–225.

Godman, G. C., Miranda, A. F., Deitch, A. D., and Tanenbaum, S. W., 1975, Action of cytochalasin D on cells of established lines. III Zeiosis and movements at the cell surface, *J. Cell Biol.* **64:**644–660.

Hall, T. G., and Bennett, V., 1987, Regulatory domains of erythrocyte ankyrin, *J. Biol. Chem.* **262**(22):10537–10545.

Harding, C. V., Susan, R. S., Lo, W. K., Bobrowski, W. F., Maisel, H., and Chylack, L. T., 1985, The structure of the human cataractous lens, in: *The Ocular Lens* (H. Maisel, ed.), Marcel Dekker, New York, pp. 367–389.

Johnson, K. R., Lampe, P. D., Hur, K. C., Louis, C. F., and Johnson, R. G., 1986, A lens intercellular junction protein, MP26, is a phosphoprotein, *J. Cell Biol.* **102:**1334–1343.

Kistler, J., and Bullivant, S., 1989, Structural and molecular biology of the eye lens membranes, *Crit. Rev. Biochem. Mol. Biol.* **24**(2):151–181.

Krohne, G., and Benavente, R., 1986, The nuclear lamins, *Exp. Cell. Res.* **162:**1–10.

Langanger, G., De Mey, J., Moeremans, M., Daneels, G. U., De Brabander, M., and Small, J. V., 1984, Ultrastructural localization of α-actinin and filamin in cultured cells with the immunogold staining (IGS) method, *J. Cell Biol.* **99:**1324–1334.

Lazarides, E., and Moon, R. T., 1984, Assembly and topogenesis of the spectrin-based membrane skeleton in erythroid development, *Cell* **37:**354–356.

Lazarides, E., and Nelson, W. J., 1982, Expression of spectrin in nonerythroid cells, *Cell* **31:**505–508.

Lien, N. D., Paroutaud, P., Dunia, I., Benedetti, E. L., and Hoebeke, J., 1985, Sequence analysis of peptide fragments from the intrinsic membrane protein of calf lens fibers MP26 and its natural maturation product MP22, *FEBS Lett.* **181**(1):74–78.

Manenti, S., Dunia, I., Le Maire, M., and Benedetti, E. L., 1988, High-performance liquid chromatography of the main polypeptide (MP26) of lens fiber plasma membranes solubilized with *n*-octyl β-D-glycopyranoside, *FEBS Lett.* **233**(1):148–152.

Moon, R. T., and Lazarides, E., 1984, Biogenesis of the avian erythroid membrane skeleton: Receptor-mediated assembly and stabilization of ankyrin (goblin) and spectrin, *J. Cell Biol.* **98:**1899–1904.

Niggli, V., and Burger, M. M., 1987, Interaction of the cytoskeleton with the plasma membrane, *J. Membr. Biol.* **100:**97–121.

Palek, J., 1987, Hereditary elliptocytosis, spherocytosis and related disorders: Consequences of a deficiency or a mutation of membrane skeletal proteins, *Blood Rev.* **1**:147–168.

Partridge, T. A., Morgan, J. E., Coulton, G. R., Hoffman, E. P., and Kunkel, L. M., 1989, Conversion of *mdx* myofibres from dystrophin-negative to positive by injection of normal myoblasts, *Nature* **337**:176–179.

Ramaekers, F. C. S., Dunia, I., Dodemont, H. J., Benedetti, E. L., and Bloemendal, H., 1982, Lenticular intermediate-sized filaments: Biosynthesis and interaction with plasma membrane, *Proc. Natl. Acad. Sci. U.S.A.* **79**:3208–3212.

Revel, J. P., Yancey, S. B., Nicholson, B. J., and Hoh, J., 1987, Sequence diversity of gap junction proteins, in: *Junctional Complexes of Epithelial Cells, Ciba Foundation Symposium 125* (G. Bock and S. Clark, eds.), John Wiley & Sons, Chichester, pp. 108–127.

4

Lens Calcium and Cataract

GEORGE DUNCAN, JULIA M. MARCANTONIO, and JULIE TOMLINSON

CALCIUM AND HUMAN CATARACT

Calcium was first implicated in the cataract process in the early years of this century when it was observed that many human cataractous lenses had very high calcium concentrations compared with clear donor lenses (Adams, 1929). The early studies had shown that not all cataractous lenses had high calcium values, and this has recently been confirmed (Duncan and Bushell, 1975). An increase in lens calcium is accompanied by white, light-scattering opacities, and these are often found in highly localized regions of the outer cortex (Duncan and Jacob, 1984). Nuclear brunescence, or nuclear cataract, appears to originate by another process, since lenses with pure nuclear cataracts have near-normal internal calcium levels. In general, it appears that sodium and calcium increase together in cortical cataract, and both ions are normally distributed in nuclear cataracts (Maraini and Mangili, 1973).

Significant advances were made in the study of the role of calcium in human cataract by the application of ion-sensitive microelectrode techniques to the lens (Duncan and Jacob, 1984). It became immediately obvious from these studies that a large fraction of the total calcium in the lens is in a sequestered or bound form. For example, the total calcium content of the normal human lens measured by atomic absorption techniques is approximately 0.2 to 0.5 mM, whereas the free calcium concentration measured by ion-sensitive microelectrodes is in the micromolar region (Duncan and Jacob, 1984). It is estimated that in the normal human lens only 1% of the total calcium is in the free state. In cortical cataract there are very significant increases in both the free and total calcium content of the lens. In lenses with localized opacities it is possible to insert ion-sensitive electrodes into adjacent clear and turbid regions, and it indeed appears that those regions scattering most light also have the highest internal free calcium content (Duncan and Jacob, 1984). During the process of cortical cataract formation the total calcium can increase from 0.1 mM to over 20 mM while the total sodium increases from 20 mM to over 200 mM.

It appears that small but significant increases in lens sodium and calcium can also occur as a result of the aging process itself (Duncan *et al.*, 1989a). Measurements of sodium content, sodium influx, potassium efflux, and electrical conductance all show a significant increase with age. Since the lens voltage also depolarizes, it is likely that some progressive increase in lens permeability is involved. In fact, the increase in lens sodium/potassium permeability ratio exactly parallels the increase with age in lens optical density (Fig. 4-1). Accompanying these changes is a small but significant rise in lens free calcium. It therefore appears that many of the ion changes that occur in cortical cataract are already under way as a result of the aging process itself.

CALCIUM AND ANIMAL CATARACT

Cataract is not uncommon in domestic animals, especially dogs (Barnett, 1982), and isolated cases have been reported in cattle (Ashton *et al.*, 1977). It has rarely been reported however, in species in the wild, and so it was surprising to find in a survey of netted salmon from the west coast of Scotland that cataract has increased recently from near-negligible levels to around 55% (Fraser *et al.*, 1989). The cata-

GEORGE DUNCAN, JULIA M. MARCANTONIO, AND JULIE TOMLINSON • School of Biological Sciences, University of East Anglia, Norwich NR4 7TJ, Great Britain.

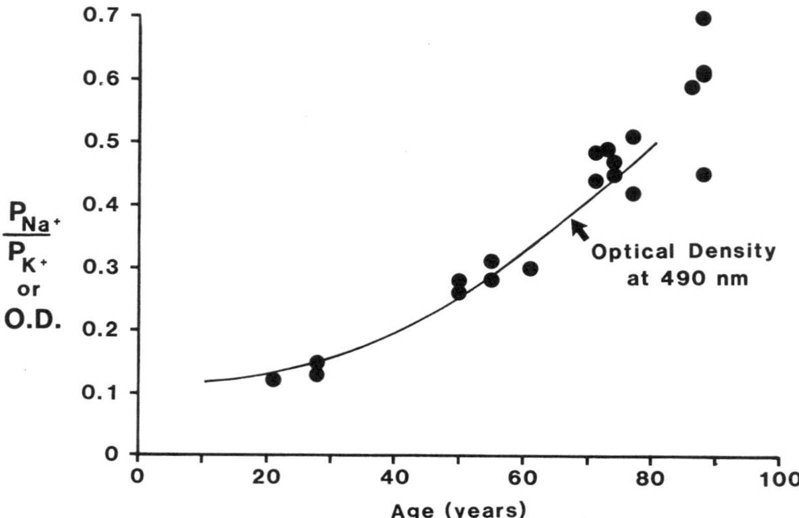

FIGURE 4-1. Relative cation permeability (P_{Na}/P_K) of human lenses as a function of age computed from measurements of ion concentration and electrical potential. The solid line shows the change with age of the mean optical density of the human lens measured at 490 nm. (See Duncan *et al.*, 1989a, for further details.)

racts could occur in one or both eyes, and in many cases only small discrete regions of the lens were affected. Ion measurements made on whole lenses demonstrated that both sodium and calcium had increased in cataractous lenses compared to clear, control salmon lenses. The increase in salmon cataract was found to parallel the exponential increase in fish farming in the west of Scotland that has occurred recently, so it is likely that some toxic pollutant is to blame.

CALCIUM IN ANIMAL MODELS

Although cataract does not normally occur in animals, a range of agents can be used to produce cataract in animal models. These include naphthalene, photosensitizing agents, selenium, etc. Cataract has also been widely studied in diabetic animals. In almost every case, cortical or osmotic cataracts are formed, and there is a general disturbance of the ion balance of the lens. In a study of cataracts produced by feeding rats with the photosensitizing agent 8-methoxypsoralen (Boutros *et al.*, 1984), it was pointed out that changes in lens calcium followed very significant changes in lens sodium and potassium and that the major changes in lens protein content and transparency appeared to accompany the calcium changes. A very recent study of the diabetic rat lens, which develops localized opacities, has also demonstrated that regions with the greatest light scatter also show the greatest increase in free calcium (Hightower *et al.*, 1989). These data are very similar to the ion-sensitive microelectrode studies carried out on human lenses (Duncan and Jacob, 1984).

An interesting animal model is provided by administering sodium selenite to young rats (Shearer *et al.*, 1987); after several days dense nuclear cataracts are produced. Again it appears that the free calcium levels are highest in those regions showing the greatest light scatter. Lens proteolysis and protein loss again occur during this type of cataract, and it has been suggested that the calcium-activated protease calpain II is responsible (Shearer *et al.*, 1987).

LENS CATARACT STUDIES *IN VITRO*

Dorothy Rose Adams (1929) was probably the first to carry out physiological experiments with isolated lenses concerning the role of calcium in cataract. She observed that opacities could be produced in rabbit lenses by incubating them in media with artificially elevated calcium levels. During the 1950s, however, the emphasis shifted to the importance of calcium in maintaining transparency when it was observed that incubating lenses in media with artificially reduced calcium levels caused them to swell and, to a certain extent, scatter light. It was also noted that medical conditions leading to hypocalcaemia were associated with an increased incidence of cataract (see review by Delamere and Paterson, 1981).

We now know that a certain level of external calcium is required to maintain the normal sodium and potassium permeability of the lens membranes (Jacob and Duncan, 1981). A reduction of external calcium below 0.5 mM (approximately one half the normal value) leads to an increase in both sodium and potassium movement across the lens membranes. Interestingly, calcium also appears to control its own permeability, as a decrease in external calcium also stimulates calcium entry into the lens (Delamere and Paterson, 1981). It is possible that a single channel may be responsible for all of these stimulated movements, as it has recently been shown by patch-clamp techniques (Jacob *et al.*, 1985; Cooper *et al.*, 1986) that reducing external calcium leads to the activation of a nonspecific cation channel in the lens membranes. It appears that sodium, potassium, and calcium can move through this channel, and, although it is largely quiescent in the normal lens, it can be activated by pressure (Cooper *et al.*, 1986) and membrane sulfhydryl oxidation (Duncan *et al.*, 1988) as well as by reducing external calcium (Jacob *et al.*, 1985). This channel could therefore provide a major route for calcium entry into the lens during the cataract process.

It is also possible that calcium enters through calcium-specific channels, and although there are several reports indicating that such channels do not provide a major route in the normal lens (Rae, 1985; Delamere and Paterson, 1985), there is one report that indicates that calcium entry into the rat lens during diabetic cataract can be prevented by administering drugs such as verapamil that are presumed to be specific blockers of calcium channels (Fleckenstein, 1983).

Human lens studies showed that, generally, sodium and calcium increase together in cataract, and so it was not possible to discriminate between the relative importance of sodium and calcium in opacification. With long-term *in vitro* studies it is possible to develop strategies in which sodium alone is allowed to increase (Marcantonio *et al.*, 1986) or where only a calcium increase is permitted (Duncan and Jacob, 1984). A sodium increase alone produces little opacification and negligible proteolysis, whereas a specific calcium increase produces marked opacification. In long-term organ culture studies it is also possible to demonstrate that lenses with a high calcium content also contain significant amounts of high-molecular-weight protein aggregates (Marcantonio *et al.*, 1986).

As previously mentioned, calpain II has for some time been implicated in crystallin proteolysis. However, recent studies have shown that the lens cytoskeleton is equally sensitive to breakdown in the presence of high internal calcium (Truscott *et al.*, 1990), and this may well explain the catastrophic loss of lens structure and transparency that occurs in cortical cataract.

CONTROL OF INTRACELLULAR CALCIUM

Calcium regulation occurs through a balance between passive influx down a large concentration gradient and active efflux mechanisms through carriers (Fig. 4-2). One carrier system is coupled directly via a calmodulin-dependent, calcium-activated ATPase, and this active process has been characterized in both fish and mammalian lenses (Iwata, 1985; Iimuro *et al.*, 1987). The requirement for an active regulatory process can clearly be seen, for example, if the glucose concentration in the medium bathing a cultured lens is lowered (Hightower and Reddy, 1981; Duncan and Jacob, 1984). This results in a large increase in both the free and total calcium in the lens, and light-scattering occurs in those regions that have suffered the greatest calcium increase (Duncan and Jacob, 1984).

A further carrier-based regulatory process, which is widely distributed in nerve and muscle membranes, is indirectly linked to the transmembrane sodium gradient through a coupled Na^+/Ca^{2+} exchange mechanism (Dipolo and Beauge, 1983). Such a system plays a critical role in regulating internal calcium in rod outer segments, for example, but previous experiments have failed to find such an exchange mechanism in the rabbit lens (Hightower *et al.*, 1980; McGahan *et al.*, 1983; Delamere and Paterson, 1985). If such a system did exist in the lens, then reducing the transmembrane sodium gradient either by decreasing the external sodium concentration or by increasing the internal level should lead to a decrease in the efflux of calcium from the lens.

The simplest way to investigate calcium efflux systems is to load a lens with ^{45}Ca and subsequently measure the ^{45}Ca efflux rate constant. In most systems, including the lens, the efflux is multiexponential in nature because of contributions from the ex-

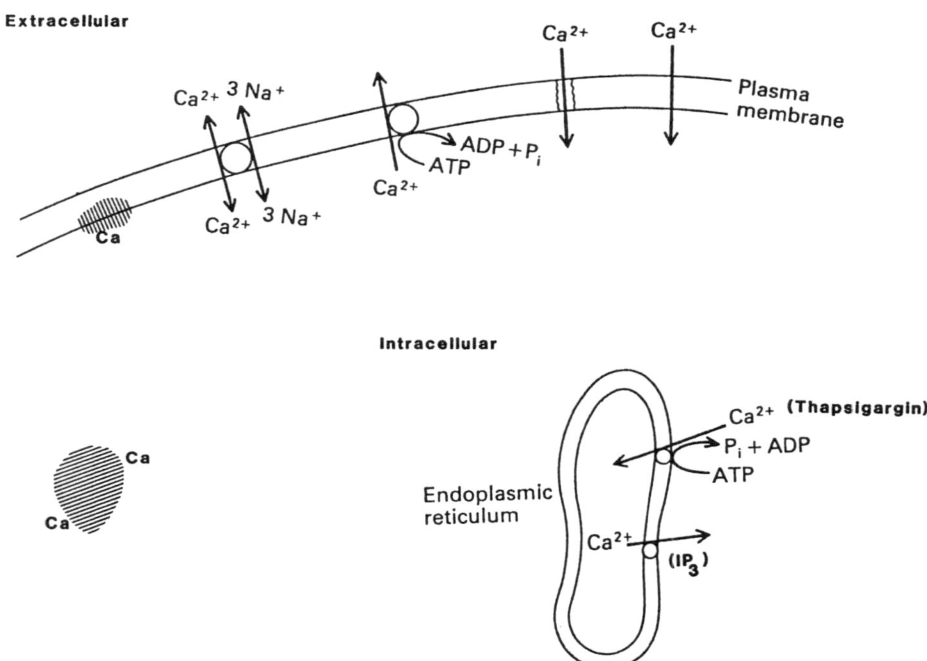

FIGURE 4-2. Outline of mechanisms for the regulation of lens internal calcium. These include influx through different types of channel and efflux via two carrier mechanisms. Calcium sequestration could occur through binding proteins in the membrane and cytosol and also by incorporation into a dynamic cytoplasmic store.

tracellular space and bound and free compartments within the lens. However, it has been found in the rat lens (Tomlinson *et al.*, 1991) that when the bound fraction is subtracted from the total calcium within the lens, then the efflux kinetics consist of only two components corresponding to the extracellular and exchangeable intracellular components. When the rate constants are calculated in this way, a rapid drop in efflux rate is seen when the external sodium is reduced (Fig. 4-3). The efflux rate returns to normal on return to the normal level of external sodium.

If the efflux is decreased and the passive influx carries on at the same rate, then the lens should fill up with calcium on exposure to a sodium-free medium. This is precisely what happens in the rat lens (Fig. 4-4). Since the lens fills up with calcium, it should scatter more light, and again the rat lens fulfills expectation (Tomlinson *et al.*, 1991).

Any agent leading to a decrease in the sodium gradient across the lens membranes should lead to an increase in internal calcium if the Na^+/Ca^{2+} exchange mechanism plays a significant role in regulating internal calcium. For example, poisoning the sodium pump by the cardiac glycoside ouabain increases internal sodium (Marcantonio *et al.*, 1986). In the rat lens (Fig. 4-5) and in the organ-cultured bovine lens (Marcantonio *et al.*, 1986), Na^+,K^+-ATPase inhibition is indeed followed by an increase in internal calcium.

Normally it appears that the inward leak of calcium into the lens is extremely low (Delamere and Paterson, 1985). Potent cataractogenic agents would include any process that increased this inward leak. Membrane sulfhydryl groups have been implicated both in aging and in cortical cataract (Alcala and Maisel, 1985), and it is interesting to find that very low concentrations of pCMPS, a specific membrane SH-complexing agent, can induce a very marked rise in internal calcium (Fig. 4-6). Accompanying this rise is a significant increase in light scatter in the lens. It is likely that calcium enters the lens in this case through nonspecific cation channels

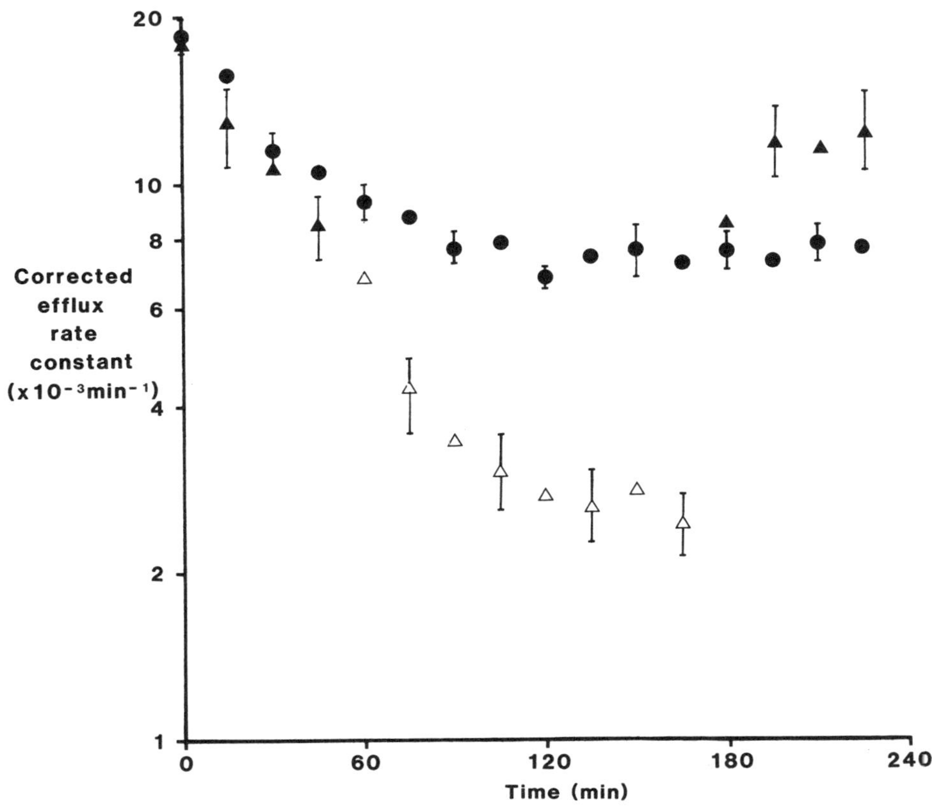

FIGURE 4-3. Efflux rate constant of exchangeable ^{45}Ca from the rat lens as a function of time. During the period in control solution (closed symbols), the rate constant is initially high because of efflux from the extracellular space; then a constant value of 7×10^{-3} min^{-1} is achieved through membrane-limited efflux. During exposure to sodium-free solution (open symbols) containing 150 mM *n*-methyl glucamine in place of sodium, the rate constant falls to below 4×10^{-3} min^{-1}. The rate constant increases dramatically on return to control solution (filled triangles).

that have been activated following membrane SH oxidation (Duncan *et al.*, 1988).

So far we have established the importance of pumps, carriers, and channels in regulating internal calcium, but there is one system, namely, the endoplasmic reticulum, that plays an important role in regulating internal calcium in many cell types, especially during stimulation by receptor systems that involve second messengers (Berridge and Irvine, 1984). Calcium is pumped into this dynamic store from the cytoplasm via a Ca^{2+}-ATPase and is released on stimulation by a range of trigger substances including IP_3 (Fig. 4-2). Either inhibiting the pump or increasing internal IP_3, for example, would be expected to lead to a loss of calcium from the store and an increase in intracellular calcium. Few strategies are available at present to increase IP_3 directly, but a novel inhibitor of the endoplasmic reticulum Ca^{2+}-ATPase has recently become available (Thastrup *et al.*, 1989). Thapsigargin, extracted from the umbelliferous plant *Thapsia garganica*, is a sesquiterpene lactone compound, and because it is extremely hydrophobic, it penetrates the cell membrane relatively rapidly.

The free calcium concentration in tissue-cultured human lens epithelial cells can be monitored using the trapped fluorescent dye fura 2 (Tsien *et al.*, 1982), and when thapsigargin (40 nM) is added to a dish of cells bathed in a calcium-free artificial aqueous humor solution, there is a rapid and transient increase in intracellular calcium corresponding to a release of calcium from intracellular stores. When the thapsigargin is removed and solution containing the normal external calcium level (1 mM) is applied, then the stores refill and can be emptied once more on reapplying thapsigargin. This time the rise in intracellular calcium is larger and more maintained (Fig. 4-7). It is likely that this dynamic store has a role to play in the normal growth and development of human lens epithelial cells.

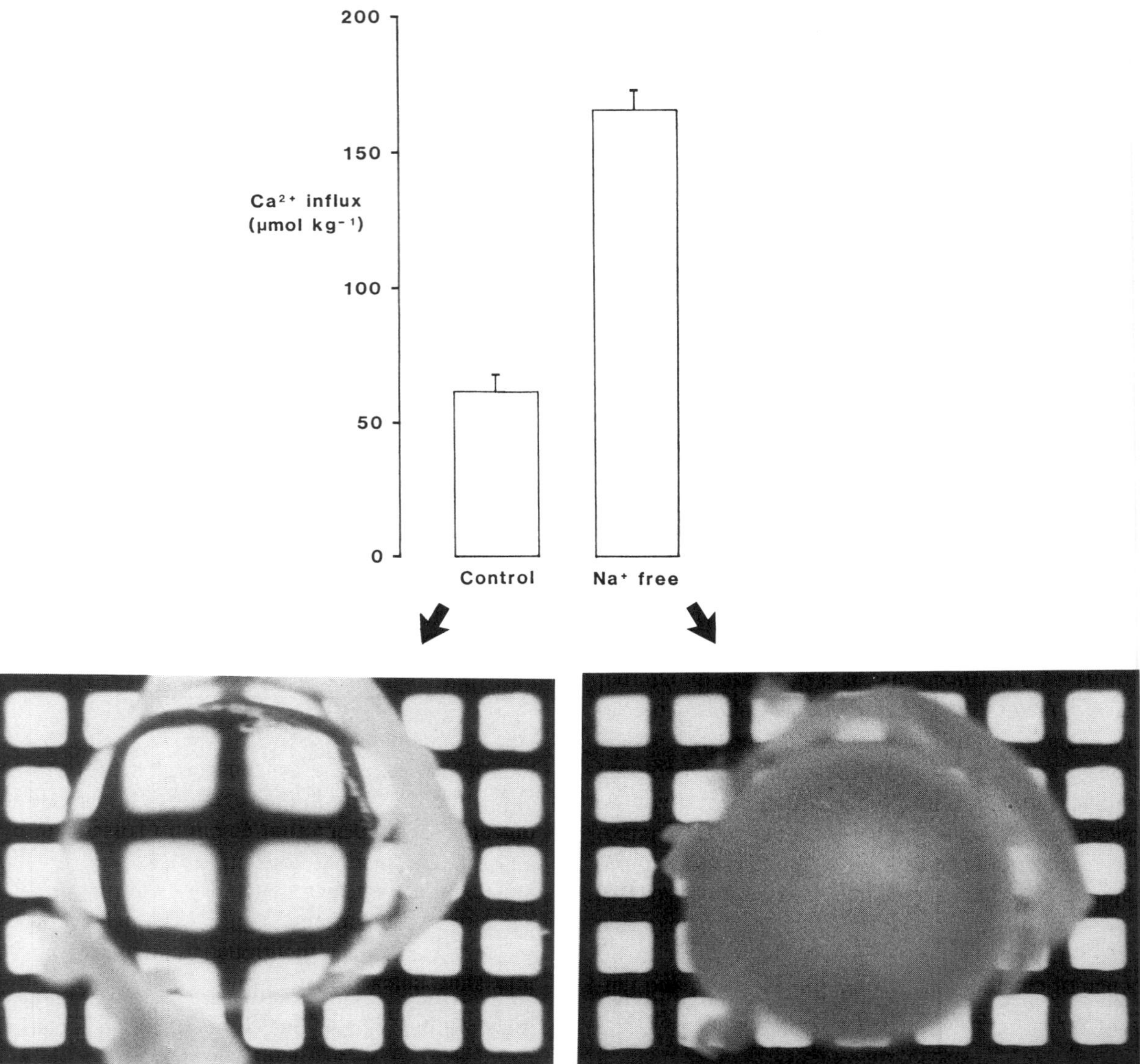

FIGURE 4-4. $^{45}Ca^{2+}$ content of rat lenses after a 5-hr influx in control or Na^+-free (*n*-methyl glucamine) medium. The data are presented as the mean ± S.E. of six lenses in each case. The lenses were also photographed against a 1-mm grid at the end of the incubation period.

CONCLUSIONS

The total calcium concentration in the normal transparent lens is in the region of 0.1 mM, whereas the free calcium in fiber cells measured by ion-sensitive microelectrodes is less than 1% of this. A large fraction of lens calcium must therefore be sequestered or bound in some way. During the development of white, light-scattering opacities, both the total sodium and calcium concentrations increase, and there is a concomitant increase in the free calcium concentration. Studies with ion-sensitive microelectrodes in both human and animal lenses have shown that the increase in free calcium is greatest in those regions that scatter most light.

The cation permeability of normal lenses also changes with age, as does the free calcium content, and it is suggested that these changes could occur as a result of the increased activation of nonspecific cation channels in the lens membranes with age. Such channels can be activated by pressure, decreasing external calcium, and membrane sulfhydryl oxidation.

The loss of transparency associated with an

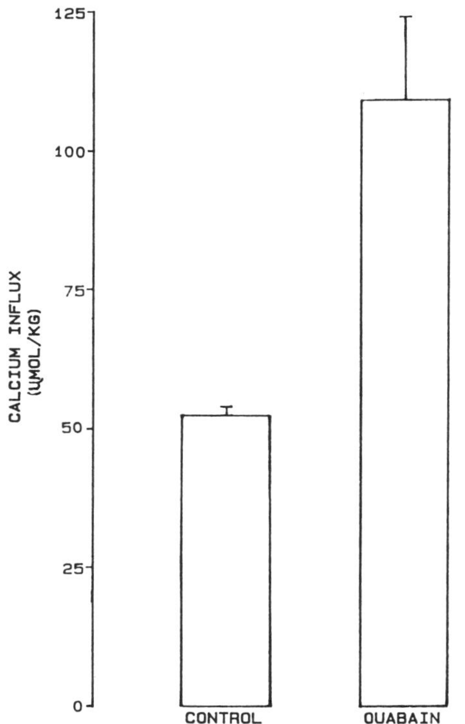

FIGURE 4-5. ^{45}Ca content of the lens after a 4-hr incubation in control medium or with ouabain (1 mM).

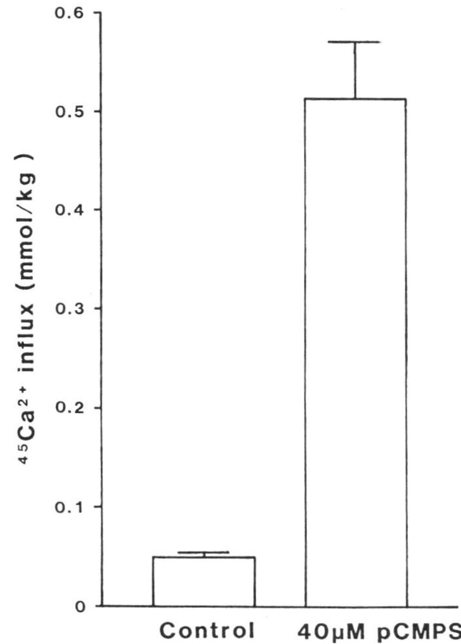

FIGURE 4-6. ^{45}Ca content of rat lenses after a 5-hr incubation in control medium or with pCMPS (40 μM).

increase in internal calcium is probably caused by the action of the protease calpain II on lens crystallins and cytoskeleton.

Intracellular lens calcium is regulated by a combination of membrane-based carrier and channel systems together with a dynamic intracellular store. A calmodulin-dependent Ca^{2+}-ATPase and an Na^{+}/Ca^{2+} exchange carrier both expel calcium from the lens, while calcium can probably enter through a number of channel systems, the most important of which in the development of cataract is likely to be a nonspecific cation channel. The free calcium in the cytoplasm of the epithelium can also be buffered by an endoplasmic reticulum store. Calcium is pumped from the cytoplasm into the store by

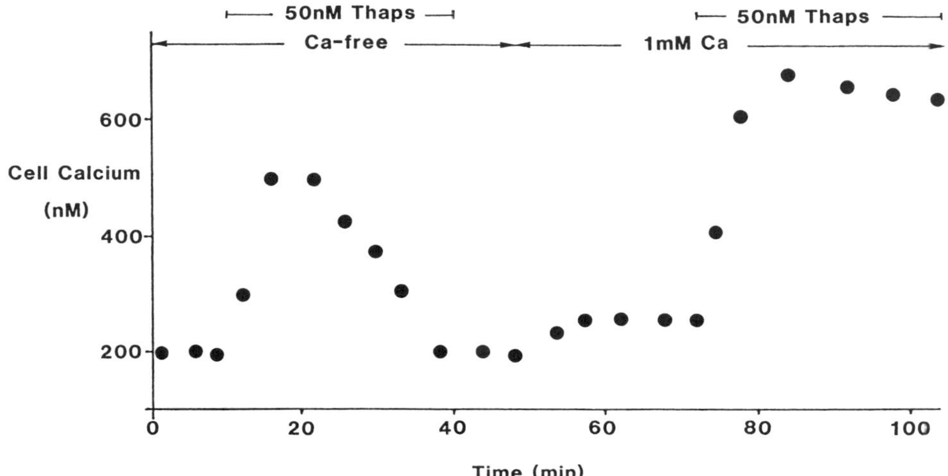

FIGURE 4-7. Calcium concentration of human tissue-cultured cells monitored by fura-2. The cells were calibrated by permeabilizing with ionophore (ionomycin) at the end of the experiment. Thapsigargin (40 nM) produces an increase in internal calcium in both the presence and absence of external calcium, and the fact that a second response can be produced shows that the dynamic store can be refilled. (See also Duncan *et al.*, 1989b.)

a thapsigargin-sensitive Ca^{2+}-ATPase, and it is suggested that emptying of the store can be triggered by second messengers such as IP_3.

REFERENCES

Adams, D. R., 1929, The role of calcium in senile cataract, *Biochem. J.* **23:**902–912.

Alcala, J., and Maisel, H., 1985, Biochemistry of lens plasma membranes and cytoskeleton, in: *The Ocular Lens* (H. Maisel, ed.), Marcel Dekker, New York, pp. 169–222.

Ashton, N., Barnett, K. C., Clay, C. E., and Clegg, F. G., 1977, Congenital nuclear cataracts in cattle, *Vet. Rec.* **100:**505–508.

Barnett, K. C., 1982, Lens opacities in the dog as models for human eye disease, *Trans. Ophthalmol. Soc. U.K.* **102:**346–349.

Berridge, M. J., and Irvine, R. F., 1984, Inositol trisphosphate, a novel second messenger in cellular signal transduction, *Nature* **312:**315–321.

Boutros, G., Koch, H. R., Jansen, R., Jacob, T. J. C., and Duncan, G., 1984, Effect of 8-methoxypsoralen on rat lens cations, membrane potential and protein levels, *Exp. Eye Res.* **38:**509–513.

Cooper, K. E., Tang, J. M., Rae, J. L., and Eisenberg, R. S., 1986, A cation channel in frog lens epithelia responsive to pressure and calcium, *J. Membr. Biol.* **93:**259–265.

Delamere, N. A., and Paterson, C. A., 1981, Hypocalcaemic cataract, in: *Mechanisms of Cataract Formation in the Human Lens* (G. Duncan, ed.), Academic Press, New York, pp. 219–236.

Delamere, N. A., and Paterson, C. A., 1985, Characteristics of ^{45}Ca uptake by the rabbit lens, *Exp. Eye Res.* **41:**11–16.

Dipolo, R., and Beauge, L., 1983, The calcium pump and sodium–calcium exchange in squid axons, *Annu. Rev. Physiol.* **45:**313–340.

Duncan, G., and Bushell, A. R., 1975, Ion analysis of human cataractous lenses, *Exp. Eye Res.* **20:**223–230.

Duncan, G., and Jacob, T. J. C., 1984, Calcium and the physiology of cataract, *Ciba Found. Symp.* **106:**132–148.

Duncan, G., Gandolfi, S. A., and Maraini, G., 1988, Diamide alters membrane Na^+ and K^+ conductances and increases internal resistance in the isolated rat lens, *Exp. Eye Res.* **47:**807–818.

Duncan, G., Hightower, K. R., Gandolfi, S. A., Tomlinson, J., and Maraini, G., 1989a, Human lens membrane cation permeability increases with age, *Invest. Ophthalmol. Vis. Sci.* **30:**1855–1859 (in press).

Duncan, G., Elliott, J. A., Webb, S. F., Dawson, A. P., Cullen, P. J., and Thastrup, O., 1989b, Calcium release from intracellular stores in tissue—cultured human lens epithelial cells, *Invest. Ophthalmol. Vis. Sci.* **30:**130a.

Fleckenstein, A., 1983, *Calcium Antogonism in Heart and Smooth Muscle,* John Wiley & Sons, New York.

Fraser, P. J., Duncan, G., and Tomlinson, J., 1989, Effects of a cholinsterase inhibitor on salmonid lens: A possible cause for the increased incidence of cataract in *Salmo solar, Exp. Eye Res.* **49:**293–298.

Hightower, K. R., and Reddy, V. N., 1981, Metabolic studies on calcium transport in mammalian lens, *Curr. Eye Res.* **1:**197–204.

Hightower, K. R., Leverenz, V., and Reddy, V. N., 1980, Calcium transport in the lens, *Invest. Ophthalmol. Vis. Sci.* **19:**1059–1064.

Hightower, K. R., Riley, M. V., and McCready, J., 1989, Regional distribution of calcium in alloxan diabetic rabbit lens, *Curr. Eye Res.* **8:**517–522.

Iimuro, A., Takenhana, M., and Iwata, S., 1987, Influence of calmodulin antagonists on Ca^{2+} transport in the lens, *Ophthalmic Res.* **19:**95–101.

Iwata, S., 1985, Calcium-pump and its modulator in the lens: A review, *Curr. Eye Res.* **4:**299–304.

Jacob, T. J. C., and Duncan, G., 1981, Calcium controls both sodium and potassium permeability of lens membranes, *Exp. Eye Res.* **33:**85–93.

Jacob, T. J. C., Bangham, J. A., and Duncan, G., 1985, The characterisation of a cation channel on the optical surface of the frog lens epithelium, *Q. J. Physiol.* **70:**403–421.

Maraini, G., and Mangili, R., 1973, Differences in proteins and in the water balance of the lens in nuclear and cortical types of senile cataract, *Ciba Found. Symp.* **19:**79–94.

Marcantonio, J. M., Duncan, G., and Rink, H., 1986, Calcium-induced opacification and loss of protein in the organ-cultured bovine lens, *Exp. Eye Res.* **42:**617–630.

McGahan, M. C., Chin, B., and Bentley, P. J., 1983, Calcium metabolism in the rabbit lens, *Exp. Eye Res.* **36:**57–65.

Rae, J. L., 1985, The application of patch clamp methods to ocular epithelia, *Curr. Eye Res.* **4:**409–424.

Shearer, T. R., David, L. L., and Anderson, R. S., 1987, Selenite cataract: A review, *Curr. Eye Res.* **6:**289–296.

Thastrup, O., Dawson, A. P., Cullen, P. J., Sharff, O., Foder, B., Bjerrum, P. J., Christensen, S. B., and Hanley, M. R., 1989, Thapsigargin, a novel molecular probe for studying intracellular calcium release and storage, *Agents Actions* (in press).

Tomlinson, J., Bannister, S. C., Croghan, P. C., and Duncan, G., 1991, Analysis of rat lens $^{45}Ca^{2+}$ fluxes: Evidence for NA^+–Ca^{2+} exchange, *Exp. Eye Res.* **52:** (in press).

Truscott, R. J. W., Marcantonio, J. M., Tomlinson, J., and Duncan, G., 1990, Opacification and proteolysis in the intact rat lens, *Invest. Ophthalmol. Vis. Sci.* **31:**2405–2411.

Tsien, R. Y., Pozzan, T., and Rink, T. J., 1982, Calcium homeostasis in intact lymphocytes: Cytoplasmic free calcium monitored with a new intracellularly trapped fluorescent indicator, *J. Cell Biol.* **94:**325–334.

5

Molecular Interactions of Crystallins in Relation to Optical Properties

C. SLINGSBY, B. BAX, R. LAPATTO, O. A. BATEMAN, H. DRIESSEN, P. F. LINDLEY, D. S. MOSS, S. NAJMUDIN, and T. L. BLUNDELL

INTRODUCTION

The transparency of the lens depends on an even distribution of protein and water over distances comparable to the wavelength of light, while the degree of refraction is controlled partly by the ability of the lens to change shape. The core regions of certain lenses such as carp and rat have an extremely high refractive index as a result of high protein concentration, which confers rigidity on that region of the lens. By contrast the outer regions of these lenses, like the complete human lens, have a lower proportion of protein to water and are malleable (van Heyningen, 1976; Philipson, 1969; Fagerholm *et al.*, 1981). Furthermore, there is an increasing protein concentration gradient from the periphery to the core of the lens, leading to a gradient of refractive index that almost abolishes spherical aberration (Fernald and Wright, 1983; Sivak, 1985).

Although lenses are made up of many different kinds of crystallin proteins, the regions of high and low protein density are characterized by different proportions of component proteins (Summers *et al.*, 1986). In the rat lens not only has a careful analysis been made of the relative proportions of α-, β-, and γ-crystallins along the refractive index gradient (Siezen *et al.*, 1988), but also the cross-linking of these molecules has been monitored spectroscopically in the intact lens as a function of maturation (Askren *et al.*, 1979). These studies clearly indicate that in a medium of high refractive index where the lens becomes hard, the solubility of crystallin proteins is drastically reduced concomitant with the protein thiols becoming oxidized. It may well be that alterations in the protein structure, in particular those that alter their potential for intermolecular interactions, result in macroscopic effects such as an increase in lens rigidity. In this chapter those aspects of crystallin protein structure that are related to forming a high packing density are emphasized, as they may be related to the loss of accommodation during the progression of presbyopia.

CONFORMATION OF γ-CRYSTALLINS

A most useful property of proteins as far as refractive index is concerned is that they can fold compactly with the exclusion of water, creating domains of high density that can further associate in a variety of ways producing water-soluble aggregates. We have shown how a small, compact domain of around 30 Å diameter is formed from some 80 amino acid resi-

C. Slingsby, B. Bax, R. Lapatto, O. A. Bateman, H. Driessen, P. F. Lindley, D. S. Moss, S. Najmudin, and T. L. Blundell • Laboratory of Molecular Biology, Department of Crystallography, Birkbeck College, London WC1E 7HX, Great Britain.

dues of a crystallin polypeptide chain by the backbone folding into two "Greek key" motifs (Blundell *et al.*, 1981; Wistow *et al.*, 1983; Summers *et al.*, 1984; Slingsby, 1985). The topology of the polypeptide chain is described as a Greek key (Richardson, 1977) to indicate that the direction of the backbone reverses three times, causing four strands to run antiparallel. The Greek key in lens proteins comprises about 40 amino acid residues, and the four strands are called *a, b, c,* and *d* in those regions where the conformation of the polypeptide chain backbone is in extended regular β structure (Fig. 5-1). Two such Greek keys intercalate in a symmetrical way around a pseudotwofold axis such that two β sheets are formed from four β strands forming a compact wedge-shaped domain (Fig. 5-2). The large loop between *a* and *b* strands is a folded β hairpin, which packs against the side of the sheet and is similar in structure in all crystallin Greek key motifs (White *et al.*, 1989). Each domain can then be described as a wedge of two β sheets stabilized by hydrogen bonds between antiparallel β-sheet strands, with the core being tightly packed with hydrophobic side chains. The molecular theme is one of motifs packing in a symmetrical way to form compact domains. The bulk of the eye lens proteins are derived from pairs of such domains organized in a symmetrical way. Whereas this high symmetry probably contributes to the stability of the system, the spacing and arrangements of these pairs of domains will determine the protein density and hence the refractive index.

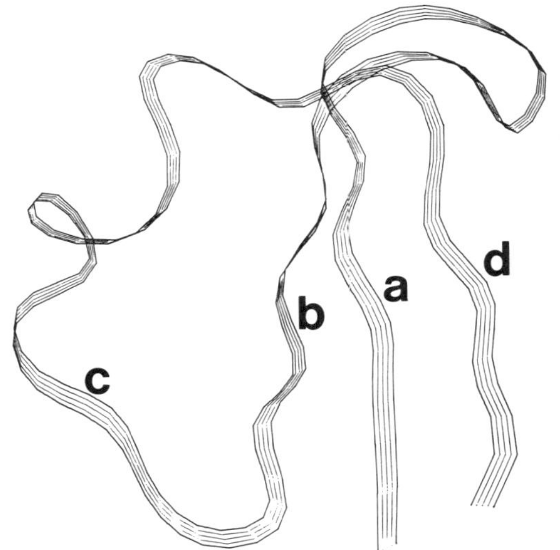

FIGURE 5-1. Ribbon diagram showing the folding of the polypeptide backbone of one crystallin "Greek key" motif; β strands (a, b, c, d) are shown where the backbone has regular β structure. Three of the β strands (b, a, d) hydrogen bond with each other together with c strand from another motif to form a β sheet. A distinctive feature of crystallin Greek keys is the folded β hairpin connecting a and b β strands.

RELATION OF β- TO γ-CRYSTALLIN

Of the three classes of lens proteins, α-, β-, and γ-crystallins, both β and γ are constructed from multiple copies of this basic domain. The γ-crystallins are a family of closely related molecules of about 174 amino acids. Each protein is formed from an N-terminal and a C-terminal domain, which associate

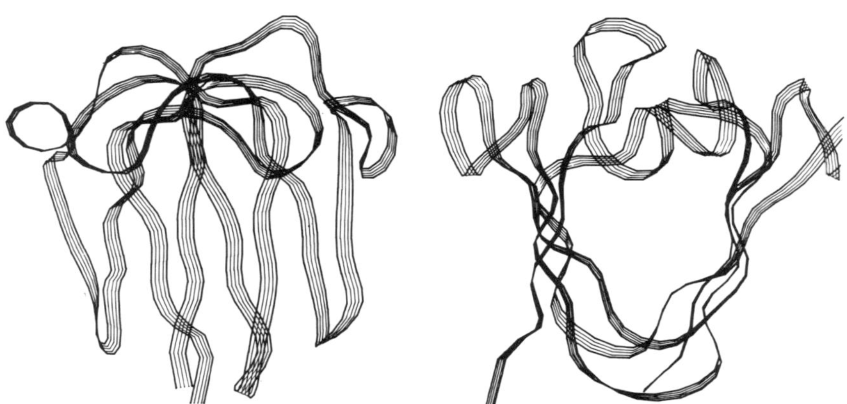

FIGURE 5-2. Ribbon diagrams of two views of a single domain made from two "Greek key" motifs. The view on the right shows how the four β strands from two Greek key motifs are aligned as two planes of β sheets stacked together. Rotation by 90° about a vertical axis results in the view on the left. The two β sheets are inclined so that the domain can be described as a wedge with two folded β hairpins at the wide end.

around an approximate twofold axis centered on a cluster of hydrophobic side chains that hold the domains together (Fig. 5-3). In a similar way β-crystallin polypeptides are comprised of two domains, but in addition they have N-terminal and sometimes C-terminal extensions. However, whereas γ-crystallins are monomers in solution, β-crystallins interact further to form oligomers (Wistow and Piatigorsky, 1988). The simplest such aggregate is βB2 homodimer consisting of four domains. We have recently solved the structure of this dimer using single crystal Xray crystallography (Bax *et al.*, 1990). Comparison of the β-crystallin structure with γ-crystallin shows that the N-terminal and C-terminal domains of both proteins are indeed very similar. However, the conformation of the connecting peptide between domains in the β-crystallin is extended, whereas in γ-crystallin it makes a sharp turn. Conformational differences in this critical region of the protein structure thus determine whether the protein is oligomeric or monomeric. These findings are surprising, since it was predicted that the β-crystallin extensions together with the hydrophobic regions on the surface of the N-terminal domain would have held the dimer together (Wistow *et al.*, 1981; Slingsby *et al.*, 1988a).

SOLUBILITY OF β- and γ-CRYSTALLINS AND RELATION TO CATARACT

The spacing of sets of domains in solution is critically dependent on a balance between their interaction potential for other protein domains and for water. If one member of the γ-crystallin family, γ-II (γB), is dissolved in water, it will form a transparent solution. In this case the solution will consist of symmetrically disposed pairs of domains, randomly oriented but evenly distributed with respect to each other. The βB2 polypeptide will also form a transparent solution, but in this case the two domains of one subunit will be tightly packed against another pair to form a compact dimer or tetramer of domains. However, if a different member of the γ-crystallin family, γ-IVa (γE), is placed in solution and then chilled, the transparent solution becomes

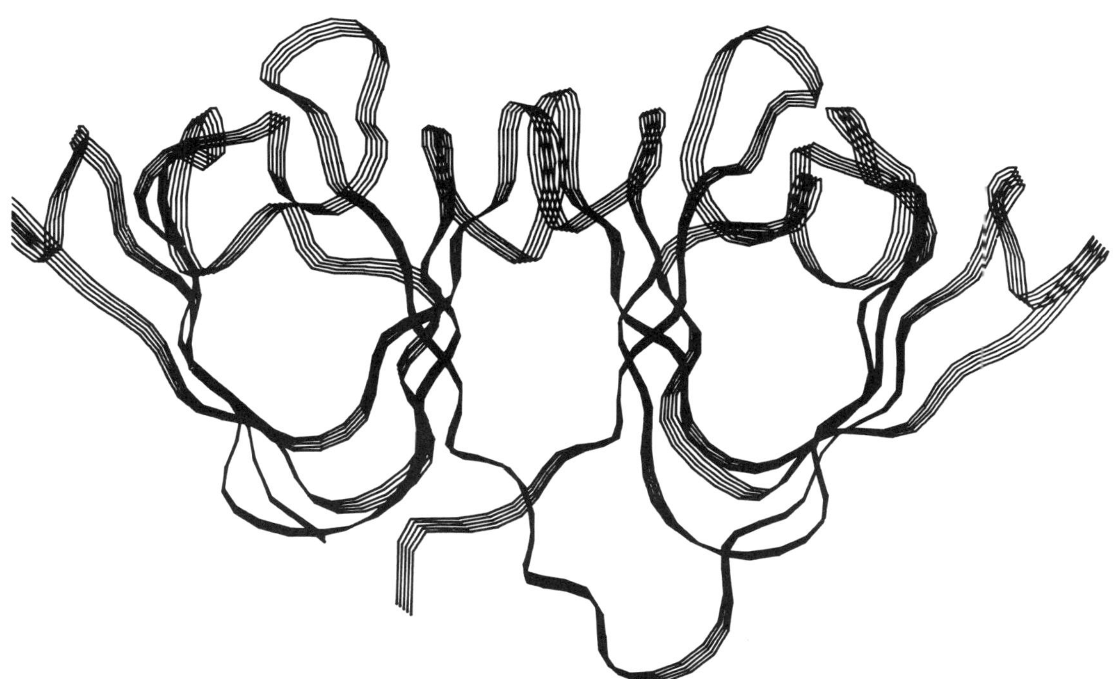

FIGURE 5-3. Ribbon diagram of a complete γ-crystallin molecule showing how it is composed of two domains connected at the bottom by one stretch of polypeptide. This view is perpendicular to the approximate plane containing the three pseudotwofold axes and shows how one molecule comprises four β sheets. The interiors of both domains are filled with hydrophobic side chains. The two domains contact each other through hydrophobic side chains.

opaque (Blundell *et al.*, 1983; Siezen *et al.*, 1985). In this case, the interaction potential exerted between pairs of γ-IVa domains is greatly increased as the temperature is lowered such that at a critical point, phase separation occurs. As the clusters of γ-IVa domains increase in size, the solution partitions into fluctuating regions of high and low protein density, causing light scattering. Similarly, when young mammalian lenses are chilled, the central regions become opaque, causing cold cataract (Benedek *et al.*, 1979). This is the region of the lens that is enriched in γ-IVa crystallin, implying that soft, immature lenses are sufficiently liquid-like to exhibit a phase separation.

We have shown that γ-IVa in the crystalline state, in which the proportion of protein to water is similar to lens core region, makes an intimate twofold interaction centered perpendicular to a β strand at the edge of a sheet in the C-terminal domain (Fig. 5-4) (White *et al.*, 1989). If the lens is cooled, interactions such as the twofold one around the C-terminal β sheet will probably become so favorable that clusters of domains build up, perhaps using the dimer as a seed, causing a local increase in protein density. Within these light-scattering clusters the γ-IVa sets of domains are as densely packed as in the crystal, and although interactions similar to those observed in the lattice will be favored, the order will probably only be short range (Bernal, 1964; Delaye and Tardieu, 1983). We have argued that interactions such as these will probably be responsible for triggering the lens cytoplasm phase separation that occurs in cold cataract (Slingsby *et al.*, 1988b; White *et al.*, 1989).

The βB2-crystallin subunits on their own form dimers and never monomers. The βB2 subunit specifically recognizes a complementary site on another molecule to give a globular homodimer. The subunit has no strong tendency to interact at other sites, so only one type of dimer forms, and it has been crystallized with a crystallographic dyad running through the dimer interface (Bax and Slingsby, 1989). Solutions of βB2 dimers show no tendency to phase separate, indicating that the globular "tet-

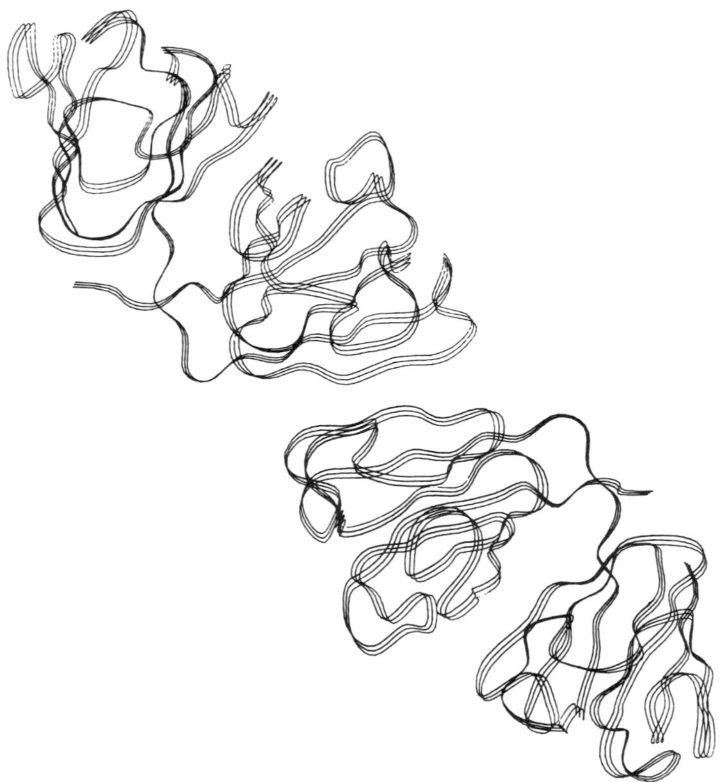

FIGURE 5-4. Ribbon diagram of two γ-IVa crystallin molecules arranged as they are in the crystal lattice. The view is perpendicular to the crystallographic twofold axis that relates the positions of the two molecules. The interaction is mediated through the edge strands of two β sheets from two molecules.

ramer" of domains interacts preferentially with water rather than other domains. By contrast, γ-IVa dimers once formed are probably unstable and continue to interact further with more protein domains. As long as the lens cytoplasm comprises a sufficient amount of water such that protein domains are surrounded by water, then these γ-crystallins will still have a restricted range of rotational and translational movements available, and consequently a limited range of fluctuating interactions with other proteins will be possible, including self-association. At this stage the lens will exhibit cold cataract. However, once the dehydration of the lens approaches that found in crystals, the domains will have fewer opportunities for exploring alternative protein interactions. So long as the packing is even, the system will remain transparent but will be more stable, as the lens will no longer turn opaque on cooling.

β-CRYSTALLIN INTERACTIONS

Human lenses maintain a high level of hydration throughout life although they are formed from similar α-, β-, and γ-crystallins. However, some of the genes coding for γ-crystallins have been turned off during human evolution, particularly those γ-crystallins that appear to be tight-packing proteins (Meakin *et al.*, 1985; Russell *et al.*, 1987; Siezen *et al.*, 1987), which leaves β-crystallins as the predominant proteins in the water-soluble phase of human lens (Zigler and Sidbury, 1973; de Jong, 1981). β-Crystallins are extremely heterogeneous, as many other subunits, all related to βB2, are involved in interactions that result in groups of differently sized oligomers displaying extensive charge polydispersity (Bindels *et al.*, 1981; Berbers *et al.*, 1982).

Although interactions stabilizing the simple βB2 dimers are beginning to be worked out, a major remaining problem is how the larger, mixed aggregates are constructed. The six genes coding for β subunits can be divided into a basic group comprising βB1, βB2, and βB3 and an acidic group βA2, βA3, and βA4 (Berbers *et al.*, 1984; Lubsen *et al.*, 1988; Wistow and Piatigorsky, 1988). The basic group has a variable N-terminal arm and a more constant C-terminal extension, whereas the acidics have only an N-terminal arm. β-Crystallin basic and acidic subunits are capable of interacting with each other in many combinations to produce a vast array of oligomers. Bovine βB2 subunit can be purified from all the other subunits by chromatographing a mixed population of β-crystallin aggregates in the presence of a denaturant (Herbrink *et al.*, 1975). The βB2 polypeptide refolds on removal of denaturant into a pair of domains that self-associate to form a homodimer (Slingsby *et al.*, 1982). In a similar way we have isolated several basic and acidic β-crystallin subunits and refolded them either on their own or in combination with other β-subunits. By monitoring the size of the reassociated oligomer, we were able to show that formation of an intermediate size class of β-crystallins depended on heterologous interactions between βB2 and an acidic subunit with a long N-terminal extension (Slingsby and Bateman, 1990).

The largest β-crystallin aggregates are probably either hexamers or octamers of β subunits containing both βB2 and acidic subunits but also, importantly, a basic subunit (βB1) characterized by an extremely long N-terminal extension (Berbers *et al.*, 1982; Siezen *et al.*, 1986). In bovine lens this extension is 58 amino acids long and has many simple repeats of Ala-Pro in the sequence (Berbers *et al.*, 1983, 1984). A major problem now is understanding the role played by these arms. It may be that the arm makes essential interactions with other domains or arms of a smaller oligomer and thus is used in the construction of the larger aggregate; or it may interact with membranes or other cytoskeletal components in the cell (Bloemendel *et al.*, 1982). One possibility is that the arms prevent intimate domain interactions between different aggregates and thus help to keep the surface of the heteroligomer surrounded largely by water. Modification of these arms would then easily alter the surface interaction potential of the heteroligomer. Such a model would invoke the principle of extensions being used as spacers for controlling the distribution of domain clusters in a way that could easily accommodate increasing dehydration.

ROLES OF γ- AND β-CRYSTALLINS IN THE LENS

A general picture emerges in which γ-crystallins make a large contribution to tight packing, whereas in more hydrated systems β-crystallin oligomers with their large selection of polypeptide extensions probably play a major role. Clearly tight packing of protein in eye lenses leads to harder lenses as exemplified in the core regions of rat and fish lenses:

the increase in refractive index is at the expense of diminished plasticity. When packing is such that approximately 50% of the mass of the lens is protein, the crystallin domains will no longer be surrounded by water and will be interacting extensively with other proteins in all directions to an extent similar to that observed in the γ-IVa crystal lattice (White *et al.*, 1989). At this level of packing in the rat lens, the once soluble proteins are almost completely transformed into an insoluble form with α- and β-crystallin oligomers being removed from the soluble phase first, followed eventually by γ-crystallins (Siezen *et al.*, 1988); a similar sequence of events was observed in the core region of bovine lens (Blundell *et al.*, 1983). As the lens proceeds to harden, therefore, the proportion of γ-crystallin in the soluble phase increases.

The surfaces of these small proteins are covered with ion-pair networks, allowing them easily to make local interactions with neighboring charged side chains and thus reducing the amount of water required to stabilize the monomeric state (Summers *et al.*, 1984, 1986). Spectroscopic measurements on intact rodent lenses have shown that the maturation of the hard rat and mouse lens is associated with total oxidation of protein thiol groups, in contrast with the soft guinea pig lens where thiols remain reduced as they do in the mature human lens (Askren *et al.*, 1979; Kuck *et al.*, 1982; Yu *et al.*, 1985). Furthermore, the oxidation begins while the γ-crystallins are still in the soluble phase and is predominantly intramolecular (Hum and Augusteyn, 1987). Although initially the oxidation could involve local clusters of thiols in the N-terminal domain (Summers *et al.*, 1984) or be between domains (Breitman *et al.*, 1984) without radically altering the domain structures, the final stages of oxidation must involve some disruption of the sheet structure caused by cross-linking of distal cysteines, which in turn will probably contribute to the protein's insolubilization.

An insight into protein close-packing systems came with the observation that haddock γ-crystallins have an extraordinarily high level of methionine (Croft, 1973). Two γ-crystallins have recently been sequenced from carp lens; there are 16 methionines on the surface of one carp γ-crystallin and eight cysteines in the domain cores of the other (Chang *et al.*, 1988). A recent review of protein subunit and domain interfaces of proteins generally concluded that aromatics and methionine sulfurs make "particularly good glue for pasting together protein units" (Argos, 1988). Another review concluded that arginine side chains are often found buried at protein interfaces (Janin *et al.*, 1988). γ-Crystallins have large numbers of conserved arginines on the surface of their domains, and the combination of numerous additional surface methionines appears to facilitate their close-packing role in the fish lens. Sulfur-containing residues are frequently involved in γ-crystallin lattice interactions (Wistow *et al.*, 1983; Sergeev *et al.*, 1988; White *et al.*, 1989), and their role in β-crystallin aggregation has been suggested (Slingsby *et al.*, 1988a). Interestingly, one of the carp γ-crystallin sequences has a particularly dense cluster of methionines (Fig. 5-5) on loops similar to the observed interface region of βB2 dimer (Bax *et al.*, 1990).

CRYSTALLIN INTERACTIONS IN CATARACT AND PRESBYOPIA

In contrast with hard lenses, the protein concentration in human lenses halts around 30% of the mass; more protein remains in solution, and little protein thiol oxidation occurs in the transparent lens. In the relatively hydrated human lens, overextensive crystallin interaction would cause local regions of high refractive index, which would contrast with the more watery surrounds. Protein oxidation and modifications leading to accelerated loss of monomeric crystallins from the soluble phase are the hallmarks of the cataractous lens (Spector, 1984; Harding and Crabbe, 1984). Presumably such alterations to the γ-crystallin structure promote extensive protein interface interactions, which are incompatible with lens transparency at this level of hydration.

In human lens the continuous cortex-to-core increase in insoluble protein becomes more apparent around the age of onset of presbyopia (Li *et al.*, 1986). This implies that the kinds of interaction that lead to protein aggregation without involving disruptive cross-links are causing increased rigidity in the presbyopic lens. Nondisruptive protein interactions have been observed in the lattices of three γ-crystallins. Point mutations on the domain surfaces appear to direct different packing arrangements, which require different amounts of water to fill in the spaces (Sergeev *et al.*, 1988; White *et al.*, 1989). However, in the complete lens, interactions will be considerably influenced by α- and β-crystallins. Now that the basic mode of interaction of β-crystallin domains has been determined (Bax *et al.*, 1990), it remains to be seen how the β-crystallin

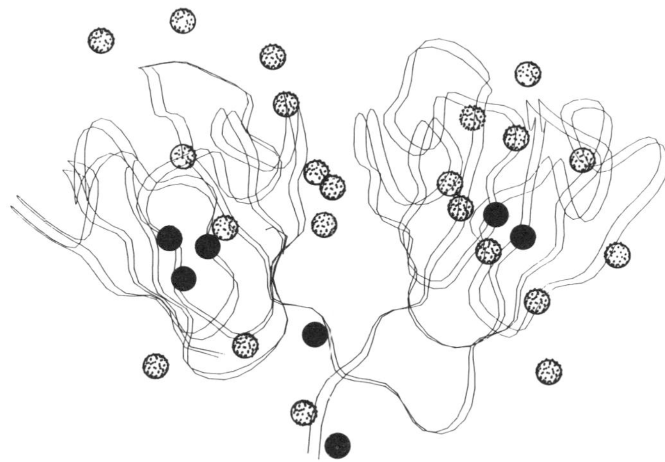

FIGURE 5-5. Ribbon diagram showing the backbone conformation of carp γ-m1 crystallin. The structure was model built using the sequence of carp γ (Chang *et al.*, 1988) and the coordinates of x-ray-determined calf γ-crystallin. Solid spheres represent cysteine sulfur atoms, and dotted spheres represent methionine sulfur atoms.

aggregates use their extensions. Because the human equivalent of βB1 comprises around 10% of the soluble protein of a young lens (Alcala *et al.*, 1988), it appears that this β-crystallin subunit with a very long N-terminal extension may have an important role in determining crystallin interactions in the human lens.

ACKNOWLEDGMENT. This work was supported by the Medical Research Council (London), U.K.

REFERENCES

Alcala, J., Katar, M., Rudner, G., and Maisel, H., 1988, Human beta crystallins: Regional and age related changes, *Curr. Eye Res.* **7**:353–359.

Argos, P., 1988, An investigation of protein subunit and domain interfaces, *Protein Eng.* **2**:101–113.

Askren, C. C., Yu, N.-T., and Kuck, J. F. R., Jr., 1979, Variation of the concentration of sulphydryl along the visual axis of aging lenses by Raman optical dissection technique, *Exp. Eye Res.* **29**:647–54.

Bax, B., and Slingsby, C., 1989, Crystallization of a new form of the eye lens protein βB2-crystallin, *J. Mol. Biol.* **208**:715–717.

Bax, B., Lapatto, R., Nalini, V., Driessen, H., Lindley, P. F., Mahadevan, D., Blundell, T. L., and Slingsby, C., 1990, X-ray analysis of βB2-crystallin and evolution of oligomeric lens proteins, *Nature* **347**:776–780.

Benedek, G. B., Clark, J. I., Serralach, E. N., Young, C. Y., Mengel, L., Sauke, T., Bagg, A., and Benedek, K., 1979, Light scattering and reversible cataracts in the calf and human lens, *Phil. Trans. R. Soc. [A]***293**:329–340.

Berbers, G. A. M., Boermann, O. C., Bloemendal, H., and de Jong, W. W., 1982, Primary gene products of bovine β-crystallin and reassociation of its aggregates, *Eur. J. Biochem.* **128**:495–502.

Berbers, G. A. M., Hoekman, W. A., Bloemendal, H., de Jong, W. W., Kleinschmidt, T., and Braunitzer, G., 1983, Proline and alanine rich N-terminal extension of the basic bovine β-crystallin B_1 chains, *FEBS Lett.* **161**:225–229.

Berbers, G. A. M., Hoekman, W. A., Bloemendal, H., de Jong, W. W., Kleinschmidt, T., and Braunitzer, G. 1984, Homology between the primary structures of the major β-crystallin chains, *Eur. J. Biochem.* **139**:467–479.

Bernal, J. D., 1964, The structure of liquids, *Proc, R. Soc. [A]* **280**:299–322.

Bindels, J. G., Koppers, A., and Hoenders, H. J., 1981, Structural aspects of bovine β-crystallins: Physical characterization including dissociation–association behaviour, *Exp. Eye Res.* **33**:333–343.

Bloemendal, H., Hermsen, T., Dunia, I., and Benedetti, E. L., 1982, Association of the crystallins with the plasma membrane, *Exp. Eye Res.* **35**:61–67.

Blundell, T., Lindley, P. F., Miller, L. R., Moss, D. S., Slingsby, C., Turnell, W. G., and Wistow, G., 1981, The molecular structure and stability of the eye lens: X-ray analysis of γ-crystallin II, *Nature* **289**:771–777.

Blundell, T., Lindley, P. F., Miller, L. R., Moss, D. S., Slingsby, C., Turnell, W. G., and Wistow, G., 1983, Interactions of γ-crystallin in relation to eye-lens transparency, *Lens Res.* **1**:109–131.

Breitman, M. L., Lok, S., Wistow, G., Piatigorsky, J., Treton, J. A., Gold, R. J. M., and Tsui, L.-C., 1984, γ-Crystallin family of the mouse lens: Structural and evolutionary relationships, *Proc. Natl. Acad. Sci. U.S.A.* **81**:7762–7766.

Chang, T., Jiang, Y.-J., Chiou, S.-H., and Chang, W.-C., 1988, Carp gamma-crystallin with high methionine content: Cloning and sequencing of the complementary DNA, *Biochim. Biophys. Acta* **951**:226–229.

Croft, L. R., 1973, Amino and carboxy terminal sequence of γ-crystallin from haddock lens, *Biochim. Biophys. Acta* **295**:174–177.

de Jong, W. W., 1981, Evolution of lens and crystallins, in: *Molecular and Cellular Biology of the Eye Lens* (H. Bloemendal, ed.), John Wiley & Sons, New York, pp. 221–278.

Delaye, M., and Tardieu, A., 1983, Short-range order of crystallin proteins accounts for eye lens transparency, *Nature* **302:**415–417.

Fagerholm, P. P., Philipson, B. T., and Linstrom, B., 1981, Normal human lens—the distribution of protein, *Exp. Eye Res.* **33:**615–620.

Fernald, R. D., and Wright, S. E., 1983, Maintenance of optical quality during crystalline lens growth, *Nature* **301:**618–620.

Harding, J. J., and Crabbe, M. J. C., 1984, in: *The Eye: The Lens: development, proteins, metabolism, and cataract,* vol. 1B, 3rd ed. (H. Davson, ed.), Academic Press, London, pp. 207–492.

Herbrink, P., van Westreenen, H., and Bloemendal, H., 1975, Further studies on the polypeptide chains of γ-crystallin, *Exp. Eye Res.* **20:**541–548.

Hum, T. P., and Augusteyn, R. C., 1987, The nature of disulphide bonds in rat lens proteins, *Curr. Eye Res.* **6:**1103–1108.

Janin, J., Miller, S., and Chothia, C., 1988, Surface, subunit interfaces and interior of oligomeric proteins, *J. Mol. Biol.* **204:**155–164.

Kuck, J. F. R., Yu, N.-T., and Askren, C. C., 1982, Total sulphydryl by Raman spectroscopy in the intact lens of several species: Variations in the nucleus and along the optical axis during aging, *Exp. Eye Res.* **34:**23–37.

Li, L.-K., Roy, D., and Spector, A., 1986, Changes in lens protein in concentric fractions from individual normal human lenses, *Curr. Eye Res.* **5:**127–135.

Lubsen, N. H., Aarts, H. J. M., and Schoenmakers, J. G. G., 1988, The evolution of lenticular proteins: The β- and γ-crystallin super gene family, *Prog. Biophys. Mol. Biol.* **51:**47–76.

Meakin, S. O., Breitman, M. L., and Tsui, L.-C., 1985, Structural and evolutionary relationships among five members of the human γ-crystallin gene family, *Mol. Cell Biol.* **5:**1408–1414.

Philipson, B., 1969, Distribution of protein within the normal rat lens, *Invest. Ophthalmol.* **8:**258–270.

Richardson, J. S., 1977, β-Sheet topology and the relatedness of proteins, *Nature* **268:**495–500.

Russell, P., Meakin, S. O., Hohman, T. C., Tsui, L.-C., and Breitman, M. L., 1987, Relationships between proteins encoded by three human γ-crystallin genes and distinct polypeptides in the eye lens, *Mol. Cell Biol.* **7:**3320–3323.

Sergeev, Y. V., Chirgadze, Y. N., Mylvaganam, S. E., Driessen, H., Slingsby, C., and Blundell, T. L., 1988, Surface interactions of γ-crystallins in the crystal medium in relation to their association in the eye lens, *Proteins Struct. Funct. Genet.* **4:**137–147.

Siezen, R. J., Fisch, M. R., Slingsby, C., and Benedek, G. B., 1985, Opacification of γ-crystallin solutions from calf lens in relation to cold cataract formation, *Proc. Natl. Acad. Sci. U.S.A.* **82:**1701–1705.

Siezen, R. J., Anello, R. D., and Thomson, J. A., 1986, Interactions of lens proteins. Concentration dependence of β-crystallin aggregation, *Exp. Eye Res.* **43:**293–303.

Siezen, R. J., Thomson, J. A., Kaplan, E. D., and Benedek, G. B., 1987, Human lens γ-crystallins: Isolation, identification, and characterization of the expressed gene products, *Proc. Natl. Acad. Sci. U.S.A.* **84:**6088–6092.

Siezen, R. J., Wu, E., Kaplan, E., Thomson, J. A., and Benedek, G. B., 1988, Rat lens γ-crystallins, *J. Mol. Biol.* **199:**475–490.

Sivak, J. G., 1985, Optics of the crystalline lens, *Am. J. Optom. Physiol. Opt.* **62:**299–308.

Slingsby, C., 1985, Structural variation in lens crystallins, *Trends Biochem. Sci.* **10:**281–284.

Slingsby, C., and Bateman, O. A., 1990, Quaternary interactions in eye lens β-crystallins: Basic and acidic subunits of β-crystallins favor heterologous association, *Biochemistry* **29:**6592–6599.

Slingsby, C., Miller, L. R., and Berbers, G. A. M., 1982, Preliminary x-ray crystallographic study of the principle subunit of the lens structural protein, bovine β-crystallin, *J. Mol. Biol.* **157:**191–194.

Slingsby, C., Driessen, H. P. C., Mahadevan, D., Bax, B., and Blundell, T. L., 1988a, Evolutionary and functional relationships between the basic and acidic β-crystallins, *Exp. Eye Res.* **46:**375–403.

Slingsby, C., Driessen, H. P. C., White, H., Mylvaganam, S., Najmudin, S., Bax, B., Bibby, M. A., Lindley, P. F., Moss, D. S., and Blundell, T. L., 1988b, Molecular interactions in relation to cataract, in: *Molecular Biology of the Eye: Genes, Vision, and Ocular Disease,* (J. Piatigorsky, T. Shinohara, and P. S. Zelenka, eds.), Alan R. Liss, New York, pp. 419–426.

Spector, A., 1984, The search for a solution to senile cataracts, *Invest. Ophthalmol. Vis. Sci.* **25:**130–146.

Summers, L., Wistow, G., Marebor, M., Moss, D. S., Lindley, P., Slingsby, C., Blundell, T., Bartunik, H., and Bartels, K., 1984, X-ray studies of the lens specific proteins: The crystallins, *Peptide Protein Rev.* **3:**147–168.

Summers, L. J., Slingsby, C., Blundell, T. L., den Dunnen, J. T., Moormann, R. J. M., and Schoenmakers, J. G. G., 1986, Structural variation in mammalian γ-crystallins based on computer graphics analyses of human, rat and calf sequences, *Exp. Eye Res.* **43:**77–92.

van Heyningen, R., 1976, Experimental studies on cataract, *Invest. Ophthalmol.* **15:**685–697.

White, H. E., Driessen, H. P. C., Slingsby, C., Moss, D. S., and Lindley, P. F., 1989, Packing interactions in the eye-lens: Structural analysis, internal symmetry and lattice interactions of bovine γIVa-crystallin, *J. Mol. Biol.* **207:**217–235.

Wistow, G. J., and Piatigorsky, J., 1988, Lens crystallins: The evolution and expression of proteins for a highly specialized tissue, *Annu. Rev. Biochem.* **57:**479–504.

Wistow, G., Slingsby, C., Blundell, T., Driessen, H., de Jong, W., and Bloemendal, H., 1981, Eye lens proteins: The three dimensional structure of β-crystallin predicted from monomeric γ-crystallin, *FEBS Lett.* **133:**9–16.

Wistow, G., Turnell, B., Summers, L., Slingsby, C., Moss, D., Miller, L., Lindley, P., and Blundell, T., 1983, X-ray analysis of the eye lens protein γ-II crystallin at 1.9 Å resolution, *J. Mol. Biol.* **170:**175–202.

Yu, N.-T., De Nagel, D. C., Pruett, P. L., and Kuck, J. F. R., Jr., 1985, Disulphide bond formation in the eye lens, *Proc. Natl. Acad. Sci. U.S.A.* **82:**207–214.

Zigler, J. S., Jr., and Sidbury, J. B., Jr., 1973, Structure of calf lens β-crystallins, *Exp. Eye Res.* **16:**207–214.

6

Biophysical Analysis of Eye Lens Transparency

ANNETTE TARDIEU and FRANÇOISE VÉRÉTOUT

LENS TRANSPARENCY*

The molecular basis of eye lens transparency has been analyzed recently using osmotic pressure and solution X-ray scattering experiments, and the fundamental contribution of protein–protein interactions to the process was described (Vérétout *et al.*, 1989). The aim of this chapter is to present the role of protein–protein interactions in a more pictorial way.

The images of transparent and opaque lenses are familiar. For most of us, however, opacity means cataract and, therefore, a pathological event that we would prefer to avoid. From a physical standpoint, as explained by Trokel as early as 1962, lens transparency is limited by two phenomena, absorption and scattering of visible light, which are functions of lens structure and composition. Both reduce the light intensity transmitted through the lens. This transmitted intensity I_t may be simply written as a function of the incident intensity I_0, of the lens thickness l and of an extinction coefficient τ:

$$I_t = I_0 \exp(-\tau l) \qquad (1)$$

When $\tau = 0$, 100% of the incident light is transmitted, and the lens is transparent. When τ increases because of absorption and/or scattering, I_t/I_0 decreases until the lens becomes opaque. In other words, age, metabolic deficiency, and so on are not *per se* responsible for cataracts. They can only be responsible for modifications of the lens structure and composition leading to increased scattering and/or absorption and possibly to opacities.

As also explained by Trokel (1962), given the lens composition, in the normal state light absorption is negligible, and lens transparency is determined by light scattering, which arises from fluctuations of the refractive index. The extinction coefficient corresponds to the integral of the scattered intensity I_s:

$$\tau = \int I_s/I_0 \qquad (2)$$

The normal lens is composed of two independent scattering systems, membranes and proteins. Membranes are quasiperiodically organized with periodicities much larger than light wavelengths. The cytoplasm is a concentrated protein solution with scatterers much smaller than light wavelengths. In a transparent lens the membrane scattering was estimated to remain negligible (a simplification not likely to be valid for a variety of cataracts), and the light scattering is essentially a result of lens-specific proteins, the crystallins. At the high protein concentration prevailing in the lens, the level of scattering is determined by the spatial organization, or local order, of the crystallin proteins (Trokel, 1962). Trokel was, however, unlucky in his appreciation of the local order. He suggested a local order approaching a paracrystalline state. Some years later, Benedek (1971) correctly pointed out that at the high protein concentrations present *in vivo*, a liquid-like organization was sufficient to account theoretically for transparency.

*The work on transparency was initiated by Mireille Delaye around 1981. We pursued it in close collaboration with Mireille until her death in 1987.

ANNETTE TARDIEU AND FRANÇOISE VÉRÉTOUT • Centre de Génétique Moléculaire, CNRS, 91198 Gif sur Yvette Cedex, France.

LENS AS A PROTEIN SOLUTION

We prefer here to present the problem in the following way. The lens cytoplasm is essentially a concentrated protein solution. The light scattered by one protein is proportional to its molecular weight. The light scattered by a protein solution is, in addition, proportional to the protein concentration. Furthermore, with concentrated solutions, the scattering is modulated because of the protein distribution.

Light scattering, I_s/I_0, of lens was first measured as a function of protein concentration by Mireille Delaye around 1981 (Delaye and Gromiec, 1983). The use of calf lens cortical cytoplasmic extracts (Clark *et al.*, 1982) made the experiments feasible, since the protein concentration could be easily varied while keeping the same protein composition. The measurements clearly showed that light scattering does not vary linearly with protein concentration but first increases and then decreases, as can be seen in Fig. 6-1a. From that, it can be inferred that the transparency first decreases and then increases with the protein concentration. Light scattering, however, does not allow us to analyze the protein distribution at the origin of such results. However, with X-ray measurements that are able to probe the local organization of the scatterers, it became possible to demonstrate from the shape of the X-ray scattering curves that the variation of the scattering as a function of c was indeed caused by a short-range, liquid-like order, as detailed by Delaye and Tardieu (1983).

That was the state of the question in 1983. At that time, we thought it would be difficult to calculate the transmitted intensity from the scattered intensity, so the quantitative calculation was not done. In fact it is quite simple to use equations 1 and 2 to calculate the corresponding percentage of light transmission (Tardieu and Delaye, 1988; Vérétout *et al.*, 1989). The result, shown in Fig. 6-1b, is really illustrative. In the physiological concentration range, for a lens 1 cm thick and green light, about 95% of the incident intensity is transmitted. Under the same conditions, with the same particles, but with a "random" protein distribution, only 50% of the incident light would be transmitted, and the lens would be turbid.

PROTEIN DISTRIBUTIONS

To find the molecular basis of eye lens transparency, it is necessary to go some steps further, and the next step may be expressed by the statement that the protein distribution is itself determined by the protein–protein interactions. To show how the protein–protein interactions can be analyzed, it is convenient to remember that, in the case of a solution of identical spherical particles, the light scattering has simple thermodynamic (Eisenberg, 1976) and structural

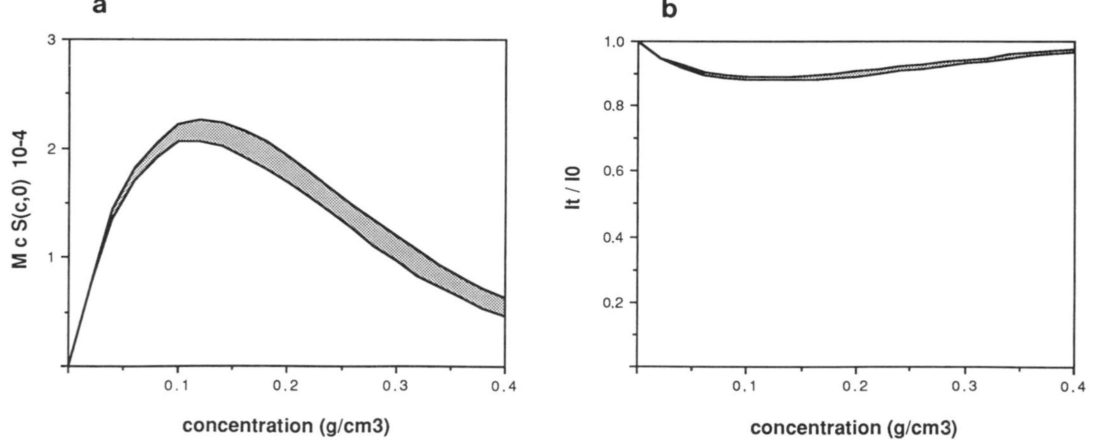

FIGURE 6-1. (a) Schematic representation of the experimental measurements of the scattering, I_s/I_0, as a function of protein concentration. The two lines enclose the data presented by Delaye and Tardieu (1983). Experiments were preformed on calf lens cortical cytoplasmic extracts using both direct light scattering and extrapolated X-ray scattering measurements. The normalization is as explained in legend to Fig. 6-2. (b) Percentage of transmitted light, I_t/I_0, as a function of protein concentration for the scattering experiments shown in a. The calculation was done using equations 1 and 2 for a 1-cm-thick lens and a wavelength of 500 nm. The coefficients needed to calculate b from a can be found in Tardieu and Delaye (1988) and Vérétout *et al.* (1989).

(Tardieu and Delaye, 1988) expressions, which are, respectively:

$$I_s/I_0 \sim cM\,[(RT/M)(\partial\Pi/\partial c)^{-1}] = cRT(\partial\Pi/\partial c)^{-1} \qquad (3)$$

$$I_s/I_0 \sim cM\,[S(c,0)] \qquad (4)$$

where Π is the osmotic pressure of the solution and $S(c,0)$ the structure factor at the origin of the solution. I_s/I_0 could be directly measured with light scattering. Π may be measured using the osmotic stress technique developed by Prouty *et al.* (1985) and Parsegian *et al.* (1986). $S(c,0)$ is conveniently obtained from extrapolation to the origin of X-ray-scattering experiments, which provide us with $S(c,s)$, $s = 2\sin\theta/\lambda$, where 2θ is the scattering angle, in a large s range. Synchrotron radiation at LURE (Orsay) was found particularly convenient for such experiments, since a large number of scattering curves could be measured in a short period of time.

As can be seen from the equations above, it is possible to go further into the analysis of transparency by using solutions made of identical particles. In order to approach this situation, α-crystallins were purified. They were chosen because, since they are the largest particles, they were considered to be mainly responsible for the scattering *in vivo*.

The osmotic pressure data are shown in Fig. 6-2a, and examples of X-ray measurements are seen in Fig. 6-2b. The concentration effect is very important. In the absence of protein–protein interactions, osmotic pressure would vary linearly with protein concentration, and, with the normalization used, the X-ray curves would remain identical whatever the concentration. The figure also shows that, as expected, the general behavior of α-crystallins is similar to the behavior of cortical cytoplasmic extracts published by Delaye and Tardieu (1983). In addition, the experiments were performed at two different ionic strengths. Comparison of the two series of experiments clearly shows an effect of charge, which is also shown in Fig. 6-2a. As explained in the

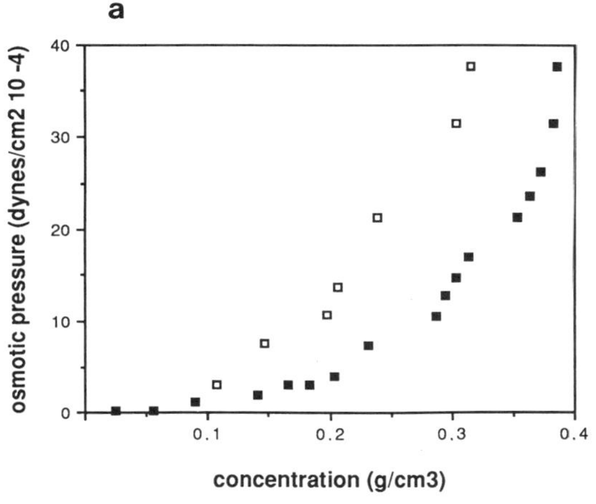

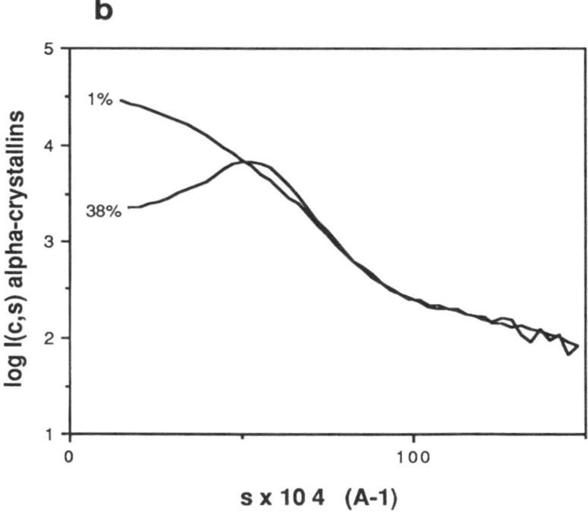

FIGURE 6-2. Examples of experimental data obtained with α-crystallin solutions. (a) Osmotic pressure experiments at 150(■) and 17 mM (□). Each point is the average of two to four experiments. In the absence of protein–protein interactions, the osmotic pressure would vary linearly with the protein concentration, and no dependence on ionic strength would be observed. (b) Solution X-ray-scattering curves recorded at 150 mM at two protein concentrations indicated on the figure, 1% and 38% (i.e., 0.010 and 0.380 g/cm^3). One X-ray experiment provides one scattering curve as a function of the scattering angle $s = 2\sin\theta/\lambda$. Such curves, $I(c,s)$, are shown here on a log scale. They are normalized to coincide at large scattering angles. With this normalization, in the absence of interactions, the scattering curves would be identical whatever the protein concentration, and they would look like the 1% curve. The low-angle decrease in intensity and the maximum at medium angles are features typical of repulsive, liquid-like, interactions. For the sake of simplicity, only two curves are shown here. About 25 such curves were measured at each ionic strength (see series in Vérétout *et al.*, 1989). At 17 mM ionic strength, the low-angle intensity is lower, and the maximum is higher. To determine the best-fit model parameters, we calculate the structure factor $S(c,s)$ according to: $S(c,s) = I(c,s)/I(c,0)$, where $I(c,0)$ is the intensity curve extrapolated to zero concentration, i.e., in practice, the 1% concentration curve (Vérétout *et al.*, 1989). The structure factor calculated for each protein concentration is then extrapolated to $s = 0$ to obtain $S(c,0)$ (equation 4). It remains to multiply this value by the protein concentration, c(g/cm^3), and the protein molecular weight, M, to construct curves like those shown in Fig. 6-1a, where the scattering is represented by $McS(c,0)$.

legend of Fig. 6-2, $S(c,0)$ is obtained from extrapolation of the X-ray curves, and comparison with the osmotic pressure data can then be done using equations 3 and 4. X-ray and osmotic pressure experiments were shown to be entirely consistent, allowing us to conclude that the macroscopic thermodynamic properties of the solution are indeed determined by the microscopic structural properties. Since the proportionality constant in equations 3 and 4 is known, I_s/I_0 may be calculated from either Π or $S(c,0)$; τ is then obtained from I_s/I_0 using equation 2, and I_t is obtained using equation 1.

PROTEIN–PROTEIN INTERACTIONS

We now know that the behavior shown in Fig. 6-2 is the result of overall repulsive protein–protein interactions, more precisely, electrostatic screened coulombic repulsive interactions. Models have been developed in liquid-state physics for such repulsive interactions (Hayter and Penfold, 1981; Hansen and Hayter, 1982). It happens that, with a solution of identical spherical particles, such interactions can be modeled with a few structural parameters. This is the last missing step leading to an understanding of the molecular basis of lens transparency.

The structural parameters are the particle specific excluded volume, φ/c, i.e., the ratio of the excluded volume fraction associated with the particles to the particle concentration, the particle diameter, σ, and the particle charge, Z. The validity of the repulsive electrostatic coulombic interaction model is easily demonstrated with the X-ray experiments. Furthermore, the three parameters φ/c, σ, and Z are almost independently determined from amplitude, position, and amplitude difference at both ionic strengths of the maximum of the scattering curves. Once the validity of the model has been shown with X-rays, two parameters, φ/c and Z, are easily determined from the curvature and the difference at both ionic strength of the osmotic pressure curves (Vérétout *et al.*, 1989). The osmotic pressure and the structure factor of a solution may now be calculated from these three parameters, and, therefore, so can the extinction coefficient and the transparency.

The values of the parameters obtained at high and low protein concentration are given in Table 6-1. It has to be emphasized that the excluded specific volume, the diameter, and the charge so obtained are best-fit model parameters that define an equivalent particle.

TABLE 6–1
Properties of Lens α-crystallins at High and Low Concentrations

Low c	High c
$\varphi/c = 2$	$\varphi/c = 1.4$
$\sigma = 190$ Å	$\sigma = 170$ Å
$Z = 60$	$Z = 40$

When compared to what is known of α-crystallin quaternary structure, the excluded volume of the equivalent particle found at low c is close to the volume of the spherical envelope of the α-crystallin, and that found at high c is close to the volume of the average envelope. This observation, coupled to unpublished results on other proteins, indicates that excluded volume and compactness of the quaternary structure are closely related. The φ/c value was found rather high, varying from 2 to 1.4 from low to high protein concentration in agreement with the type of noncompact structure proposed for α-crystallins (Tardieu *et al.*, 1986). The charge, although important, is almost completely screened at *in vivo* ionic strength.

The table therefore shows that the parameter relevant for transparency *in vivo,* in addition to concentration and molecular weight, is essentially the φ/c ratio (Fig. 6-3). We were thus led to analyze in more detail the role of compactness for transparency. So far in the literature, this parameter has not been considered to play a role, and the excluded volume is always considered to be equal to either the dry or the hydrated volume of the protein (Bettelheim and Siew, 1983). With compact globular particles (i.e., with $\varphi/c = 0.75$ and the excluded volume equal to the dry volume), the scattering is maximum for a protein concentration of about 0.150 g/cm^3 whatever the particle molecular weight. Changing the compactness has the effect of shifting the maximum to lower protein concentration values, as shown in Fig. 6-3b. Also, at high protein concentrations, it can be seen that for a given molecular weight, changing the φ/c ratio by a factor of 2 may change the scattering and therefore the extinction coefficient by a factor of 10, which may be particularly important for the lens. We can now calculate from the scattering that, as illustrated in Fig. 6-4, given their molecular weight, compact α-crystallins would be turbid whereas α-crystallins as a pile of subunits are perfectly compatible with transparency. The result is entirely consistent with the model pro-

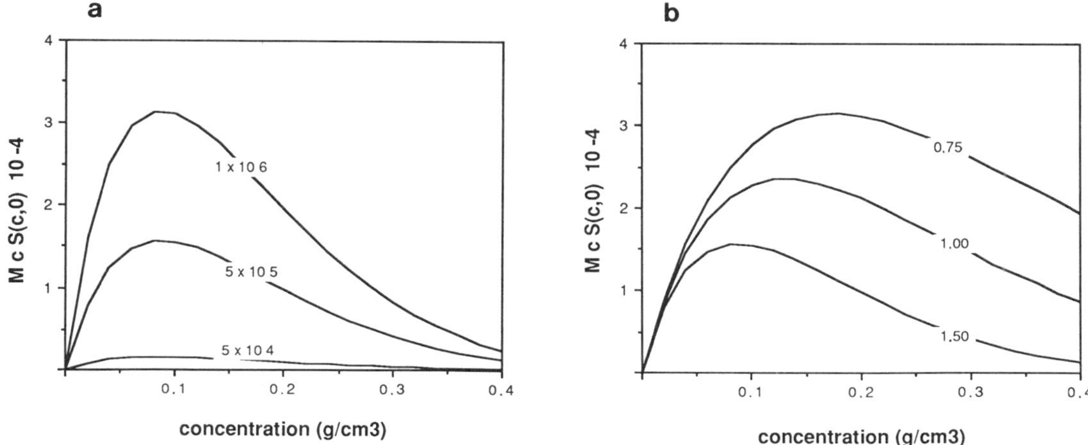

FIGURE 6-3. Variation as a function of protein concentration of the scattering, I_s/I_0. The scattering is represented by the product $McS(c,0)$, according to equation 4. The proteins are supposed to interact through screened electrostatic repulsive interactions. (a) The proteins have a specific excluded volume φ/c of 1.50. The scattering was calculated for different protein molecular weight values as indicated on the figure. (b) The protein molecular weight is 500,000. The scattering was calculated for different φ/c values as indicated in the figure.

posed for α-crystallin quaternary structure (Tardieu *et al.*, 1986).

One of the major result of our study is therefore to show that, by adequate combination of molecular weight and compactness, transparency may be achieved in many ways and that the constraint for transparency might not be as severe as thought previously. As far as transparency is concerned, many different proteins might have done as well. So the question remains why the crystallins were selected through evolution, and in particular why high-molecular-weight α-crystallins. Part of the answer probably lies in the stability as a function of age of such structures and/or in their ability to incorporate modified subunits. One other answer could be the charge. The charge does not seem to play a role

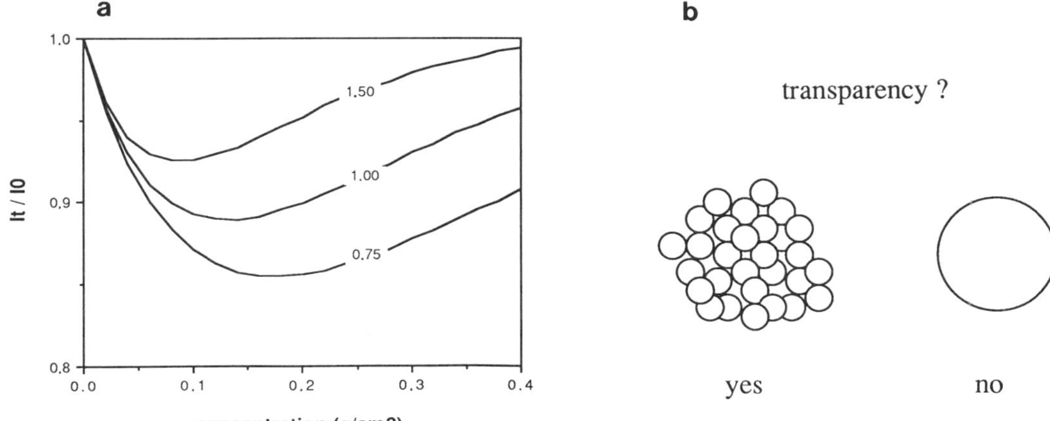

FIGURE 6-4. (a) Percentage of transmitted light, I_t/I_0, as a function of protein concentration for the scattering experiments shown in Fig. 6-3b. The calculation was done, as for Fig. 6-1b, using equations 1 and 2 for a 1-cm-thick lens and a wavelength of 500 nm. More extensive calculations can be found in Vérétout *et al.* (1989). (b) Pictorial equivalent of a. Consider a concentrated protein solution for which the protein molecular weight is of the order of 500,000 and the protein–protein interactions are repulsive. Proteins with a quaternary structure like a pile of subunits, i.e., with excluded specific volumes φ/c of the order of 1.50, are compatible with transparency, whereas proteins with compact structures, i.e., with excluded volume close to the dry volume and φ/c close to 0.75, would be slightly turbid.

directly, since at the ionic strength prevailing *in vivo* the charge is almost completely screened. It could play an important role, however, in preventing aggregation. In this context, oligomers are a good way of forming a globular particle with an evenly distributed charge density. Polydispersity may, in addition, help to avoid crystallization. Small crystals would efficiently scatter light and produce localized opacities. Nevertheless, obviously, small proteins with the same type of repulsive interactions as those observed with α-crystallins would have been better for transparency. This result certainly means that there must be other types of constraints that account for the protein mixture and protein distribution observed *in vivo*. One advantage of a mixture of different sized proteins might be to increase the maximum concentration attainable. Another one is, again, that size heterogeneity is an efficient way to prevent crystallization.

In our opinion, however, the reasons for a protein mixture are to be found elsewhere. One could imagine that the three main crystallin classes present the same type of interactions and that, therefore, the lower the molecular weight, the higher the transparency. This is not true at all. Electrostatic repulsive interactions are also found for the β-crystallins, yet with a different φ/c ratio and a different charge, and γ-crystallins present attractive interactions (different macroscopic behavior had already been shown, e.g., Siezen *et al.*, 1985; Thomson *et al.*, 1987). In fact, one of the results of our recent work was to show that different proteins may display different types of interactions. The effect on scattering of attractive interactions, illustrated in Fig. 6-5, is strikingly different from the one observed previously with α-crystallins (Fig. 6-2b). Since the γ-crystallin molecular weight is so small compared to that of α-crystallins, such different interaction behavior has no drastic effect on the overall transparency of the cytoplasmic extracts, as can be seen from the comparison in Fig. 6-5 of the variation of the cortical extract scattering and of the α-crystallin scattering as a function of protein concentration.

Such experiments led us, however, to hypothesize that the protein concentration gradient within the eye lens might originate from constant osmotic pressure coupled to differential interactive properties of crystallins (Vérétout and Tardieu, 1989). Indeed, in the physiological concentration range, for a given osmotic stress, the equilibrium concentration is found to be higher in nuclear than in cortical extracts (it seems that similar measurements were done independently by Magid *et al.*, 1989; see also Reiff, 1986).

CONCLUSIONS

The scattering experiments that have been discussed here are, of course, also relevant to understanding opacity in cases where the latter could be caused by

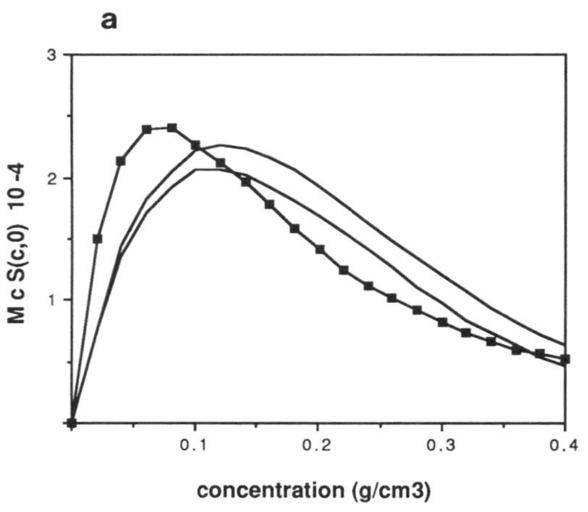

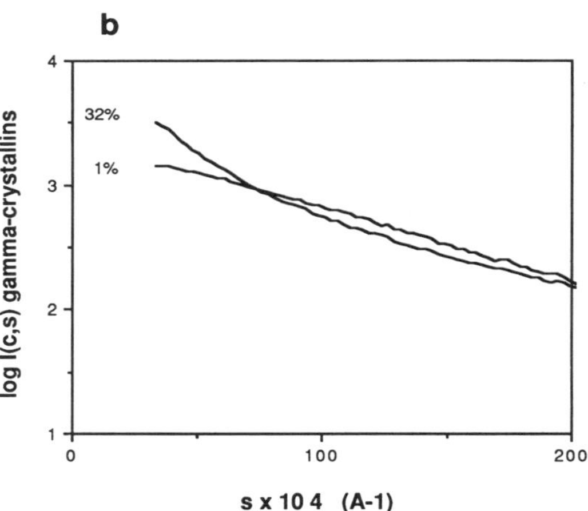

FIGURE 6-5. (a) Comparison of the light-scattering variation with protein concentration of cortical cytoplasmic extracts (the two lines as in Fig. 6-1a) and of α-crystallin solutions (■). Both curves are close to each other in the physiological concentration range, in agreement with the hypothesis that in the normal state, lens transparency is essentially limited by the α-crystallin scattering. (b) X-ray-scattering curves recorded at low (1%) and high (32%) concentration with γ-crystallins (experiments on γ-crystallins were started in collaboration with C. Slingsby). Comparison with the α-crystallin curves in Fig. 6-2b immediately shows that the protein–protein interactions are totally different.

protein aggregation and/or changes in the protein concentration. It would take too long to develop the topic here. It is simple, however, to see that the formalism used to calculate the scattering of a protein solution only needs to be slightly modified to estimate, given the size of an aggregate, the expected opacity. For such calculations, it is sufficient to assume that the normal cytoplasm behaves as a continuous background for the aggregates and to neglect, in a first approximation, the interactions between the aggregates. The scattering is then calculated as a function of contrast between aggregates and background. As already shown, a few percent macromolecular aggregates are sufficient to induce opacities (Delaye *et al.*, 1987). In the same way, it can be shown that even a small contrast difference caused by a small protein concentration difference between neighboring cells may have drastic effects because of the cell size as compared to the wavelength. What exactly happens, however, in the various cataracts remains to be established. For a more general discussion of cataracts, including the role of membranes and ions, we find it better to refer the reader to Duncan *et al.* (Chapter 4, this book).

Finally, to come back to the molecular basis of transparency, the next step will be to relate the observed differences in the interactive properties of α- and γ-crystallins to differences in their three-dimensional structures, a challenge for crystallographers already under way (Slingsby, 1985; Slingsby *et al.*, Chapter 5, this book).

REFERENCES

Benedek, G. B., 1971, Theory of transparency of the eye, *Appl. Opt.* **10**:459–473.

Bettelheim, F. A., and Siew, E. L., 1983, Effect of change in concentration upon lens turbidity as predicted by the random fluctuation theory, *Biophys. J.* **41**:29–33.

Clark, J. I., Delaye, M., Hammer, P., and Menge, L., 1982, Preparation and characterization of native lens cell cytoplasm, *Curr. Eye Res.* **1**:695–704.

Delaye, M., and Gromiec, A., 1983, Mutual diffusion of crystallin proteins at finite concentrations: A light scattering study, *Biopolymers* **22**:1203–1221.

Delaye, M., and Tardieu, A., 1983, Short range order of crystallin protein accounts for eye lens transparency, *Nature* **302**:415–417.

Delaye, M., Danford-Kaplan, M. E., Clark, J. I., Krop, B., Gulik-Krzywicki, T., and Tardieu, A., 1987, Effect of calcium on the calf lens cytoplasm, *Exp. Eye Res.* **44**:601–616.

Eisenberg, H., 1976, *Biological Macromolecules and Polyelectrolytes in Solution,* Clarendon Press, Oxford.

Hansen, J. P., and Hayter, J. B., 1982, A rescaled MSA structure factor for dilute charged colloïdal dispersions, *Mol. Phys.* **46**:651–656.

Hayter, J. B., and Penfold, J., 1981, An analytic structure factor for macroion solutions, *Mol. Phys.* **42**:109–118.

Magid, A. D., McIntosh, T. J., and Simon, S. A., 1989, Osmotic pressure of bovine lens crystallins, *Biophys. J.* **55**:21a.

Parsegian, V. A., Rand, R. P., Fuller, N. L., and Rau, D. C., 1986, Osmotic stress for the direct measurement of intermolecular forces, *Methods Enzymol.* **127**:400–416.

Prouty, M. S., Schechter, A. N., and Parsegian, V. A., 1985, Chemical potential measurements of deoxyhemoglobin S polymerization: Determination of the phase diagram of an assembling protein, *J. Mol. Biol.* **184**:517–528.

Reiff, T. R., 1986, A colloid osmotic model of macromolecular aggregation to explain tissue water loss in aging, *Exp. Gerontol.* **21**:267–276.

Siezen, R. J., Fisch, M. R., Slingsby, C., and Benedek, G. B., 1985, Opacification of γ-crystallin solutions from calf lens in relation to cold cataract formation, *Proc. Natl. Acad. Sci. U.S.A.* **82**:1701–1705.

Slingsby, C., 1985, Structural variation in lens crystallins, *Trends Biochem. Sci.* **10**:281–284.

Tardieu, A., and Delaye, M., 1988, Eye lens proteins and transparency: From light transmission theory to solution x-ray structural analysis, *Annu. Rev. Biophys. Biophys. Chem.* **17**:47–70.

Tardieu, A., Laporte, D., Licinio, P., Krop, B., and Delaye, M., 1986, Calf lens α-crystallin quaternary structure: A three-layer tetrahedral model, *J. Mol. Biol.* **192**:711–724.

Thomson, J. A., Schurtenberger, P., Thurston, G. M., and Benedek, G. B., 1987, Binary liquid phase separation and critical phenomena in a protein/water solution, *Proc. Natl. Acad. Sci. U.S.A.* **84**:7079–7083.

Trokel, S. L., 1962, The physical basis for transparency of the crystalline lens, *Invest. Ophthalmol.* **1**:493–501.

Vérétout, F., and Tardieu, A., 1989, The protein concentration gradient within eye lens might originate from constant osmotic pressure coupled to differential interactive properties of crystallins, *Eur. Biophys. J.* **17**:61–68.

Vérétout, F., Delaye, M., and Tardieu, A., 1989, The molecular basis of eye lens transparency: Osmotic pressure and X-ray analysis of α-crystallin solutions, *J. Mol. Biol.* **205**:713–728.

7

Nonenzymic Posttranslational Modification of Lens Proteins in Aging

JOHN J. HARDING

INTRODUCTION

Many proteins are modified after translation by reactions catalyzed and controlled by enzymes. Phosphorylation, glycosylation, acetylation, hydroxylations, and other changes are well known to biochemists in different fields. Apart from these changes proteins are thought of as relatively inert, surviving unchanged until removal by proteolysis and replacement, but proteins consist of amino acids, many of which have reactive groups in their side chains: thiols, amides, indoles, imidazoles, phenols, thioethers, and amino and guanidino groups. Furthermore, they exist in an environment containing many reactive small molecules including sugars, reactive metabolites, cyanate derived from urea, corticosteroids, acetaldehyde, and other aldehydes. All of these can attack proteins *in vivo* in reactions that do not require enzymes. These nonenzymic reactions of active groups on proteins with surrounding molecules are the subject of this chapter. The emphasis is on lens proteins and on changes in aging, although the same modifications are important in many diseases and toxic effects including the long-term complications of diabetes and renal failure and the toxic effects of alcohol and other xenobiotics (Harding, 1985).

The first clue to the existence of some nonenzymic modifications was the appearance of new components separable by electrophoresis, isoelectric focusing, or ion-exchange chromatography. These components were invariably more negatively charged (or less positively charged). Many were tabulated by Robinson and Rudd (1974) when they were seeking evidence for deamidation of proteins (Table 7-1). They used such anodic shifts as evidence for deamidation, but most of the possible nonenzymic modifications that occur *in vivo* would cause anodic shifts (Table 7-2) because they eliminate the positive charges of amino groups where the reactions occur. Some such as glucose 6-phosphate also introduce negative charges, thus enhancing the anodic shift. Some changes such as thiol oxidation and racemization do not produce a change in charge and yet progress with age.

NONENZYMIC MODIFICATIONS

Most nonenzymic posttranslational modifications are thought to be deleterious (Harding, 1985). They progress slowly, so long-lived proteins may be the most seriously affected. The proteins in the lens nucleus are perhaps the most long-lived proteins in the body, with no detectable synthesis (see Harding and Crabbe, 1984), and so they are ideal candidates for the accumulation of damage caused by slow nonenzymic attack. These proteins give a fuzzier electrophoretic pattern, with more anodic bands appearing, and they become more difficult to keep in solution when isolated from older lenses.

The thiols and disulfides of clear lenses can be studied noninvasively by Raman spectroscopy, and

JOHN J. HARDING • Nuffield Laboratory of Ophthalmology, Oxford University, Oxford OX2 6AW, Great Britain.

TABLE 7–1
The Appearance of Anodic Fractions of Proteins with Time[a]

Protein	Source	Evidence[b]
Albumin	Bovine plasma	E,F
Aldolase	Rabbit muscle	C,F
α-Amylase	Human saliva	C,E,F
Aminopeptidase B	Rat liver	E,F
L-Asparaginase	*E. coli*	E
L-Asp aminotransferase	Porcine heart	F
Carbonic anhydrase B	Human erythrocyte	E,F
Carbonic anhydrase C	Human erythrocyte	E,F
Catalase	Bovine and murine liver	C,E
	Horse and human erythrocyte	C,E,F
Colicin E3	*E. coli*	C,F
α-Corticotropin	Ovine	C,E
β-Corticotropin	Porcine	C
α-Crystallin	Bovine and human lens	C,E,F
Cytochrome C	Bovine and rat heart	C,E
	Rat kidney	E
	Rat liver	E
Enterotoxin C_2	*Staphylococcus*	F
Esterase	Human	E
α_2-Globulin	Rat serum	E,F
Glu–Asp aminotransferase	Porcine heart	E
L-Glycerol 3-phosphate dehydrogenase	Rat liver and muscle	E,F
Growth hormone	Human	E
γG-immunoglobulin	Mouse serum	C,E,F
Insulin	Bovine	C
	Porcine	E
Lysozyme	Hen egg white	C,E
Myoglobin	Bovine	F
Phosphoryl carrier protein	*E. coli*	E
Prolactin	Bovine	C,E
	Ovine	C,E,F
Transferrin	Human	E

[a]Adapted from Robinson and Rudd (1974).
[b]C, chromatography; E, electrophoresis; F, isoelectric focusing.

this shows a progressive loss of thiol and appearance of disulfide with age in mouse and rat lens (Kuck *et al.*, 1982) but not in guinea pig or human lens (Yu *et al.*, 1985). The apparent loss of all thiol from the rat lens as indicated by Raman spectroscopy is surprising because some thiol is buried within the protein structures and would require extensive unfolding of proteins to make it available. Chemical analyses do not support the view that all the protein thiol is lost with age (Perry *et al.*, 1987; Hum and Augusteyn, 1987).

Other nonenzymic modifications of lens proteins that increase with age are nonenzymic glycosylation (Chiou *et al.*, 1981) and racemization of aspartyl residues (Masters *et al.*, 1977; Muraoka *et al.*, 1987). Cataract is a strongly age-dependent disease and in the older population has been called senile cataract. In Oxford the mean age of patients at the time of cataract surgery is 73 years. Many age-related changes to the lens are found to be more pronounced in cataract, and this is true of nonenzymic modification of proteins. Glycation (nonenzymic glycosylation), carbamylation, oxidation of methionine, addition of gluthathione, and addition of corticosteroids have all been identified in at least some human cataracts.

Cerami *et al.* (1987) have suggested that nonenzymic glycosylation may play a role in aging, but many nonenzymic changes have a similar effect, and therefore I suggest that many nonenzymic modifications of proteins may play a role in aging and in slowly developing age-related diseases such as cataract.

One modification we have studied is that pro-

TABLE 7–2
Nonenzymic Modifications of Proteins That Produce Anodic Shifts

Reaction	Reactant	Charge change per site
Glycosylation (glycation)	Glucose	1
	Galactose	1
	Fructose	1
	Glucose 6-phosphate	3
	Glyceraldehyde	1
Carbamylation	Cyanate	1
With aldehydes and ketones	Acetaldehyde	1
	Pyridoxal	1
	Prednisolone	1
Acetylation	Aspirin	1
Oxidation of methionine	—	<1
Deamidation	—	1
Addition of glutathione	Glutathione	1

duced by cyanate (carbamylation). Cyanate is a small reactive ion derived from a simple equilibrium with urea present in plasma and other tissues (Fig. 7-1). Blood urea levels are about 5 mM in the young but increase with age. Cyanate reacts and binds covalently with lens proteins (Harding and Rixon, 1980). It modifies the protein surface and provokes conformational changes (Beswick and Harding, 1984) that lead to exposure of thiol groups and their oxidation to disulfides. In a more subtle way the changes to the surface cause a phase separation opacity, like a "cold cataract," in incubated rat lenses (Crompton *et al.*, 1985). We have since shown that unfolding of lens proteins can be achieved not only by reaction with cyanate but by reaction with glucose 6-phosphate (Beswick and Harding, 1987). In other studies we have shown that a variety of chemical modifiers can induce the increase in phase separation temperature. These include methyl isocyanate, the toxic compound released in the Bhopal disaster (Harding and Rixon,

$$\begin{array}{l} NH_2 \\ | \\ C{=}O \rightleftharpoons HNCO + NH_3 \\ | \\ NH_2 \end{array}$$

FIGURE 7-1. Equilibrium between urea, on the left, and cyanate (isocyanate shown as isocyanic acid).

1985), thiocyanate, glucose 6-phosphate, and glucosamine (R. Ajiboye, K. C. Rixon, and J. J. Harding, unpublished results).

The nonenzymic modifications described are not specific either with respect to modifiers, many of which occur *in vivo,* or to proteins, most of which have the same reactive groups. If such modifications are important in cataract, they should also be relevant to other age-related conditions. Indeed, nonenzymic glycosylation is thought to be important in the long-term sequelae of diabetes, reaction with acetaldehyde in liver cirrhosis, with cyanate in renal failure, and so on.

CLINICAL FINDINGS

Peripheral neuropathy, like cataract, occurs with increasing frequency in the older population, and posttranslational modifications of proteins have been associated with both cataract and neuropathy, especially peripheral neuropathy (Table 7-3). If posttranslational modification plays an important role in both cataract and peripheral neuropathy, one would predict that these two conditions would be associated. To test this idea we incorporated a question to assess peripheral neuropathy into the questionnaire for a case-control study of cataract (Harding *et al.*, 1989).

In this study 423 cataract patients and 608 con-

TABLE 7–3
Nonenzymic Posttranslational Modifications of Proteins Associated with Neuropathy and Cataract[a]

Modification or modifier	Neuropathy	Cataract
Glycosylation	+	+
Acetaldehyde (from ethanol)	+	+
Pyridoxal	+	
γ-Diketones	+	
Corticosteroids		+
Naphthoquinone		+
Cyanate	+	+
Urea (renal failure)	+	+
Disulfiram	+	
Methionine oxidation		+
Racemization		+

[a]See Harding (1985).

trols were interviewed in their own homes. The study was designed to identify risk factors for cataract, to confirm the protective effect of aspirin-like analgesics, and in particular to identify the dose of aspirin-like drugs associated with a protective effect. The question relating to peripheral neuropathy asked if the subjects experienced tingling or numbness in hands or feet. As a result, peripheral neuropathy appeared as a risk factor for cataract (Table 7-4). This result could have been caused by the known link between diabetes and both peripheral neuropathy and cataract, or it might have been caused by a link between glaucoma and cataract, glaucoma being a risk factor for cataract, and peripheral neuropathy can be caused by drugs used to treat glaucoma. As a result of multifactorial analysis, log-linear analysis, to study these interactions, it became clear that peripheral neuropathy was an independent risk factor for cataract. This is not to suggest that neuropathy causes cataract. This epidemiological finding is consistent with the notion that some cataracts are caused by factors that also cause peripheral neuropathy. These factors include, in addition to diabetes, a known risk factor for both conditions. We suggest that both cataract and peripheral neuropathy are partly caused by posttranslational modification of the proteins of the lens and of the peripheral nerves, respectively.

TABLE 7–4
Peripheral Neuropathy Appears as a Risk Factor for Cataract in a Case-Control Study of Cataract in Oxfordshire[a]

	Controls	Cases	Total
Without neuropathy	451	281	732
With neuropathy	157	142	299
Total	608	423	1031
Percentage with neuropathy	25.8	33.6	

[a]Relative risk is 1.45 with a 95% confidence interval 1.11 to 1.90; $P = 0.007$ ($\chi^2 = 7.27$). From Harding *et al.* (1989).

REFERENCES

Beswick, H. T., and Harding, J. J., 1984, Conformational changes induced in bovine lens α-crystallin by carbamylation. Relevance to cataract, *Biochem. J.* **223**:221–227.

Beswick, H. T., and Harding, J. J., 1987, Conformational changes induced in lens α- and γ-crystallins by modification with glucose 6-phosphate. Implications for cataract, *Biochem. J.* **246**:761–769.

Cerami, A., Vlassara, H., and Brownlee, M., 1987, Glucose and aging, *Sci. Am.* **256**:90–96.

Chiou, S.-H., Chylack, L. T., Tung, W. H., and Bunn, H. F., 1981, Nonenzymic glycosylation of bovine lens crystallins. Effect of aging, *J. Biol. Chem.* **256**:5176–5180.

Crompton, M., Rixon, K. C., and Harding, J. J., 1985, Aspirin prevents carbamylation of soluble lens proteins and prevents cyanate-induced phase separation opacities *in vitro:* A possible mechanism by which aspirin could prevent cataract, *Exp. Eye Res.* **40**:297–311.

Harding, J. J., 1985, Nonenzymatic covalent post-translational modification of proteins *in vivo, Adv. Protein Chem.* **37**:247–334.

Harding, J. J., and Crabbe, M. J. C., 1984, The lens: Development, proteins, metabolism and cataract, in: *The Eye,* 3rd ed., Vol. 1B (H. Davson, ed.), Academic Press, London, pp. 207–492.

Harding, J. J., and Rixon, K. C., 1980, Carbamylation of lens proteins a possible factor in cataractogenesis in some tropical countries, *Exp. Eye Res.* **31**:567–571.

Harding, J. J., and Rixon, K. C., 1985, Lens opacities induced in rat lenses by methyl isocyanate, *Lancet* **1**:762.

Harding, J. J., Harding, R. S., and Egerton, M., 1989, Risk factors for cataract in Oxfordshire: Diabetes, peripheral neuropathy, myopia, glaucoma and diarrhoea, *Acta Ophthalmol.* **67**:510–517.

Hum, T. P., and Augusteyn, R. C., 1987, The nature of disulphide bonds in rat lens proteins, *Curr. Eye Res.* **6**:1103–1108.

Kuck, J. F. R., Yu, N.-T., and Askren, C. C., 1982, Total sulfhydryl by Raman spectroscopy in the intact lens of several species: Variations in the nucleus and along the optic axis during aging, *Exp. Eye Res.* **34**:23–27.

Masters, P. M., Bada, J. L., and Zigler, J. S., 1977, Aspartic acid racemization during aging and in cataract formation, *Nature* **268**:71–73.

Muraoka, S., Fujii, N., Tamanoi, I., and Harada, K., 1987, Characterization of a protein containing D-aspartic acid in aged mouse lens, *Biochem. Biophys. Res. Commun.* **146**:1432–1438.

Perry, R. E., Swamy, M. S., and Abraham, E. C., 1987, Progressive changes in lens crystallin glycation and high-molecular-weight aggregate formation leading to cataract development in streptozotocin-diabetic rats, *Exp. Eye Res.* **44**:269–282.

Robinson, A. B., and Rudd, C. J., 1974, Deamidation of glutaminyl and asparaginyl residues in peptides and proteins, *Curr. Top. Cell. Regul.* **8**:247–295.

Yu, N.-T., DeNagel, D. C., Pruett, P. L., and Kuck, J. F. R., 1985, Disulfide bond formation in the eye lens, *Proc. Natl. Acad. Sci. U.S.A.* **82**:7965–7968.

8

Lenticular Senescence and the Retina

ROBERT WEALE

INTRODUCTION

Ever since the time of Helmholtz, research workers have concerned themselves with why and how presbyopia occurs (Bito, 1988; Brückner, 1959). However, the matter has only rarely been approached from an evolutionary point of view (cf. Bito and Miranda, 1987). The outstanding question in this context is not so much why accommodation is lost in midlife as why it is well over 10 D in the very young. In what circumstances does anyone have to form detailed retinal images of objects a mere 8 or 10 cm from the eyes?

A POSSIBLE CAUSE OF THE MAGNITUDE OF EARLY ACCOMMODATION

It may be instructive to invert the problem and to examine it from the point of view of modern biological theories of aging (Warner *et al.*, 1987). Presbyopia is multifactorial (Weale, 1963) and depends on the properties of the capsule, the lens matrix and its component fibers, on the ciliary muscle, on the position of the lens in the eye, perhaps on properties of the iris and its musculature, etc. An approach based on simple evolutionary concepts may therefore be more immediately revealing than one leaning on the molecular biology of the relevant tissues. It is not intended to minimize the importance of the latter in connection with the problem. But it is probably true to say that accommodation and presbyopia developed as a result of certain biological pressures, the response to which can be explained in terms of molecular biology. The pressures define the need to which a feasible response has evolved as determined by modern scientific methods. My immediate object is to define the need.

It does not appear likely that presbyopia can bear a direct relation to the human life span (Brückner *et al.*, 1987); its theoretical maximum is about 115 years (Lestienne, 1988). From the point of view of vision, the interests of the propagation of the species are served if the eyes are at their best for a period approximately twice as long as the interval between birth and puberty or birth and the commencement of an independent (family) life. This suffices for the adequate care and protection of the next generation. If accommodation of 3–4 D is biologically useful and were to be maintained for anything like a lifetime—whether this be 70 or 115 years—the biological cost would be enormous. The standard error of measurements of accommodation is 0.7–0.8 D (Duane, 1912; McBrien and Millodot, 1986). If man is to be guaranteed an amplitude of 3–4 D—i.e., if this is a minimum—then the normal amplitude has to be 5–6 D higher; the reason is that then the odds will be about 1 in 10^{10} (which corresponds to the world population) against the value dropping below the minimum.

The biological cost of developing and maintaining such a system for a lifetime will therefore be even more expensive than the system based on just 3–4 D. It can be shown (Weale, 1990) that the most economic system meeting the minimum requirements is a linear one (Fig. 8-1) starting with an amplitude of accommodation of ~15 D and decreasing by approximately 0.25 D per year in conformity with observation.

VISUAL ACUITY AND PRESBYOPIA

Another type of pressure militating against the maintenance of a high accommodative power is to

ROBERT WEALE • Age Concern Institute of Gerontology, King's College London, University of London, London SE1 8TX, Great Britain.

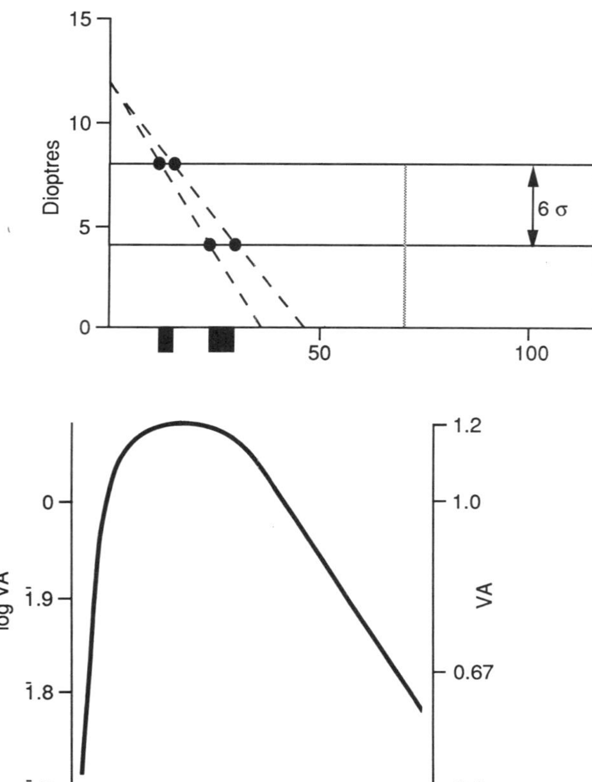

FIGURE 8-1. A high accommodative amplitude in infancy is of economic advantage in the devolution of accommodation. The left-hand black oblong indicates an estimate of the spread of the age of puberty; the right-hand one corresponds to twice that age spread and is placed at double the age of puberty; historically it represents an estimate of the approximate upper age limit up to which vision might be expected to be optimal. The straight lines, which agree well with accommodative devolution, were obtained on the premises that a minimum of, say, 4 D accommodation is required up to the age at which an individual ceases to have to look after his/her offspring and that the biological cost of providing this should be a minimum.

FIGURE 8-2. The variation of visual acuity with age (after Weale, 1982). Note that there is an apparent temporal link between the cessation of effective accommodation and the age-linked onset in the reduction of visual acuity.

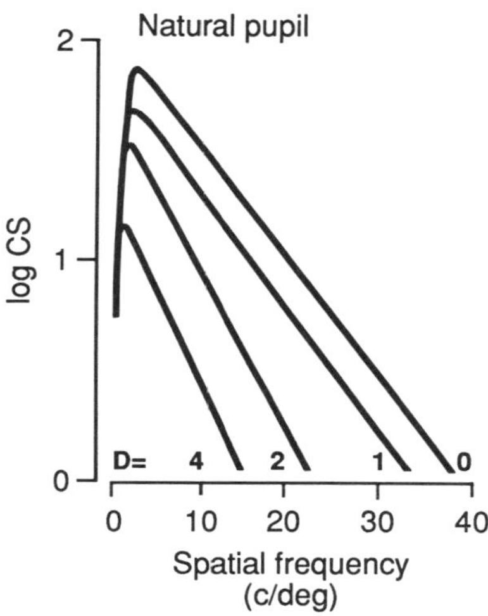

FIGURE 8-3. The variation of contrast sensitivity (CS) with spatial frequency. Parameter: ocular refraction (after Kay & Morrison, 1987). Note that visual acuity as defined by minimum CS drops by about 50% when defocus is of the order of 3 D. Or could the reverse be true? (Cf. Figs. 8-1 and 8-2.)

be found in the senescence of the retina. Between the ages of 40 and 80 years, contrast sensitivity decreases (Sloane *et al.*, 1988), and visual acuity (Fig. 8-2) drops to about 50% of its maximum (cf. Weale, 1982). But a similar drop in contrast sensitivity for high spatial frequencies is observed (Kay and Morrison, 1987) when there is a loss of focus corresponding to that resulting from the disappearance of accommodation (Fig. 8-3). Although it is true that this decrement is not a function of viewing distance, the fact that the neural grain coarsens so as to impede the perception of fine detail is hardly going to constitute a biological pressure for the long-term maintenance of a focusing mechanism serving the optical optimization for near and far.

THE LENS AS A FILTER

The apparent symbiosis between lens and retina is not, however, confined to image quality control. It has been reported from more than one laboratory that the young human retinal pigment epithelium accumulates one of the most ubiquitous of age pig-

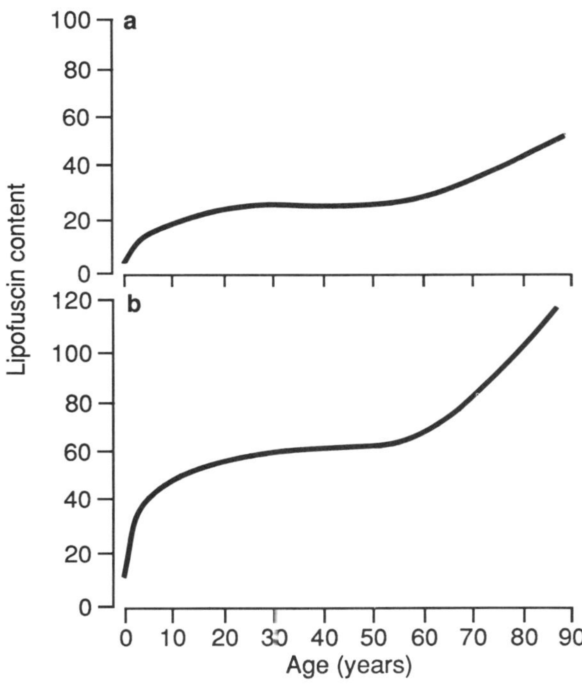

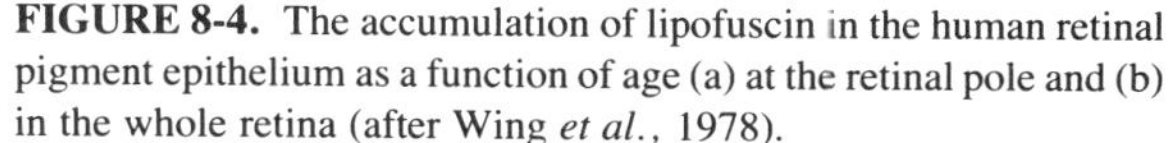
FIGURE 8-4. The accumulation of lipofuscin in the human retinal pigment epithelium as a function of age (a) at the retinal pole and (b) in the whole retina (after Wing *et al.*, 1978).

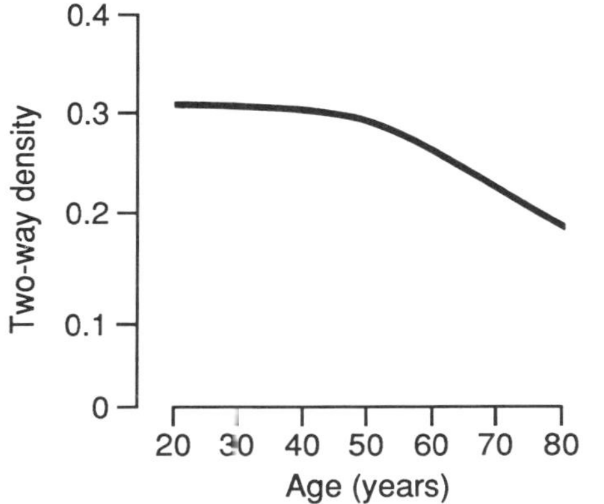

FIGURE 8-6. The variation of photolysable cone pigment concentration with age (after Keunen *et al.*, 1987). The data from Kilbride *et al.* (1986) do not show the plateau but do not differ greatly from the curve shown in that the data for their youngest group show considerable variability. On the other had, Keunen *et al.* found that the data of a large fraction of their subjects were unacceptable and had to be discarded. Note that the plateau in this figure appears to be linked temporally to that shown in Fig. 8-4.

ments, namely, lipofuscin (Feeney-Burns *et al.*, 1984; Wing *et al.*, 1978). The appearance of large amounts of an age pigment (Fig. 8-4) early in life is amazing and has been tentatively attributed to the influence of short-wavelength radiation (Weiter *et al.*, 1986). After a rapid rise during the first decade or two, there occurs a plateau extending for some two or three decades up to the age of about 50 years. Thereafter the rise resumes.

The slowdown in accumulation has been linked to the increase in lenticular absorbance (Weiter *et al.*, 1986), which progresses in an exponential manner (Weale, 1988a): the more the lenticular nucleus absorbs short-wavelength radiations, the fewer of them will reach the retina, perhaps causing a slowing in the accumulation of lipofuscin. The time constants of the two processes do not, however, entirely agree (Weale, 1988b), and more than one mechanism is probably involved in the process of lipofuscin formation. It may, for example, be mediated via epithelial vitamin A and photoreceptor activity, respectively.

Looked at from the point of view of evolution, the young eye receives a large amount of light, presumably so as to vouchsafe to the individual information needed for successful development and survival. The yellowing of the lens seems to put a brake on the concomitant formation of lipofuscin, which may harbor, or be the result of, a hazard (Harman, 1985; Sohal and Wolfe, 1986). In this manner, optimal function of the retina coexists with the period necessary for the growth of offspring.

LENS GROWTH AND RETINAL EXPOSURE TO RADIATION: LOSS OF VISUAL PIGMENT

However, the rise of lipofuscin concentration is resumed (Fig. 8-4), a fact often overlooked. A variety of possible causes spring to mind, but the continued growth of the lens has recently been advanced as a possible explanation (Weale, 1988b). In the young lens, the light-absorbing nucleus is near the pupil, as the cortex is narrow; hence, the pupil is effectively stoppered, short-wavelength radiation entry into the eye being controlled by nuclear transmissivity. But, because of continual lenticular growth, the cortex widens, a gap forms between the nucleus and the iris (Fig. 8-5), and, if light can traverse it without being intercepted by the ciliary system, it will reach the retina primarily in an equatorial belt. Following reflection, some of it will reach the posterior pole.

The second phase of lipofuscin accumulation is accompanied by the aforementioned decrease in contrast sensitivity and visual acuity. Their cause appears to lie in progressive neural attrition, notably in the diminution of foveal cones. Both Kilbride *et al.* (1986) and Keunen *et al.* (1987) have reported a decrease in the concentration of foveal visual pigment as a function of age (Fig. 8-6), which can be linked quantitatively to the synchronous drop in visual acuity. This provides another example of the eye departing from its optimal physiological function once the period required for the survival of the individual of the species has ended.

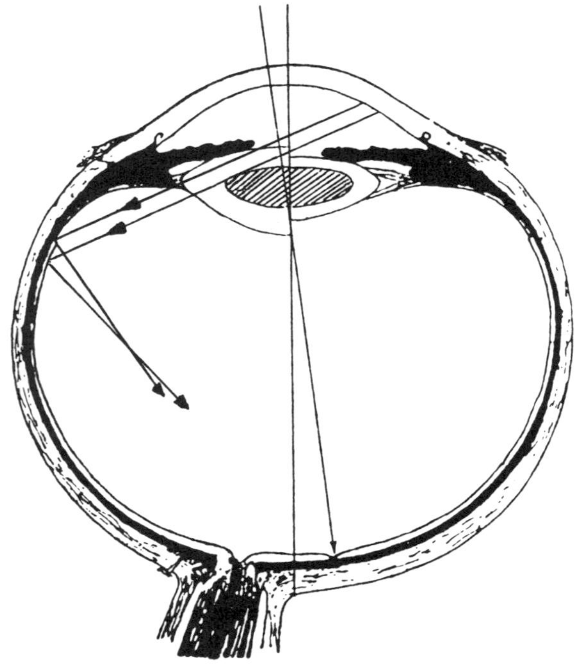

FIGURE 8-5. How the lenticular nucleus appears to control the entry of short-wavelength radiation into the human eye (Weale, 1988b).

CONCLUSION

It would seem to follow that the lens plays an organizing role in ocular senescence. No matter whether presbyopia is primarily lenticular or ciliary in origin (Bito, 1988), it results from two evolutionary principles, namely, those of biological economy and of the gradualness of changes needed to achieve this (which is also a matter of cost). It is not clear whether the rate of accommodative devolution is ultimately controlled by the retina–brain through the reduction in contrast responses. Lenticular absorbance appears to protect the retina from radiational insults during the early decades when potential damage and intake of information have to be balanced. The need for this protection is diminished after midlife, perhaps contributing to retinal devolution, a process that, in turn, reduces biological costs.

REFERENCES

Bito, L. Z., 1988, Presbyopia, *Arch. Ophthalmol.* **106:**1526–1527.

Bito, L. Z., and Miranda, O. C., 1987, Presbyopia, the need for a closer look, in: *Presbyopia* (L. Stark and G. Obrecht, eds.), Fairchild Publications, New York, pp. 411–429.

Brückner, R., 1959, Über Methoden longitudinaler Alternsforschung am Auge, *Ophthalmologica* **138:**59–75.

Brückner, R., Batschelet, E., and Hugenschmidt, F., 1987, The Basel longitudinal study on aging (1955–1978), *Docum. Ophthalmol.* **64:**235–310.

Duane, A., 1912, Normal values of the accommodation at all ages, *J.A.M.A.* **59:**1010–1013.

Feeney-Burns, L., Hildebrand, E. S., and Eldridge, S., 1984, Aging human RPE: Morphometric analysis of macular, equatorial and peripheral cells, *Invest. Ophthalmol. Vis. Sci.* **25:**195–200.

Harman, D., 1985, Role of free radicals in aging and disease, in: *Relations between Normal Aging and Disease* (A. H. Johnson, ed.), Raven Press, New York, pp. 45–84.

Kay, C. D., and Morrison, J. D., 1987, A quantitative investigation into the effects of pupil diameter and defocus on contrast sensitivity for an extended range of spatial frequencies in natural and homatropinized eyes, *Ophthalmol. Physiol. Opt.* **7:**21–30.

Keunen, J. E. E., Van Norren, D., and Van Meel, G., 1987, Density of foveal cone pigments at older age, *Invest. Ophthalmol. Vis. Sci.* **28:**985–991.

Kilbride, P. E., Hutman, L. P., Fishman, M., and Read, J. S., 1986, Foveal cone pigment density difference in the aging human eye,

Vision Res. **26:**321–325.

Lestienne, R., 1988, On the thermodynamical and biological interpretation of the Gompertzian mortality rate distribution, *Mech. Ageing Dev.* **42:**197–214.

McBrien, N. A., and Millodot, M., 1986, Amplitude of accommodation and refractive error, *Invest. Ophthalmol. Vis. Sci.* **27:**1187–1190.

Sloane, M. E., Owsley, C., and Alvarez, S. L., 1988, Aging, senile miosis and spatial contrast sensitivity at low luminance, *Vision Res.* **28:**1235–1246.

Sohal, R. S., and Wolfe, L. S , 1986, Lipofuscin: Characteristics and significance, *Prog. Brain Res.* **70:**171–183.

Warner, H. R., Butler, R. N., Sprott, R. L., and Schneider, E. L. (eds.), 1987, *Modern Biological Theories of Aging,* Raven Press, New York.

Weale, R. A., 1963, New light on old eyes, *Nature* **198:**944–946.

Weale, R. A., 1982, *A Biography of the Eye—Development, Growth, Age,* H. K. Lewis, London.

Weale, R. A., 1988a, Age and the transmittance of the human crystalline lens, *J. Physiol. (Lond.)* **395:**577–587.

Weale, R. A., 1988b, The senescent lens and the retinal pigment epithelium, *Eye* **2:**157S–163S.

Weale, R. A., 1990, Evolution, age, and ocular focus, *Mech. Aging Dev.* **53:**85–89.

Weiter, J. J., Delori, F. C., Wing, G. L., and Fitch, K. A., 1986, Retinal pigment epithelium lipofuscin and melanin and choroidal melanin in human eyes, *Invest. Ophthalmol. Vis. Sci.* **27:**145–152.

Wing, G. L., Blanchard, G. C., and Weiter, J. J., 1978, The topography and age relationship of lipofuscin concentration in the retinal pigment epithelium, *Invest. Ophthalmol. Vis. Sci.* **17:**601–607.

Part I

Conclusions

GEORGE DUNCAN

I used to have a sinking feeling returning home at about 6-monthly intervals during my Ph.D., when my old grandmother used to ask, "Well, what do you know now that you didn't know 6 months ago?" I had great difficulty in answering her with 6 months to think about it and I had similar missgivings when asked to review Part I of the meeting, with about 6 hours warning! In the first talk of the session we gained a tremendous insight into the regulation of α-crystallin gene expression from Ana Chepelinsky's work. Using a combination of highly sophisticated techniques, including the bacterial chloramphenicol acetyltransferase gene as the reporter, and developing a transgenic mouse system, she has found by point mutation techniques which of the bases in the 5′ flanking sequences are essential for gene promoter activity. This basic information is vital if we are to understand how the lens-specific αA genes are regulated, and this is now of even greater importance when we realize that the αB-chain gene is active in other tissues such as heart and appears to be switched on in certain diseases quite unrelated to cataract.

The regulation of growth and development of the whole lens was admirably dealt with by Yves Courtois, whose team over the past few years has been instrumental in expanding our knowledge of the interaction between the lens and retina in the growth and development of the lens. Furthermore, through gene cloning and *in situ* hybridization techniques, he has demonstrated that the receptors for fibroblast growth factors (FGF) are found not only in the lens epithelial cells but also in the capsule surrounding the lens. His most recent work has demonstrated that the lens cells can also synthesize growth factors, and so there are many mechanisms available by which FGF could regulate lens growth.

Undoubtedly, it now appears that arachidonic acid metabolism is involved in one of these. One of the key markers in lens fiber cell differentiation is in fact MP26, the main intrinsic protein of 26 kDa, which is absent in lens epithelial cells but is the most abundant protein in lens fiber membranes. Lucio Benedetti's work allows us to understand how MP26 is arranged in the lens fiber membranes. The cytoskeleton appears to play a key role in their structural organization, which can in fact be at a very high level, as junctional plaques are often formed by this protein, and Lucio very elegantly outlined the immense amount of work involving combined freeze-fracture and immunogold labeling that had to be undertaken to prove that MP26 was involved in junctional communication in the lens and that spectrin, vinculin, and α-actin all had a role to play in this organization. Calcium plays a key role in communication in many systems, and the next chapter (Duncan *et al.*) went on to analyze the importance of calcium in producing highly localized opacities. I have also demonstrated that the lens free calcium and sodium increase as the result of the aging process, and I have suggested that one of the processes involved with membrane aging is the acquisition of nonspecific cation channels that allow sodium, calcium, and potassium to pass through the membranes.

Chapter 5 presents a beautifully illustrated account of protein packing and lens fiber cells by Christine Slingsby. She literally took us on a Moroccan magic carpet tour of the lens crystallins. In this tour de force she summarized her work over a number of years, which we can now see resulting in immense benefit. The high-resolution structures of γ-crystallin, obtained by x-ray diffraction, have given a great stimulus to lens molecular biologists and physical chemists alike. We know that γ-crystallin is a bilobal structure made up of two similar domains, and this proved to be the key, the Greek key as it turned out, to unlocking the structures of the various β-crystallins as well, since they, with great economy on part of the lens, form a super-

George Duncan • School of Biological Sciences, University of East Anglia, Norwich NR4 7TJ, Great Britain.

family with the γ-crystallins. We can now begin to understand how the γs, βs, and αs can fit together to form either tight, high-density structures, as they do in the rat, or rather loose structures, as they do in the human case. Obviously, any complete molecular explanation of cataract will be based on the breakdown of this packing arrangement. Annette Tardieu took up this packing theme when she reviewed her work with Mireille Delaye on the physical basis for lens transparency, and crystallin interactions are vital in this. Her most recent work concerns a thermodynamic interpretation of transparency, and through this she has elucidated the importance of protein quaternary structure in transparency.

Protein interactions were also dealt with by John Harding, who pointed out that lens proteins become more negatively charged with age. He reviewed several posttranslational mechanisms that account for this, the major ones of which appeared to be glycation and cyanation. Diabetes and reduced glucose tolerance with age would give rise to glycation, for example, and urea, a major source of cyanate in the body, increases with age. The concentration of urea also unfortunately increases with severe diarrhea, which has already been shown to be a very high risk factor in cataract. However, aspirin, John pointed out, prevents cyanation.

Robert Weale brought Part I to a very stimulating end with a typically thoughtful yet provocative account of lens–retinal interaction in aging. Robert's thesis is that our vision is geared to be optimal around 18 to 36 years of age, and that is why we start out with an unphysiologically high value for accommodation of about 14 diopters. This gradually drops from 14 to 8 at a rate that minimizes the biological cost. Thereafter, aging is a battle between light-induced retinal damage and loss of visual acuity. A perfectly transparent lens will permit photo-induced loss of cone pigment, whereas production of screening pigment within the lens would degrade visual acuity.

Part II

VISUOMOTOR ADAPTATION AND GAZE ADAPTATION

Part II

Introduction

JOHN FINDLAY

A very common expression used in an optician's practice must be "You'll soon get used to them" as an individual is given a new prescription for spectacles or lenses. This symposium concerns the emerging science that enables us to see what is involved in such a deceptively simple statement. We can now claim some considerable understanding of the adaptive processes that occur when an individual wears a new pair of spectacles. This understanding has come about through the studies of physiologists, psychologists, engineers, and optometrists within the discipline of neuroscience.

Perhaps surprisingly, the route to this understanding takes us through territory that involves quite fundamental science. Space and time are very basic concepts indeed, and we shall be studying phenomena that have a spatial and a temporal dimension. Vision is preeminently a spatial modality, and adaptation is preeminently a temporal process. There is a familiar conceptual framework of space and time within which science, especially physics, operates. But when we are trying to understand how the brain deals implicitly with space and time in visual processing, we need to be able to step outside this framework. In the study of vision it is necessary for many purposes to abandon the familiar three-dimensional Cartesian configuration. Visual space is quite different; it is radially organized around the visual axis with one small foveal region having crucial importance. Several of the contributors contrast the roles of central and peripheral vision in contributing to adaptation. The contrast between physical time and perceptual time is no less marked. Our concept of chronology as derived from our experience of clocks is not the same as the biological or psychological devolution of time. But even more crucially, there is a whole aspect of vision in which space and time are integrated at an early stage in a way that proved difficult to assimilate for many years. Only relatively recently and relatively gradually has the importance of motion as a fundamental feature of visual stimulation been accepted. Because motion cuts across the traditional spatial and temporal dimensions, it was in the past frequently assigned a secondary importance. Fortunately, this balance has now been redressed, and several of the contributors emphasize the way in which visual motion is involved in adaptation.

This volume also reflects the fusion, in an exciting way, of two rather separate scientific traditions that have been concerned with vision. One tradition places emphasis on mechanism, and the other places emphasis on process. It is necessary to appreciate both traditions in order to understand fully the adaptation phenomena under consideration.

The first scientific tradition derives ultimately from physics and is typified by the field of physiological optics. The message here is that vision—seeing—is a process that may be understood with reference to specifiable physical and physiological mechanisms. Our comprehension of vision and visual phenomena is enhanced by drawing on the physicists' understanding of optics, as shown last century by Helmholtz and this century by Campbell. Within this tradition, rigorous modeling of the processes of vision is given prominence.

Let me introduce just one example from my own particular research area of saccadic eye movements—rapid movements of the eye used to redirect attention. We are aware from our understanding of the image-forming process that the expected consequence of relative motion of retina and retinal image would be blurring. The logic of the mechanistic tradition tells us that the extensive blur accompanying saccadic eye movements would not be useful for visual purposes. This leads to a consideration of how the visual system copes with this situation. Many demonstrations have shown that vision of the blurred image during the course of a saccade is pre-

JOHN FINDLAY • Department of Psychology, University of Durham, Durham DH1 3LE, Great Britain.

vented. Some suppressive process takes place. The importance of this suppression is indicated by the fact that more than one separate mechanism seems to contribute to it. Of course this isn't all the story. Vision is too large and impressive a process to be totally encompassed by simple analogies. We also perceive objects in motion, and they don't generally appear blurred. There are fascinating reasons for this that relate to a topic mentioned earlier—the status of motion within visual perception.

The second tradition has roots in biology and psychology. Biological evolution has favored organisms that can survive in changing circumstances. Adaptability to different environments is a ubiquitous feature of biology. We encounter numerous adaptational processes. Some of these processes are very specific and specialized. An example from visual science is the process of dark adaptation, which depends, although in a far from simple way, on the regeneration of retinal rhodopsin. Here adaptation is dependent on photochemistry. However, much of the adaptation that forms the topic of this symposium concerns processes within the nervous system. Adaptability and plasticity are ubiquitous features of nervous systems. Even very simple nervous systems possess some adaptive capacity for rudimentary habituation and learning. One of the major scientific developments of the 1980s has been the realization and detailed specification of the more complex forms of adaptation and learning possessed by different types of neural networks.

Fusion of these traditions can lead to potential conflict. Exact modeling of a system that is capable of adaptive change is a challenge. To understand what is going on it is often necessary to know not only about the properties of the observer's visual system but also about the properties of the visual environment within which that visual system has evolved. It is also vital for experimental workers never to forget that the subjects of their experiments—whether animals, student volunteers, or clinical patients—are likely to shift their behavioral responses over the course of time. But when these pitfalls are avoided, fusion of the traditions can lead to some of the most exciting current research in neuroscience, as this book demonstrates.

9

Adaptations to Transformations of the Optic Array

IAN P. HOWARD

INTRODUCTION

People adapt to visual distortions that are experienced, for instance, when they first wear spectacles. The changes underlying adaptation can occur in the oculocentric, or retinotopic, system, which codes the positions and movements of images on the retina. They can occur in the headcentric system, which codes the positions of objects with respect to the head, or in the bodycentric system, the system that allows us to point to a seen object with an unseen hand. Finally, they can occur in the exocentric system, which allows us to judge the position of an object with respect to an external frame of reference, for instance, the orientation of a line relative to vertical.

I discuss adaptations to various types of visual distortion and in each case try to show which frame of reference is involved. One general principle is that adaptive changes occur in the oculocentric system only if the distortion is visually apparent.

ADAPTATIONS TO VISUAL CURVATURE AND TILT

If a person views the world through lenses that cause objects to appear curved or tilted, the objects appear less curved or less tilted after some time. In normal visual experience there are as many lines tilted or curved to the right as there are lines tilted or curved to the left. Therefore, an overall optical tilt or curvature of the visual field can be detected purely visually, and this means that the changes responsible for adaptation to this type of distortion can, in theory, occur in the oculocentric system.

A sense of the oculocentric position of an object depends on the *local-sign mechanism* of the visual system. This is the mechanism in which, for a given position of the eye, each region of the visual field has a stable mapping onto the retina (the optical part) and visual cortex (the neural part). A sensory system of this type, in which information is conveyed along spatially distinct channels, is a *labeled-line coding system*. When the spatial ordering of the channels on the sensory surface is also important, as it is in the local-sign system, we have a *topographic labeled-line system*. Approximately one million ganglion cells convey spatial information to the visual cortex, and, beyond the first few months of life, the connections these cells make are fixed. Furthermore, the fovea provides a well-defined reference point, or *Kernpunkt*, for the retinal local-sign system, and this too is built in and therefore not subject to modification. If the system is anatomically determined, how are we to account for the fact that people adapt to distorted visual inputs?

I think the answer lies in the fact that not all the million ganglion-cell axons that go from eye to brain are independent channels of spatial information. Each ganglion cell, at least outside the foveal region, typically receives excitatory inputs from a local group of receptors (its excitatory center) and inhibitory inputs from a surrounding group of receptors (the inhibitory surround). Even in the foveal region, where there is only one receptor for each

IAN P. HOWARD • Human Performance in Space Laboratory, York University, North York, Ontario M3J 1P3, Canada.

ganglion cell, even the smallest image falls across several receptors. In each location of the visual field, the receptive fields of neighboring ganglion cells overlap. A sensory system in which the tuning function of each neural channel overlaps that of every other neural channel is a *metameric system*. The local-sign system is locally metameric; that is, it is metameric within each small region of the retina where a group of receptive fields overlap. By contrast, the receptor stage of the color system is wholly metameric, since it has only three overlapping channels.

Any sensory system, such as an eye, an ear, or the skin, that has receptors distributed over a sensory membrane devotes its local-sign system to only one sensory feature, which I shall refer to as the local-sign feature. Since the local-sign feature in a given sensory system is the only one coded topographically, it may also be referred to as the topographic feature for that sensory system. Position is the local sign or topographic feature of both vision and touch; for audition it is frequency. Local sign is a primary feature, since distinct channels are formed at the receptor level. Color is also a primary feature, but nontopographic. Luminance and temporal intermittency are temporally coded primary features and therefore also nontopographic. Color, luminance, and intermittency are primary features, and none of them is derived from local sign. All other visual features such as motion, orientation, and depth are secondary features because they are spatial or spatiotemporal derivatives. This means that the formation of distinct channels for these features can be deferred until inputs reach the visual cortex. This is a help because there is more room in the brain than in the eye for the detectors required for channeling these features. The packing problem can also be alleviated by limiting the number of channels for each nontopographic feature. This also has the advantage that it makes it possible to have a complete set of detectors for each nontopographic feature at each location in the local-sign system. Thus, we can see the color, motion, orientation, and depth of stimuli at each location in the visual field. But if there are only a few channels to cover a sensory dimension, their tuning functions must overlap; if they did not, there would be no continuity of fine discrimination across the sensory dimension. Nontopographic features are therefore highly metameric.

Metameric systems exhibit the following properties.

Metameric Systems Have Poor Resolution

Two stimuli are resolved when they are detectable as two stimuli. In order to be resolved, two stimuli must excite two detectors at a discriminably higher level than they excite a detector in an intermediate position. Another way of saying this is that an ordered set of detectors can resolve a periodic stimulus only if the spatial frequency of the detectors is at least twice that of the stimulus (the Nyquist limit). Since the color system has only three channels with tuning functions that overlap, our capacity to resolve wavelengths is zero. No matter how many wavelengths are present in a patch of light, we see only one color. The color we see depends on the relative extent to which the different receptors are excited, and if two lights with different wavelength components excite the receptors in the same ratios, those lights appear identical, even though the wavelength components are discriminably different when presented one at a time. The lights are said to be metameric matches.

Metamerism is particularly evident in the skin. Békésy (1967) and others have shown that two or more points applied to a local area of skin feel as if they are one point at an intermediate position that depends on the relative strengths of the stimuli. Békésy called this metameric process funneling.

The visual local-sign system is metameric only locally. It has many independent channels, which, at the theoretical limit, allow it to resolve black-and-white gratings in which the bars are as narrow as the diameter of receptors. In practice, for a given optical system, grating acuity is limited by the mean size of receptive fields, by the extent to which their excitatory centers and inhibitory surrounds overlap, and by the noise in detectors. This can be summed up by saying that resolution is limited by the ability of the visual system to detect local differences in luminance contrast. Note that, although the spatial structure of a receptive field does not allow a ganglion cell to detect the position of stimuli within its receptive field, it does render it optimally sensitive to a particular distribution of light (typically a point of light in a dark surround).

Because the local-sign system is locally metameric, two or more visual stimuli falling wholly within an area of the retina where the excitatory regions of neighboring receptive fields overlap should appear as one stimulus in a position that depends on their relative luminances. This occurs when two short parallel lines are seen together within an area of about 2′ of arc, which is about the size of the smallest receptive fields in the retina (Badcock and Westheimer, 1985; Watt *et al.*, 1983). When the lines are presented separately (one above the other or successively) their distinct positions can be discriminated.

Metameric effects are difficult to detect in secondary features because all such features are spatiotemporal derivatives and are therefore confounded with the local-sign system for visual direction. Two short lines in the same location but at slightly different orientations should metamerize their orientations (they should both appear at an intermediate orientation), and this should happen before the lines perceptibly fuse into one line, because the sensory units that code orientation have larger receptive fields than those that code location. There seem to be no experimental investigations of this possibility, but perhaps metameric processes within excitatory regions of neighboring orientation detectors can explain why parallel lines in a grating near the limit of resolution appear wavy or why the orientation of a short, briefly presented line is difficult to detect (Andrews, 1967). For similar reasons, two superimposed displays of dots moving at similar but different speeds should appear to move at an intermediate velocity. There are no reports of this effect, but there is a report that directions of motion metamerize. Watamaniuk *et al.* (1989) found that the discriminability of the direction of motion in a display of random dots was the same when all the dots moved in the same direction as when they moved in different directions about the same mean direction. They explained their results in terms of metameric processes within the motion-detection system.

Metameric Systems Show Good Discrimination

Two stimuli are discriminated if one is detectably different from the other, given that they have been resolved as two stimuli. A metameric system with poor resolution can have exquisite discrimination. For instance, the human eye can discriminate many hundreds of spectral colors (or their metamers) as long as they are presented sequentially (resolved in time) or to distinct regions of the retina (spatially resolved).

Resolution depends mainly on the noisiness and width of tuning functions of sensory channels. Discrimination depends on the extraction of a difference signal from neighboring detectors. It therefore does not depend on the width of tuning functions of channels but on the steepness of their tuning functions, that is, on the relative rate of change of difference signals as the stimulus is moved over the stimulus continuum, as illustrated in Fig. 9-1 for orientation discrimination. The noisiness of the individual channels seems also to be less important for discrimination (Bowne, 1990). Discrimination sensitivity in a metameric system should show local maxima at places where the tuning functions of detectors intersect, because this is where the difference signal changes most rapidly. The well-known hue discrimination function is an example. If the local-sign system is locally metameric, it too should show undulating discrimination functions. There has been some contention about whether sensitivity to changes in separation of lines shows peaks and troughs as the distance between the lines is varied (Hirsch and Hylton, 1982; Westheimer, 1984). Wilson (1986) interpreted some data from Klein and Levi (1985) as showing peaks and troughs analogous to those in the hue discrimination function. Regan and Price (1986) found undulations in sensitivity to changes in line orientation as the line was varied in orientation, as shown in Fig. 9-2. The highest sensitivity to changing orientation occurred at vertical and horizontal, which suggests that the tuning functions of orientation detectors overlap at these values.

This raises the interesting possibility that tuning functions of neighboring detectors intersect where sensitivity to change is required to be highest. Another illustration of this point is the fact that sensitivity to changes in binocular disparity (stereoscopic depth) is greatest about zero disparity, which is where the tuning functions of disparity detectors seem to overlap (see Regan, 1989, for a discussion of this issue).

The fine discrimination in metameric systems explains hyperacuity, an example of which is the ability to detect vernier offsets several times smaller than the mean spacing of receptive fields. The distinction between resolution and discrimination (hyperacuity) in a locally metameric spatial modality can be vividly illustrated on the skin. If the skin of the back is prodded simultaneously by two pencils about 1 cm apart, they metamerize; that is, they feel like one object at an intermediate position. But if the pencils are presented sequentially with the same

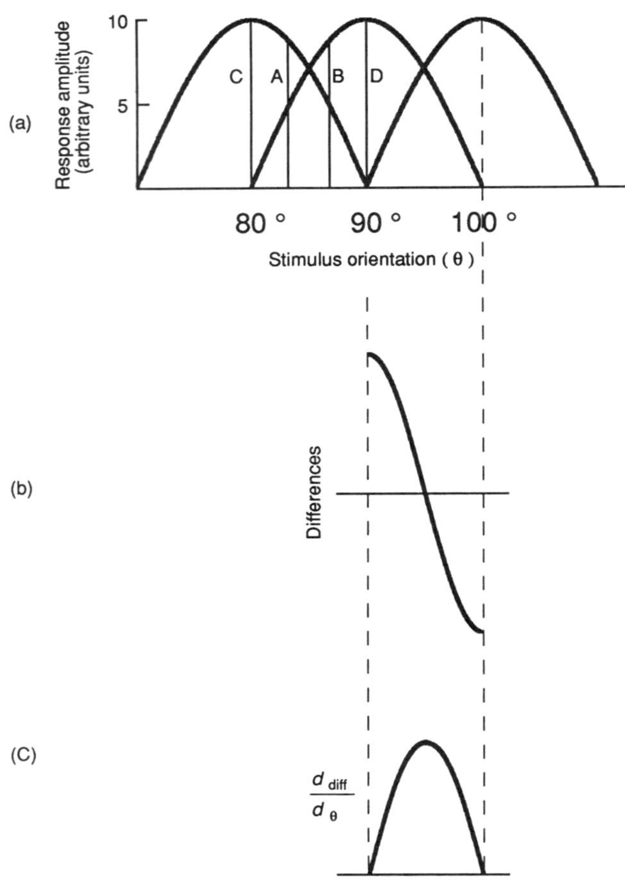

FIGURE 9-1. (a) Hypothetical tuning functions of three metameric orientation detectors in the visual cortex. (b) The signed difference in firing rate of the three detectors as a line is rotated from 80° to 100°. (c) Orientation discrimination function derived by differentiating the difference function.

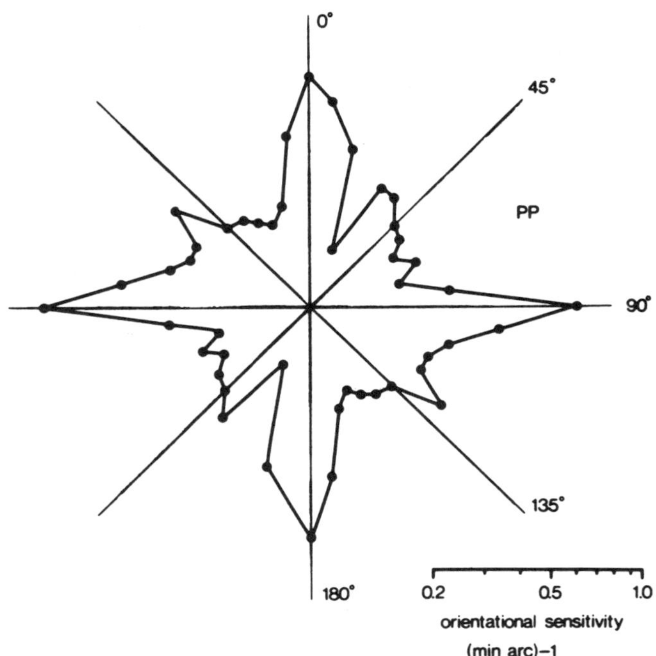

FIGURE 9-2. Orientation discrimination function in one subject. It can be seen that sensitivity to changes in the orientation of the test grating is greatest at the vertical and horizontal positions (from Regan and Price, 1986).

separation, their distinct positions can be discriminated (Loomis and Carter, 1978).

Metameric Systems Show Assimilation and Contrast

Neighboring stimuli presented to a metameric system are subject to simultaneous or successive interactions because of the overlap between the tuning functions of neighboring detectors. For instance, two parallel lines separated by a gap slightly greater than is needed to resolve them appear to be closer than they are. This is an assimilation effect. When parallel lines are slightly further apart than about 4′ of arc, they appear to repel each other (Badcock and Westheimer, 1985). This is a contrast effect. Successive interactive effects in visual direction are known as figural aftereffects. Assimilation is probably a result of interactions between the excitatory centers of the receptive fields on which the images of the lines fall. Contrast is probably created by interactions between the inhibitory regions of neighboring receptive fields. These effects should be no larger than the diameter of the region within which receptive fields overlap, which, in the foveal region, should be no more than a few minutes of arc. This is the typical size of a figural aftereffect (Kohler and Wallach, 1944).

There are contrast effects in motion (Nakayama and Tyler, 1978) and in motion in depth (Regan and Beverley, 1978). In orientation, contrast reveals itself as geometric illusions and successive tilt contrast, examples of which are shown in Fig. 9-3. Tilt contrast has a typical magnitude of about 2°, but effects up to about 5° have been reported when the test stimulus is presented for durations as short as 100 msec (Wolfe, 1984).

Since contrast effects depend on the extent to which sensory channels overlap, they provide a powerful psychophysical procedure for measuring the physiological properties of metameric systems, such as the number and bandwidth of channels coding a given feature. Contrast effects serve to enhance signals associated with changes in stimulation relative to signals associated with less important regions of steady stimulation.

Metameric Systems Typically Exhibit Opponency

An opponent system is one in which inputs from different detectors are subtracted to produce a difference signal. In the color opponent system inputs from red and green and from blue and yellow (red plus green) detectors are combined in a see-saw fashion. Opponent mechanisms provide signals that are independent of variations in stimulus intensity (Howard, 1982, p. 111). Thus, the color opponency mechanism produces signals that vary with hue, independently of changes in luminance. There is evidence that orientation discrimination is independent of contrast over a wide range of contrasts (Regan and Beverley, 1985), and a similar independence of contrast has been reported for spatial frequency discrimination (Regan *et al.*, 1982), speed discrimination (McKee *et al.*, 1986), and temporal frequency discrimination (Bowne, 1990).

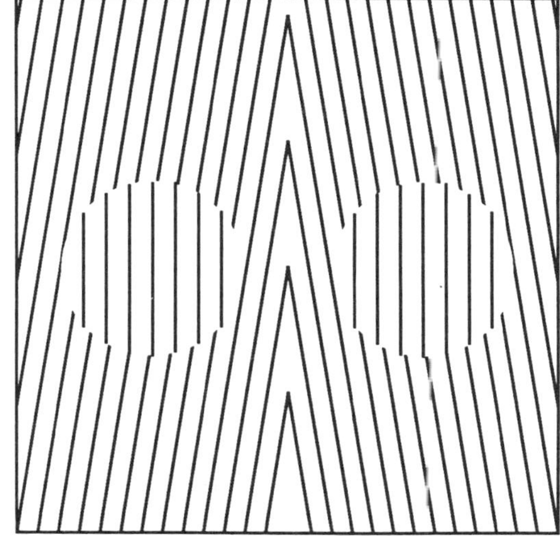

FIGURE 9-3. A geometric illusion that illustrates tilt contrast. The lines in both central disks are really vertical but appear tilted because of the contrasting tilt of the background lines.

In many sensory dimensions there are natural balance points or *norms*. For instance, "vertical" is a norm for orientation, "equidistance" is a norm for relative depth, and "stationarity" is a norm in a scale of movement one way or the opposite way. Such stimulus dimensions are known as oppositional. It seems to be a universal characteristic of oppositional stimulus dimensions that inspection of a stimulus that is off the norm causes that stimulus to appear more like the norm and other values of the feature to be displaced accordingly. Thus, if one looks at a tilted line, it gradually appears more vertical, and a subsequently seen vertical line looks tilted in the opposite direction, a phenomenon known as the tilt aftereffect. Although tilt contrast is a component of the tilt aftereffect, there is also a component specifically related to the vertical (Templeton *et al.*, 1965). To take another example. If one looks at a moving display, it appears to slow down, and a stationary display appears to move in the opposite direction, a phenomenon known as the motion aftereffect. Curved lines come to look more straight, and objects at different distances come to appear more equidistant (Howard and Templeton, 1963).

Persistent asymmetric stimulation with respect to a norm serves to inform the visual system that there is a systematic distortion of the visual input and provides an automatic mechanism for adjusting the system to the disturbance. For example, in the natural world there are as many lines slanting or curving one way as there are lines slanting or curving another way. Any consistent slant or curvature over a period of time signifies that the visual system is distorted and is in need of correction. After a period of time the tilt or curvature diminishes because the visual opponency systems for the detection of orientation or curvature automatically adjust themselves. It is an automatic calibration system or error-correcting mechanism.

A plausible physiological explanation for normalization can be provided in terms of adaptation within opponent mechanisms. It is simply necessary to assume that, in a system that exhibits normalization, the tuning functions of detectors overlap at the position of the norm. For instance, after a line just off the vertical has been inspected for some time, it should appear displaced toward the vertical because the detector on that side of vertical will have become adapted or fatigued relative to the one on the other side of vertical. Note that a line at vertical would not normalize, since it falls in a region where it excites two detectors by the same amount. I suggest that the tuning functions of systems in which normalization is an advantage evolve or grow so that they overlap at the position of the natural norm. This has the added advantage that sensitivity to change in the stimulus feature will be greatest at this position.

The adaptation process can be an elevation of threshold, lasting a few minutes and involving a process akin to fatigue. But there are also adaptation processes that can last days or even weeks. The best-known of these are the contingent aftereffects. For example, a period of exposure to red vertical lines alternating with green horizontal lines produces a long-lasting aftereffect in which vertical lines look green and horizontal lines look red (McCollough, 1965). These effects can be understood as mechanisms for correcting for cross-talk between feature-detection systems. The orientation–color contingent effect may be a mechanism the human visual system uses to null the perceptual effects of chromatic aberration in the optical system of the eye or in lenses people habitually use. Long-lasting adaptation effects have also been reported for simple visual features presented singly, such as visual motion alone (Favreau, 1979) and visual tilt alone (Wolfe and O'Connell, 1986). These long-lasting effects are strongest when test trials are not given immediately following the induction stimulus. However, these seemingly simple long-lasting aftereffects may be specific to incidental features of the induction stimulus such as its size or location, and if so, they may be best described as contingent aftereffects.

All the oculocentric effects I have mentioned are small, presumably because naturally occurring distortions of the visual system are also small. I do not know of any evidence that visual experience can induce large, long-lasting effects in either the visual local-sign system or in secondary visual features.

ADAPTING TO VISUAL DISPLACEMENT

A coherent displacement of the whole visual scene cannot be detected visually. Therefore, adaptations to visual displacement should not occur within the oculocentric system. A person can detect a visual displacement when an attempt is made to reach for a visual object with an unseen hand. It is well known that people very rapidly learn to correct their pointing errors when wearing displacing prisms. Many experiments have revealed that this visuomotor learning results mainly from a change in the felt position of the arm on the body or of the head on the body or of the eyes in the head but not from any change in the visual local-sign system (Howard, 1982). The adaptation occurs in the headcentric or bodycentric systems, not in the oculocentric system.

Several claims have been made that small distortions of oculocentric visual space can be induced by pointing with a hidden hand to visual targets that are optically displaced, even though the displacement is not visually evident (Cohen, 1966; Held and Rekosh, 1963; Courjon *et al.*, 1981). Others have claimed that such effects are caused by uncontrolled asymmetries in the visual input. We are left with no convincing evidence that shifts in the visual local-sign system can be induced purely by a mismatch between the seen position of objects and goal-directed arm movements (see Howard, 1982, 501; Scheuhammer and Timney, 1982).

A person may also detect a displacement of the visual scene by comparing the visual direction of an object with the direction of a sound that it makes. People immediately adapt to intersensory discordances of this type by making their nonvisual judgments conform to their visual judgments. One experiences this process when watching a ventriloquist and feeling convinced that the voice comes out of the mouth of the dummy.

ADAPTING TO POSTURAL DISPLACEMENTS

After the eyes have been held in an eccentric position for a few minutes, a visual target must be displaced several degrees in the direction of the eccentric gaze to be perceived as straight ahead. Apparent displacements of visual objects with respect to the head arise from a misregistration of the position or movement of the eyes. They therefore occur in the headcentric system. The magnitude of the deviation of apparent straight ahead has been shown to depend on the duration of eye deviation and to be a linear function of the eccentricity of gaze (Hill, 1972; Morgan, 1978; Paap and Ebenholtz, 1976).

Similar deviations of bodycentric visual direction occur during and after holding the head in an eccentric posture (Howard and Anstis, 1974). It has never been settled whether these effects are caused by changes in afference or by changes in efference associated with holding the eyes or the head in a given posture (see Howard, 1982, for a discussion of this issue). Whatever the cause of these effects, it is evident that the headcentric and bodycentric systems are much more labile than the oculocentric system. Headcentric visual tasks require information from the retina to be related to information regarding the position and motion of the eyes. The relative calibrations of these systems must be subject to modification because the head changes size with age and muscular systems are subject to fatigue and aging.

ADAPTING TO VISUAL MINIFICATION

The visual image may be minified by lenses. If the whole visual scene consists of unfamiliar objects, minification produces no detectable change in the retinal image. Adaptations of the oculocentric system are therefore not to be expected. However, uniform minification affects relationships between vision and touch, between accommodation and convergence, between binocular disparity and relative depth, and between the vestibular–ocular response and visual motion. These topics are discussed in other chapters of this volume.

If we look at familiar scenes through minifying lenses, we can detect minification in terms of the remembered sizes of objects, given that there are cues to distance present. The evidence suggests that people adapt to the new perceived sizes of familiar objects but that adaptation is short-lived. It consists of learning a new set of expectations regarding the perceived sizes of objects in relation to their perceived distances (Rock, 1965). This is an oculocentric effect, but it does not occur at the level of the local-sign system but at a higher level. The symptom of micropsia, or minified vision, experienced by people with damage to the right parietal lobe can be understood as a disturbance of this size–distance scaling mechanism. Micropsia can be experienced by simply misconverging on a regular display of dots or parallel lines. The resulting mismatch between accommodation and convergence induces a sensation of reduced size of the dots or lines if the eyes are overconverged or of increased size if the eyes are underconverged.

ADAPTING TO ROTATED VISUAL SCENES

Adaptations to optically induced tilts of the visual field of up to about 2° can be accounted for in terms of normalization processes within the visual system, as I have already explained. But adaptation to rotations of the visual scene through a large angle must be accounted for in another way (Rieser and Banks, 1981). Obviously, rotations through 90° or 180° cannot be accounted for in terms of visual normalization because these rotations of the visual field leave the main lines of visual space still predominantly vertical and horizontal.

It has been reported by several investigators that after a long period of wearing inverting spectacles the world comes to appear the right way up. There has been much confusion in the literature on this point (see Howard and Templeton, 1966). The answer lies in realizing that there are two distinct questions. The first has to do with perceiving which of two points is "up" and which is "down" relative to gravity. This is not an oculocentric task because the part of the retina on which a light must fall to be gravitationally "up" varies with the tilt of the eye. It is an exocentric task because the frame of reference is the direction of gravity as sensed by the vestibular and somesthetic systems. Any adaptation must therefore involve a recalibration of the polarity of visual space relative to information from the vestibular and somesthetic systems. An astronaut who has just arrived in space is in a peculiar position because his vestibular and somesthetic systems give him no sense of orientation. Usually what is under the feet is perceived as the floor and what is above the head as the ceiling, but sometimes the astronaut feels that he is inverted and hanging down from the ceiling rather than standing on a floor.

The second question has to do with the recognizability of familiar objects that are normally seen in one orientation, for example, trees, print, and faces. Such objects are said to be monooriented. Our ability to recognize such objects depends on their orientation on the retina, not on their orientation with respect to gravity (Rock and Heimer, 1957). For instance, a page of print viewed with the head in an upside-down orientation is difficult to read because it is inverted on the retina even though it is upright with respect to gravity. This is therefore an oculocentric phenomenon. With continued experience in viewing the world through inverting spectacles, a person learns to recognize upside-down objects. Several investigators, such as Kohler (1955), have concluded that this implies a radical inversion of the oculocentric coding of visual direction and have even claimed that one part of the visual field can come to be reversed with respect to another. This conclusion is totally unjustified. Learning to read inverted print does not involve any change in the local-sign system but merely a change in the way meaning is ascribed to visual patterns in particular retinal orientations. It is an oculocentric effect, but it occurs at a high level of processing and does not involve any shift in local sign. One can see how the idea arose that one part of the visual field might appear upside-down while another appears upright. Imagine that you are wearing inverting lenses and spend a lot of time learning to read but do not look at any faces. After a while, gravitationally upright print would look normal in the sense of looking familiar and easy to read, but faces would still look upside-down in the sense of being difficult to recognize.

SUMMARY

Visual adaptation to certain types of visual distortion, such as curvature and tilt, may be accounted for in terms of oculocentric changes in the visual system. However, the changes are not in the gross topography of the local-sign system but rather in localized metameric regions of the local-sign system or in specialized mechanisms that code secondary features such as orientation and motion. There is no purely visual adaptation to visual displacement, since this cannot be visually detected. One can adapt accuracy at pointing to displaced visual targets, and this involves changes in the felt position of eyes, head, or arms. For similar reasons adaptation to visual minification of unfamiliar objects cannot be purely visual, but such adaptation can lead to recalibrations of the relationships between accommodation and vergence or between disparity and relative depth or between head rotation and the gain of the vestibuloocular response.

Changes in the felt position of the eyes or the head can affect the sense of visual direction. If familiar objects are minified, people adapt to the unfamiliar relationship between perceived size and perceived distance. This is an oculocentric form of adaptation but not retinotopic, since it occurs at a high level of neural processing. Learning to recognize what is "up" in a visual field that has been optically rotated through a large angle involves remapping visual space in terms of information from the vestibular and somesthetic systems. This is an exocentric task. Learning to recognize inverted familiar objects is an oculocentric task, but it does not involve changes in the local-sign system or in visual feature-detection systems. It is simply a matter of learning to recognize familiar patterns in an unfamiliar orientation.

REFERENCES

Andrews, D. P., 1967, Perception of contour orientation in the central fovea. Part I: Short lines, *Vision Res.* **7:**975–997.

Badcock, D. R., and Westheimer, G., 1985, Spatial location and hyperacuity: The centre/surround location contribution function has two substrates, *Vision Res.* **25:**1259–1267.

Békésy, G. von, 1967, *Sensory Inhibition*, McGraw, Princeton.

Bowne, S. F., 1990, Contrast discrimination cannot explain spatial frequency, orientation or temporal frequency discrimination, *Vision Res.* **30:**449–461.

Cohen, H. B., 1966, Some critical factors in prism adaptation, *Am. J. Psychol.* **79:**285–290.

Courjon, J. H., Jeannerod, M., and Prablanc, C., 1981, An attempt at correlating visuomotor-induced tilt aftereffect and ocular cyclotorsion, *Perception* **10**:519–524.

Favreau, O. E., 1979, Persistence of simple and contingent motion aftereffects, *Percept. Psychophys.* **26**:187–194.

Held, R., and Rekosh, J., 1963, Motor–sensory feedback and geometry of visual space, *Science* **141**:722–723.

Hill, A. L., 1972, Direction constancy, *Percept. Psychophys.* **11**:175–178.

Hirsch, J., and Hylton, R., 1982, Limits of spatial-frequency discrimination as evidence of neural interpolation, *J. Opt. Soc. Am.* **72**:1367–1374.

Howard, I. P., 1982, *Human Visual Orientation,* Wiley, Chichester.

Howard, I. P., and Anstis, T., 1974, Muscular and joint-receptor components in postural persistence, *J. Exp. Psychol.* **103**:167–170.

Howard, I. P., and Templeton, W. B., 1963, The effect of steady fixation on the judgement of relative depth, *Q. J. Exp. Psychol.* **16**:193–203.

Howard, I. P., and Templeton, W. B., 1966, *Human Spatial Orientation,* Wiley, Chichester.

Klein, S. A., and Levi, D. M., 1985, Hyperacuity thresholds of 1.0 second: Theoretical predictions and empirical validation, *J. Opt. Soc. Am.* **A2**:1170–1190.

Kohler, I., 1965, Experiments with prolonged optical distortions, *Acta Psychol.* **11**:176–178.

Kohler, W., and Wallach, H., 1944, Figural aftereffects: An investigation of visual processes, *Proc. Am. Phil. Soc.* **88**:269–357.

Loomis, J. M., and Carter, C. C., 1978, Sensitivity to shifts of a point stimulus: An instance of tactile hyperacuity, *Percept Psychophys.* **24**:487–492.

McCollough, C., 1965, Color adaptation of edge-detectors in the human visual system, *Science* **149**:1115.

McKee, S. P., Silverman, G. H., and Nakayama, K., 1986, Precise velocity discrimination despite random variations in temporal frequency and contrast, *Vision Res.* **26**:609–619.

Morgan, C. L., 1978, Constancy of egocentric visual direction, *Percept. Psychophys.* **23**:6[illegible]–68.

Nakayama, K., and Tyler, C. W., 1978, Relative motion induced between stationary lines, *Vision Res.* **18**:1663–1668.

Paap, K. R., and Ebenholtz, S. M., 1976, Perceptual consequences of potentiation in the extraocular muscles: An alternative explanation for adaptation to wedge prisms, *J. Exp. Psychol.* **2**:457–468.

Regan, D., 1989, *Human Brain Electrophysiology,* Elsevier, New York.

Regan, D., and Beverley, K. I., 1978, Illusory motion in depth: Aftereffect of adaptation to changing size, *Vision Res.* **18**:209–212.

Regan, D., and Beverley, K. I., 1985, Postadaptation orientation discrimination, *J. Opt. Soc. Am.* **2A**:147–155.

Regan, D., Bartol, S., Murray, T. J., and Beverley, K. I., 1982, Spatial frequency discrimination in normal vision and in patients with multiple sclerosis, *Brain* **105**:735–754.

Regan, D., and Price, P., 1986, Periodicity in orientation discrimination and the unconfounding of visual information, *Vision Res.* **26**:1299–1302.

Rieser, J. J., and Banks, M. S., 1981, The perception of verticality and the frame of reference of the visual tilt aftereffect, *Percept. Psychophys.* **29**:113–120.

Rock, I., 1965, Adaptation to a minified image, *Psychonom. Sci.* **2**:105–106.

Rock, I., and Heimer, W., 1957, The effect of retinal and phenominal orientation on the perception of form, *Am. J. Psychol.* **70**:493–511.

Scheuhammer, J., and Timney, B., 1982, Adaptation to optically reduced size, *Perception* **11**:139–152.

Templeton, W. B., Howard, I. P., and Easting, G., 1965, Satiation and the tilt aftereffect, *Am. J. Psychol.* **78**:656–659.

Watamaniuk, S. N. J., Sekuler, R., and Williams, D. W., 1989, Direction perception in complex dynamic displays: The integration of direction information, *Vision Res.* **29**:47–59.

Watt, R. J., Morgan, M. J., and Ward, R. M., 1983, Stimulus features that determine the visual location of a bright bar, *Invest. Ophthalmol. Vis. Sci.* **24**:66–71.

Westheimer, G., 1984, Line-separation discrimination curve in the human fovea: smooth or segmented? *J. Opt. Soc. Am.* **1A**:683–684.

Wilson, H. R., 1986, Responses of spatial mechanisms can explain hyperacuity, *Vision Res.* **26**:453–469.

Wolfe, J. M., 1984, Short test flashes produce large tilt aftereffects, *Vision Res.* **24**:1959–1964.

Wolfe, J. M., and O'Connell, M., 1986, Fatigue and structural change: Two consequences of visual pattern adaptation, *Invest. Ophthalmol. Vis. Sci.* **27**:538–543.

10

Adaptation to Anamorphosing Lenses: Perceptive Responses and Visuomotor Coordination

VALÉRIE CORNILLEAU-PERES and JACQUES DROULEZ

INTRODUCTION

The control of the interactions between the body and its environment requires from the central nervous system a capacity to perceive three-dimensional (3-D) metric parameters such as lengths, angles, and curvatures. However, primary visual information consists in bidimensional (2-D) images formed on the retina and coded retinotopically by the first neural networks in the visual pathways. Motion parallax is one of the depth cues that allow the visual system to perform the inference of 3-D structure from 2-D images. This is demonstrated by the kinetic depth effect (Wallach and O'Connell, 1953) where the shadows of 3-D objects are projected on a screen (Fig. 10-1). As the objects move in the 3-D space, an observer can perceive their structure by looking at the screen. In order to account for this type of perceptual phenomenon, Gibson (1966) developed an approach leading to the scheme presented in Fig. 10-2. The 2-D image changes are interpreted in terms of motion of the objects in the 3-D space by use of a principle of spatial constancy. The computational approach that followed Gibson's work expresses the mathematical relationships between the 2-D retinal parameters and the 3-D characteristics of structure and motion and shows that the problem is underdetermined if no constraint is added concerning the nature of the 3-D environment. The assumption of rigidity of the 3-D objects, which is the mathematical expression of spatial constancy, allows a complete resolution of the problem in most cases and has been extensively used in theoretical studies of the perception of structure from motion (Ullman, 1979; Longuet-Higgins and Prazdny, 1980; Waxman and Ullman, 1985). It should be pointed out that the physiological reality of the rigidity assumption has not been demonstrated and that the visual perception of deformable objects remains to be understood.

The computational approach has proved to be a very powerful tool for the comprehension and modeling of the visual processes. However, it uses an implicit hypothesis that seems to us of fundamental importance and was stated explicitedly by Droulez (1985), namely, that a visual metric is required in the perception of structure from motion. In the first part of this chapter we discuss the existence of a visual metric and its possible coding in the visual pathways. The second part raises the question of the plasticity of this visual metric by referring to different types of adaptive changes already evidenced in the perceptual or sensorimotor systems. In order to demonstrate the adaptability of the visual system to a metric perturbation, we exposed subjects to anamorphosing lenses yielding a selective lengthening in the vertical direction (Droulez and Cornilleau,

VALÉRIE CORNILLEAU-PERES AND JACQUES DROULEZ • Essilor International, Laboratory of Physiological Optics, 94000 Creteil, France, and Laboratory of Neurosensory Physiology, CNRS, 75270 Paris Cedex 06, France.

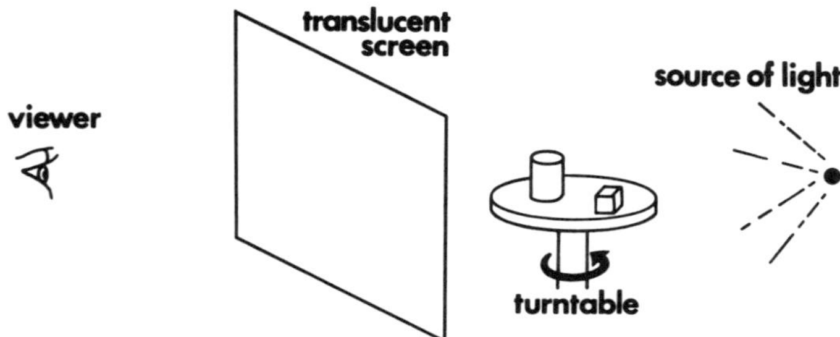

FIGURE 10-1. The kinetic depth effect. A set of objects is lying on a table that rotates around a vertical axis. The images of the moving objects are projected on a translucent screen. By looking at these projections, an observer can perceive the 3-D shape of the objects.

1986). The results of this experiment are exposed in the third section and demonstrate the existence of both perceptual and visuomotor adaptation to our lenses. Finally, the biological need for plasticity of the visual metric is discussed, and we give an interpretation of our results that accounts for the adaptability of the visual metric.

THE HYPOTHESIS OF A VISUAL METRIC

Consider the example shown in Fig. 10-3A. An observer is viewing monocularly a square that is translating along a sagittal axis toward him. The corresponding 2-D pattern, as viewed by the observer, is presented in Fig. 10-3B. The computational approach demonstrates how a global expansion of the visual scene can be interpreted, under the rigidity assumption, as reflecting a 3-D translation in depth. However, it does not state how the image expansion can be computed by the visual system. Clearly, the detection of the translation in depth of a square requires the use of some type of metric operators that enable the visual system (1) to estimate the lengths of the sides and the angles of the square and compare the measures performed in different parts of the retinal image at a given instant (the ability of an observer to discriminate a frontal square from a rectangle is demonstrated by the experimental results detailed below); (2) to measure the temporal variations of length and angle in the retinal images and compare these variations in different parts of the visual field; and (3) to deduce the 3-D direction of the translation from these metric measurements.

The differential geometry shows that such operations can be performed if the visual system is a metric space. It is then provided with metric operators that allow the calculus of the scalar product of two vectors (i.e., the calculus of angles and lengths) in every point of the retinal images and the comparison of the measures performed in different points of these images (i.e., the measurement of a spatial variation of metric parameters).

Our physical space is euclidian, which means that the metric operators do not vary with the loca-

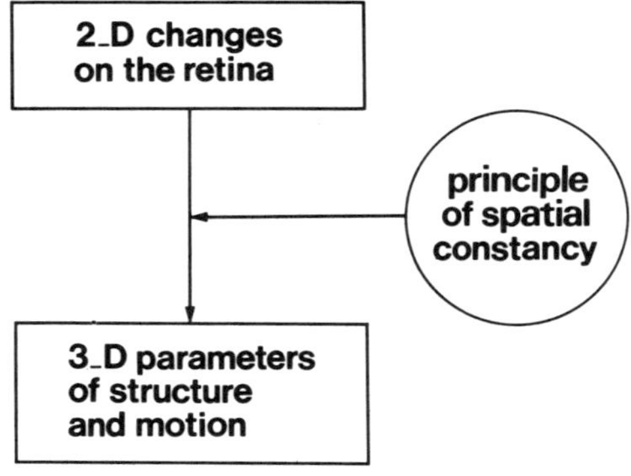

FIGURE 10-2. The Gibsonian approach to motion and structure perception. During relative motion of an observer and his environment, the 2-D patterns of luminous intensity fluctuate on his retina. This flow of 2-D information, as associated with the principle of spatial constancy, specifies 3-D motion and structure.

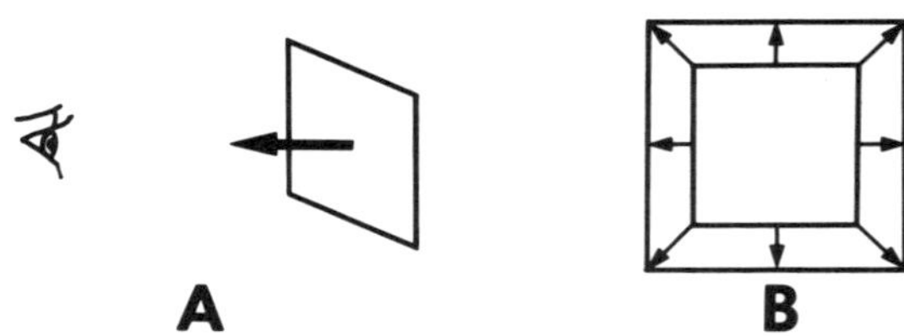

FIGURE 10-3. Perception of an approaching square. An observer is viewing a frontal square that translates toward him along a sagittal axis (A). This motion corresponds to expansion of the image of the square on the retina of the observer (B).

tion where the metric operations are performed. These operators can be represented by a ruler and a protractor. However, metric spaces do not verify this property in general. For instance, the measurement of lengths on a rugby ball requires rulers of different curvature according to the position and direction of the measure. Since there is no *a priori* reason for the visual system to be euclidean, the metric operators have to be defined locally in every point of the retinal image. The neurophysiological interpretation of these operators might then be the following: Consider the retinotopic coding of velocity by a retinal map, as was found in the striate visual cortex of the cat (Hubel and Wiesel, 1968) and monkey (for instance, DeValois *et al.*, 1982). Retinian velocity is coded in every point of this map by a set of neurons with directional selectivity. The scalar product of two velocity vectors can be obtained by the weighted product of the outputs of different neurons, the weights being a characteristic of the metric.

It should be mentioned that Pellionisz and Llinas (1979) have postulated the existence of such metric operators in the central nervous system to account for the transformation between sensory and motor maps.

ADAPTATION TO OPTICAL PERTURBATIONS

Let us assume the existence of geometric operators encoded in the visual pathways. The question arises whether these operators are rigidly determined, by genetics for instance, or can undergo adaptive changes to keep the internal representation of space consistent with its physical properties.

Since the original experiments reported by Helmholtz (1866) and Stratton (1897), many investigators have exposed subjects to optical perturbations and reported visual adaptation. We here present an overall view of these studies and of the typical time constants of adaptation they have exhibited.

A first type of studies on visuomotor adaptation concerns eye movements. When human subjects or animals are exposed to optical devices that magnify or diminish the visual scene, their vestibuloocular reflex (VOR) exhibits plastic changes (Gauthier and Robinson, 1975; Istl-Lenz *et al.*, 1985; Miles and Fuller, 1974) that become substantial after a few days (typically 5). The use of left–right reversing prisms also yields adaptive changes in the VOR tending to compensate for the reversal (Gonshor and Melvill-Jones, 1973). Melvill-Jones *et al.* (1988) analyzed the modifications in eye–head tracking over a 6-hr experience with active reversed vision and showed that several characteristics of these adaptive changes are detectable within the first 30 sec of exposure.

Another way of measuring visuomotor adaptation involves eye–hand coordination. Mandelbrojt *et al.* (1984) exposed subjects to magnifying (×2) lenses and trained them to point in a closed-loop condition. With the same pointing task performed in open loop, they could measure significant adaptive changes after 10 min of exposure. In view of this result, we can wonder what the role is of visuomotor training in the adaptation process. This problem has been extensively discussed in many studies concerning adaptation to displacing prisms and is reviewed by Howard (1982). Three points emerge from this discussion:

1. The adaptation, as measured by the test procedure, depends on the similarity between the test procedure and the training task.
2. Self-produced movements performed during the exposure period tend to accelerate the adaptive processes.
3. The time course of adaptation and the site of the recalibration (sense of position of the eye, sense of position of the arm) vary among subjects.

All these studies involve visuomotor test procedures (pointing with the hand, indicating the straight-ahead direction) and thus necessarily measure a visuomotor adaptation. There exist, to our knowledge, very few experiments concerning purely visual adaptation, which we define as a perceived change of the structure of the environment, as measured independently of the perceived position and orientation of the body. The perceptual adaptation thus defined cannot be evidenced with reversing or displacing prisms, since they perturb the relationship between the environment and the body but not the apparent internal structure of the environment. Rock (1966) attempted to demonstrate a perceptual adaptation by exposing subjects to optical minification. The test procedure consisted in the estimation of the length of a test line as compared to the length of a memorized standard line. After 30 min of exposure and active visuomanual training, the subject had compensated for 23% of the total minification. However, Rock did not control for the fact that the perceived distance of the test lines and the memorized length of the standard line had remained constant in the course of the experiment. Moreover, this adaptation might as well not be purely perceptual, since we have no idea of the relationship between a memorized standard line and the mental representation of the body. Therefore the existence of a purely perceptual adaptation remains to be clearly demonstrated, as well as its correlation with visuomotor adaptation.

ADAPTATION TO ANAMORPHOSING LENSES

Goal of the Experiment

The goal of this experiment was to evidence a purely perceptual adaptation to a perturbation of the visual metric and to relate it to the changes in the visuomotor coordination. In order to keep the horizontal direction as a reference, we used lenses that lengthened the visual scene only in a vertical direction.

Methods

LENSES

Three pairs of cylindrical lenses with the concave side facing the eye were used in the experiments. Their optical power was null, their axes were in the horizontal plane of the eyes, and their lengthening effect was almost exclusively vertical with respect to the subject (see Fig. 10-4). The total lengthening effect was 5, 8, or 25%.

Because of the selective lengthening in the vertical direction, the rotation of an object around the

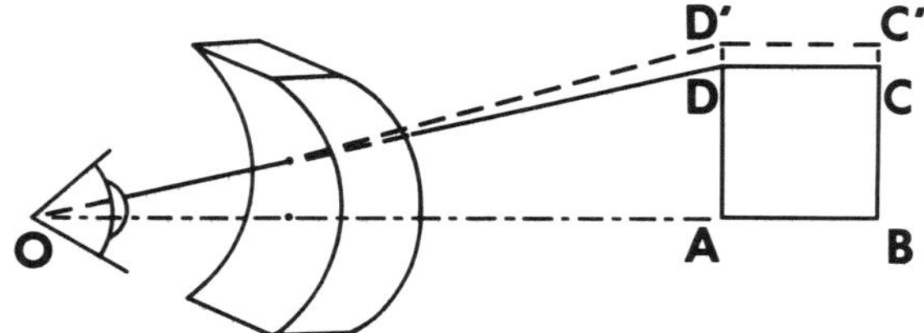

FIGURE 10-4. Optical effect of the anamorphosing lenses. The faces of the lens are horizontal cylinders with the concavity facing the eye. The square ABCD viewed by the observer *O* who wears the lenses appears as a vertical rectangle ABCD. [Reprinted with permission from Droulez and Cornilleau (1986), copyright 1986, Pergamon Press PLC.]

visual axis of the observer induced a perceived deformation of this object (Fig. 10-5).

SUBJECTS

The 11 subjects were 20–26 years old, had normal uncorrected vision, and were right-handed. All of them were naive and wore the lenses continuously for durations of 1 to 9 days while living normally. Their perceptual and motor responses were measured before the experiment ("control tests"), during adaptation ("adaptation tests"), and after lens removal ("postadaptation tests").

PERCEIVED SQUARE TEST

Computer-generated rectangles were presented on a display 57 cm in front of the subject (Fig. 10-6A,B). The subject's head was held still by a chin rest, and a black tunnel prevented lateral vision.

The rectangles presented a constant area (0.01 m^2), but their height/width ratio (HWR) ranged from 0.75 to 1.25 and varied during the experiment. A forced-choice method with two alternatives ("vertical rectangle" or "horizontal rectangle") was applied with a convergent procedure. This allowed the measure of the HWR characterizing the rectangle that is perceived as a square by the subject. This HWR was averaged over 40 trials during a given session. Its value should ideally decrease 25% relative to the control value when the subject wears the 25% lenses for the first time.

POINTING TEST

The subject was seated in front of a digitizing tablet were he could point to visual targets with an electronic pen. An array of nine targets was fixed under the top side of a box and could be seen by the subject through a mirror placed horizontally in the box at equal distance between the top side and the digitizing tablet (Fig. 10-6C,D). Therefore, the targets were perceived as lying on the pointing surface by the subject who could not see his hands and arms. The head of the subject was held still by a chin rest, and the central target was seen in a straight-ahead direction, about 50 cm from the eyes. After pointing at the central target, which was checked by the experimenter, the subject had to point to one of the eight peripheral targets randomly chosen by the computer. A session consisted of 15 pointings at

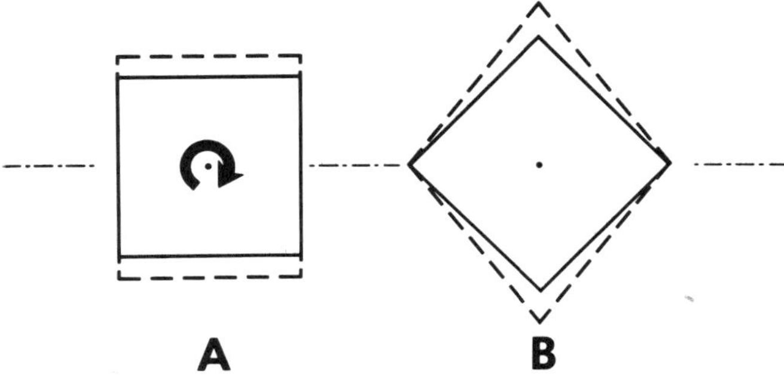

FIGURE 10-5. Apparent deformation of the rotating square through the cylindrical lenses. In its initial position the square (A, continuous line) is seen through the lenses as a vertical rectangle (dashed line). After a rotation of 45°, the square looks like a lozenge elongated vertically (B). Therefore, for a rotation around the sagittal axis, a deformation of the square is perceived by a subject who wears the lenses for the first time.

the eight targets. The mean ordinates Y of targets 1, 3, 5, and 7, and the mean abscissas X of targets 2, 4, 6, and 8, as pointed to by the subject, served in the calculus of the following ratios:

$$\text{HWR1} = (Y_1 - Y_3) / (X_2 - X_4)$$
$$\text{HWR2} = (Y_5 - Y_7) /(X_6 - X_8)$$

which quantify the amplitude of horizontal pointing relative to vertical pointing. HWR1 and HWR2 characterize the pointing movements of small and larger amplitude, respectively. The mean of these two ratios is written HWRM. The effect of our lenses should ideally result in a 25% increase of HWR1 and HWR2 when the subject first puts the lenses on.

Results

PERCEIVED SQUARE TEST

Table 10-1 displays the results obtained for the 11 subjects (column 1) wearing the 5%, 8%, or 25% lenses (column 2) and for adaptation periods ranging

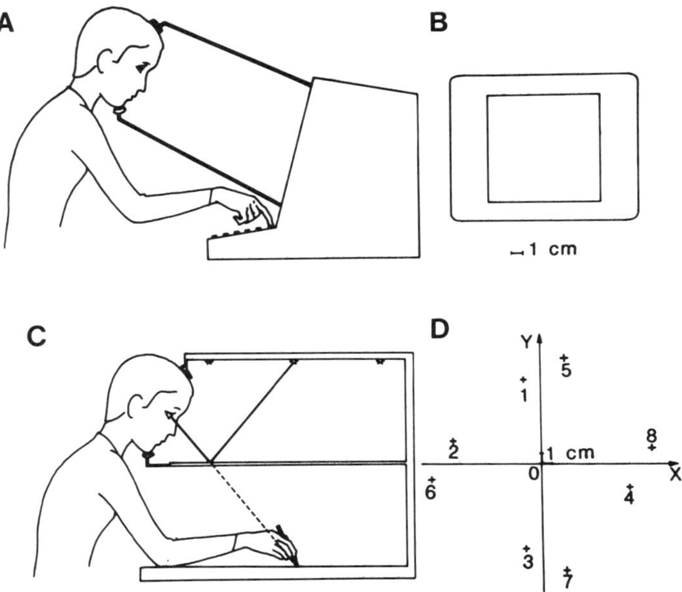

FIGURE 10-6. Experimental setup. (A) Perceived square test. The subject's head is held by a chin rest, and a black tunnel blocks out peripheral vision. (B) An example of a rectangle displayed on the screen. (C) Pointing test. The subject looks at a set of targets through a horizontal mirror and points to the target images, which lie in the plane of a digitizing tablet. (D) Positions of the eight targets around the central point O. [Reprinted with permission from Droulez and Cornilleau (1986), copyright 1986, Pergamon Press PLC.]

TABLE 10–1
Perceived Square Test

1 Subject	2 Lenses effect (%)	3 Exposure duration (days)	4 Control (± S.D.)	5 Initial decrease (%)	6 Final increase (%)	7 Adaptation (%; significance)	8 Aftereffect (%)	9 Adaptation rate (% per day)
E.V.	25	1	1.008 (± 0.015)	25.2	28.2	− 0.7 (N.S.)	0.9 (N.S.)	0
A.L.	25	1	0.968 (± 0.035)	31.0	27.9	5.5 ($P < 0.01$)	2.2 ($P = 0.02$)	3.8
C.L.	25	1	0.979 (± 0.012)	27.1	26.3	3.1 ($P < 0.01$)	2.8 ($P < 0.01$)	3.0
I.D.	25	1	1.060 (± 0.035)	23.0	24.1	0 (N.S.)	1.0 (N.S.)	0
X.M.	25	3	0.968 (± 0.012)	23.5	25.8	0.20 (N.S.)	3.7 ($P < 0.01$)	0.7
L.B.	25	4	1.054 (± 0.020)	18.6	22.2	4.6 ($P < 0.01$)	6.0 ($P < 0.01$)	1.3
C.C.	25	7	0.958 (± 0.010)	22.8	24.5	4.5 ($P < 0.01$)	6.9 ($P < 0.01$)	0.8
A.B.	8	7	0.920 (± 0.015)	8.7	4.9	6.8 ($P < 0.01$)	3.0 ($P < 0.01$)	0.7
G.C.	5	7	0.940 (± 0.015)	6.1	6.6	2.9 ($P < 0.01$)	3.4 ($P < 0.01$)	0.5
C.R.	8	7	1.025 (± 0.015)	8	8.3	7.3 ($P < 0.01$)	7.6 ($P < 0.01$)	1.1
T.J.	8	9	0.970 (± 0.015)	9.5	6.2	4.0 ($P < 0.01$)	1.0 ($P < 0.01$)	0.3

from 1 to 9 days (column 3). Column 4 indicates the average of the control HWR (i.e., before adaptation) and its standard deviation. The intrasubject variability of the responses is very low (currently less than 2%, with a maximum of 3.5%), but the mean HWR varied significantly according to the subject.

Figures 10-7 and 10-8 present the subjects' responses as a function of time for long-term experiments (3 to 9 days) with the 5, 8, and 25% lenses. The marked decrease of HWR from the optical effect of the lenses, observed in the first test of the exposure period, is followed by a slow increase of HWR. In some cases this increase is noticeable after 24 hr of continuous wearing.

In Table 10-1, columns 5 and 6 present the response variation obtained when the subject puts the lenses on ("initial decrease") and when he takes them off ("final increase"). These values are well correlated with the optical effect of the lenses.

The adaptation is quantified as the difference between the HWR measured just before lens removal and that in the beginning of the adaptation period (Table 10-1, column 7). All adaptation measures were significant except for subject X.M. The modifications in the visual processes can also be characterized by the aftereffect, which we measured from the difference between the control value and the HWR value observed just after lens removal (Table 10-1, column 8). All the subjects in the long-

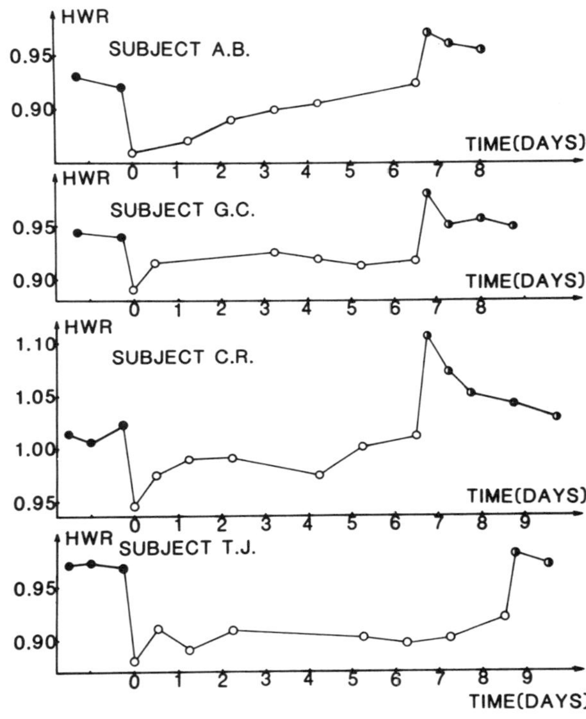

FIGURE 10-7. Perceived square test. Collected results for the subjects wearing the 8% lenses (A.B., C.R., T.J.) and the 5% lenses (subject G.C.). Abscissa, time of exposure (days) starting at the beginning of lens wear; ordinate, mean height/width ratio (HWR) of the rectangle that is perceived as a square by the subject before (solid circles), during (open circles), and after (half-solid circles) lens wear. [Reprinted with permission from Droulez and Cornilleau (1986), copyright 1986, Pergamon Press PLC.]

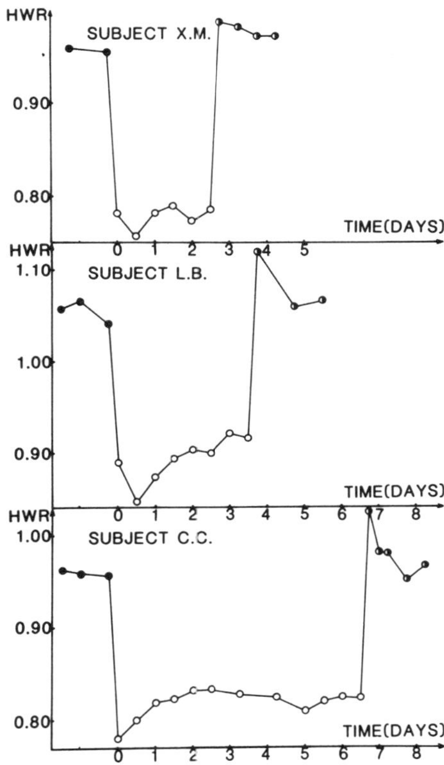

FIGURE 10-8. Perceived square test. Collected results from subjects wearing the 25% lenses for 3 days (X.M.), 4 days (L.B.), or 7 days (C.C.). Abscissa and ordinates as in Fig. 10-7. [Reprinted with permission from Droulez and Cornilleau (1986), copyright 1986, Pergamon Press PLC.]

term experiments (3 to 9 days) exhibited a significant aftereffect.

An adaptation rate was obtained by averaging the adaptation and the aftereffect measures and dividing the result by the number of exposure days (Table 10-1, column 9). This adaptation rate ranged from 0 to 3.5% per day (mean 1.1% per day). For two of the four subjects who wore the lenses only 8 hr, no adaptation could be exhibited.

In conclusion, we found that the perceptual adaptation was highly variable between subjects and could not always be shown in the 8-hr experiments. Moreover, in long-term experiments, the rate of adaptation was slightly higher for the strong lenses (mean 0.93% per day) than for the weak lenses (mean 0.65% per day), but because of the high intersubject variability, this difference was not statistically significant.

POINTING TEST

The three subjects who wore the 25% lenses for more than 3 days performed this test. Figure 10-9 presents HWR1, HWR2, and HWRM, as measured before, during and after lens wear. HWR2, which corresponds to large arm movements, is systematically higher than HWR1. The mean ratio HWR2/HWR1 averaged for all sessions of the three subjects is 1.066 (standard deviation 0.06).

The adaptation curves of Fig. 10-9 can be compared to those of Fig. 10-8, which were obtained for the same subjects completing the perceived square test. In Fig. 10-9, the responses of subjects X.M. and L.B. present complex variations. The initial increase and final decrease are smaller than the optical lengthening (between 4.2% and 14%). Moreover, because the responses of subject L.B. did not return to the initial control values by 3 days after lens removal, a control value of HWR could not be defined

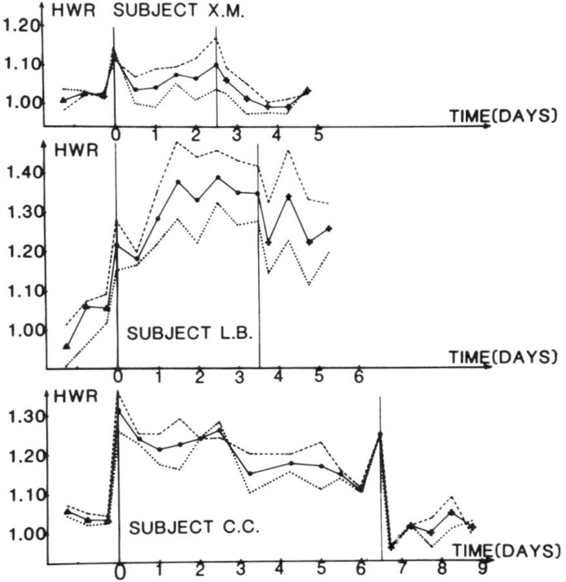

FIGURE 10-9. Open-loop pointing test. Collected results for the three subjects who wore the 25% lenses for 3 days (X.M.), 4 days (L.B.), or 7 days (C.C.). Abscissa, time of exposure (days) starting at the beginning of lens wear; ordinate, the averaged HWR1 and HWR2 and their mean, as computed from the pen position at the end of the pointing movements. HWR1 (small dots) was calculated for the four targets close to the central point. HWR2 (dashed line) was calculated for the four targets far from the central point. The mean value of HWR1 and HWR2 is shown as the continuous line. The two light vertical bars indicate the beginning and end of lens wear. [Reprinted with permission from Droulez and Cornilleau (1986), copyright 1986, Pergamon Press PLC.]

clearly. Therefore, no adaptation could be demonstrated for these two subjects.

In contrast, subject C.C. exhibited a clear adaptation profile, which can be characterized by the following values: initial increase, 26.9%; final decrease, 28.6%; adaptation measure, 6.0%; aftereffect measure, 7.5%; mean adaptation rate, 0.96% per day. It should be noted that this adaptation rate is very close to the rate obtained with the perceived square test (0.8% per day).

VERBAL REPORTS

When they first put the lenses on, the subjects experienced three types of perceptual phenomena: objects and faces appeared lengthened in the vertical direction; large and planar surfaces were seen as concave; head movements induced an apparent motion of the peripheral visual field and a slight feeling of loss of equilibrium. However, visuomotor coordination was maintained, and the subjects could live normally with the lenses on. The main disturbance came from the reduction of the visual field in the case of the 25% lenses, since they were fixed on masking goggles.

After 1 day of exposure, the subjects were still aware of the lengthening effect and of the distortions but claimed that these effects did not bother them. For those who had an exposure period of 3 days or more, the opposite effects were experienced just after lens removal.

Discussion

Our results lead to the following conclusions: (1) the subjects could compare vertical and horizontal length with good precision, and this perceptual task showed low intrasubject variability; (2) the perceptive responses obtained during exposure to the anamorphosing lenses exhibit long-term plastic adaptive changes, although the resulting adaptation rate is rather slow (less than 1% per day) and seems independent of the initial degree of optical distortion; (3) the variability of the responses obtained in the visuomotor task of pointing in open-loop conditions was about one order of magnitude higher than in the purely perceptual task; and (4) despite this variability, a slow process of visuomotor adaptation was also shown for one subject who presented a strong negative correlation ($r = -0.92$) between perceptual and visuomanual adaptation.

Because the subjects were immediately aware of the optical effect of the lenses (faces and familiar objects were lengthened vertically), we cannot exclude a systematic correction of the responses of a "cognitive" type (i.e., related to an *a priori* knowledge of the environment). However, the slowness of the adaptation process and the correlation between the response changes and the optical effect at the beginning and at the end of the exposure period argue against this hypothesis.

Rock (1966) obtained a significant adaptation to a minification of the visual scene in 30 min, and Mandelbrojt *et al.* (1984) obtained high rates of adaptation to a magnification of the visual scene within 20 min. Compared to these results, the adaptation reported in our experiment is very slow. This might result from our use of a selective lengthening of the visual scene, which could imply more complex processes than those yielded by an isotropic magnification or minification. The difference in the experimental conditions might also account for this discrepancy. The changes in length estimation reported by Rock in the course of adaptation might be caused by a change in depth perception. During the adaptation period Mandelbrojt *et al.* trained the subjects with a task that is the same as the task dedicated to adaptation measure, except that it is performed in closed loop. Therefore, the measured adaptation might be specific for the test and involve only a partial reorganization of the sensory and motor maps.

A POSSIBLE SCHEME FOR THE ADAPTATION OF THE VISUAL METRIC

The goal of sensorimotor adaptation is clear, whereas the biological necessity for plasticity of the visual metric seems less obvious. What is the discrepancy that induces perceptual adaptation in our experiment? The classical interpretation of visuomotor adaptation as a result of a recalibration of the motor commands or of the proprioceptive signals remains unable to explain perceptual adaptation. If the visual metric adapts, it must be that perceiving a square as a square is a necessity for the visual system. This necessity for agreement between the physical and the perceived geometry of the environment becomes clearer when we consider that the observer is in continual motion relative to his environment and that an optical perturbation such as the vertical lengthening elicited by our lenses is associated to apparent distortions of the environment (Fig. 10-5).

A first adaptation scheme directly issued from the computational approach consists of a search for the maximum rigidity of the environment. The visual system would then adjust its metric so as to maximize the apparent rigidity of objects. However, we cannot present any physiological arguments in favor of this scheme. We cannot even argue that the visual system always chooses the rigid interpretation of a visual scene, since it has been shown that 3-D rigid objects in motion may be perceived as deforming (Ames, 1951; Adelson, 1985).

However, this scheme can be improved, and the need for plasticity of the visual metric may be explained, in physiological terms, as required for the apparent stability of the world during egomotion. Numerous studies on vection (the illusory perception of self-motion induced by visual stimulation) indicate that a part of the visual scene is chosen as a stable frame of reference relative to which the subject perceives his self-motion. The apparent stability of this frame of reference, as well as the coherence of the different types of sensory signals from which it is defined, seems to be of considerable importance for the organism: conflicting vestibular and visual information may yield sensations of loss of equilibrium and nausea. Indeed, our subjects were subject to such conflicts during lens wear.

Therefore, a second scheme for perceptual adaptation consists of the minimization of deformation of the environment during self-motion. It may be considered to be based on an intersensory discrepancy if we distinguish the proprioceptive and exteroceptive roles of visual perception: as a global motion of the visual scene indicates a self-motion, the optic flow analysis fails to confirm any rigidity in the visual scene. Other sensory modalities involved in the perception of egomotion may then be coupled with the proprioceptive visual signals in conflict with exteroceptive visual information.

This interpretation contradicts the results presented by Rock, since no apparent deformation of the visual scene is elicited by a minification of the visual scene. This contradiction might appear because Rock did not measure a perceptual adaptation but rather a visuomotor adaptation (for instance, the subject might have related the length of the standard line to the length of his finger).

REFERENCES

Adelson, E. H., 1985, Rigid objects that appear highly non-rigid, *Invest. Ophthalmol. Vis. Sci. [Suppl.]* **26:**56.

Ames, A. J., 1951, Visual perception and the rotating trapezoidal window, *Psychol. Monogr.* **65**(7):324.

De Valois, R. L., William Yund, E. W., and Hepler, N., 1982, The orientation and direction selectivity of cells in macaque visual cortex, *Vision Res.* **22:**531–544.

Droulez, J., 1985, Neurosensory processing of moving images, Submitted to *Biol. Cybernet.* (in press).

Droulez, J., and Cornilleau, V., 1986, Adaptive changes in perceptual responses and visuomanual coordination during exposure to visual metrical distortion, *Vision Res.* **26**(11):1783–1792.

Gauthier, G. M., and Robinson, D. A., 1975, Adaptation of the human vestibulo-ocular reflex to magnifying lenses, *Brain Res.* **92:**331–335.

Gibson, J. J., 1966, *The Senses Considered as Perceptual Systems,* Boston: Houghton Mifflin.

Gonshor, A., and Melvill-Jones, G., 1973, Changes of human vestibulo-ocular response by vision reversal during head rotation. *J. Physiol. (Lond.)* **234:**102–103.

Helmholtz, H. von, 1866, *Handbuch der Physiologischen Optik,* Vos, Leipzig.

Howard, I. P., 1982, *Human Visual Orientation,* John Wiley & Sons, New York.

Hubel, D. H., and Wiesel, T. N., 1968, Receptive fields and architecture of monkey striate cortex, *J. Physiol. (Lond.)* **195:**215–243.

Istl-Lenz, Y., Hydén, D., and Schwartz, D. W. F., 1985, Response of the human vestibulo-ocular reflex following long term 2× magnified visual input, *Exp. Brain Res.* **57:**448–455.

Longuet-Higgins, H. C., and Prazdny, K., 1980, The interpretation of a moving retinal image, *Proc. R. Soc. Lond. [Biol.]* **208:**385–397.

Mandelbrojt, P., Gauthier, G. M., Vercher, J. L., Ouaknine, M., and Obrecht, G., 1984, Ensemble expérimental pour l'étude des propriétés adaptatives du système visuo-manuel chez le sujet nouvellement équipé de corrections optiques, *J. Fr. Ophtalmol.* **7:**157–165.

Melvill Jones, G., Guitton, D., and Berthoz, A., 1988, Changing patterns of eye–head coordination during 6 h of optically reversed vision, *Exp. Brain Res.* **69:**531–544.

Miles, F. A., and Fuller, J. H., 1974, Adaptive plasticity in the vestibulo-ocular responses of the rhesus monkey, *Brain Res.* **80:**512–516.

Pellionisz, A., and Llinas, R., 1979, Brain modeling by tensor network theory and computer simulation. The cerebellum: Distributive processor for predictive coordination, *Neuroscience* **4:**323–348.

Rock, I., 1966, *The Nature of Perceptual Adaptation,* Basic Books, New York.

Stratton, G., 1897, Upright vision and the retinal image, *Psychol. Rev.* **4:**182–187.

Ullman, S., 1979, *The Interpretation of Visual Motion,* Cambridge, MA, MIT Press.

Wallach, H., and O'Connell, D. N., 1953, The kinetic depth effect, *J. Exp. Psychol.* **45**(4):205–217.

Waxman, A. M., and Ullman, S., 1985, Surface structure and three dimensional motion from image flow kinematics, *Int. J. Robot. Res.* **4**(3):72–94.

11

Adaptive Control of Saccade Metrics

HEINER DEUBEL

INTRODUCTION

Applying optical aids distorts the relationship established between the visual signals impinging on the retina and our "model of the outside world" by which we plan directed motor activity such as hand or head movements toward an object. So, putting on magnifying or reducing spectacles raises a serious problem for the performance of the vestibuloocular reflex (VOR), which takes a nonvisual signal derived from the semicircular canals to compensate for retinal slip otherwise induced by head turns. The VOR gain becomes inappropriate, and this results, perceptually, in perceived instability of the visual world, eventually causing nausea. Fortunately, these initial problems disappear rapidly, within a few hours or days. An important factor for this kind of habituation is the capacity of the VOR gain for adaptive recalibration, as has been investigated in numerous studies (e.g., Miles and Fuller, 1974; Miles and Eighmy, 1980).

The saccadic system, on the other hand, senses the eccentricity of targets by retinal information and programs rapid eye turns to bring the fovea to the target. It is immediately conceivable that a magnification or a reduction of the retinal image by spectacles mounted to the head does not bring about inappropriate saccadic performance, since the angle under which the target is seen before the eye movement matches the required rotation of the eye to acquire the target. So, with spectacles, there is no need for saccadic adaptation, at least as long as the induced effects are conjugate for both eyes and interaction with head movements is not considered. Contact lenses that move with the eye, however, alter the apparent eccentricity of targets while the required amount of rotation of the line of sight is unchanged. Hence, contact lenses in fact cause saccadic dysmetria. Here it should be considered that, fortunately, the visual magnification induced by contact lenses is rather small, about 1% per 5 D. Nevertheless, it seems important to ask to what extent the saccadic system might be able to cope with these extraneous changes.

Ample evidence for the ability of the saccadic system to compensate for even much more profound dysmetrias is provided from two basically different lines of research. On the one hand, long-term effects of altered properties of the peripheral oculomotor system have been investigated in patients as well as by means of animal experiments. It has been found, for instance, that patients suffering from partial eye muscle paralysis manage to adjust saccadic gain within a few days to account for the effect of the disease (Kommerell *et al.*, 1976; Abel *et al.*, 1978). Optican and Robinson (1980) surgically weakened the medial and lateral recti muscles of one eye in rhesus monkeys. Normal visual experience with the weakened eye then led to a gradual recovery of saccadic accuracy, provided the cerebellum was intact. Interestingly, this adaptive process included not only metric adjustment of saccade size but also the gradual compensation of postsaccadic drift eventually induced by muscle weakness. Thus, not only saccade metrics but also the time course of the eye movement are under adaptive control.

In the second principal line of investigations, saccadic dysmetria has been induced psychophysically by consistent target shifts while the subject tries to acquire the target with a saccade (e.g., McLaughlin, 1967; Henson, 1978; Miller *et al.*, 1981; Deubel *et al.*, 1986; Deubel, 1987). Using a paradigm of this kind, the first part of this paper

HEINER DEUBEL • Max-Planck-Institute for Behavioral Physiology D-8130 Seewiesen, Germany

demonstrates basic properties of the adaptive adjustment of saccade metrics. In the second part, an experiment is described that emphasizes the important role of sensory visual processing in adaptive control.

ADAPTATION OF SACCADE METRICS INDUCED BY MODIFIED VISUAL FEEDBACK

In the laboratory, adaptive changes of saccadic reactions can be induced by systematically altering visual feedback. A "classical" paradigm that has been used in most of our experiments is shown in Fig. 11-1. Subjects (or trained rhesus monkeys) have to follow a small target projected on a large screen that steps in rapid succession. Each trial begins with a stepwise target displacement in the range of 6° to 12°. When the saccade to acquire this target is on the way, the stimulus is systematically shifted by a small amount in a direction consistent with the initial step. The target position after the second step then serves as starting position for the next double-step stimulus. In the example given in Fig. 11-1, the intrasaccadic target step systematically goes in the opposite direction of the initial step. Given this kind of intrasaccadic shift, the primary saccade at first consistently overshoots the target and is usually followed by a large secondary corrective saccade. In other words, this paradigm mimics a situation that occurs when putting on positive (magnifying) contact lenses. It should also be mentioned that in the

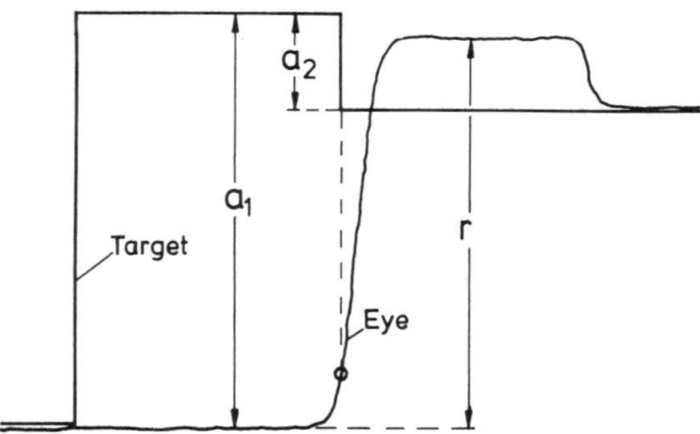

FIGURE 11-1. Stimulus movement and resulting oculomotor response. The amplitude of the first target step a_1 was varied in the range of 6° to 12°. Triggered by the primary saccade, a second target step was induced that consistently occurred in the same direction (gain-increase condition) or, as in this example, in the opposite direction (gain-decrease condition) of the initial step. Step size a_2 was 30% of the initial step amplitude a_1. Saccadic gain is defined as $g = r/a_1 \times 100\%$.

unadapted state saccadic gain is normally slightly smaller than 1.

It is interesting to note that, provided the intrasaccadic shifts are not too large, the subjects completely fail to perceive them (Bridgeman *et al.*, 1975). The oculomotor control system, however, becomes aware of the systematic postsaccadic errors and reacts, as can be seen clearly in Fig. 11-2. The open circles in this figure reveal the time course of the gain of subsequent saccades in such an adaptation paradigm. Obviously, subjects soon begin to compensate by gain reduction and, within 200–300 trials, are finally acquiring the displaced target with normal accuracy. In fact, we found that saccades of all directions are susceptible to these fast adaptive gain changes (Deubel, 1987).

The speed of the adaptive process is asymmetric: the fast time constants indicated above only hold for experimental situations where the adaptive system is required to reduce saccadic gain. Gain increases, on the other hand, occur considerably more slowly, as can be seen from the crosses in Fig. 11-2, representing data from an experimental session in which the intrasaccadic target shifts occurred in the same direction as the initial steps. Also, adaptation under this condition generally tends to be incomplete. Note that gain increases are induced by an experimental paradigm that initially leads to large saccadic undershoots. This result seems to indicate that systematic undershoots are largely tolerated by the adaptive controller, whereas the occurrence of saccadic overshoots leads to immediate gain reduction. Therefore, compensation for positive contact lenses should occur considerably faster than for negative ones.

The capability of the adaptive mechanism to cope with a variety of demands resulting from the application of optical devices as well as from central and peripheral diseases, fatigue, aging, and so forth is closely related to the question of how specifically the adaptive compensation can react to specific stimulation properties. In other words, the question arises to what extent this kind of fast adaptation can be specific to the amplitude and the direction of the saccadic eye movement, specific to the position of the eye in the orbit, or even specific to the eye that was exposed to the training procedure.

A series of experiments (Deubel *et al.*, 1986) gave strong evidence that induced adaptation is indeed not specific to the saccadic amplitude that was conditioned in the training phase. To demonstrate this basic finding more clearly, we recently per-

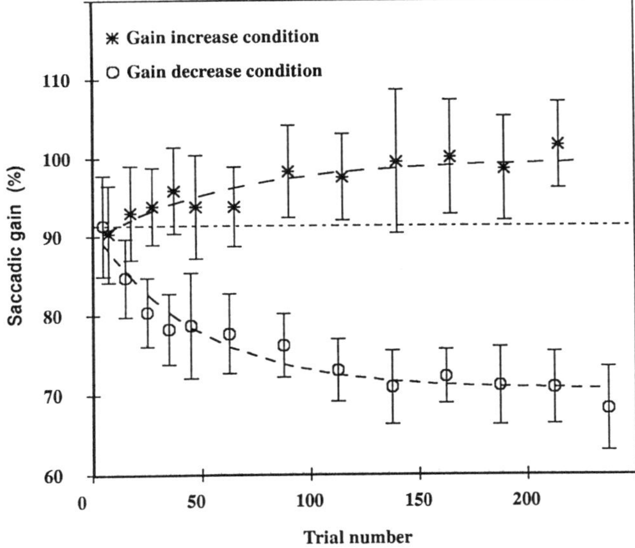

FIGURE 11-2. Mean percentage gain as a function of subsequent trial number. Data for the gain-increase condition are displayed as crosses, and those for the gain-decrease condition as circles.

formed additional investigations in which subjects were given horizontal target steps ranging in size from 3° up to 20°. Again, for adaptation, a second intrasaccadic target shift directed opposite to the initial step was provided, but only for the saccades to 8° target steps; for the other step sizes normal visual feedback (i.e., without intrasaccadic displacement) was given. Results from this type of experiment are presented in Fig. 11-3. The open and filled circles in Fig. 11-3 compare pre- and postadaptive undershoot of the primary saccades with respect to the initial target position as a function of step sizes. Although the conditioning stimulus was only delivered for the 8° saccades, saccades of all sizes are equally affected by the adaptation, as becomes obvious from the increased amount of undershoot for all step amplitudes. This means that the gain reduction induced for the 8° steps transfers to other amplitudes and indicates that gain control is indeed parametric with a single parameter determining saccadic amplitude in response to all target step sizes. Very similar results are obtained in cases where the intrasaccadic shift systematically went in the same direction as the primary saccade, calling for an adaptive gain increase (Fig. 11-3, crosses and triangles, respectively). It should be mentioned that here, in order to obtain sufficient adaptation, only initial target steps of 8° (plus conditioning intrasaccadic target shifts) were given in the adaptation phase.

It is concluded that for both gain decrease and increase, gain changes cannot be made specific to the amplitude of the movement. Conflicting results indicating some ability for amplitude-specific adaptation (Miller *et al.*, 1981; Semmlow *et al.*, 1989) may be attributed to the fact that in these studies,

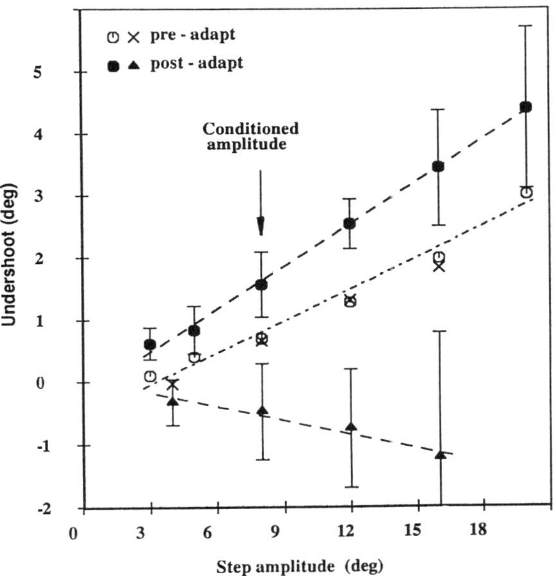

FIGURE 11-3. Mean saccadic undershoot as a function of target step amplitude a_1. Open circles and crosses represent the data from the preadaptive control experiment. Filled circles represent mean postadaptive undershoot for the gain-decrease condition, and triangles those for the gain-increase condition. For adaptation, a conditioning intrasaccadic target shift was only provided for the 8° initial target steps.

spatial loci for initial fixation position and final target positions were not sufficiently varied. Therefore, these results might reflect the effect of position-specific motor learning rather than indicating amplitude specificity of saccadic gain control.

Do the above findings mean that only one single "gain" value determines the amplitudes of all possible saccades? This is indeed not the case; in fact, it has long been known for horizontal eye movements that, say, adaptation of the gain of leftward saccades does not transfer to rightward saccades (Abel *et al.*, 1978; Miller *et al.*, 1981). This kind of directional selectivity was recently investigated more closely, now for oblique saccades with both horizontal and vertical components (Deubel, 1987).

To that end, initially sufficient gain adaptation of saccades in a certain direction was induced with a paradigm as described in Fig. 11-1. Subsequently, the amount of transfer of induced adaptation to saccades into other movement directions was tested. Figure 11-4 gives a typical example of the results, here for a case in which all vertical, i.e., 90° upward, saccades underwent adaptation. The left diagram of the plot shows mean radial gain and direction of saccades in a polar plot. Open circles denote data from the unadapted state; filled circles represent the data from the postadapted state. The fat arrow denotes the saccadic direction (here the 90° upward direction) that underwent adaptation.

As expected, the largest gain change is revealed for the movement direction that was previously adapted. Thus, in this example the gain of the 90° upward saccade is reduced from the nominal value of 86% to about 56% after adaptation. Adjacent directions also yield reduced saccadic gains. However, the effect decreases steeply with angular distance from the adapted direction. This becomes still more apparent in the right diagram of Fig. 11-4, giving the percentage change of radial gain for the different movement directions. Evidently, the adaptation is tightly tuned; gain can be adjusted independently for saccadic directions with only moderate angular distance. It must be concluded that a high directional selectivity of the induced gain modification exists.

Next I addressed the question of whether not only the gain but also the direction of the saccade can be adaptively modified. For this purpose an experimental paradigm as in Fig. 11-5 (left) was used.

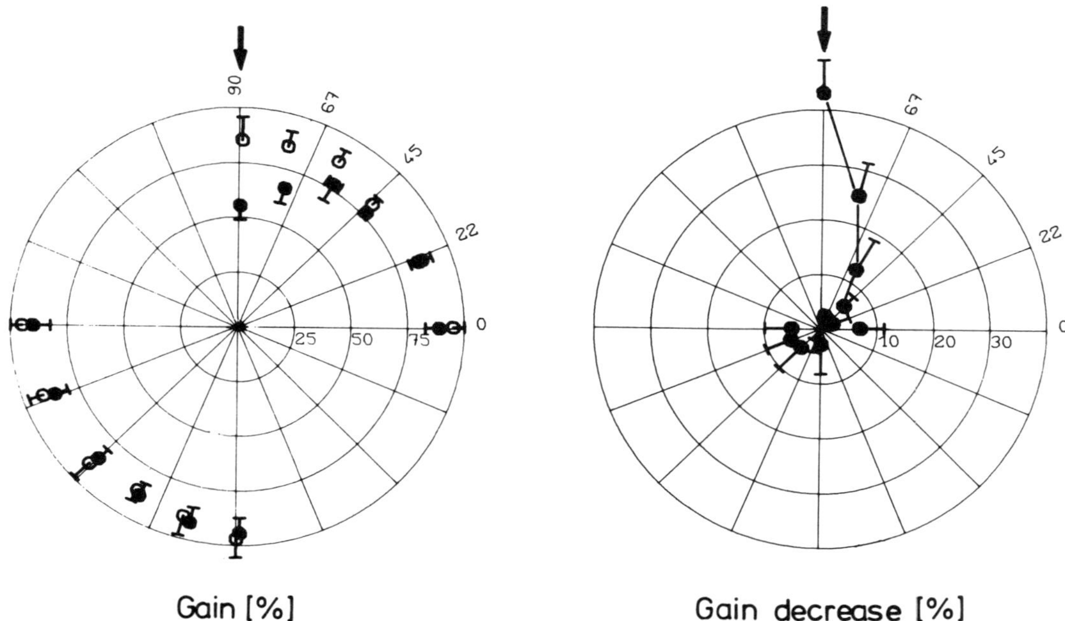

FIGURE 11-4. Left: Means of radial saccadic gain and direction in a polar plot. Open circles represent data from the unadapted state; filled circles represent data from the postadaptive test phase. The arrow denotes the target angle for which conditioning double-step stimuli were given during adaptation (here amplitude reduction for the 90° upward steps). Radial bars represent standard deviations. Right: Percentage changes of saccadic gain resulting from adaptation. Radial bars represent the 99.5% significance level for the difference between pre- and postadaptive means.

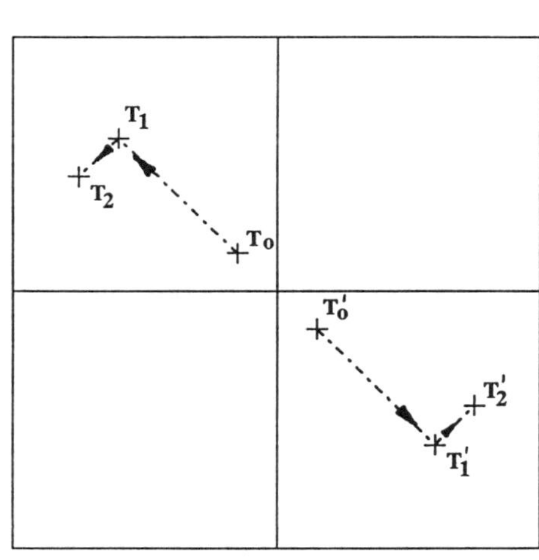

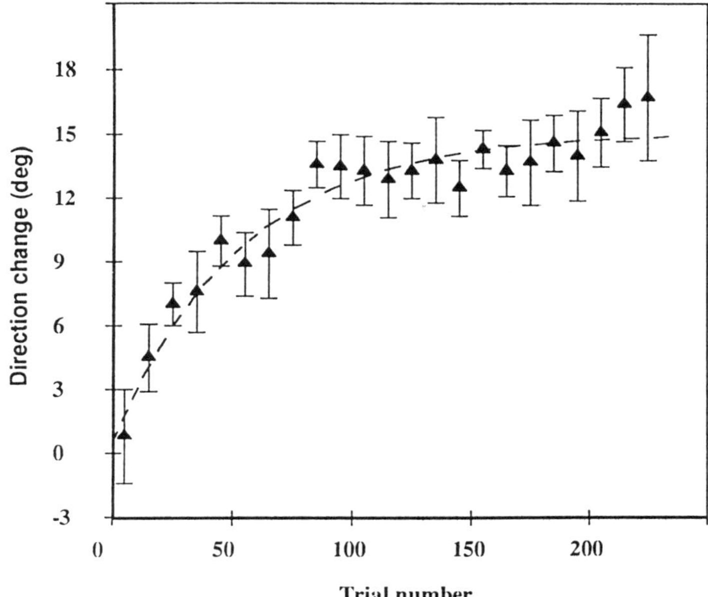

FIGURE 11-5. Left: Target movements for the induction of adaptive changes of saccade direction. Right: Typical time course of adaptive changes of saccadic direction.

A target jumped from a random position T_0 to T_1. Here again, the catch-up saccade triggered a consistent second, intrasaccadic target displacement to position T_2. In this case, however, the conditioning target shift occurred orthogonally to the direction of the initial step. It is clear that the adaptive controller is now invited to modify saccade direction.

Intuitively, we tended to believe that the direction of the eye movement is programmed more rigorously than its amplitude and therefore expected these adaptive changes to occur considerably more slowly. The general finding, however, was that saccadic direction is in fact as well and easily adaptable as saccadic gain. The diagram on the right of Fig. 11-5 shows the change of saccadic direction during adaptation as a function of time. In general, the time constants for small direction changes were in the range of 30–200 trials.

After adaptation, the spread of direction modification on saccades into adjacent directions can be tested. Figure 11-6 shows the differences of saccadic direction between the pre- and postadapted phase. Obviously, the angular change is highest for the adapted direction but decreases steeply with angular distance, revealing that only saccadic directions close to the adapted one are affected. The tuning width of the adaptation effect is very similar to the results from the previous gain-change experiments. Evidently, then, the direction of voluntary saccades to visible targets can be easily modified. Furthermore, this type of adaptation is also highly directionally selective.

In summary, these findings demonstrate that modification of visual feedback by consistent intrasaccadic target displacements induces profound changes of saccade metrics. Amplitude as well as direction of saccades to a visual target adapt quickly, within a few minutes of training. Time constants for

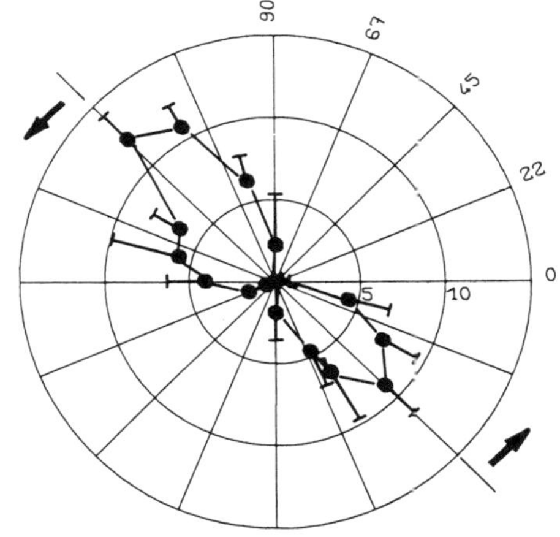

FIGURE 11-6. Mean angular change of saccadic direction caused by adaptation of the −45° oblique saccades, given as a function of the directions under which the targets were presented.

gain reduction as well as for direction modification are in the range of 30 to 200 trials; gain increases need 400 or more training movements. These adaptive changes are surprisingly specific to the direction of the saccadic eye movement. Induced adaptation is, however, not specific to the size of the saccade. In other words, a single gain parameter determines saccadic gain for all amplitudes in a specific movement direction.

In further experiments not presented here we found that induced adaptation is also not specific to the orbital eye position; i.e., training in a certain oculomotor range transfers completely to other areas in the motor field. Also, for this kind of short-term adaptation, adaptive changes are conjugate in both eyes, obeying Hering's law of equal innervation. Finally, adaptation does not change the well-known amplitude/velocity relationship of saccadic eye movements ("main sequence"; Bahill *et al.*, 1975).

The above properties were found for the fast adaptation processes induced experimentally in the laboratory, within a session, by systematic target shifts. The question arises if these results still hold when the subject is given considerably more time to adapt. Indeed, there is evidence for slower, long-term adaptation processes (occurring over several days or weeks) that allow the saccadic system to adapt still more specifically to various demands. It has recently been shown for patients with ocular muscle weakness, for instance, that adaptive changes observed over a period of days or weeks may vary with orbital position (Optican and Miles, 1985). This finding is indeed to be expected if the adaptive controller should be able to account for the inherent position-dependent nonlinearities of the oculomotor plant. Also, in contrast to our finding with short-term adaptation, there are recently accumulating indications that saccades (and VOR) might have the capacity for monocular adaptive changes (Virre *et al.*, 1988; Gleason and Schor, 1989). Again, the experimental data suggest that this capacity is considerably more limited than for conjugate changes and occurs only with a long period of training.

ROLE OF SENSORY PROCESSING IN SACCADIC ADAPTATION

So far, it has been demonstrated how motor responses are changed by conditioning stimuli. It should be kept in mind, however, that understanding such an adaptive system requires knowledge not only about parameters of the motor subsystem but also about the sensory mechanisms that continuously have to provide an error signal indicating saccade performance. In fact, although our understanding of the motor control part advances very rapidly, very little is yet known about how such a signal could be evaluated from the visual input.

In the ecological situation outside our laboratories, a saccade is usually made to only one item within a visually highly complex scene. Knowledge that the saccade was inaccurate requires some form of "recognition" of the object appearing on a different retinal location after the saccade, a problem that is somewhat equivalent to the well-known "correspondence problem" in motion perception and stereovision. That this type of correlation of presaccadic information from the retinal periphery with the postsaccadic foveal or parafoveal input indeed exists, and that it is not necessarily bound to a conscious recognition process, was the conclusion of a series of experiments in which complex random stimuli without semantic meaning were used as saccade targets (Deubel *et al.*, 1984).

In these experiments, the subject initially fixated a prominent fixation line on an extended pseudonoise pattern of vertical bars generated on a large display, similar to the upper pattern (I) in Fig. 11-7. The "target" was then defined by an abrupt dark/bright inversion of a limited area in the periphery. This means that now the stimulus looked like pattern (II) in the figure. On questioning, the subjects frequently interpreted this change as a "sort of short movement" and had no problems initiating horizontal saccades to this area. Triggered by the saccades, the computer shifted the whole scene systematically by a small amount, consistently in the direction opposite the saccade. In order to keep the subjects from becoming familiar with the specific stimuli, the form of the pattern was changed with each trial.

The rationale for selecting this kind of experimental approach was that, after the primary saccade, the subjects had completely lost track of the position of the "target." Since target area and background had similar structure, the subjects were unable perceptually to reidentify the target amid the surrounding background. Also, as with the dot stimuli, they did not perceive the intrasaccadic displacement, as was tested in forced-choice tests. Surprisingly, however, saccadic gain also adapted in this

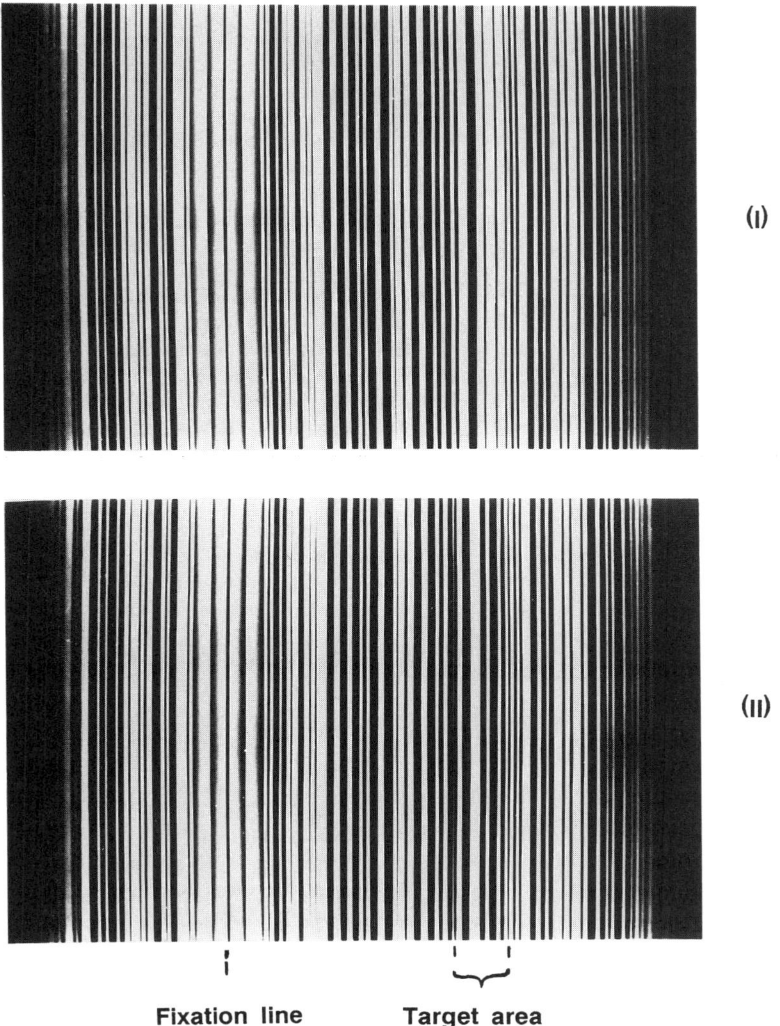

FIGURE 11-7. Stimulus pattern. The stimuli were calculated by means of a random process keeping effective spatial frequencies in the range of 0.5–3 c/°. The patterns were presented on a fast display and subtended a visual angle of about 10° width and 8° height. In this trial, pattern I was presented initially, and the subject had to fixate on the exterior fixation line. Then, a limited area in the visual periphery was abruptly inverted, leading to pattern II. Triggered by the saccadic eye movement directed to this "target," the whole modulated pattern was shifted by 1° opposite to the direction of the saccade.

visually very complex situation and became normometric with respect to the final target position after 150–200 trials, as demonstrated by the apparent shift of the amplitude distributions in Fig. 11-8.

This finding indicates that saccadic adaptation is dependent on a low-level automatic error-detection mechanism based on visual input but is independent of conscious perception. This mechanism is supposed to store information about the peripheral target structure and, after the saccade, to compare this information with the foveal reafference in order to find the target and determine an error signal that indicates saccadic performance. This experiment emphasizes the importance of visual processing capabilities for saccadic adaptation. It is to be expected that blur and distortion of the visual signals will abolish the underlying matching process. To what extent adaptive performance depends on sufficient capabilities of foveal and peripheral vision appears to be a question of high interest indeed.

In summary, analysis of adaptation in the saccadic system discloses a highly efficient mechanism for the plastic mapping of sensory signals into motor reactions sophisticated enough to compensate for the minor effects of optical aids. On the other hand, the effect of reduced vision on the underlying error-

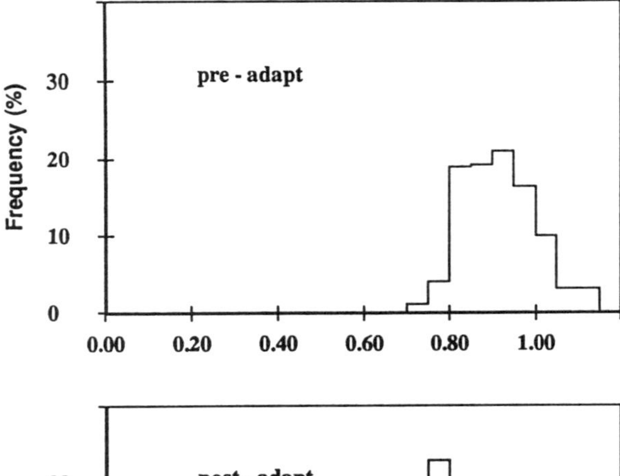

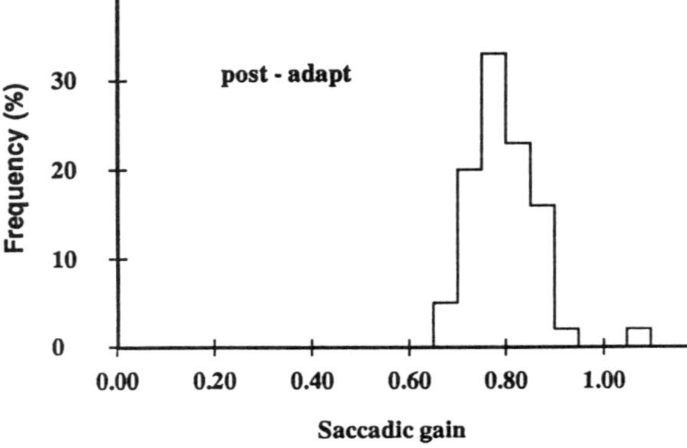

FIGURE 11-8. Distribution of normalized saccadic landing positions before and after 160 training sequences, as explained in Fig. 11-7.

detection mechanism and thereby on adaptive performance is still largely unknown.

REFERENCES

Abel, L. A., Schmidt, D., Dell'Osso, L. F., and Daroff, R. B., 1978, Saccadic system plasticity in humans, *Ann. Neurol.* **4**:313–318.

Bahill, A. T., Clark, M. R., and Stark, L., 1975, The main sequence, a tool for studying human eye movements, *Math. Biosci.* **24**:191–204.

Bridgeman, B., Hendry, D., and Stark, L., 1975, Failure to detect displacement of the visual world during saccadic eye movements, *Vision Res.* **15**:719–722.

Deubel, H., 1987, Adaptivity of gain and direction in oblique saccades, in: *Eye Movements: From Physiology to Cognition* (J. K. O'Regan and A. Levy-Schoen, eds.), Elsevier/North-Holland, Amsterdam, pp. 181–190.

Deubel, H., Wolf, W., and Hauske, G., 1984, The evaluation of the oculomotor error signal, in: *Theoretical and Applied Aspects of Eye Movement Research* (A. G. Gale and F. W. Johnson, eds.), Elsevier/North-Holland, Amsterdam, pp. 55–62.

Deubel, H., Wolf, W., and Hauske, G., 1986, Adaptive gain control of saccadic eye movements, *Hum. Neurobiol.* **5**:245–253.

Gleason, G., and Schor, C., 1989, Selective differential binocular adaptation of vertical saccades and pursuits, *Invest. Ophthalmol. Vis. Sci.* **30**(3)(Suppl.):185.

Henson, D. B., 1978, Corrective saccades: Effect of altering visual feedback, *Vision Res.* **18**:63–67.

Kommerell, G., Olivier, D., and Theopold, H., 1976, Adaptive programming of phasic and tonic components in saccadic eye movements, *Invest. Ophthalmol.* **15**:657–660.

McLaughlin, S. C., 1967, Parametric adjustment in saccadic eye movements, *Percept. Psychophys.* **2**:359–361.

Miles, F. A., and Eighmy, B. B., 1980, Long-term adaptive changes in primate vestibuloocular reflex. I. Behavioural observations, *J. Neurophysiol.* **43**:1406–1425.

Miles, F. A., and Fuller, J. H., 1974, Adaptive plasticity in the vestibuloocular responses of the rhesus monkey, *Brain Res.* **80**:512–516.

Miller, J. M., Anstis, T., and Templeton, W. B., 1981, Saccadic plasticity: Parametric adaptive control by retinal feedback, *J. Exp. Psychol. Hum. Percept. Perform.* **7**:356–366.

Optican, L. M., and Miles, F. A., 1985, Visually induced adaptive changes in primate oculomotor control signals, *J. Neurophysiol.* **54**:940–958.

Optican, L. M., and Robinson, D. A., 1980, Cerebellar-dependent adaptive control of primate saccadic system, *J. Neurophysiol.* **44**:1058–1076.

Semmlow, J. L., Gauthier, G. M., and Vercher, J. L., 1989, Mechanisms of short-term saccadic adaptation, *J. Exp. Psychol. Hum. Percept. Perform.* **15**:249–258.

Virre, E., Cadera, W., and Vilis, T., 1988, Monocular adaptation of the saccadic system and vestibulo-ocular reflex, *Invest. Ophthalmol. Vis. Sci.* **29**(8):1339-1347.

12

Short-Term and Long-Term Adaptative Changes in Eye–Head Movement Coordination Resulting from Reduced Peripheral Vision

G. M. GAUTHIER, J. L. SEMMLOW, J. L. VERCHER, C. PEDRONO, and G. OBRECHT

VISUAL DISCRIMINATION

Fine visual discrimination of an object requires stabilization of the image on the fovea. If the head is immobile and the target in motion, the smooth pursuit system and the saccadic system cooperate first to move the eyes in the direction of the object then to maintain its image in the fovea. In such a task, visual discrimination may start after a delay of the order of 300 msec for a target presented and set into motion 10° from the fovea. Indeed, the saccadic and the smooth pursuit systems have a reaction time of 200 and 130 msec, respectively, whereas a 10° saccade lasts about 100 msec (Robinson, 1981). When the head and the eyes move in the direction of the target, the final position of the eyes in space is the result of the linear summation of the position of the eyes in the orbits and the position of the head in space. To reach such final positions both the eyes and the head move according to particular patterns showing definite eye and head coordination (Bizzi *et al.*, 1972).

G. M. Gauthier and J. L. Vercher • Sensorimotor Control Laboratory, University of Provence, 13397 Marseilles Cedex 13, France. J. L. Semmlow • Department of Biomedical Engineering, Rutgers University, Piscataway, New Jersey 08855. C. Pedrono and G. Obrecht • Essilor International, Laboratory of Physiological Optics, 94000 Creteil, France.

EYE–HEAD COORDINATION

At least four different coordination patterns have been identified by Zangemeister and Stark (1982) and classified in terms of eye-to-head delays. The most common pattern observed in eye–head tracking of stepping targets consists of a sequence of events in which the head starts to move at the same time as or slightly after the eyes. Because of its larger inertia, the head has not significantly moved by the time the eyes have reached the target. Although the head is still moving, visual discrimination may start as soon as the eyes reach and remain stable on the target. The stability of the eyes in space while the head is moving is provided by the vestibuloocular reflex (VOR), which activates fully as soon as the eyes reach the target. This reflex, when activated, induces a movement of the eyes in the direction opposite to that of the head. The VOR is usually defined through its gain, that is, the ratio of eye over head movement velocities. When the target to be fixated is fairly far away from the eyes (so that convergence bias can be neglected), perfect stability of the eyes in space requires VOR gain to be unity. This condition, usually satisfied in eye and head tracking of eccentric targets, leads to perfect counterrolling of the eyes in the orbit from the time the eyes reach the target until the end of head motion. In

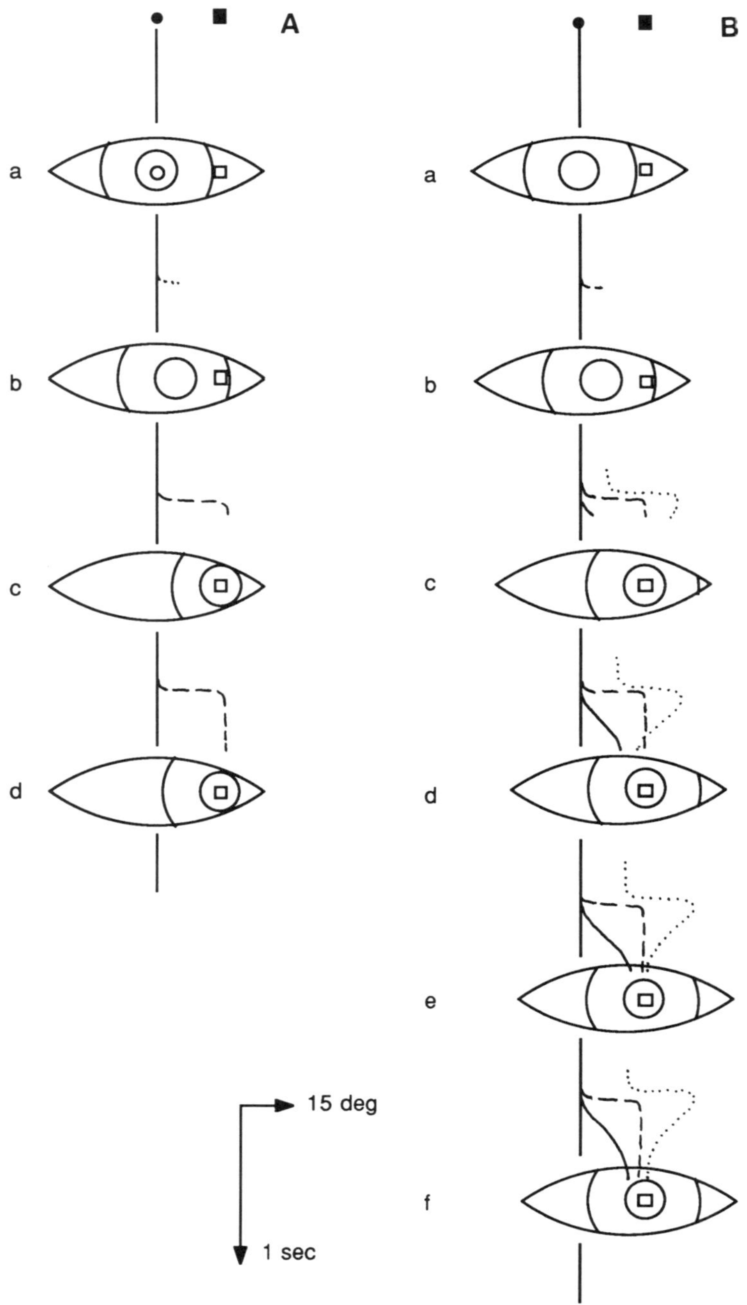

FIGURE 12-1. Eye and head movements in normal target-tracking condition. A: When the head is prevented from moving, displacement is handled by a saccade that brings the eyes to the target in 300 msec including saccadic delay and movement time (a–c). B: When the eyes and the head move, a saccade is first triggererd that brings the eyes on the target (a, b). Gaze remains stable in space through the VOR, which provides perfect counterrolling of the eyes while the head slowly rotates toward the target (d–f).

a typical eye–head tracking sequence, such as shown in Fig. 12-1, 40% to 70% of gaze displacement is provided by head rotation, while the remaining 60% to 30% result from angular displacement of the eyes in orbit.

As described above, the VOR plays an essential role in the coordination of motor events during eye–head tracking. The VOR system is also involved in perception and particularly in the generation of perceptive correlates of motor activities.

Indeed, when the eyes and the head are concomitantly involved in the tracking of a smoothly and slowly moving target, the brain determines—and we sense—the velocity of the target in space by comparing head velocity, as measured by the vestibular system, to eye velocity in orbit. The difference between these two velocities yields the angular velocity of the target in space. It is still not clear how the eye velocity in orbit is calculated. Several hypotheses are proposed, among which the most plausible ones require information derived from either the smooth pursuit system internal model, the eye muscle activation (efferent copy), or ocular muscle proprioception (inflow). Nonetheless, reliable sensing of target velocity requires precise monitoring of both eye movement and vestibular activation. Besides, if the vestibular system is to be used as a reference to calculate target velocity, its function must remain fairly stable in time. This might not be the case, since it has been shown that the VOR is an open-loop system. Fortunately, nature has provided the system with long-term adaptative control.

VESTIBULOOCULAR REFLEX CALIBRATION

Experiments carried out on man and various animals have shown that the VOR system modifies itself in order to maintain or restore a stable input–output relationship to compensate for internal and external changes. Among internal modifications, the most drastic ones relate to morphological changes occurring through childhood such as in head size and interocular distance as well as changes affecting, naturally or pathologically, the nervous and muscular elements of the VOR system. Alteration of the VOR input–output relationship also occurs as a result of optical systems used to correct refraction (spectacles) or increase visual discrimination (telescopic lenses). In fact, reversing prisms (dove prisms) (Gonshor and Melvill-Jones, 1976) and magnifying lenses (Gauthier and Robinson, 1975) have been used to demonstrate and quantify the adaptative properties of the VOR. As will be shown later, VOR adaptation is essential to recover appropriate motor and sensory functions in individuals being fitted with corrective glasses, particularly with progressive lenses.

To better approach the complex and multifaceted problems raised by progressive lenses, problems that are successfully solved by the CNS, we now examine the sensorimotor adaptive reactions in eye–head coordination resulting from magnifying lenses and from lenses with peripherally limited vision.

ADAPTATION TO MAGNIFYING LENSES

If a subject is suddenly fitted with magnifying lenses having, for example, a magnification factor of 2 (×2), the objects in the visual field will appear bigger and closer to him, but no oculomotor problem should arise. To fixate an eccentric target the eyes will simply jump to the correct orbital position leading to the capture of the image by the fovea. The saccade will be twice the size of the saccade produced in the control condition. No perceptual alteration will result. However, localization error will occur if the subject attempts to reach the target with his hand. Besides, when the subject slowly moves his head right and left while fixating a stationary target, he will perceive an illusory motion of the target in the direction opposite to head motion with an amplitude about equal to head rotation amplitude. This illusion may be readily explained by considering some functional aspects of the smooth pursuit and vestibular systems involved in this task.

Indeed, if the head rotates slowly, the eyes will remain essentially on target. Because of the ×2 lenses, this implies that they rotate in the orbit over an angle twice the amplitude of the head rotation. The overall eye rotation is the result of the combined action of the smooth pursuit system and the VOR. Assuming a VOR gain equal to 1, it follows that the smooth pursuit system is responsible for one half of the counterrotation of the eyes. The CNS, which monitors both head rotation through the vestibular system output and the smooth pursuit system through inflow and/or outflow information, detects that the motion of the eyes in orbit is in excess of head rotation, and assumes that the target has moved in space; hence, the illusion.

If a subject is requested to wear ×2 magnifying lenses for a long period of time, adaptive changes will progressively take place leading to an increase of VOR gain and a concomitant decrease of the illusory sensation of motion. At a behavioral level, it appears that within minutes some changes take place, but they are weak, labile, and fairly difficult to quantify. Nonetheless, they tend toward the restoration of normal function. (Melvill-Jones *et al.*, 1988).

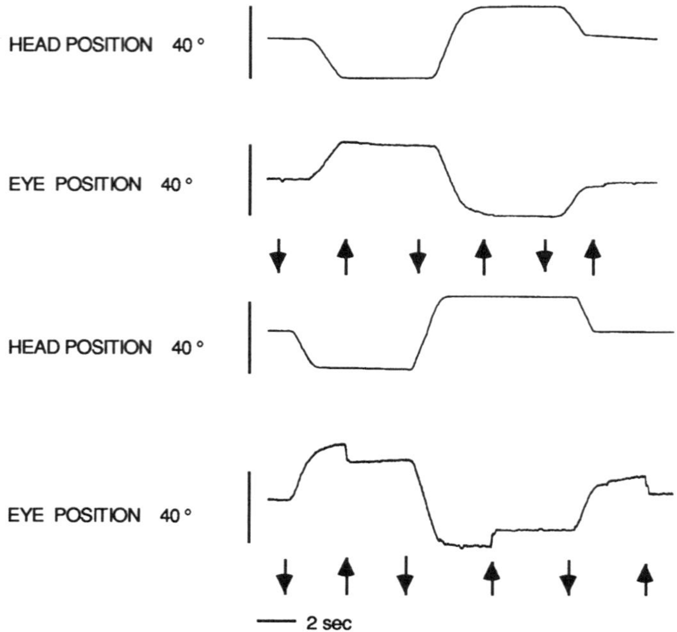

FIGURE 12-2. Vestibuloocular reflex adaptation to ×2 magnifying lenses. Subject's ability to look at the remembered position of a fixed target when rotated 20° or 40° in the dark before (upper two traces), and after (lower two traces) wearing magnifying lenses. Chair (head) and eye positions are shown in each test. At the down arrows the target disappears, and the subject is rotated while attempting to keep fixating at the target location. At the up arrows, the target reappears, and the subject refixates it. After lens adaptation, the subject mislocates the target. The amplitude of the saccade necessary to refixate the target is a measure of the increase in gain (from Gauthier and Robinson, 1975).

Experiments have shown that continuous wearing of ×2 lenses for 4 days is likely to bring an increase of VOR gain of 70%, as tested in the dark (Gauthier and Robinson, 1975). Figure 12-2 shows the increase of VOR measured with transient chair rotations while the subject fixates an imaginary target. Longer wearing of dove prisms, as tested by Gonshor and Melvill-Jones (1976), suggests that the time constant of the adaptive processes leading to major changes of the VOR gain is of the order of 3 to 4 weeks. Still, complete adaptation may require a much longer period and may involve central nervous system processes different from those responsible for the VOR modifications observed in the first few weeks. From these time-constant observations one may conclude that at least three mechanisms are sequentially involved in the overall VOR adaptation phenomenon: one very labile with a short time constant of the order of a few trials or a few seconds, one with a time constant of the order of minutes or hours, and a much longer time constant of a few weeks.

EYE–HEAD TRACKING AND VISION DISCRIMINATION WITH REDUCED PERIPHERAL VISION

In order to comprehend the more complex problem raised by progressive lenses one may first analyze visual discrimination and eye–head coordination changes observed in subjects with artificially reduced peripheral vision. One experimental paradigm uses spectacles in which a vertical slit of clear central vision has been preserved while the peripheral hemifields have been blurred by a thin layer of silicone grease. Such an arrangement allows precise location of a target presented eccentrically but prevents its discrimination anywhere except through the clear slit. Let's examine this experimental procedure in detail.

In such an experiment, visual performance is quantified in terms of the amount of time required to correctly identify a three-digit number. Under conditions where peripheral vision is restricted, previous experiments have shown a constant relationship between the ability to discern a peripheral image and the time available to view this image (Gauthier *et al.*, 1987). In particular, short presentations of a three-digit random number, appearing in unpredictable peripheral locations, produced a consistent psychophysical "identification function," which is the frequency of correct identifications as a function of stimulus presentation time. This frequency function showed that the percentage of correct identifications increased monotonically over a 100- to 300-msec period. As expected, the stimulus viewing time for correct identification is greater for more peripheral targets. This "acquisition" time also increases when the peripheral field is restricted (Gauthier *et al.*, 1987).

To identify successfully a peripheral image viewed through peripherally limited vision lenses, the position of the eyes in space (i.e., gaze), must move to coincide with target position. As mentioned above, in the case of unrestricted peripheral vision, the shift in gaze position can be accomplished entirely by conjunctive eye movements but is generally shared by both the head and the eyes. In this situation, a series of experiments (Semmlow *et al.*, 1990) has shown that the primary determinant of successful target acquisition was eye position error (which would be the same as gaze error under these circumstances) when the target presentation ended. Although variations were found between different subjects, targets presented ±24° peripherally were correctly identified provided the eye position was within 2.0° to 3.0° of the target when the stimulus presentation ended. Though not rigorously investigated, the viewing time during which eye position was within this error boundary did not appear significant. That is, position accuracy, not viewing time, determines if a stimulus is correctly identified.

When the peripheral field is restricted, the range of possible eye positions for successful stimulus acquisition is limited: hence, gaze position must be established by a change in head position. Typically, an initial eye movement does occur, but the eye returns to the central position in coordinated compensation of the head movement. Figure 12-3 illustrates the sequence of events leading to centering of both eyes and head in the direction of the target. When peripheral vision is restricted, the acquisition time increases, as shown in the identification function of Fig. 12-4. In addition, the slope of the identification function decreases, particularly for very restricted peripheral fields.

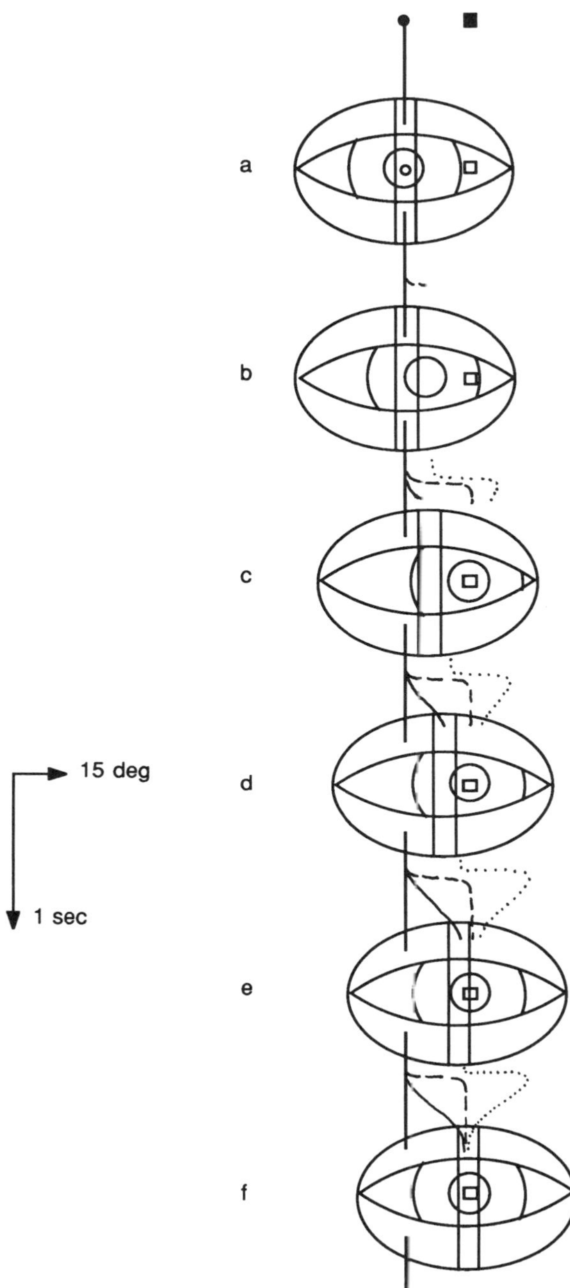

FIGURE 12-3. Eye and head movements in target tracking with slit glasses. The eyes reach the target in about 300 msec (a,b), but fine foveal discrimination can only start when the head has rotated toward the target to the point where the target can be seen through the slit (e). Head angle is then about equal to target eccentricity, and the eyes are recentered in their orbits (f).

Since both eye and head movements are required, and since both must be fairly precise, successful stimulus identification could depend on a number of factors. For example, either head or eye

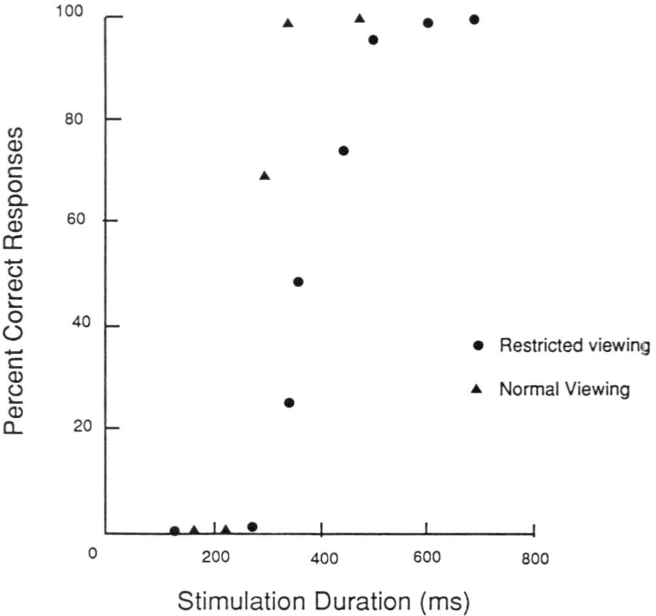

FIGURE 12-4. Identification function. In the normal condition, 100% of the targets are correctly identified if the presentation duration is 300 msec or more (▲). With peripherally limited viewing, 500 msec presentation duration is necessary for 100% correct identification (●).

movements could be subject to amplitude errors. Acquisition times would also be strongly influenced by the response time of these movements. In turn, response times are dependent on movement dynamics as well as response latencies.

To determine which of these several eye and head movement parameters influence the acquisition time, a series of experiments was performed in which head and eye positions were continuously monitored during the acquisition task.

The identification curves were obtained using a specially designed stimulus/response monitor (Semmlow *et al.*, 1986) similar to that described by Pedrono *et al.* (1987). Four numerical displays of three digits each were placed at 12° and 24° left and right of a central fixation stimulus at a viewing distance of 52 cm. Similarly, the viewing time, or stimulus duration, was varied over a range of values.

During the stimulus presentations, both head and eye movements were simultaneously recorded. Eye movements were monitored using the infrared reflection technique while head movement recording relied on a specially designed ultrasonic head movement monitor (Rollero *et al.*, 1990). A typical experimental run consisted of 100 individual head and eye movement recordings.

Eye movement was calibrated following the stimulus presentation and head calibration using a known step change between two target positions (2.0 deg.).

Typical response records are shown after calibration adjustment in Fig. 12–5. The dotted line is the eye movement response, the dashed line is the head movement, and the solid line the gaze movement computed from the two movements. The vertical arrow indicates the extent of the stimulus duration and the horizontal bar indicates the stimulus position; hence, Fig. 12-5 presents a response to a target positioned 24.0° left of center. A small adjustment in head position can be seen during the head alignment period as well as the subsequent small calibrating saccadic eye movements. Analysis consisted of identifying the critical features of both head and eye responses from the calibrated recordings.

The data indicate that although the overall acquisition time is a combination of head response latency and movement dynamics, the variation in identification performance is primarily caused by the response latency. Thus, whereas the intercent of the identification function is determined by both latency and movement dynamics, the slope is determined primarily by latency. (If the time characteristics of all head movements were always the same, identification outcomes would only depend

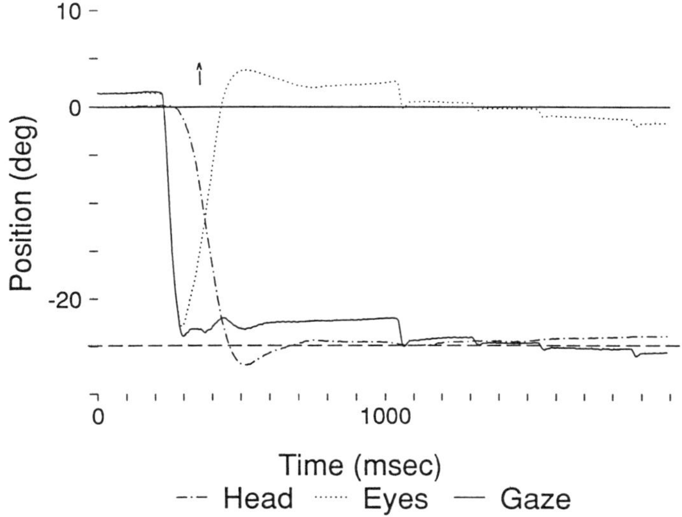

FIGURE 12-5. Eye and head movements in target tracking with slit glasses. Targets are presented 24° to the left of a central fixation target with restricted peripheral viewing (8° slit). Head (dashed line), eye (dotted line), and gaze positions (solid line) are shown (vertical arrow). The horizontal dashed line indicates the target position.

on the stimulus duration and the slope of the identification function would be vertical.)

To confirm the roles established for each of the head movement components in the identification task, we have developed a computer-simulated model of the identification task. The model was constructed under the assumption that identification outcome is solely dependent on the head position when the stimulus period ends; that is neither gaze position nor the eye movements that determine gaze position influence performance in our task. An implicit assumption is that the identification task can occur with negligible delay provided the head position is within a given limit.

Figure 12-6 presents a block diagram of the model. The first element generates the response latency of a head movement. This element utilizes a random number generator in conjunction with the appropriate distribution function to produce a distribution of response latencies similar to that of a given subject. The second element combines the response latency with the stimulus duration to produce the relative stimulus duration, that is, the stimulus viewing time remaining once the head movement has started. The third element converts relative stimulus duration to head position error using the relationship between these variables empirically determined from actual data points. The model has an optional gaussian noise source to represent the scatter of data points. The output of the third element is head error, which, after passing through a threshold element, determines if a given trial results in a correct or incorrect response.

The model can be used to predict overall performance, as described by the identification function, by tabulating the outcomes of a large number

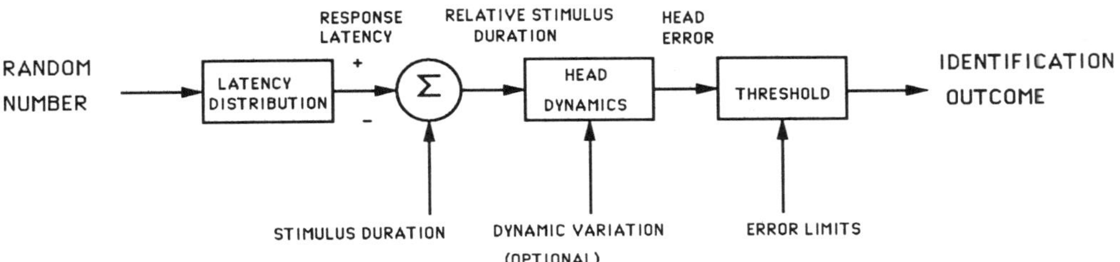

FIGURE 12-6. Model. Block diagram of a model to represent the task of identifying eccentric targets when peripheral vision is restricted (From Semmlow *et al.*, 1990).

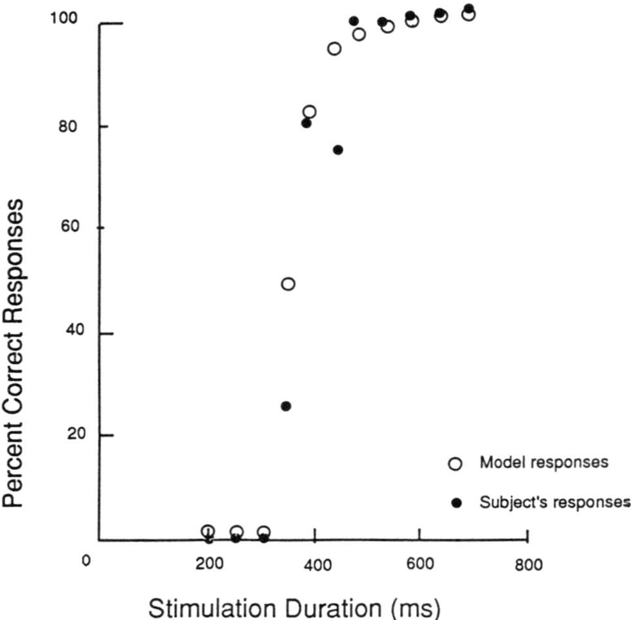

FIGURE 12-7. Identification function. Identification function obtained experimentally (●) and from computer simulation (○) of the model. Target positions were 24° left and right of center, and the slit restricting peripheral vision was 4°.

of simulated trials. The identification function of Fig. 12-7, obtained experimentally (●), shows the overall performance of one subject in the identification task when the peripheral vision is restricted with a 4° slit. The simulation results (○) are from a model in which the head error threshold for a correct response was limited to $\pm 2.5°$. The model is in agreement with the experimental results, and the discrepancies probably arise from the difficulty in obtaining sufficient data over a reasonable experimental period.

From such series of experiments one may conclude that the identification of peripheral targets when peripheral vision is restricted requires precise, coordinated motor responses from the head and the eyes. Psychophysical experiments demonstrate that a consistent performance measurement of this task can be constructed by tabulating the number of correct responses over a range of stimulus viewing times or stimulus durations. Recordings of dynamic head and eye movements during this task show that performance is solely determined by head movement characteristics. Although the acquisition time depends on both the head movement response latency and the duration, the variation in performance results primarily from the variability in response latency. The computer-simulated model of the identification task produces results closely resembling those obtained experimentally and confirms the role that head movement parameters play in determining overall performance.

VISION AND ADAPTATION THROUGH PROGRESSIVE LENSES

When a subject wearing progressive lenses tracks a slowly moving target or attempts to foveate a target through the intermediate zone of the lens, head movements must be generated in the direction of the target any time the eyes reach the limit of the clear vision zone. If this zone is narrow the head movement amplitude yielding foveal fixation is nearly equal to the target eccentricity. Figure 12-8 illustrates the sequence of eye and head motor events resulting in the capture of a target presented 30° to the right of a central fixation dot.

In spite of a fairly large intermediate zone of clear vision (10°), the final head position is clearly directed toward the target, and the eye is recentered in the orbit. Comparison of the sequence of eye and head motor events applying to normal viewing (Fig. 12-1), viewing through a slit (Fig. 12-3), and viewing through a progressive lens (Fig. 12-8) shows that discrimination time will definitely be increased in the two modified conditions. Progressive lenses re-

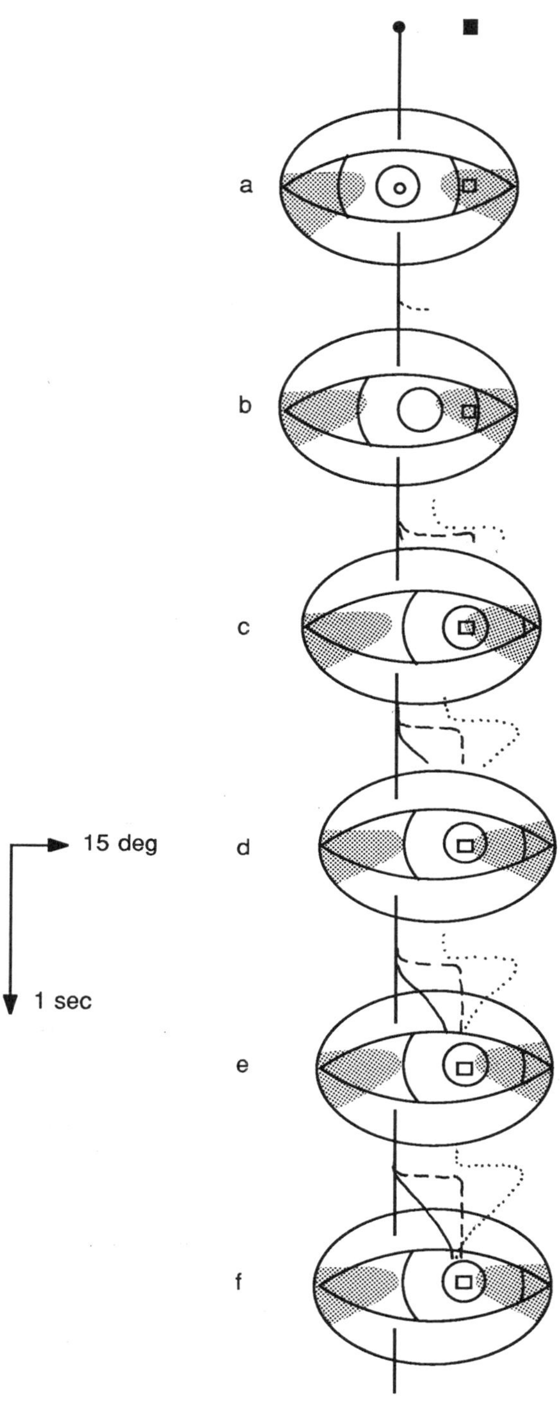

FIGURE 12-8. Eye and head movements in target tracking with progressive lenses. The sequence is similar to that observed in Fig. 12-4 for slit viewing except that head movement amplitude does not have to be equal to target eccentricity to permit foveal discrimination.

quire a time intermediate between normal and slit viewing conditions.

In a task involving discrimination of a three-digit number, data show that the average discrimination time is around 1 sec in normal viewing, around 1.5 sec with progressive lenses, and longer than 2.1 sec with 4-mm slit lenses (Gauthier *et al.,* 1987).

Detailed analysis of the data shows that some adaptive changes take place during a target discrimination experiment with a subject wearing progressive lenses. Head movement delay becomes shorter, and head velocity larger, than that recorded in normal vision. In addition, both head movement stability and precision increase markedly within a few responses. These changes are similar, in terms of time and amplitude, to those observed in subjects performing a tracking task while wearing a helmet designed to increase head inertia (Gauthier *et al.,* 1986).

Long-term wearing of progressive lenses brings about further adaptive changes as demonstrated by discrimination tasks performed over 1 to 3 weeks. The improvements are the result of further decreases in eye-to-head delay: eye and head movements become practically synchronous and show a higher stabilization of gaze (Fig. 12-9). One may suspect that these improvements result from appropriate changes in the VOR gain similar to the changes described above in experiments involving long-term wearing of magnifying lenses (Gauthier and Robinson, 1975), reducing lenses (Miles and Fuller, 1974), or reversing prisms (Gonshor and Melvill-Jones, 1976). Still, this hypothesis has yet to be demonstrated, since, as we discuss later, the VOR gain must be adjusted in orbital coordinates to fit the magnification optics, which vary throughout the lens surface.

A more efficient use of the lenses is suggested by the observation that in most eye–head movement sequences, the head movement amplitude was no more than 80% of target eccentricity, in contrast to amplitudes about equal to target eccentricity during the early trials on the first day (Fig. 12-9). It follows that adaptive optimization of eye–head coordination sequences is accompanied by a sensory adaptation that permits the use of the marginal zone of less than clear optics.

PERCEPTUAL ADAPTATION

As commonly observed and reported by presbyopic subjects wearing progressive lenses, a few days to a

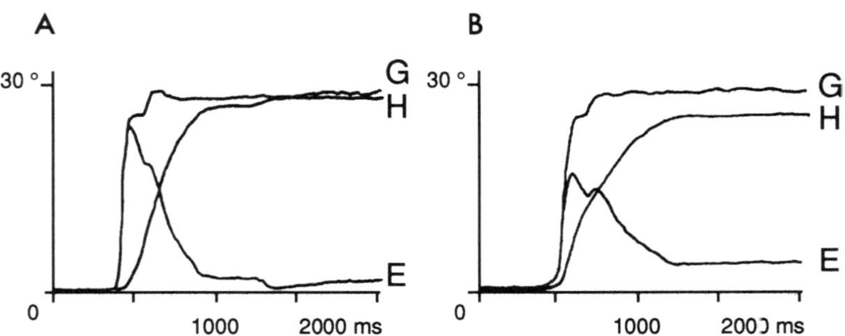

FIGURE 12-9. Adaptated eye and head movements in target tracking with progressive lenses. At first, head movement amplitude, about equal to target eccentricity (A), is fairly unstable. Adaptation brings about a shortening of head delay and an increase of movement and fixation stability. Targets can now be discriminated with a head movement shorter than target eccentricity (B).

few weeks is necessary to recover close to normal sensory, motor, and perceptual visuomotor functions. Apparently, perceptual adaptation is concomitant with motor adaptation (most likely the former is more or less conditioned by the latter). Sensorimotor and perceptual processes have approximately equivalent time constants.

It is worth noting that the major perceptual change experienced by a subject wearing progressive lenses for the first time results in a fairly uncomfortable sensation of illusory instability of the visual field. This instability has the same origin as that described above for magnifying lenses but is more troublesome. Indeed, because of the geometry of the lens, the upper and lower parts of the lens—not to mention the lateral parts of the intermediate zone—may have a large power difference resulting in different magnification factors. It follows that on head rotation (as mentioned above for magnifying lenses), the subject will sense illusory motion of the visual field at different velocities through the upper and lower fields of view and within hemifields as well. This usually results in a "sensation of rocking," which may induce nausea and vertigo. These sensations soon decrease and finally vanish in 2 to 3 weeks.

One may wonder about, and marvel at, the sophistication of the various adaptive controls handling the observed changes in response to a somewhat constraining alteration of the visual input. We discussed earlier the adaptive changes of VOR gain resulting from wearing ×2 lenses. The same changes are necessary throughout the visual field if one assumes a homogeneous magnification factor throughout the lens. A more complex adaptation is necessary in a subject wearing bifocals or progressive lenses. With bifocals, one VOR gain applies to viewing conditions through the upper part of the lens and another through the lower part. With progressive lenses, an ideal setting will require a distinct gain value for each direction of view, or at least for each viewing zone. One may question how this is accomplished, since the essential factor is not head position in space (which conditions the input to the semicircular canals) that sets the appropriate gain but the actual direction of the eyes through the lens. Although the sensing of this position may be achieved either through eye muscle activation (efferent copy) or through ocular muscle proprioception, the demonstration that such fine adaptation occurs will probably puzzle the most fervent advocates of sensory motor adaptation. Still, one may speculate that the CNS may well execute such elaborate control if one considers that the necessary process is essentially an extention of the coordinate-dependent motor adaptation seen in double-step saccadic adaptation (Wolf *et al.*, 1983; Semmlow *et al.*, 1989; Deubel Chapter 11, this book) and in perceptual adaptation observed in subjects fitted with unhomogeneous lens optics (Droulez and Cornilleau, 1986).

SUMMARY

In a visual target-tracking task with the head free, the eyes usually reach a peripheral target while the head is still moving. The vestibuloocular reflex stabilizes gaze to allow for fine foveal discrimination. For targets in the 10° to 20° eccentricity range, head rotation contributes weakly (20–50%) to gaze displacement. If peripheral vision is altered by an optical system attached to the head, such as progressive lenses, the beginning of visual discrimination is likely to be delayed until the head has rotated to almost equal target eccentricity. The eyes, after counterrotation, are then recentered in their orbits.

Previous studies (Gauthier *et al.*, 1987; Pedrono *et al.*, 1987) have described sensory and motor changes resulting from lenses with peripherally modulated vision that progressively recover to close to normal visuomotor performance.

The question remains as to the nature, amplitude, and timing of the changes in eye–head movement coordination induced by limited peripheral vision. We have quantified these short- and longer-term changes in normal subjects wearing lenses with a narrow, vertically oriented slit of clear vision. Peripheral vision was made homogeneously blurry with a film of silicone grease. The subjects were requested to discriminate three-digit targets randomly presented for various durations at 12° or 24° to the right or to the left of a central fixation target.

The basic data show that visual discrimination may only start when gaze is stabilized through the zone of the clear vision (slit). It follows that discrimination time is largely related to the width of the clear-vision slit.

The changes in eye–head coordination, analyzed in light of current knowledge on sensorimotor adaptation, should help to explain observations carried out on presbyopic subjects wearing progressive lenses.

REFERENCES

Bizzi, E., Kalil, R. E., and Morasso, P., 1972, Two modes of active eye–head coordination in monkeys, *Brain Res.* **40:**45–48.

Droulez, J., and Cornilleau, V., 1986, Adaptive changes in perceptual responses on visuomanual coordination during exposure to visual metrical distorsion, *Vision Res.* **26:**1783–1792.

Gauthier, G. M., and Robinson, D. A., 1975, Adaptation of the human vestibuloocular reflex to magnifying lenses, *Brain Res.* **92:**331–335.

Gauthier, G. M., Martin, B. J., and Stark, L. W., 1986, Adapted head-and-eye movement responses to added head inertia, *Aviat. Space Environ. Med.* **57:**336–342.

Gauthier, G. M., Obrecht, G., Pedrono, C., Vercher, J.-L., and Stark, L., 1987, Adaptative optimization of eye–head coordination with degraded peripheral vision, in: *From Physiology to Cognition* (J. K. O'Regan and A. Lévy-Schoen, eds., Elsevier/North-Holland, Amsterdam, pp. 201–210.

Gonshor, A., and Melvill-Jones, G., 1976, Extreme vestibulo-ocular adaptation induced by prolonged optical reversal of vision, *J. Physiol. (Lond.)* **256:**381–414.

Melvill–Jones, G., Guitton, D., and Bertthez, A., 1988, Changing patterns of eye–head coordination during 6 h of optically reversed vision, *Exp. Brain Res.*, **69:**531–544.

Miles, F. A., and Fuller, J. H., 1974, Adaptive plasticity in the vestibulo-ocular responses of the rhesus monkey, *Brain Res.* **80:**512–516.

Pedrono, C., Obrecht, G., and Stark, L., 1987, Eye–head coordination with laterally "modulated" gaze field, *Am. J. Optomet. Physiol. Optics* **64:**853–860.

Robinson, D. A., 1981, Control of eye movements, in: *Handbook of Physiology. The Nervous System* (V. B. Brooks, ed.), American Physiological Society, Bethesda, pp. 1275–1336.

Rollero, R., Vercher, J. L., Semmlow, J. L., and Gauthier, G. M., 1990, Ultrasonic two-axis rotation detector, *IEEE Trans. Biomed. Eng.* **37:**450–457.

Semmlow, J., Gauthier, G., and Vercher, J. L., 1986, *Instrumentation for Determining Acquisition Times, Interim Report*, Essilor International, Creteil, France.

Semmlow, J. L., Gauthier, G. M., and Vercher, J. L., 1989, Short term adaptive modification of saccadic amplitude, *J. Exp. Psychol.* **15:**249–258.

Semmlow, J. L., Gauthier, G. N., and Vercher, J. L., 1990, Identification of peripheral visual images in a laterally restricted gaze field, *Percept. Motor Skills* **70:** 175–194.

Wolf, W., Deubel, H., and Hauske, G., 1983, Properties of parametric adjustment in the saccadic system, in: *Theoretical and Applied Aspects of Eye Movement Research* (G. Stelman and P. Vroon, eds.), North-Holland, Amsterdam, pp. 79–86.

Zangemeister, W., and Stark, L., 1982, Types of gaze movement: Variable interactions of eye and head movements, *Exp. Neurol.* **77:**563–577.

13

The Disruptive Effects of Optical Aids on Retinal Image Stability during Head Movements

FREDERICK A. MILES

EYE MOVEMENTS FUNCTION TO IMPROVE VISION

Eye movements exist to aid vision by directing gaze toward objects of particular interest and, should those objects move, by tracking them. There are two basic kinds of eye movements: *saccadic,* which serve to shift fixation and thereby bring pertinent retinal images into the fovea where vision is most acute, and *smooth,* which serve to keep those images in the fovea by compensating for movements of the object or the observer. Visual acuity begins to deteriorate significantly when retinal images drift at more than a few degrees per second (Westheimer and McKee, 1975), and it is the smooth eye movements that operate to minimize such drift.

Earlier this century, Ronne (1923) pointed out that patients wearing newly prescribed spectacles experience movement of their visual world when they turn their heads; after wearing the lenses for a few days their world becomes stable again, but, if the spectacles are now removed, then motion is again experienced. Ronne correctly attributed this problem to the vestibuloocular reflex (VOR), which is concerned with keeping the eyes stable during head turns.

In this review I examine the effect of various spectacle lenses on the operation of this reflex, emphasizing the precise nature of the optical interference and the nervous system's ability to compensate for it. Since there have been few clinical studies of these problems, I rely heavily on relevant laboratory studies. However, most of this research addresses only issues related to *rotational* disturbances of gaze. *Translational* disturbances, such as occur during walking, are potentially just as disruptive and create problems that are sufficiently different from rotational ones to require the nervous system to make special provision for them in the form of a second vestibuloocular reflex. I make a sharp distinction between these two entirely separate reflexes, referring to the one as the *rotational vestibuloocular reflex* (RVOR) and the other as the *translational vestibuloocular reflex* (TVOR), and consider them quite separately. Of course, under normal circumstances these two reflexes will frequently be acting in concert. Since the mechanisms compensating for translations have attracted relatively little attention, I have much less to say about them, but this should not be taken to indicate that they are less important. Indeed, the optical devices that interfere with the operation of the RVOR also affect the TVOR, and some of the visual problems previously attributed solely to interference with the operation of the RVOR probably also involve the TVOR.

THE ROTATIONAL VESTIBULOOCULAR REFLEX

It is instructive to begin by examining some eye movement recordings in order to show the RVOR in

FREDERICK A. MILES • Laboratory of Sensorimotor Research, National Eye Institute, National Institutes of Health, Bethesda, Maryland 20892.

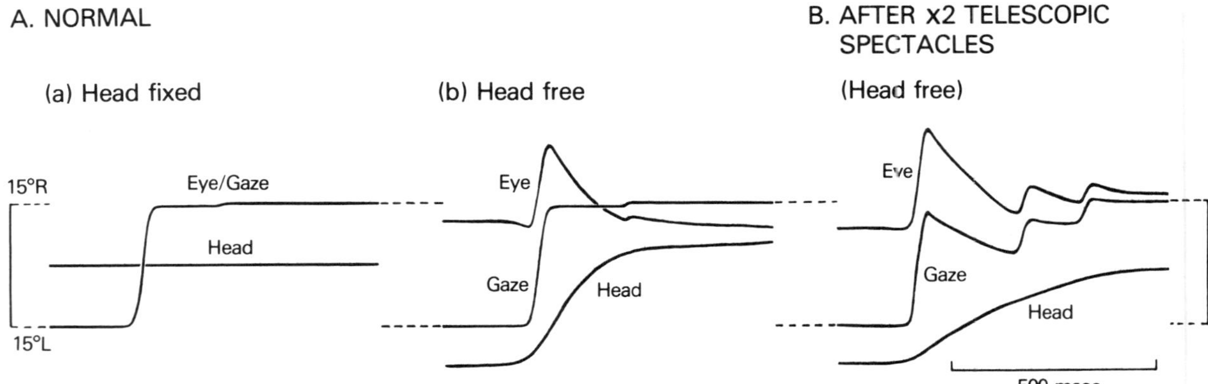

FIGURE 13-1. Effect of optically induced changes in the RVOR on the transfer of gaze during head turns in the monkey. (A) Whether the head is stationary as in a or moving as in b, the transfer of gaze between two targets normally proceeds with an almost identical time course because the head movements are offset by compensatory eye movements that are known to be generated largely by the RVOR. (B) After wearing ×2 telescopic spectacles for a few days, the compensatory eye movements are now much too large, reflecting a much increased RVOR gain, and gaze fails to acquire the new fixation target until the head movement is virtually complete. Note that the spectacles are removed for the test, which is performed in total darkness except for the small fixation target lights. (After Miles and Fuller, 1974.)

operation. Figure 13-1 shows the transfer of fixation between two targets, the first 15° to the left and the second 15° to the right, under three different conditions (Miles and Fuller, 1974). In Figure 13-1A(a) the head was fixed in position, and the transfer was achieved solely by means of a rightward saccadic eye movement. It is evident that such movements are executed very rapidly, thereby minimizing the period during which retinal images are disturbed by such necessary transfers of fixation. However, under normal conditions, these saccadic eye movements are accompanied by turning of the head. In order to follow the shift of fixation under these circumstances, we must record not only the movements of the eyes with respect to the head but also of the head with respect to the surroundings and then sum the two together. When this is done, the result is a plot of the movement of the eyes with respect to the surroundings—often termed "gaze"—and it then becomes apparent that the transfer of gaze is practically unaffected by the head movement [see Fig. 13-1A(b)]. The reason for this is that the head movement is offset by compensatory movements of the eyes, which are largely generated by the RVOR. Closer scrutiny of Fig. 13-1A(b) indicates that the motor sequence is initiated by a rightward head turn that lasts for several hundred milliseconds. At first, the eyes move leftward solely to compensate for this head movement so that gaze is almost undisturbed, but at some point the subject makes a rapid rightward saccadic eye movement that shifts the gaze to the new target. This saccade only lasts for a few tens of milliseconds and so is complete long before the head has stopped moving. After the saccade is complete, the eyes once more move leftward solely to compensate for the continuing head movement, so that gaze is now maintained at its new location. These data were recorded from a monkey, but very similar records can be obtained from human subjects. In summary, saccades serve to select and acquire new images, whereas the RVOR is solely concerned with generating eye movements to compensate for any rotation of the head and thereby maintains fixation on the selected object of regard when the head turns.

The RVOR receives its input from the semicircular canals, which are embedded in the base of the skull and act as angular speedometers, sensing rotations of the head. The canals initiate neural activity that results in compensatory eye movements in a mere 14 msec (Lisberger, 1984). Such rapidity is imperative if the high speeds achieved during normal head turns are to be adequately compensated. However, the RVOR lacks any immediate error-correction capability: the semicircular canals that sense the input (head rotation) are unaffected by the output (eye rotation) and hence "unaware" whether the output is correct or not; i.e., the reflex operates open loop. If these compensatory eye movements do not have the correct magnitude, then fixation will be disturbed. The adequacy of reflexes such as the RVOR is often assessed from their *gain,* which is given by the amplitude ratio of the output and the input. Clearly, open-loop reflexes perform their functions best when appropriately calibrated, i.e., have a gain close to unity, and it is this aspect of RVOR function that will be our major concern here because it is very susceptible to optical interference.

RVOR Adaptation

Many studies on a variety of species including man have demonstrated that there is a special neural mechanism responsible for ensuring that the gain of the RVOR is appropriately calibrated. For recent reviews see Miles (1983, 1986, 1987). These studies indicate that if the visual input associated with head turns is artificially altered with optical devices such as reading spectacles, then, over a period of days, the gain undergoes changes that are said to be adaptive because they compensate for the imposed optical challenge. I start by briefly considering current ideas about the adaptive mechanism before concentrating on the effects of various optical devices.

Accurate calibration of the RVOR requires reliable "error" information about its performance. Any shortcomings in the gain of the RVOR will result in persistent retinal slip during head turns, and the system might, in principle, use the associated visual input to diagnose the gain error. Thus, if the gain is too low the world will be seen to drift in the direction opposite to that in which the head is moving, and if the gain is too high then the reverse will occur. Clearly, the directional relationship between retinal slip and head movement reliably indicates the direction in which the gain needs to change to improve image stabilization, and all early models of the adaptive mechanism assumed that RVOR gain errors were sensed by matching retinal slip and vestibular signals directly (Ito, 1972; Davies and Melvill Jones, 1976; Gonshor and Melvill Jones, 1976; Robinson, 1976). However, in theory, the system is not obliged to use raw visual and vestibular signals to deduce the gain error—any reliable central correlates of the same would suffice (Miles and Lisberger, 1981). It is probably also safe to assume that the system is extremely resourceful and makes use of whichever of the several possible cues happens to be available.

Some two decades have elapsed since various theoreticians suggested that the cerebellum was an important site for motor learning, drawing a clear distinction between the two major types of input to the cerebellar cortex: *mossy fibers*, which were assumed to carry the bulk of the information used in the moment-by-moment computations whereby the cerebellum coordinates complex patterns of motor behavior, and *climbing fibers*, which were assumed to exert a long-term modulatory influence on the mossy-fiber pathway through the cortical networks (Marr, 1969; Grossberg, 1969; Albus, 1971). Cerebellar lesions, particularly those that involve the vestibulocerebellum (flocculus, nodulus, uvula), have been shown to cause severe deficits in adaptive gain control of the RVOR in all species that have been tested (Robinson, 1976; Schairer and Bennett, 1980; Nagao, 1983; Lisberger *et al.*, 1984; Michnovicz and Bennett, 1987). Similar deficits also follow lesions of the dorsal cap of the inferior olive, which normally supplies the climbing-fiber inputs to the flocculus (Haddad *et al.*, 1980; Ito and Miyashita, 1975).

There is general agreement that the cerebellum plays a critical role in the adaptive regulation of the RVOR. The current wisdom is that the flocculus is particularly important, and two hypotheses have been proposed to account for its involvement. One is based on the earlier ideas about the cerebellum's involvement in motor learning, invoking the flocculus as a side loop of the RVOR that relays vestibular signals (borne by mossy-fiber inputs) through cerebellar cortical networks containing modifiable synapses whose long-term efficacy is regulated by climbing-fiber inputs (Ito, 1972). In this model, errors in RVOR gain are assumed to be signaled directly by the retinal image slip associated with head rotations and sensed by the climbing fibers. Since the modifiable elements underlying plasticity in this model are assumed to be located within the cerebellar cortex, it has been termed the "cerebellar" hypothesis. The second hypothesis suggests that the flocculus is solely concerned with providing the complex error information needed to calibrate the RVOR gain and is assumed to bring about appropriate changes in the efficacy of modifiable synapses that are located in the brainstem RVOR pathways (Miles *et al.*, 1980; Miles and Lisberger, 1981). Since the modifiable elements underlying plasticity in this model are located within the brainstem, it has been termed the "brainstem" hypothesis. At present, conclusive evidence for either of these two hypotheses is lacking.

Telescopic Spectacles

In reviewing the effects of various optical devices on the operation of the RVOR, I start with the extreme challenge of telescopic spectacles. A major reason for this is that much of the research on the adaptability of the RVOR has been done using these devices, which elicit large changes in the gain of this reflex; I then go on to consider the more modest problems created by the commonly prescribed spectacle aids. That the RVOR can be altered by optical devices can be seen in Fig. 13-1B, which shows the eye movements of a monkey that has worn ×2 telescopic spectacles for several days: when the spectacles are removed, the animal has great difficulty maintaining stable gaze during head turns. (Note that these tests were performed in total darkness except for the small fixation target lights.) It is evident from the gaze trace in Fig. 13-1B that the animal successfully acquires the fixation target with his initial saccade but then overcompensates for the accompanying head movement. The animal's problem is that his RVOR gain has increased to be in accordance with the requirements of the ×2 spectacles and hence is now too large for the normal viewing situation. In fact, the animal makes a number of catch-up saccades in a vain attempt to maintain fixation on the newly acquired target, and it isn't until the head has come to a stop that he is successful. This serves to demonstrate not only the adaptive capability of the RVOR but also how critical it is that the RVOR should have a gain that is appropriate for the viewing conditions. If the telescopic spectacles are turned around they become ×0.5 minifying spectacles and gradually elicit an appropriate decrease in the gain of the RVOR.

A critical point here is that the telescopic spectacles only interfere with the functioning of the oculomotor system *during head movements*. Thus, acquiring and tracking targets are not affected by the telescopes when the head is still: a target that appears to be 10° off to one side when seen through the ×2 spectacles can be acquired with a 10° saccade despite the fact that it is actually only 5° off to one side; likewise, a target that is seen to be moving at 10°/sec can be tracked successfully with a matching eye velocity of 10°/sec despite the fact that the target is actually moving at 5°/sec. Only when the head moves is the oculomotor system confronted with a new challenge: if the head rotates at, say, 10°/sec, the world seen through the spectacles will move at twice that rate, and, in order to keep the images on its retina stable, the animal would have to generate compensatory eye movements that have twice their normal velocity.

The magnification factors typically used to demonstrate the adaptive capability of the RVOR in the laboratory are quite large (×2 and ×0.5), and during the initial adaptation period the subject is hampered because he is unable to keep his retinal images stable when he turns his head. Although there are visually driven reflexes that would help the subject to track the moving images seen through the spectacles, these have severely limited dynamic capabilities and so are never as effective as an appropriate RVOR: there is no substitute for calibrating the reflex.

Spherical Spectacle Lenses

After the prescription for their spectacles has been changed, patients often report slight disorientation, sometimes accompanied by nausea, during head turns. This generally lasts no more than a few days as the RVOR (and other visuomotor systems) gradually adapts to the slight change in optical magnification. As already mentioned, this problem was beautifully described more than half a century ago and correctly attributed to the RVOR (Ronne, 1923). As usually fitted (15 mm in front of the cornea), spectacle lenses have only a weak magnifying effect, given by the expression $40/(40 - D)$, where D is the power of the lens in diopters (Rubin, 1974). This means that positive lenses increase the magnitude of the compensatory eye movements required to stabilize the images seen through the spectacles during head turns by about 3% per diopter, whereas negative lenses decrease the required eye movements by about 2% per diopter. In view of the adaptive capability of the RVOR, it is not surprising that, among spectacle wearers, myopes tend to have lower RVOR gains than hyperopes, though the differences are quite small (Cannon *et al.*, 1985).

Although the extreme challenge of telescopic spectacles leads to rather gradual changes in RVOR gain that generally take days to complete, more modest stresses such as that caused by 5-D reading spectacles, for example, can be largely compensated within an hour if the subject diligently turns his head from side to side repeatedly to maximize the visuovestibular interaction necessary for adaptation (Collewijn *et al.*, 1983). Of course, most patients might be expected to curtail their head movements after acquiring new spectacles, particularly if they experience nausea, and this might be expected to slow their rate of adaptation. Thus, even though they have the potential for compensating within hours, it seems that it often takes several days to complete.

It is well known that contact lenses generally do not cause such severe orientation problems, and, of course, one reason is that contact lenses have a much smaller effect on magnification. However, even if this were not so, the contact lenses would still not affect visual stability during head turns because they move with the eyes. Thus, if a person wearing contact lenses rotates his head 10°, then, as usual, a compensatory eye rotation of 10° will keep that person's gaze aligned on the fixation target regardless of any magnifying effect of the lenses (compare A and B in Fig. 13-2). The critical factor that causes spectacle lenses to interfere with the operation of the RVOR is that they move with the head, making it necessary for the subject to view the images of the fixation target through eccentric regions of the lenses where there is a prismatic effect (see Fig. 13-2C). That contact lenses do not alter the

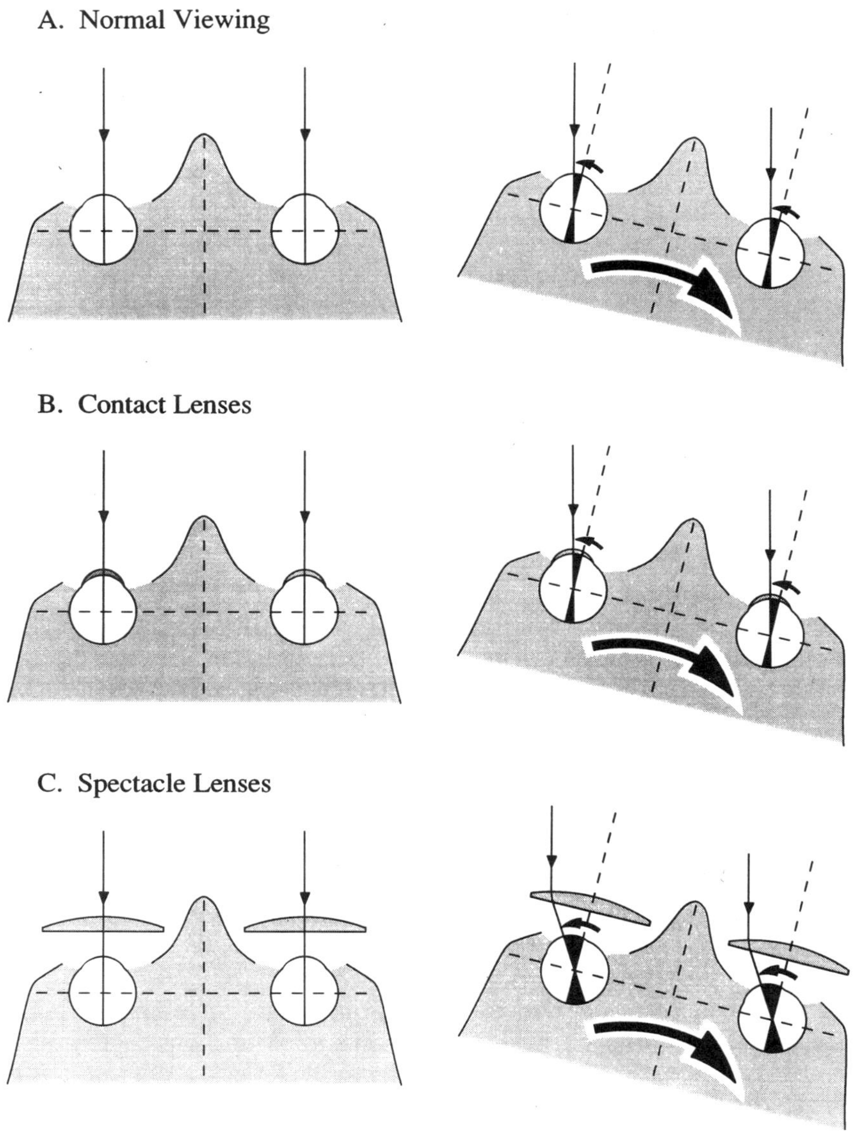

FIGURE 13-2. The effect of optical devices that move with the head or eyes on the eye movements required to compensate for head turns. In the normal situation, compensatory rotation of the eyes must simply match the rotation of the head (A), and this is also true for the subject wearing contact lenses (B). However, spectacle lenses move with the head, and the required compensatory eye movements are altered by the prismatic effect associated with eccentric viewing through the lenses (C). In the case of the positive lenses shown here, compensation must be increased by about 3% per diopter.

compensatory eye movements required to keep retinal images stable during head turns and, therefore, do not affect the operation of the RVOR is one clear way in which they are superior to spectacle lenses.

The need of the presbyopic patient for increasingly positive lenses with advancing age raises the issue of whether the RVOR retains its adaptive capability throughout life. Several studies have looked into this, and all have failed to find any significant deterioration with age in the capacity to adapt (Gonshor and Melvill Jones, 1976; Sekine, 1983; Demer *et al.*, 1989), though it would not be surprising if the rate of adaptation were affected.

Cylindrical Spectacle Lenses

The optical challenge offered by spherical lenses is largely independent of the direction or axis of the head rotation. In these circumstances it is appropriate for the RVOR gain to adapt uniformly for head rotations about any axis. However, the same would not be true of the cylindrical lenses used to correct astigmatism. The effective magnification will now depend on the angle between the axis of the cylinder and the axis of the head rotation (θ), and the relevant lens power is given by the expression, $D_{sph} + (D_{cyl}\cos\theta)$, where D_{sph} and D_{cyl} are the powers of the spherical and cylindrical corrective lenses in diopters. Thus, head rotations about the axis of the cylinder will require additional compensation to the extent of the full power of the cylinder, whereas head rotations about an axis that is orthogonal to this will require no supplementary compensation.

To date, there have been no studies to see if the adaptive mechanism in the RVOR can respond appropriately to the asymmetric challenge of cylindrical lenses. However, the system's response to other devices such as reversing prisms suggests that it should be quite capable of this. Thus, left–right reversing dove prisms present a highly asymmetric challenge to the RVOR, which is most readily understood by considering the visual consequences of head rotations about the three principal orthogonal axes: in addition to reversing the visual input associated with horizontal (yaw) rotations of the head (occurring about a vertical axis and stressing the horizontal RVOR), they also reverse the visual input associated with rolling the head from side to side (head rotating about a front-to-back horizontal axis, stressing the torsional RVOR) but do not alter the visual input associated with pitching the head backwards and forwards (head rotating about a side-to-side horizontal axis, stressing the vertical RVOR). The human nervous system is able to respond appropriately to this asymmetric optic challenge, selectively attenuating the compensatory eye movements associated with head rotations about the yaw and roll axes while leaving those concerned with the rotations about the pitch axis unchanged (Gonshor and Melvill Jones, 1976; Berthoz *et al.*, 1981).

Such directional selectivity in the adaptive response is evident in an even more extreme form in experiments on cats using cross-axis stimulation. In these experiments the animal was oscillated vertically about the pitch axis while the visual surroundings were oscillated horizontally, and this resulted in the gradual emergence of a horizontal component in the compensatory eye movements associated with subsequent test rotations about the pitch axis in the dark (Schultheis and Robinson, 1981). Such orthogonal components are also evident in the vestibuloocular responses of human subjects whose visual input has been rotated 30° in the frontal, or roll, plane using a pair of rotated dove prisms (Callan and Ebenholtz, 1982). All of these observations suggest that the adaptive mechanism in the RVOR should be sufficiently versatile to deal with the asymmetric challenge offered by cylindrical spectacle lenses.

Multifocal Spectacle Lenses

In discussing the modifiability of the RVOR I have stressed the adaptive nature of the system whereby changes in gain are always directed toward the goal of improving retinal image stability during head turns. Another important facet of the system is that it seems to have memory, and, although its performance is modifiable—when inappropriate—over time, it has the ability to retain the modified state for long periods in the absence of feedback. Thus, animals that have adapted to telescopic or reversing prism spectacles retain the adapted state if kept in darkness or if their head is immobilized, both being procedures that prevent the CNS from obtaining feedback information about the RVOR gain (Robinson, 1976; Miles and Eighmy, 1980). For this reason the system is often termed *plastic*, signifying that it can be molded to a new "form" and then retains this "form" until actively remolded. In fact, it is this enduring quality that has led to the suggestion that changes in the gain of the reflex must result from semipermanent changes in the synaptic relays within the CNS and has resulted in the RVOR becoming one of the model systems for studying "modifiable synapses" in the central nervous system. This all implies that the gain of the RVOR in the short term is fixed and immutable, which raises a problem for patients wearing bi- or trifocal spectacles, since the magnification—and hence the required ocular compensation—varies from one sector of the lens to another. There have been no direct studies of this problem, so that it is not known if the system can deal with the several different magnifications on a moment-by-moment basis. However, there are data that suggest that the system might be able to rise to this challenge by modulating the gain of the RVOR.

The compensatory eye movements generated by humans in the usual RVOR test situation—involving oscillations about a vertical axis in total darkness—are very sensitive to the instructions given to the subject (Barr *et al.*, 1976; Baloh *et al.*, 1984; McKinley and Peterson, 1985). For example, if the subject is asked to imagine that he is trying to fixate a target that is moving with him in the dark, his compensatory eye movements are considerably reduced in magnitude; if the imagined target is stationary, then his compensatory eye movements increase. Clearly, humans can adopt mental strategies that can have a profound influence on the apparent "gain" of the RVOR, and one must always consider this cognitive contribution before concluding that the gain of the basic reflex has been altered by exposure to spectacles. Presumably, these immediate changes that can be brought about by cognitive strategies involve modulation rather than modification of the RVOR. This suggests that, in the short term, there are some aspects of the reflex that are fixed and immutable while other aspects can be altered in accordance with the immediate visual challenge. From our present standpoint, this ability to modulate the performance of the RVOR rapidly seems to offer a solution to the problem of bi- and trifocal spectacles. Although there have been no direct studies of this, there are some data from deep-sea divers, who have a related problem, that might offer some insights.

Professional deep-sea divers spend considerable time viewing the world through goggles with an air–glass–water interface that magnifies the image approximately 35%. When their RVOR is tested in darkness, these divers appear to generate compensatory eye movements that are at times appropriate for terrestrial viewing and at other times more suited to underwater viewing (Gauthier and Robinson, 1975). Presumably, this bistable behavior results from a learned strategy that enables the subject to modulate the performance of the RVOR according to the prevailing optical conditions. It would be interesting to know if patients with multifocal lenses show compensatory eye movements appropriate for the multimagnifications.

Before leaving the problem of adapting to optical conditions that vary from moment to moment, I should mention the recent experiments of Baker *et al.* (1987) who used the cross-axis RVOR adaptation paradigm mentioned earlier, combining vertical (pitch) vestibular stimulation with horizontal visual motion, but with the animal (cat) lying on its side. These workers added a further variable to the adaptation paradigm by alternating the animal's position between left-ear-down and right-ear-down at regular intervals and by arranging for the visual motion to be in opposite directions for the two body positions. During adaptation, matters were arranged so that when the animal was on its left side, nose-up head motion was combined with leftward visual motion, and when the animal was on its right side, nose-up head motion was combined with rightward visual motion. In this situation, the orthogonal (horizontal) "compensatory" eye movements that gradually developed during vertical (pitch) canal stimulation with the animal on, say, its left side were 180° out of phase with those when the animal was on its right side. This suggests that the RVOR can be adapted to produce two different (oppositely directed) responses depending on the animal's posture. Once more the system demonstrates the ability to alter its performance rapidly to suit the circumstances. Whether this postural dependence derives from otolith input or some other source is not known.

Corrected Anisometropes

Where the two eyes have different spectacle corrections, as in the anisometrope, the RVOR is confronted with a challenge that is not equal for the two eyes. No one has measured the RVOR in such patients, but Collewijn *et al.* (1983) have asked the question, "Can the gain of the RVOR be adjusted independently for each eye?" These workers examined the effect of spectacles with a +5-D lens for one eye and a −5-D lens for the other. One of the two subjects investigated had appreciable binocular blurring, and his RVOR showed a very slight conjugate drop with almost no differential adaptation. The second subject could use accommodation to achieve a sharp retinal image through the negative lens, and his RVOR gain also showed a conjugate decrease that in this case was appropriate only for the eye with good vision. Before concluding from this that anisometropes who wear spectacles with different power lenses for the two eyes may not be able to achieve the appropriate differential adaptation, some important limitations of this study should be pointed out. First, only two subjects were investigated. Second, it is possible that the 10-D difference between the optical challenge to the two eyes may simply have exceeded the system's capacity for correction, and more modest demands might be manageable. Third, the subjects were only followed for 24 hr. (It should be said that 24 hr was quite sufficient for these same subjects to show good adaptation to +5-D or −5-D lenses when nonconflicting for the two eyes.)

There is evidence that monkeys can adapt their RVOR selectively for each eye: merely patching one eye causes small changes in the RVOR of that eye only, and, following removal of the patch, that same eye shows selective, i.e., monocular, recovery (Viirre *et al.*, 1987). Further evidence for such selectivity comes from experiments on monkeys in which the horizontal recti for one eye were detached (Snow *et al.*, 1985). This procedure temporarily eliminates the horizontal RVOR in that eye, but the muscles gradually reattach themselves to the globe, and the RVOR recovers. If during this recovery period the normal eye is patched, then it too shows some increase in RVOR gain, indicating that some part of the recovery of the operated eye must be through an increase in the gain of the RVOR within the CNS—something that affects both eyes. However, this is of less interest to us than the fact that the animal now has one eye with normal RVOR gain—the operated eye—and one with a higher than normal RVOR gain—the unoperated eye. Subsequent removal of the patch, providing the animal with binocular vision, results in selective nonconjugate changes that reestablish a normal RVOR gain for both eyes (Snow *et al.*, 1985).

In view of these findings in monkeys it would be interesting to reexamine the issue of differential adaptation of the RVOR for each eye in human subjects. One might start by measuring the RVOR gain in both eyes of anisometropes who have been corrected with spectacle lenses: the eye with the more positive lens would be expected to have the higher RVOR gain. It would also be pertinent for such a study to include astigmats who have cylinders that differ for the 2 eyes: differences in either power or axis would require different directional asymmetries in the two eyes.

Wedge Prisms

Wedge prisms such as those sometimes used to compensate for comitant deviations, paralytic strabismus, and so forth do not have any effect on magnification and, therefore, would not be expected to alter the magnitude of the compensatory eye movements required to offset head rotations. Accordingly, wedge prisms would not be expected to have any direct influence on the operation of the RVOR. It is perhaps not surprising, therefore, that there have been no studies of the effects of such prisms on the RVOR. Nonetheless, wedge prisms can have a disorienting effect during head turns. Some of this might be caused by the well-known optical distortions produced by the prisms, but I suggest that some may also be an indirect consequence of their effect on the convergence angle of the two eyes. In order to understand this it is necessary to consider the optical consequences of head turns in more detail.

Since the axis of rotation of the head lies some distance behind the eyes, head rotations cause slight translation of the eyes. As a consequence, full compensation for head turns requires eye movements to offset both the rotation of the head and the translation of the eyes. The translational component of head turns is only of significance with near viewing—a factor that I deal with more fully later when I discuss the TVOR—and there is good evidence that the compensatory mechanism actually does take this translation into consideration: the gain of the RVOR increases with the proximity of the object of regard (Biguer and Prablanc, 1981; Blakemore and Donaghy, 1980; Gresty and Bronstein, 1986; Gresty *et al.*, 1987; Hine and Thorn, 1987; Viirre *et al.*, 1986). One attractive hypothesis is that this effect of viewing distance is achieved by direct modulation of the RVOR by some central correlate of the vergence angle between the two eyes. Support for this idea comes from the experiments of Hine and Thorn (1987), who showed that wedge prisms altered the gain of the compensatory eye movements in proportion to their effect on the vergence angle of the two eyes; in contrast, changing the level of accommodation with lenses had no effect on compensatory eye movements. This sensitivity of the RVOR to the vergence angle of the two eyes could account for some of the disorienting effects experienced by patients who move about their environment wearing wedge prisms: base-out prisms would be expected to cause the subject to overcompensate for head turns, whereas base-in prisms would be expected to cause undercompensation. Of course, these disorientation problems tend to ameliorate in a few days, suggesting that the RVOR can adapt its sensitivity to viewing distance, but, once more, there have been no formal studies of this.

THE TRANSLATIONAL VESTIBULOOCULAR REFLEX

Like rotational disturbances of the head, translational ones also excite the vestibular apparatus, but a different part: the otolith organs. These sensors, which respond to translational accelerations and are insensitive to rotations (Fernandez *et al.*, 1972; Fernandez and Goldberg, 1976), also activate central neural pathways that give rise to compensatory eye movements at short latency.* A recent study estimates the latency of the TVOR of human subjects to be 34 msec (Bronstein and Gresty, 1988). Like the RVOR, the TVOR operates open loop, and hence its performance is very sensitive to the gain of its central neural pathways. An important functional consideration when assessing the performance of the TVOR arises from the fact that the magnitude of the eye movements required to compensate for translational disturbances of the head depends critically on the three-dimensional structure of the visual scene. This effect is most readily appreciated during pure lateral translation as when looking out from a fast moving train, and I have attempted to illustrate this in Fig. 13-3. From this it can be seen that the retinal image motion caused by translation of the observer's eye depends crucially on the proximity of the external objects and includes complex image shear as the nearby objects move across the field of view much more rapidly than the more distant ones: motion parallax. It follows that the compensatory eye movements required to stabilize the retinal image of any particular object in the scene must depend on how near that object is to the observer: for appropriate operation the gain of the TVOR should be modulated inversely with the viewing distance. Indeed, if the object of interest is at a considerable distance, as in the case of the mountains in Fig. 13-3, negligible compensatory eye movements would be required (at least in the short term), and a TVOR response would be disruptive.

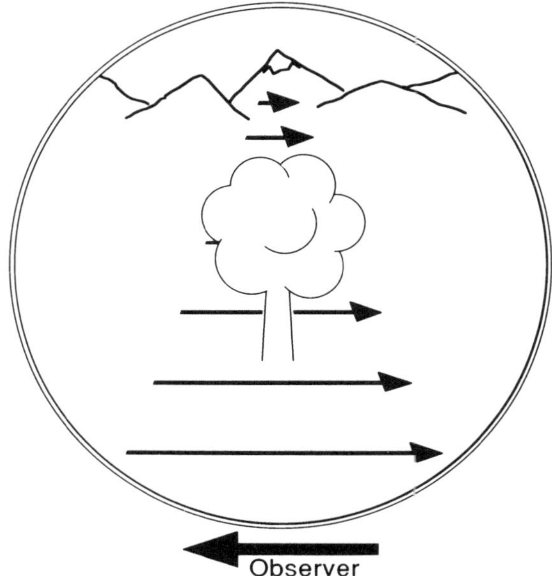

FIGURE 13-3. The visual motion experienced by an observer during lateral translation (as when looking from a train) depends on the proximity of the object. If the observer wishes to scrutinize a foreground object, then he must track it with his eyes, thereby compensating for the motion of the train. The magnitude of the required compensatory eye movement would clearly depend on how close the object of interest was to the railroad tracks: compensation for translational disturbances is a function of the proximity of the object.

The older reports in the literature suggested that the TVOR was rather weak in humans (Jongkees and Phillipszoon, 1962; Niven *et al.*, 1966; Buizza *et al.*, 1980). However, these studies did not specify proximal viewing, and it is possible that the TVOR defaulted to a low gain appropriate for distant viewing. It is now known that somewhat better responses are obtained if the subject attempts to fixate an imagined target that is nearby (Baloh *et al.*, 1988). A dependence on proximity is also suggested by a recent study on monkeys that found that the gain of the TVOR was correlated with the vergence angle of the two eyes (Paige *et al.*, 1988). Other recent experiments on monkeys indicate that the TVOR responses to lateral translation of the head are linearly related to the viewing distance expressed in reciprocal meters (Schwarz and Miles, 1989). How-

*Not surprisingly, the otolith organs are influenced by changes in orientation with respect to the earth's gravity. This can give rise to ocular counterrolling, an otolith–ocular reflex that acts to maintain the orientation of the eyes. We shall not be concerned with this reflex here, which is very weak in primates (Collewijn *et al.*, 1985).

ever, the TVOR generally overcompensated with distant viewing and undercompensated with near viewing. Further, using wedge prisms to dissociate vergence and accommodation, these workers found that the TVOR gain is just as sensitive (if not more so) to the accommodative state as it is to the vergence angle of the eyes (Schwarz and Miles, 1989). These data suggest that the system uses some internal measure of both the vergence and the accommodative states to modulate the TVOR gain in accordance with the viewing distance. Putative sources of absolute range-finding information that might be used to modulate the TVOR include neurons in the CNS that discharge in relation to vergence and/or accommodation (Mays, 1984; Mays *et al.*, 1986; Judge and Cumming, 1986).

When translational motion is intentionally combined with rotational (by rotating the subject about an eccentric axis some distance behind the head), then the compensatory eye movements increase exactly as required by the additional translation of the eyes (Viirre *et al.*, 1986; Gresty and Bronstein, 1986; Gresty *et al.*, 1987). Indeed, the compensation for the translational motion is much greater in this situation than in the purely translational one (Gresty and Bronstein, 1986). It is not clear at present why the TVOR is much better when there is concurrent activation of the RVOR.

It has been reported that translational acceleration in the fore–aft direction in total darkness produces vergence, though again the movements are rather small (Smith, 1985). Interestingly, the compensatory eye movements during forward motion are gaze dependent, operating to increase any eccentricity of the eyes: if the observer's gaze is directed downwards during the forward motion, then his/her compensatory eye movements will be downward, whereas if gaze is directed to the right then the compensatory eye movements will be rightward, and so forth (Paige *et al.*, 1988). This is exactly in accordance with the optical requirements of the situation and would seem to require information related to the position of the eyes relative to the direction of travel, i.e., gaze rather than orbital position of the eyes.

TVOR Adaptation

The evidence for adaptive gain control in the TVOR is limited to one recent study on humans, which showed that the TVOR responses approximately doubled in magnitude in two of three subjects following 20 min of continuous sinusoidal translational motion in the light (Baloh *et al.*, 1988). However, the underlying mechanism is unclear, and more experiments are required in order to know whether the changes result from alterations in the basic reflex pathway or some modulatory process.

Formal studies of the effects of optical aids on visual experience during translational disturbances are sparse. Most of my comments, therefore, are based solely on a consideration of the optical geometry.

Spectacle Lenses

In considering the effects of optical aids on visual stability during rotational disturbances of the head, I made a sharp distinction between devices that are mounted in spectacle frames and so move with the head and others such as contact lenses that move with the eyes. The same distinction must be made when considering translational disturbances of the head: contact lenses do not affect the magnitude of the eye movements required to offset translational movement of the head (compare A and B in Fig. 13-4), but spectacle lenses do because of the prismatic effect of eccentric viewing (Fig. 13-4C). However, there is a most important caveat: this effect of spectacle lenses depends on the proximity of the object of regard. In fact, in the short term no compensatory eye movements are required when the object is very distant irrespective of whether the observer is wearing spectacles. Thus, the effects of spectacles on the visual experience associated with translational motion should be most marked for images of nearby objects. Of course, positive lenses would be expected to increase the magnitude of the compensatory eye movements required to maintain fixation on objects at any given proximity, whereas negative lenses would be expected to have the converse effect. It follows that cylindrical lenses would also be expected to interfere with the operation of the TVOR, the extent depending on the orientation of the cylinder with respect to the direction of the head movement (cf. the earlier discussion of the RVOR). There have been no formal studies of any of these problems, and it is not known whether, over a period of time, the TVOR would adapt successfully.

Wedge Prisms

It has been shown in monkeys that base-out prisms bring about an immediate increase in the response of the TVOR, whereas base-in prisms have the con-

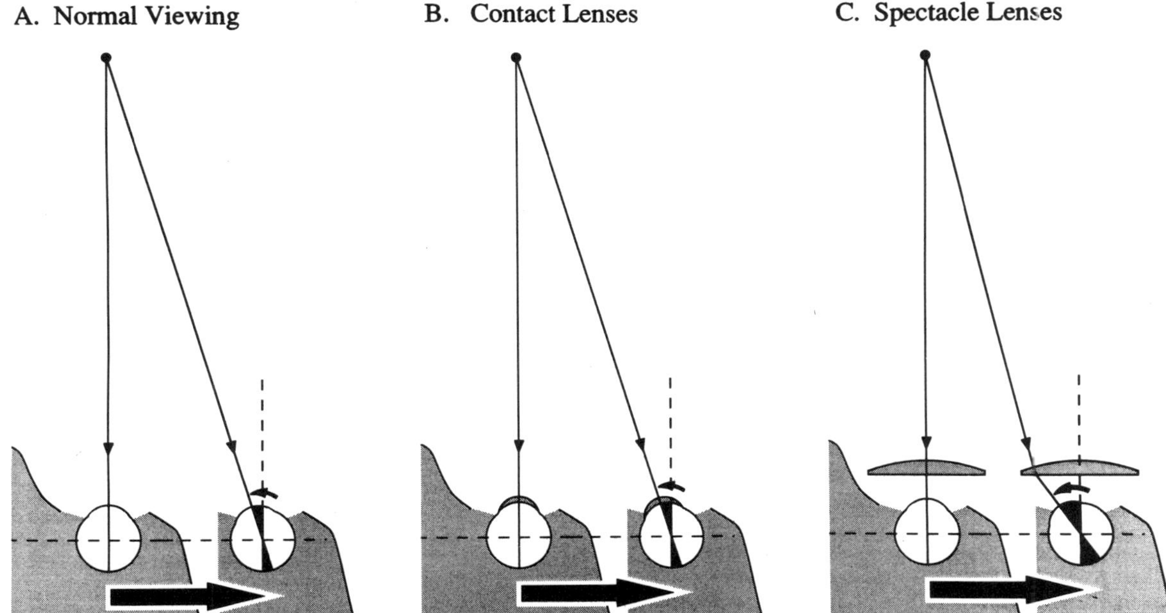

FIGURE 13-4. The effect of optical devices that move with the head or eyes on the eye movements required to compensate for lateral translation of the head. In the normal situation (A), the compensatory rotation of the eyes required to offset the observer's motion depends on the proximity of the object of regard, and this is unaffected by wearing contact lenses (B). However, spectacle lenses move with the head, and the required compensatory eye movements are altered by the prismatic effect associated with eccentric viewing through the lenses (C).

verse effect (Schwartz and Miles, 1989). Presumably, these effects of prisms are secondary to their effects on the vergence angle between the two eyes, some central neural correlate of which is used, we assume, to modulate the gain of the reflex. Thus, base-out prisms would be expected to cause the TVOR to overcompensate for head translations, and base-in prisms the converse. This would explain the disruptive effects of wedge prisms on visual stability during lateral motion of the head. Whether the TVOR would adapt to the prisms if they were left in place for a prolonged period is not known.

SUMMARY

Images that move across the retina are often more difficult to recognize than stationary ones, and visual acuity begins to deteriorate if retinal images drift at more than a few degrees per second. One of the major functions of the oculomotor system is to keep retinal images reasonably stable. Head movements pose a serious threat to this stability but are dealt with by two independent reflexes that generate compensatory eye movements to offset rotations and translations and are referred to here as the rotational vestibuloocular reflex (RVOR) and the translational vestibuloocular reflex (TVOR), respectively. The success of the RVOR depends entirely on how well the compensatory eye movements match the rotation of the head. On the other hand, the compensatory eye movements required to offset translation of the head also depend on the proximity of the object of regard as well as the head movement, and the TVOR response must be—and is—a function of the inverse of the viewing distance. Unfortunately, although the RVOR has been studied intensively in man and animals, the TVOR has received very little attention.

The optical aids commonly used to compensate for presbyopia, myopia, astigmatism, and so forth can seriously interfere with the operation of these reflexes. For example, spherical lenses, when mounted in spectacle frames secured to the head, alter the magnitude of the compensatory eye movements required to offset the head movements in proportion to their magnifying effect on the visual image. As a consequence, subjects with new spectacle lenses may experience visual instability during head movements. A critical factor here is that the optical devices move with the head and affect the stability of retinal images only during head movements. If the devices move with the eyes, as in the case of contact lenses, then, regardless of whether they affect magnification, they do not alter the compensatory eye movements required to keep retinal images stable and hence do not interfere with the operation of the vestibuloocular reflexes. A further critical factor in the case of the TVOR is that the effect of spectacles depends on the proximity of the object: images of distant objects are little affected (in the short term) by translation, regardless of whether the subject is wearing spectacle lenses. Thus, the problem here is particularly acute when viewing nearby objects. The TVOR's sensitivity to viewing distance derives at least in part from central correlates of the vergence and accommodative states, and hence the operation of this reflex is also disrupted by wedge prism spectacles.

Fortunately, subjects successfully adapt to most optical aids, a process that can take several days and, in the case of the RVOR, is known to involve central recalibration of the reflex. It seems likely that the TVOR is also recalibrated centrally, but there have been no studies of this. Adaptive regulation of the RVOR has been studied intensively in animals and humans, and it is clear that it functions to adjust the reflex so as to minimize retinal slip during head turns. The adaptive process involves changes in the central nervous system and is known to be severely compromised by damage to the cerebellum, especially to the floccular/ parafloccular lobules. Two major hypotheses, which are not mutually exclusive, have been advanced to explain the role of the cerebellum in this adaptive process: one suggests that the cerebellum is a side loop of the RVOR pathway and contains the modifiable synapses that are generally thought to mediate the changes in performance; the other suggests that the cerebellum provides the error information to bring about appropriate changes in the efficacy of synapses located elsewhere in RVOR pathways (presumably in the brainstem).

REFERENCES

Albus, J. S., 1971, A theory of cerebellar function, *Math. Biosci.* **10**:25–61.

Baker, J. F., Perlmutter, S. I., Peterson, B. W., Rude, S. A., and Robinson, F. R., 1987, Simultaneous opposing adaptive changes in cat vestibuloocular reflex direction for two body orientation, *Exp. Brain Res.* **69**:220–224.

Baloh, R. W., Lyerly, K., Yee, R. D., and Honrubia, V., 1984, Voluntary control of the human vestibulo-ocular reflex, *Acta Otolaryngol.* **97**:1–6.

Baloh, R. W., Beykirch, K., Honrubia, V., and Yee, R. D., 1988, Eye movements induced by linear acceleration on a parallel swing, *J. Neurophysiol.* **60**:2000–2013.

Barr, C. C., Schultheis, L. W., and Robinson, D. A., 1976, Voluntary, nonvisual control of the human vestibulo-ocular reflex, *Arch. Otolaryngol.* **81**:365.

Berthoz, A., Melvill Jones, G., and Bégué, A. E., 1981, Differential visual adaptation of vertical canal-dependent vestibulo-ocular reflexes, *Exp. Brain Res.* **44**:19–26.

Biguer, B., and Prablanc, C., 1981, Modulation of the vestibulo-ocular reflex in eye–head orientation as a function of target distance in man, in: *Progress in Oculomotor Research* (A. F. Fuchs and W. Becker, eds.), Elsevier/North-Holland, New York, pp. 525–530.

Blakemore, C., and Donaghy, M., 1980, Co-ordination of head and eyes in the gaze changing behaviour of cats, *J. Physiol. (Lond.)* **300**:317–335.

Bronstein, A. M., and Gresty, M. A., 1988, Short latency compensatory eye movement responses to transient linear head acceleration: A specific function of the otolith–ocular reflex, *Exp. Brain Res.* **71**:406–410.

Buizza, A., Leger, A., Droulez, J., Berthoz, A., and Schmid, R., 1980, Influence of otolithic stimulation by horizontal linear acceleration on optokinetic nystagmus and visual motion perception, *Exp. Brain Res.* **39**:167–176.

Callan, J. W., and Ebenholtz, S. M., 1982, Directional changes in the vestibular ocular response as a result of adaptation to optical tilt, *Vision Res.* **22**:37–42.

Cannon, S. C., Leigh, R. J., Zee, D. S., and Abel, L. A., 1985, The effect of the rotational magnification of corrective spectacles on the quantitative evaluation of the VOR, *Acta Otolaryngol.* **100**:81–88.

Collewijn, H., Martins, A. J., and Steinman, R. M., 1983, Compensatory eye movements during active and passive head movements: Fast adaptation to changes in visual magnification, *J. Physiol. (Lond.)* **340**:259–286.

Collewijn, H., Van der Steen, J., Ferman, L., and Jansen, T. C., 1985, Human ocular counterroll: Assessment of static and dynamic properties from electromagnetic scleral coil recordings, *Exp. Brain Res.* **59**:185–196.

Davies, P., and Melvill Jones, G., 1976, An adaptive neural model compatible with plastic changes induced in the human vestibulo-ocular reflex by prolonged optical reversal of vision, *Brain Res.* **103**:546–550.

Demer, J. L., Porter, F. I., Goldberg, J., Jenkins, H. A., and Schmidt, K., 1989, Adaptation to telescopic spectacles: Vestibulo-ocular reflex plasticity, *Invest. Ophthalmol. Vis. Sci.* **30**:159–170.

Fernandez, C., and Goldberg, J. M., 1976, Physiology of peripheral neurons innervating otolith organs of the squirrel monkey. I. Response to static tilts and to long-duration centrifugal force, *J. Neurophysiol.* **39**:970–1008.

Fernandez, C., Goldberg, J. M., and Abend, W. K., 1972, Response to static tilts of peripheral neurons innervating otolith organs of the squirrel monkey, *J. Neurophysiol.* **35**:978–997.

Gauthier, G. M., and Robinson, D. A., 1975, Adaptation of the human vestibuloocular reflex to magnifying lenses, *Brain Res.* **92**:331–335.

Gonshor, A., and Melvill Jones, G., 1976, Extreme vestibulo-ocular adaptation induced by prolonged optical reversal of vision, *J. Physiol. (Lond.)* **256**:381–414.

Gresty, M. A., and Bronstein, A. M., 1986, Otolith stimulation evokes compensatory reflex eye movements of high velocity when linear motion of the head is combined with concurrent angular motion, *Neurosci. Lett.* **65:**149–154.

Gresty, M. A., Bronstein, A. M., and Barratt, H., 1987, Eye movement responses to combined linear and angular head movement, *Exp. Brain Res.* **65:**377–384.

Grossberg, S., 1969, On learning of spatiotemporal patterns by networks with ordered sensory and motor components. 1. Excitatory components of the cerebellum, *Stud. Appl. Math.* **48:**105–132.

Haddad, G. M., Demer, J. L., and Robinson, D. A., 1980, The effect of lesions of the dorsal cap of the inferior olive on the vestibuloocular and optokinetic systems of the cat, *Brain Res.* **185:**265–275.

Hine, T., and Thorn, F., 1987, Compensatory eye movements during active head rotation for near targets: Effects of imagination, rapid head oscillation and vergence, *Vision Res.* **27:**1639–1657.

Ito, M., 1972, Neural design of the cerebellar motor control system, *Brain Res.* **40:**81–84.

Ito, M., and Miyashita, Y., 1975, The effects of chronic destruction of the inferior olive upon visual modification of the horizontal vestibulo-ocular reflex of rabbits, *Proc. Jpn. Acad.* **51:**716–720.

Jongkees, L. B. W., and Phillipszoon, A. J., 1962, Nystagmus provoked by linear accelerations, *Acta Physiol. Pharmacol. Neerl.* **10:**239–247.

Judge, S. J., and Cumming, B. G., 1986, Neurons in the monkey midbrain with activity related to vergence eye movement and accommodation, *J. Neurophysiol.* **55:**915–930.

Lisberger, S. G., 1984, The latency of pathways containing the site of motor learning in the monkey vestibulo-ocular reflex, *Science* **225:**74–76.

Lisberger, S. G., Miles, F. A., and Zee, D. S., 1984, Signals used to compute errors in monkey vestibuloocular reflex: Possible role of flocculus, *J. Neurophysiol.* **52:**1140–1153.

Marr, D., 1969, A theory of cerebellar cortex, *J. Physiol. (Lond.)* **202:**437–470.

Mays, L. E., 1984, Neural control of vergence eye movements: Convergence and divergence neurons in midbrain, *J. Neurophysiol.* **51:**1091–1108.

Mays, L. E., Porter, J. D., Gamlin, P. D. R., and Tello, C. A., 1986, Neural control of vergence eye movements: Neurons encoding vergence velocity, *J. Neurophysiol.* **56:**1007–1021.

McKinley, P. A., and Peterson, B. W., 1985, Voluntary modulation of the vestibuloocular reflex in humans and its relation to smooth pursuit, *Exp. Brain Res.* **60:**454–464.

Michnovicz, J. J., and Bennett, M. V. L., 1987, Effects of rapid cerebellectomy on adaptive gain control of the vestibulo-ocular reflex in alert goldfish, *Exp. Brain Res.* **66:**287–294.

Miles, F. A., 1983, Plasticity in the transfer of gaze, *Trends Neurosci.* **6:**57–60.

Miles, F. A., 1986, Parametric adjustments in the oculomotor system, in: *The Oculomotor and Skeletalmotor System. Progress in Brain Research,* Vol. 64 (H.-J. Freund, U. Buttner, B. Cohen, and J. Noth, eds.), Elsevier, Amsterdam, pp. 367–380.

Miles, F. A., 1987, The role of the cerebellum in adaptive regulation of the vestibulo-ocular reflex, in: *Cerebellum and Neuronal Plasticity* (M. Glickstein, C. Yeo, and J. Stein, eds.), Plenum Press, New York, pp. 293–311.

Miles, F. A., and Eighmy, B. B., 1980, Long-term adaptive changes in primate vestibuloocular reflex. I. Behavioral observations, *J. Neurophysiol.* **43:**1406–1425.

Miles, F. A., and Fuller, J. H., 1974, Adaptive plasticity in the vestibulo-ocular responses of the rhesus monkey, *Brain Res.* **80:**512–516.

Miles, F. A., and Lisberger, S. G., 1981, Plasticity in the vestibulo-ocular reflex: A new hypothesis, *Annu. Rev. Neurosci.* **4:**273–299.

Miles, F. A., Braitman, D. J., and Dow, B. M , 1980, Long-term adaptive changes in primate vestibuloocular reflex. IV. Electrophysiological observations in flocculus of adapted monkeys, *J. Neurophysiol.* **43:**1477–1493.

Nagao, S., 1983, Effects of vestibulocerebellar lesions upon dynamic characteristics and adaptation of vestibulo-ocular and optokinetic responses in pigmented rabbits, *Exp. Brain Res.* **53:**36–46.

Niven, J. I., Hixson, W. C., and Correia, M. J., 1966, Elicitation of horizontal nystagmus by periodic linear acceleration, *Acta Otolaryngol.* **62:**429–441.

Paige, G. D., Tomko, D. L., and Gordon, D. B., 1988, Visual–vestibular interactions in the linear vestibulo-ocular reflex (VOR), *Invest. Ophthalmol. Vis. Sci. [Suppl.]* **29:**342.

Robinson, D. A., 1976, Adaptive gain control of vestibuloocular reflex by the cerebellum, *J. Neurophysiol.* **39:**954–969.

Ronne, H., 1923, False movements appearing during vision through spectacle glasses; their significance with respect to experience in wearing spectacles and their connection with the vestibular apparatus, *Acta Ophthalmol.* **1:**55–62.

Rubin, M., 1974, *Optics for Clinicians,* Triad Scientific Publishers, Gainesville, FL, pp. 247–248.

Schairer, J. O., and Bennett, M. V. L., 1980 Cerebellectomy in goldfish prevents adaptive gain control of the VOR without affecting the optokinetic system, in: *The Vestibular System: Functions and Morphology* (T. Gualtierotti, ed.), Springer-Verlag, New York, pp. 463–477.

Schultheis, L. W., and Robinson, D. A., 1981, Directional plasticity of the vestibulo-ocular reflex in the cat, *Ann. N.Y. Acad. Sci.* **374:**504–512.

Schwartz, U., and Miles, F. A., 1989, Translational vestibulo-ocular (TVOR) responses of monkey are a linear function of the inverse of the viewing distance, *Soc. Neurosci. Abstr.* **15**(2):783.

Sekine, S., 1983, Age changing effect on the vestibulo-ocular reflex in humans, *Pract. Otol.* **76:**1471.

Smith, R., 1985, Vergence eye-movement responses to whole-body linear acceleration stimuli in man, *Ophthal. Physiol. Opt.* **5:**303–311.

Snow, R., Hore, J., and Vilis, T., 1985, Adaptation of saccadic and vestibulo-ocular systems after extraocular muscle tenectomy, *Invest. Ophthalmol. Vis. Sci.* **26:**924–931.

Viirre, E., Tweed, D., Milner, K., and Vilis, T., 1986, A reexamination of the gain of the vestibuloocular reflex, *J. Neurophysiol.* **56:**439–450.

Viirre, E., Cadera, W., and Vilis, T., 1987, The pattern of changes produced in the saccadic system and vestibuloocular reflex by visually patching one eye, *J. Neurophysiol.* **57:**92–103.

Westheimer, G., and McKee, S. P., 1975, Visual acuity in the presence of retinal-image motion, *J. Opt. Soc. Am.* **65:**847–850.

14

Neurophysiology of Vergence and Accommodation

STUART J. JUDGE

INTRODUCTION

Although much is known about the behavior of accommodation and vergence from studies of human performance (reviewed by Alpern, 1969; Schor and Ciuffreda, 1983; Toates, 1972), comparatively little is known about the neural pathways and mechanisms that support the behavior (Mays, 1983). Our main interest has been in studying these neural mechanisms by means of single-unit recording in awake trained monkeys (*Macaca mulatta*), but we have also carried out purely behavioral studies of accommodation and vergence behavior in the monkey, partly to test to what extent monkey behavior is similar to that of humans and partly to attempt to clarify issues not completely resolved in the human literature. The results of these behavioral experiments are described first and then followed by the results of our own and others' neurophysiological experiments. Finally some consideration is given to open questions for future research.

BEHAVIOR OF VERGENCE AND ACCOMMODATION IN THE MONKEY

Accommodation

The latency of accommodation responses to sudden target movement in depth is considerably shorter in the monkey than in man, averaging 230 msec in three monkeys (Cumming, 1985) in comparison with 370 msec in humans (Tucker and Charman, 1979). Whether or not the difference in latency arises primarily from differences in the motor periphery is unknown, since we do not have the appropriate data for human subjects, but in the monkey peripheral factors account for some 120 msec of the latency since stimulation of the Edinger–Westphal nucleus results in accommodation responses with this latency (Judge and Cumming, 1986).

DYNAMICS IN MONOCULAR VIEWING

We have studied the accommodation responses induced by sinusoidal changes in monocular accommodation demand (with various amplitudes up to ±2 D) in five monkeys. Gain decreases and phase lag increases monotonically with frequency above 0.2 Hz, with phase lags typically about 90° with ±2 D 1-Hz stimulation. As in humans, responses are not invariant with respect to stimulus amplitude, and the responses to random rather than sinusoidal changes in accommodation demand are not identical (Cumming, 1985).

SIZE CHANGE

In our standard condition, the target is viewed monocularly via a Badal lens, so there is no change in target angular size as the target moves in depth. In one monkey we have studied the change in accommodation produced with the Badal lens removed (i.e., with a size-change cue). Gains were higher and phase lags shorter in such "direct" viewing compared with Badal viewing. Similar effects have been demonstrated by Kruger and Pola (1986, 1987) in human subjects.

CHROMATICITY

In our standard condition the target is polychromatically illuminated, but we have, in two monkeys, compared the accommodation responses

STUART J. JUDGE • University Laboratory of Physiology, University of Oxford, Oxford OX1 3PT, Great Britain.

in white and monochromatic (green) illumination with target luminance matched. The responses with monochromatic illumination generally show lower gain and longer phase lags (Flitcroft and Judge, 1988). Moreover, if the stability of accommodation is examined, it is seen that in monochromatic illumination accommodation shows greater instability, with the rise in fluctuation power most prominent in the low-frequency (<0.5 Hz) region (Flitcroft and Judge, 1988).

FLUCTUATIONS

We have shown (Flitcroft, 1988) that the monkey's accommodation, like the human's, is subject to prominent fluctuations whose power falls into two bands, one near 0 Hz and the other at approximately 2 Hz, i.e., not at a higher frequency than in man, as one might have expected from the shorter latency of accommodation in the monkey if the fluctuations were the result of too high a gain in the overall feedback loop. As in man, the 2-Hz fluctuations increase as accommodation increases. We also find that the fluctuations at 2 Hz are somewhat smaller in binocular viewing than in monocular viewing and essentially absent in accommodation open-loop binocular viewing (achieved by using the measured accommodation response to control target position so as to eliminate blur).

This suggests that steady-state accommodation in binocular viewing is supported largely by blur but that vergence accommodation makes a contribution. This was an unexpected finding in view of the well-known potency of vergence accommodation in steady-state viewing (e.g., Fincham and Walton, 1957). The fluctuations in accommodation are accompanied by correlated changes in accommodative vergence. Wilson's (1973) claim to the contrary in humans is undermined by methodological problems in the analysis of his data. Schor and Kotulak (1986) have suggested that accommodative vergence is not driven by all of the monocular cues to accommodation. However, we have shown that chromatic cues and monochromatic cues contribute to accommodative vergence in the monkey in the same proportion as they do to accommodation (Flitcroft, 1988).

BINOCULAR VIEWING

In binocular viewing the dynamic responses of accommodation are superior to those in monocular viewing. By using the instantaneous measured accommodation response of the right eye to control the position of the targets (see Fig. 14-1), we were able to deprive the monkeys of blur cues and study the dynamics of vergence- (or disparity-) induced accommodation. In four of the five monkeys studied this was excellent, superior to that of blur-driven accommodation, suggesting that the dynamics of accommodation in binocular viewing are largely determined by the disparity-driven input (Cumming and Judge, 1986; B. G. Cumming and S. J. Judge, unpublished data). Figure 14-2 shows (in a Bode plot) the superior dynamics of such vergence–accommodation for one monkey.

BINOCULAR RIVALRY IN ACCOMMODATION DEMANDS

It has long been known that the accommodation response is very similar in the two eyes, but it does not appear ever to have been investigated how this *consensual response* is constructed from stimuli that must, in general, not be identical in the two eyes. Flitcroft (1988) and Flitcroft and Judge (1989) have investigated, in both man and monkey, how the accommodation system responds to dynamic conflict between the accommodation demands in the two eyes. If the targets seen by the two eyes move sinusoidally in antiphase, remarkably little accommodation occurs, suggesting linear summation. On the other hand, if only one target moves in depth while the other is stationary, again very little accommodation response occurs, whichever eye views the stationary target. If the two targets move in quadrature (i.e., with a phase difference of 90°), the phase of the accommodation response is close to that expected on the basis of linear summation, but the gain is much lower than expected. When rivalrous stimuli were presented to the two eyes (square-wave gratings of orthogonal orientation), moving in depth in antiphase, the accommodation response followed the perceptually dominant eye. This suggests that the site of the interaction of the blur signals from the two eyes lies after the site of binocular rivalry.

Vergence

The latency of vergence in the monkey seems to be very similar to that in man (about 160 msec). Moreover, if a disparity step is imposed and then maintained by electronic feedback, the vergence system responds by producing a (delayed) ramp, just as in

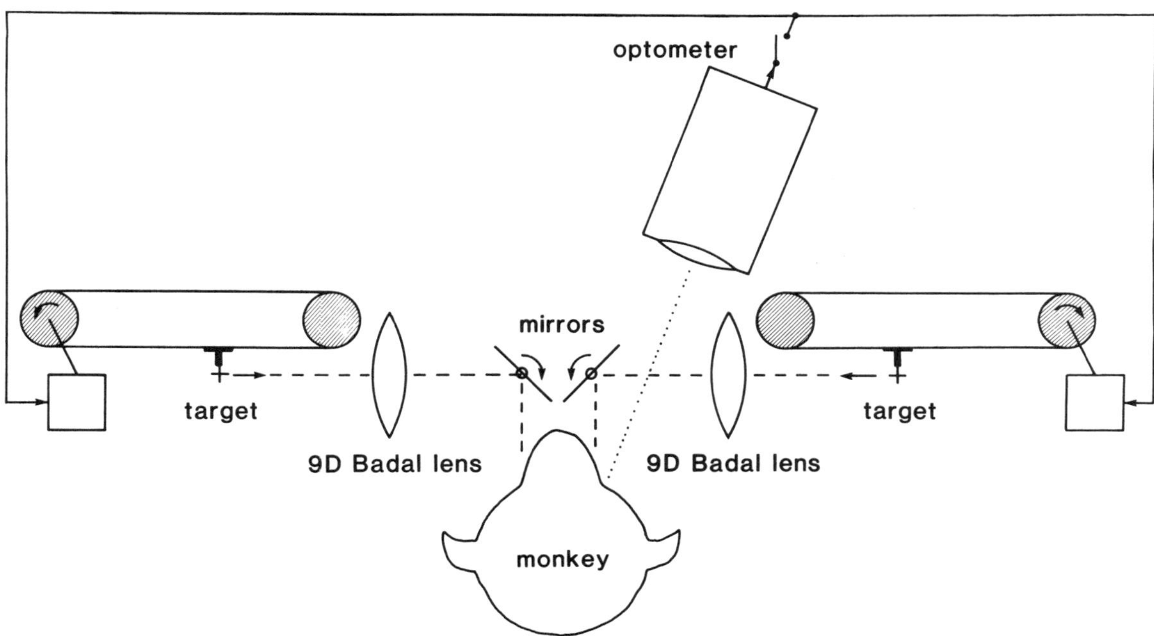

FIGURE 14-1. Experimental setup. The target was presented haploscopically. The two eyes viewed identical targets via mirrors mounted on angular servomotors and 9D Badal lenses (lenses placed so that their near focal plane was at the principal plane of the eye). The targets were mounted on linear servos. The positions of the targets and the angles of the mirrors were controlled by a PDP 11/34 computer. The accommodation of the right eye was measured with a Scheiner principle infrared optometer, and the position of each eye was measured using implanted magnetic search coils (not shown). In some experiments the signal from the optometer was used to drive the target servos.

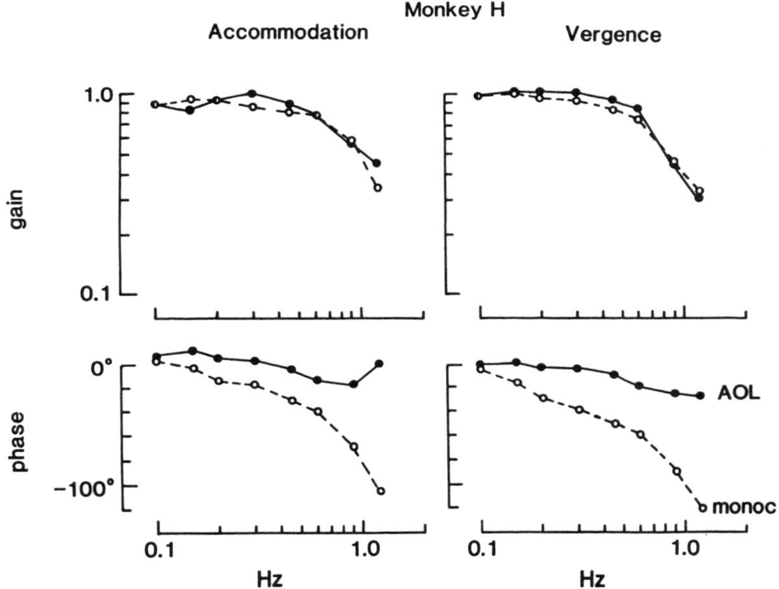

FIGURE 14-2. Bode plots of accommodation and vergence performance for monkey H in response to 4D (MA) peak-to-peak sinusoidal accommodation and vergence stimuli at frequencies from 0.1 to 1.2 Hz. Open circles show responses during monocular viewing; filled circles show responses during binocular viewing with accommodation open loop (AOL). Each point is based on at least three cycles of data (average nine cycles). Standard errors are smaller than the data points (typically ±0.03° for gain, ±2° for phase) unless explicitly shown The accommodation and vergence responses show better dynamics in binocular AOL viewing than in monocular viewing. The performance in binocular AOL viewing is similar to that in normal binocular viewing (not shown).

humans (Cumming and Judge, 1986; Rashbass and Westheimer, 1961). The relationship between step size and vergence velocity is also similar to that in humans, with vergence velocities of 5–10°/sec per degree of disparity. As in humans, however, the phase lags of the response of the vergence system to sinusoidal changes in disparity are shorter than one would expect from a system whose open-loop transfer function consists simply of an integrator and a delay of 160 msec. Interestingly, in at least one monkey we did see a slight "resonance" (i.e., gain greater than unity) in the responses to sinusoidal stimulation with a frequency of 1.5 Hz (Cumming, 1985), as one would expect from a system with the above characteristics. As with accommodation, responses were not invariant with respect to amplitude.

Accommodation vergence, as in man, had inferior dynamics to disparity-driven vergence.

Dynamically Conflicting Accommodation and Vergence Demands

We have been interested in the extent to which monkeys can dissociate accommodation and vergence. Initially we found it difficult to get monkeys to alter accommodation dynamically (sinusoidally) while vergence demand was static, but more recently we have found that at least some monkeys can do such "relative accommodation" tasks, albeit imperfectly. Varying vergence demand sinusoidally while keeping accommodation fixed ("relative vergence") seems to be an easier task, but again one that is not performed perfectly, accommodation tending to follow vergence. Judge and Cumming (1986) used a related conflict task in which the amplitudes of the accommodation and vergence demands were in a 2:1 ratio to partially dissociate accommodation and vergence responses. More re-

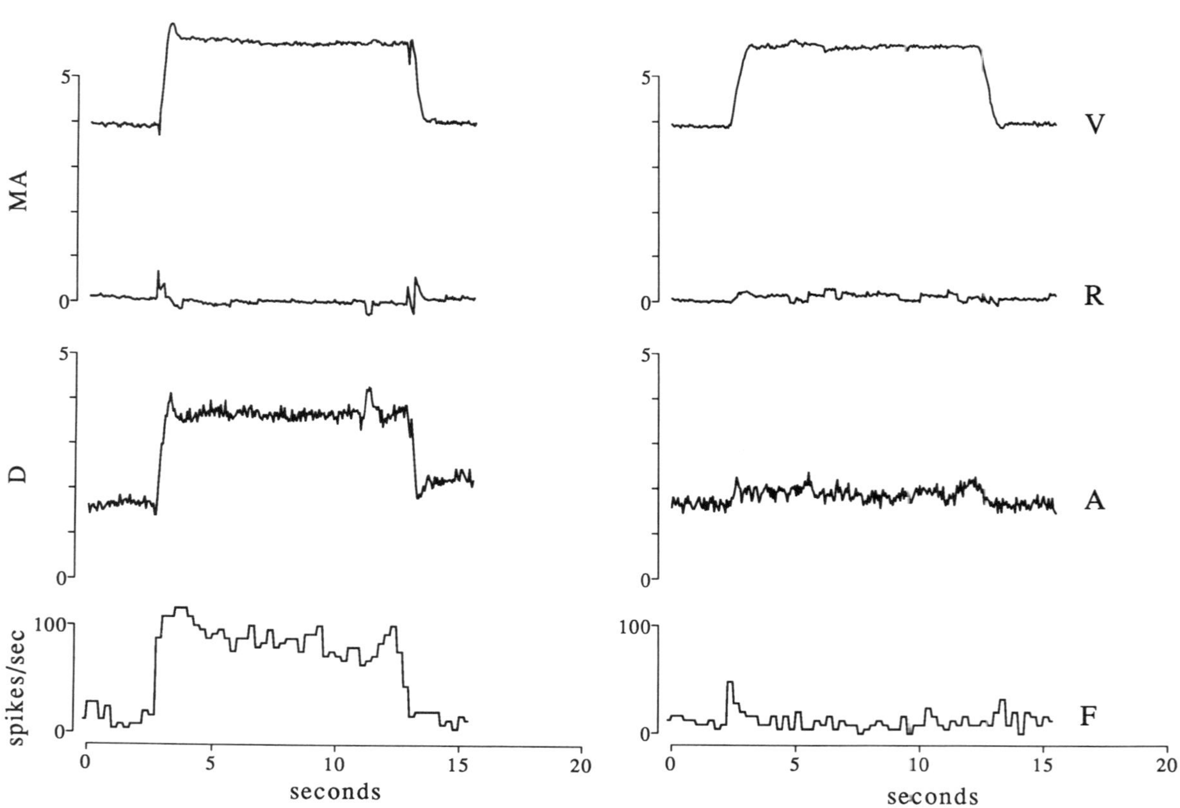

FIGURE 14-3. Comparison of activity of a midbrain accommodation-related neuron during normal binocular viewing in which the target stepped briefly toward the monkey (left-hand column) with activity when vergence demand was briefly increased by the same amount (2 meter angles) without changing accommodation demand (right-hand column). V, vergence response; R, horizontal component of right eye position; A, accommodation response; F, firing rate of neuron. Note the lack of neural response when accommodation is not called for.

cently, we (Morley *et al.*, 1986) used a square-wave relative vergence task (i.e., one in which vergence demand is briefly augmented, usually by two meter-angles, as though a prism had been briefly placed in front of one eye) to dissociate vergence and accommodation (see Fig. 14-3). In the earlier dynamic conflict task the monkeys generally behaved as one might have expected on the basis of linear combination of the responses to blur and disparity cues alone, but we have not yet examined whether this is true in the newer conflict task.

Phoria Adaptation and the Sites of the Cross-Links Between Accommodation and Vergence

We have studied phoria adaptation in several monkeys, and also the adaptability of the AC/A ratio. In general, under the conditions we have used, monkeys seem not to adapt their phoria fully to base-out prisms (Judge, 1987; Morley *et al.*, 1988). We have not systematically studied the important question of whether monocular viewing at a closer distance produces phoria adaptation, although it certainly did not in the one monkey studied (Judge, 1987). The issue is important because if the cross-link from accommodation to vergence feeds in before the postulated "slow integrator" (Schor, 1979) responsible for phoria adaptation to prisms, then monocular viewing should produce vergence adaptation. Schor (1979) originally found that monocular viewing did not induce phoria adaptation, but more recently Schor and Kotulak (1986) claim that vergence adaptation does occur in monocular viewing, and therefore in their "heuristic model" the output of the accommodative vergence cross-link drives the vergence slow integrator. Furthermore, Schor and Kotulak (1986) found that some subjects have a tonic component in *accommodation* that can be altered by viewing targets at near distances, even in blur-free binocular viewing, so the "heuristic model" also includes a slow integrator in the accommodation pathway, into which the vergence accommodation cross-link feeds. However, according to C. M. Schor (personal communication), such adaptable tonic accommodation is seen in no more than 25% of subjects. If the phenomenon were comparably uncommon in the monkey, it would be very time consuming to follow up neurophysiologically, and so we have not yet attempted to do so.

NEUROPHYSIOLOGY

Because much more is known about the neurophysiology of accommodation and vergence near the motor periphery, I will describe the neurophysiology "backwards," i.e., beginning with motoneurons.

Motor Output

Although the extraocular muscles are notable for containing a wide variety of different types of muscle fiber, the general view is that there are not special sets of neurons for different types of eye movement (Robinson, 1981). Certainly Keller and Robinson (1972) and Keller (1973) found that all motoneurons they recorded from in the abducens nucleus and the medial rectus subdivision of the oculomotor nucleus carried vergence signals. Their claim that individual motoneurons carry matched version and vergence signals seems not to have been borne out by later studies by Mays and Porter (1984), who found a wide variation in the relative strength of the version and vergence signals on motoneurons. Büttner-Ennever and Akert's (1981) suggestion that the separate "C" subgroup of medial rectus motoneurons (which innervate the small, presumed tonic, fibers in the orbital layer of the muscle) might play a special role in vergence seems at present to be only an interesting speculation. Gamlin and Mays (1986) have also shown that medial rectus motoneurons carry a vergence velocity signal in addition to a vergence position signal.

The ciliary muscle (and also the iris) is innervated by fibers from the ciliary ganglion. Because painting the ganglion with nicotine abolished the pupil response but not the accommodation response to brainstem stimulation, Westheimer and Blair (1973) suggested that at least some of the fibers innervating the ciliary muscle pass straight through the ganglion without synapsing there, but this was not rigorously proven. In the anesthetized cat, Bando *et al.* (1984a) recorded from 30 cells in the brainstem whose activity was correlated with spontaneous fluctuations of accommodation. Eleven of these cells were shown to be Edinger–Westphal cells because they could be antidromically activated by ciliary ganglion stimulation. (Curiously, the changes in firing rate of such cells preceded accommodation fluctuations by 270 msec, much longer than the comparable latency in the monkey.) Very recently Gamlin *et al.* (1989) have recorded in the awake monkey from a small number of cells just dorsal to the oculomotor nucleus that were antidromically activated by stimulation of the ipsilateral oculomotor nerve. Since the Edinger–Westphal nucleus lies in this position and the axons of its neurons run in the oculomotor nerve, it seems a reasonable assumption that these cells are Edinger–Westphal neurons. The activity of these neurons was characterized by a low spontaneous activity (mean about 12 spikes/sec, range 5–20 spikes/sec) and a mean relationship to accommodation of about 9 spikes/sec per diopter (range 5–14 spikes/sec per diopter).

Midbrain

NEAR-RESPONSE NEURONS

Some years ago Schiller (1970) found several neurons dorsolateral to the caudal part of the oculomotor nucleus whose discharge appeared to be proportional to the angle of convergence. More recently, Mays (1984) investigated such cells systematically, showing that the majority of the cells increase in firing rate by a similar extent for the same amount of convergence, whether induced by disparity-driven or accommodative vergence. A small proportion of neurons decreased firing rates for increases in vergence. Mays tentatively called all these cells "vergence cells," although he was careful to say that he could not exclude the possibility that the cells might have been related to accommodation (which was not measured in these experiments, and therefore it remained a presumption that the monkeys did not accommodate as vergence demand was changed while the accommodation demand remained fixed).

We set out to test this assumption by recording from the same area in monkeys whose accommodation was measured (in the right eye) with a purpose-built Scheiner-principle infrared optometer. We recorded from neurons dorsal and dorsolateral to the third nerve nucleus of the monkey whose discharge rates modulated when the monkey tracked targets moving in depth but not when it tracked targets moving from side to side. The neurons' activity modulated equally well whether the target moved directly toward one eye or the other. For most neurons the amplitude of modulation was similar whether the monkey tracked monocularly (blur cue alone), binocularly with accommodation open–loop (disparity cue alone), or in normal binocular viewing. By comparing the activity of neurons in normal binocular viewing and in binocular viewing in which accommodation and vergence demands were in conflict, we (Judge and Cumming, 1986) were able to show that some midbrain near-response neurons were indeed related exclusively to vergence, whereas others were related to accommodation. Figure 14-3 shows an example of an accommodation-related cell identified in this way. Although we initially thought that all the latter neurons were Edinger–Westphal neurons, this now seems unlikely for two reasons. First, such accommodation neurons can be recorded some distance from the midline, whereas Edinger–Westphal neurons are all very close to the midline; secondly, not all accommodation neurons have the discharge characteristics of Gamlin's "identified" Edinger–Westphal cells mentioned above.

If one examines the discharge of either vergence or accommodation neurons over a range of target distances, only some neurons maintain a rigorously linear relationship to vergence or accommodation, with many neurons whose firing increases as the target moves closer having curvilinear rate functions whose slope decreases at higher levels of accommodation or vergence. Such nonlinear rate functions mean that one has to be careful in interpreting the results of the conflict method of distinguishing accommodation and vergence cells.

Very recently L. E. Mays (personal communication) has been able to show that at least some identified vergence cells can be antidromically activated by stimulation of the ipsilateral medial rectus subdivision of the oculomotor nucleus, showing as one might have expected that these neurons are premotor cells.

"BOTH" NEURONS

Judge and Cumming (1986) found that not every near-response cell could be classified as purely related to accommodation or vergence responses. If one forced the firing rate of such cells to be fitted by a linear combination of accommodation and vergence response terms, then *both* terms needed to be nonzero. By calling such cells "both" cells, we did not necessarily imply that they excited both medial rectus motoneurons and Edinger–Westphal cells (though that would be a possibility). We favor instead the possibility that such cells are ones that receive imbalanced inputs from blur- and disparity-driven inputs. We have, for example, seen one cell that was completely silent when vergence demand exceeded accommodation demand or in accommodation open-loop binocular viewing but that discharged vigorously in monocular or normal binocular viewing. This firing pattern is what one would expect of a cell related exclusively to the blur-driven component of accommodation. Other cells of a similar but less extreme kind respond more strongly to monocular stimulation than binocular and behave in what at first seems to be a paradoxical way when vergence demand but not accommodation demand is increased (i.e., "relative vergence" is called for), when they alter their firing in the opposite direction to normal; i.e., a cell that (like most) increases its firing rate as accommodation and vergence increase decreases its rate when vergence demand increases but accommodation demand doesn't. (Since the vergence response in this situation is essentially accurate, whereas accommodation increases to a level closer than the target, the accommodation error is in the opposite sense to normal, and it is therefore appropriate for firing rate to fall.)

VERGENCE VELOCITY NEURONS

The above cells typically show a tonic firing pattern to maintained accommodation and vergence, although some cells have a disproportionately large change in discharge rate as accommodation and vergence demands change suddenly. There is also a category of near-response cells (or one end of a continuum) that have almost entirely transient responses (Cumming, 1985; Mays *et al.*, 1986), and Mays *et al.* (1986) have argued that these cells encode vergence velocity and could constitute the input to tonic vergence cells—but this interpretation needs more evidence to be established. Judge and Cumming (1986) examined the discharge of a small number of identified vergence neurons while the monkey tracked targets whose vergence demand varied sinusoidally over a range of frequencies and showed that the lead of firing rate over vergence response was much larger than for typical motoneurons (Skavenski and Robinson, 1973), suggesting that vergence neurons might selectively activate motoneurons with larger time constants.

ACTIVITY OF MIDBRAIN NEAR-RESPONSE NEURONS IN PRISM ADAPTATION

Both Mays' group (Tello and Mays, 1984) and my own have been interested in whether the midbrain vergence cells carry a tonic adaptable component in their discharge rate that accounts for (rapid) prism adaptation. In our studies (Morley *et al.*, 1986, and others in preparation) we induced adaptation not by using real prisms but by rotating the mirrors of a stereoscope so as to increase the vergence demand. If this was done with a slow ramp waveform, the monkeys could tolerate vergence increases (i.e., equivalent base-out prisms) of 5–7 meter angles, and a minute or two of such experience produced clear-cut increases in phoria. It is essential in these experiments to employ an optometer and to be able to distinguish accommodation- from vergence-related cells. Identified accommodation-related cells maintain a consistent relationship with accommodation before and after prism adaptation and therefore

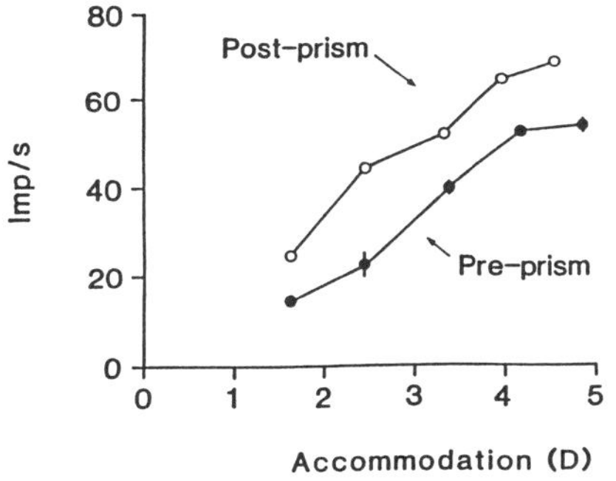

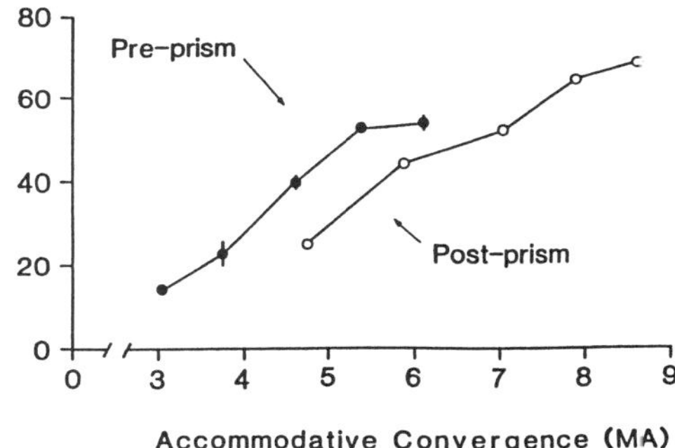

FIGURE 14-4. The response (impulses/second) of a typical midbrain vergence-related neuron plotted as a function of accommodation response (left-hand column) and the same firing rate data plotted as a function of accommodative vergence (right-hand column), both before (filled circles) and after (open circles) prism viewing. The monkey viewed targets at 1, 2, 3, 4, and 5 D monocularly with the right eye. Note that after prism viewing the phoria is increased (in right-hand column, open circles are displaced about 2-meter angles to right) and that firing rate increases after prism viewing but not by as much as needed to maintain a consistent relationship to vergence before and after prism viewing.

do *not* maintain a consistent relationship to vergence. A small proportion of identified vergence cells maintained a consistent relationship to vergence before and after prism adaptation, but the more common finding was that after prism adaptation the firing rate of vergence cells was intermediate between what one would expect if they carried no tonic adaptable signal and if they carried the complete signal. Figure 14-4 shows an example of the behavior of one such identified vergence cell before and after prism adaptation.

Pretectum

Both Mays *et al.* (1986) and Judge and Cumming (1986) noticed near-response-related cells in a more dorsal area, probably in the pretectum. In our hands these cells generally were rather weakly modulated, with a preponderance of cells whose firing rate decreased as the target approached. Their discharges were also more erratic than those of the paraoculomotor cells. The more dorsal cells lay near cells related to the pupil light response (S. J. Judge and J. W. Morley, unpublished observations). It is possible that the dorsal near-response cells are in the afferent pathway to the premotor cells; certainly, lesions that encroach on the pretectum produce a striking loss of ability to look at nearby targets (Cowey *et al.*, 1984; Cowey, 1985; Lawler, 1981).

Cerebellum

In the cat, Hosoba *et al.* (1978) found that stimulation of the deep nuclei (especially the interpositus nucleus) or the cortex of the cerebellum invoked accommodation changes. Miles *et al.* (1980) also noted that the discharge rate of a third of flocculus Purkinje cells was strongly modulated by fixation distance. Taking advantage of the existence of monkeys flucculectomized for other reasons, Judge (1987) was able to show that flocculectomized monkeys are still able to adapt their phoria (and also their AC/A ratio). More recently Luebke and Hain (1988) have shown that the majority of patients with cerebellar lesions are able to adapt phoria in response to prism viewing, thus calling into question the earlier claim by Milder and Reinecke (1983) that cerebellar patients show defective phoria adaptation.

Cortex

At least for vergence, the cortex is necessary because Westheimer and Mitchell (1969) showed that a patient whose corpus callosum had been sectioned in the attempt to contain his epileptic disorder could not make vergence eye movements to targets straddling the midline. Although there is, as far as we know, no direct evidence that the cortex is necessary for supplying the signals to drive accommodation, this seems a reasonable assumption. In the cat, Bando *et al.* (1984b) have discovered that stimulation of part of the lateral suprasylvian cortex induces accommodation and that some of the cells in this region discharge in relation to spontaneous fluctuations of accommodation.

DISPARITY PROCESSING

It seems a reasonable assumption that the cortical cells selectively responsive to particular disparities first described by Barlow *et al.* (1967) and Pettigrew *et al.* (1968) play a role in supplying signals to drive vergence as well as in supporting depth perception, but the pathways over which such information might travel are unknown. Some degree of disparity selectivity has been described for cells in a large number of different visual areas (V1, V2, V3, MT, and MST; see Felleman and Van Essen, 1987, for a review, and also Poggio, 1988, and Komatsu *et al.*, 1988), but there is as yet no indication of where the interface between sensory processing of disparity information and the vergence system might lie. The recent discovery by Wurtz's group (Komatsu and Wurtz, 1988a,b; Newsome *et al.*, 1988) that area MST plays such a role in the smooth pursuit system holds out the hope that a similar area for vergence may soon be discovered, perhaps nearby.

PROCESSING OF OTHER DEPTH CUES

It has long been known that many cues other than binocular (horizontal) disparity contribute to the perception of depth and that other cues influence vergence and accommodation. For example, vergence is influenced by size change (Erkelens and Regan, 1986; McLin *et al.*, 1988) and even perspective (Enright, 1987). Very little is known about the neural processing of these cues, although a start has been made on the search for the neural mechanisms discriminating motion in depth from velocity ratios in the two eyes (Beverley and Regan, 1973; Cynader and Regan, 1978; Zeki, 1974; Poggio and Talbot, 1981).

It is currently in vogue to argue that not only stereopsis but depth sensation in general is colorblind or largely so (see Livingstone and Hubel, 1987, for a review), and if this is true then the neural substrate of depth perception should be found outside the "color" pathway of parvocellular LGN, cytochrome oxidase staining blobs in V1, and thin cytochrome oxidase staining stripes in V2, V4, etc. (Hubel and Livingstone, 1987).

IS THERE A NEAR-RESPONSE "CENTER"?

Some years ago Jampel (1960) claimed to have located a cortical "near-response" area (surrounding the superior temporal sulcus), stimulation of which in the lightly anesthetized macaque produced all three components of the near response, but the study left much to be desired and has yet to be successfully replicated in the alert monkey despite attempts to do so (Cumming, 1985). One difficulty with replicating the Jampel study is that the crucial area is very poorly defined, apparently including cortex both posterior and anterior to the sulcus, although these are now known to be completely different functional areas. If the key area were anterior to the STS, then it might be in the part of the posterior parietal cortex where Sakata *et al.* (1980, 1983) found cells related to fixation at different distances or to tracking in depth. Certainly the parietal cortex has long been thought to play a role in perception of space (Andersen, 1987).

One can question whether the available evidence requires us to believe that there needs to be a "near-response center." It has long been common to think of vergence and accommodation as governed by interacting negative feedback systems. If this were so, then there is no need for a "center" at all. A center only becomes necessary if the manner in which the various cues that affect vergence and accommodation combine shows that the brain makes some kind of overall computation of the position of three-dimensional targets rather than simply letting the various cues (and their associated feedback and forward drives) fight it out. Erkelens *et al.* (1989a,b) have shown very recently that the vergence responses to real target motion are much better than in the impoverished stimulus situations more typical of previous experiments. It will be interesting to see if further studies of such effects show that a near-response center must after all exist.

ACKNOWLEDGMENTS. This work was supported by Medical Research Council Programme Grants to C. Blakemore and by a Wellcome Trust Major Equipment Grant.

REFERENCES

Alpern, M., 1969, Types of eye movement, in: *The Eye*, Vol. 3, 2nd ed. (H. Davson, ed.), Academic Press, New York, pp. 65–174.

Andersen, R. A., 1987, The role of the inferior parietal lobule in spatial perception and visual–motor integration, in: *The Handbook of Physiology, Section 1: The Nervous System, Volume V, Higher Functions of the Brain,* Part 2 (F. Plum, V. B. Mountcastle, and S. R. Geiger, eds.), American Physiological Society, pp. 483–518.

Bando, T., Tsukuda, K., Yamamoto, N., Maeda, J., and Tsukahara, N., 1984a, Physiological identification of midbrain neurons related to lens accommodation in cats, *J. Neurophysiol.* **52:**870–878.

Bando, T., Yamamoto, N., and Tsukahara, N., 1984b, Cortical neurons related to lens accommodation in posterior lateral suprasylvian area in cats, *J. Neurophysiol.* **52:**879–891.

Barlow, H. B., Blakemore, C., and Pettigrew, J. D., 1967, The neural mechanism of binocular depth discrimination, *J. Physiol. (Lond.)* **193:**327–342.

Beverley, K. I., and Regan, D., 1973, Evidence for the existence of neural mechanisms selectively sensitive to the direction of motion in space. *J. Physiol. (Lond.)* **235:**17–29.

Büettner-Ennever, J. A., and Akert, K., 1981, Medial rectus subgroups of the oculomotor nucleus and their abducens internuclear input in the monkey, *J. Comp. Neurol.* **197:**17–27.

Cowey, A., 1985, Disturbances of stereopsis by brain damage, in: *Brain Mechanisms and Spatial Vision* (D. J. Ingle, M. Jeannerod, and D. N. Lee, eds.), Nato ASI Series, Martinus Nijhoff, The Hague, pp. 259–278.

Cowey, A., Smith, B., and Butter, C. M., 1984, Effects of damage to superior colliculi and pre-tectum on movement discrimination in rhesus monkeys, *Exp. Brain Res.* **56:**79–91.

Cumming, B. G., 1985, *The Neural Control of Convergence Eye Movements and Accommodation,* unpublished D.Phil. thesis, University of Oxford.

Cumming, B. G., and Judge, S. J., 1986, Disparity-induced and blur-induced convergence eye movement and accommodation in the monkey, *J. Neurophysiol.* **55:**896–914.

Cynader, M., and Regan, D., 1978, Neurons in cat parastriate cortex sensitive to the direction of motion in three dimensional space, *J. Physiol. (Lond.)* **274:**549–569.

Enright, J. T., 1987, Perspective vergence: Oculomotor responses to line drawings, *Vision Res.* **27:**1513–1526

Erkelens, C. J., and Regan, D., 1986, Human ocular vergence movements induced by changing size and disparity, *J. Physiol. (Lond.)* **379:**145–169.

Erkelens, C. J., Van der Steen, J., Steinman, R. M., and Collewijn, H., 1989a, Ocular vergence under natural conditions. I. Continuous changes of target distance along the median plane, *Proc. R. Soc. Lond. [Biol.]* **236:**417–440.

Erkelens, C. J., Steinman, R. M., and Collewijn, H., 1989b, Ocular vergence under natural conditions. II. Gaze shifts between real targets differing in distance and direction, *Proc. R. Soc. Lond. [Biol.]* **236:**441–465.

Felleman, D. J., and Van Essen, D. C., 1987, Receptive field properties of neurons in area V3 of macaque monkey extrastriate cortex, *J. Neurophysiol.* **57:**889–920.

Fincham, E. F., and Walton, J., 1957, The reciprocal actions of accommodation and vergence, *J. Physiol. (Lond.)* **137:**488–508.

Flitcroft, D. I., 1988, *Sensory Control of Ocular Accommodation,* unpublished D.Phil. thesis, University of Oxford.

Flitcroft, D. I., and Judge, S. J., 1988, The effect of stimulus chromaticity on ocular accommodation in the monkey, *J. Physiol. (Lond.)* **398:**36P.

Flitcroft, D. I., and Judge, S. J., 1989, Binocular interaction between stimuli to accommodation, studied in man and the rhesus monkey (*Macaca mulatta*), *J. Physiol. (Lond.)***413:**37P.

Gamlin, P. D. R., and Mays, L. E., 1986, Medial rectus motoneurons carry a vergence velocity signal in addition to a vergence position signal, *Soc. Neurosci. Abstr.* **12**(1):460.

Gamlin, P. D. R., Zhang, Y., and Mays, L. E., 1989, Behavior of identified Edinger–Westphal neurons during ocular accommodation, *Soc. Neurosci. Abstr.* **15**(1):241.

Hosoba, M., Bando, T., and Tsukahara, N., 1978, The cerebellar control of accommodation of the eye in the cat, *Brain Res.* **153:**495–505.

Hubel, D. H., and Livingstone, M. S., 1987, Segregation of form, color, and stereopsis in primate area 18, *J. Neurosci.* **7:**3378–3415.

Jampel, R. S., 1960, Convergence, divergence, pupillary reactions and accommodation of the eyes from faradic stimulation of the macaque brain, *J. Comp. Neurol.* **115:**371–399.

Judge, S. J., 1987, Optically-induced changes in tonic vergence and AC/A ratio in normal monkeys and monkeys with lesions of the flocculus and ventral paraflocculus, *Exp. Brain Res.* **66:**1–9.

Judge, S. J., and Cumming, B. G., 1986, Neurons in the monkey midbrain with activity related to vergence eye movement and accommodation, *J. Neurophysiol.* **55:**915–930.

Keller, E. L., 1973, Accommodative vergence in the alert monkey. Motor unit analysis, *Vision Res.* **13:**1565–1575.

Keller, E. L., and Robinson, D. A., 1972, Abducens unit behaviour in the monkey during vergence eye movements, *Vision Res.* **12:**369–382.

Komatsu, H., and Wurtz, R. H., 1988a, Relation of cortical areas MT and MST to pursuit eye movements. I. Localization and visual properties of neurons, *J. Neurophysiol.* **60:**580–603.

Komatsu, H., and Wurtz, R. H., 1988b, Relation of cortical areas MT and MST to pursuit eye movements. III. Interaction with full-field visual stimulation, *J. Neurophysiol.* **60:**621–644.

Komatsu, H., Roy, J. P., and Wurtz, R. H., 1988, Binocular disparity sensitivity of cells in area MST of the monkey, *Soc. Neurosci. Abstr.* **14**(1):202.

Kruger, P. B., and Pola, J., 1986, Stimuli for accommodation: blur, chromatic aberration and size, *Vision Res.* **26:**957–971.

Kruger, P. B., and Pola, J., 1987, Dioptric and non-dioptric stimuli for accommodation: target size alone and with blur and chromatic aberration, *Vision Res.* **27:**555–567.

Lawler, K. A., 1981, *Aspects of Spatial Vision after Brain Damage,* unpublished D.Phil Thesis, University of Oxford.

Livingstone, M. S., and Hubel, D. H., 1987, Psychophysical evidence for separate channels for the perception of form, color, movement and depth, *J. Neurosci.* **7:**3416–3468.

Luebke, A. E., and Hain, T. C., 1988, Phoria adaptation in patients with cerebellar dysfunction, *Invest. Ophthalmol. Suppl.* **29:**137.

Mays, L. E., 1983, Neurophysiological correlates of vergence eye movements, in: *Vergence Eye Movements: Basic and Clinical Aspects* (C. M. Schor and K. J. Ciuffreda, eds.), Butterworths, Boston, pp. 647–670.

Mays, L. E., 1984, Neural control of vergence eye movements: Convergence and divergence neurons in midbrain, *J. Neurophysiol.* **51:**1091–1108.

Mays, L. E., and Porter, J. D., 1984, Neural control of vergence eye movements: activity of abducens and oculomotor neurons, *J. Neurophysiol.* **52:**743–761.

Mays, L. E., Porter, J. D., Gamlin, P. D. R., and Tello, C. A., 1986, Neural control of vergence movements: Neurons encoding vergence velocity, *J. Neurophysiol.* **56:**1007–1021.

McLin, L. N., Schor, C. M., and Kruger, P. B., 1988, Changing size (looming) as a stimulus of accommodation and vergence, *Vision Res.* **28:**883–898.

Milder, D. G., and Reinecke, R. D., 1983, Phoria adaptation to prisms, *Arch. Neurol.* **49:**339–342.

Miles, F. A., Fuller, J. H., Braitman, D. J., and Dow, B. M., 1980, Long-term adaptive changes in primate vestibuloocular reflex. III. Electrophysiological observations in flocculus of normal monkeys, *J. Neurophysiol.* **43:**1437–1476.

Morley, J. W., Lindsey, J. W., and Judge, S. J., 1986, Changes in activity of brainstem near-response neurons induced by prism-adaptation, *Soc. Neurosci. Abstr.* **12**(1):9.460.

Morley, J. W., Lindsey, J. W., and Judge, S. J., 1988, Prism adaptation in a strabismic monkey, *Clin. Vis. Sci.* **3:**1–8.

Newsome, W. T., Wurtz, R. H., and Komatsu, H., 1988, Relation of cortical areas MT and MST to pursuit eye movements. II. Differentiation of retinal from extraretinal inputs, *J. Neurophysiol.* **60:**604–620.

Pettigrew, J. D., Nikara, T., and Bishop, P. O., 1968, Binocular interaction on single units in cat striate cortex: Simultaneous stimulation by single moving slits with receptive fields in correspondence, *Exp. Brain Res.* **6:**391–410.

Poggio, G. F., and Talbot, W. H., 1981, Mechanisms of static and dynamic stereopsis in foveal cortex of the rhesus monkey, *J. Physiol. (Lond.)* **315:**469–492.

Poggio, G. F., Gonzalez, F., and Krause, F., 1988, Stereoscopic mechanisms in monkey visual cortex: Binocular correlation and disparity selectivity, *J. Neurosci.* **8:**4531–4550.

Rashbass, C., and Westheimer, G., 1961, Disjunctive eye movements, *J. Physiol. (Lond.)* **159:**339–360.

Robinson, D. A., 1981, Control of eye movements, in: *Handbook of Physiology, The Nervous System,* Section 1, Vol. 2 (J. M. Brookhart and V. B. Mountcastle, section eds.; V. B. Brooks, volume ed.), American Physiological Society, Bethesda, pp. 1275–1320.

Sakata, H., Shibutani, H., and Kawano, K., 1980, Spatial properties of visual fixation neurons in posterior parietal association cortex of the monkey, *J. Neurophysiol.* **43:**1654–1672.

Sakata, H., Shibutani, H., and Kawano, K., 1983, Functional properties of visual tracking neurons in posterior parietal cortex of the monkey, *J. Neurophysiol.* **49:**1364–1380.

Schiller, P. H., 1970, The discharge characteristics of single units in the oculomotor and abducens nuclei of the unanaesthetized monkey, *Exp. Brain Res.* **10:**347–362.

Schor, C. M., 1979, The relationship between fusional vergence eye movements and fixation disparity, *Vision Res.* **19:**1359–1367.

Schor, C. M., and Ciuffreda, K. J., 1983, *Vergence Eye Movements: Basic and Clinical Aspects,* Butterworths, Boston.

Schor, C. M., and Kotulak, J. C., 1986, Dynamic interactions between accommodation and convergence are velocity sensitive, *Vision Res.* **26**:927–942.

Skavenski, A. S., and Robinson, D. A., 1973, Role of abducens neurons in vestibuloocular reflex, *J. Neurophysiol.* **36**:724–738.

Tello, C. A., and Mays, L. E., 1984, Activity of mesencephallic convergence eyes during vergence adaptation. *Soc. Neurosci. Abstr.* **10**(2):988.

Toates, F. M., 1972, Accommodation function of the human eye, *Physiol. Rev.* **52**:828–863.

Tucker, J., and Charman, W. N., 1979, Reaction and response times for accommodation, *Am. J. Optom. Physiol. Opt.* **56**:490–503.

Westheimer, G., and Blair, S. M., 1973, The parasympathetic pathways to the internal eye muscles, *Invest. Ophthalmol.* **12**:193–197.

Westheimer, G., and Mitchell, D. E., 1969, The sensory stimulus for disjunctive eye movements, *Vision Res.* **9**:749–755.

Wilson, D., 1973, A centre for accommodative vergence motor control, *Vision Res.* **13**:2491–2503.

Zeki, S. M., 1974, Cells responding to changing image size and disparity in cortex of the rhesus monkey, *J. Physiol. (Lond.)* **242**:827–841.

15

Oculomotor Adaptation to Induced Vergence Demands

DAVID B. HENSON

INTRODUCTION

Several researchers have reported on the way the oculomotor system adapts to an induced heterophoria or fixation disparity (Carter, 1965; Henson and North, 1980; Ogle *et al.*, 1967; Schor, 1979). Their results show that although an ophthalmic prism placed before one eye initially induces a change in the habitual phoria and fixation disparity, continued fusion through the prism results in an adaptation of the oculomotor system and a loss of the induced change.

The data of Henson and North (1980) show that on average the phoria generated by 2 prism diopters base up before the right eye is reduced by 1.5 prism diopters with just over 3 min of binocular visual experience. The time course of the adaptation is exponential, with a time constant dependent on the amplitude of the prism (Sethi and North, 1987), the base direction (Henson and North, 1980), and the binocular status (North and Henson, 1981, 1982; Ogle *et al.*, 1967). These findings have been interpreted as indicating that the oculomotor system is capable of adapting the output of the center that normally maintains binocular alignment. Schor (1979) has presented a model to explain this adaptation, hypothesizing that the output of the fusion mechanism is responsible for adaptation.

HYPOTHESIS

Henson and North (1980) hypothesized that a process of adaptation similar to that described by them was essential to the visual system to enable it to compensate for the normal changes in orbital mechanics that occur with age. The gradual enophthalmos that results from a loss of orbital fat (Weale, 1963) and the alterations in muscle structure (Miller, 1975) are just two of the normally occurring changes that would require major alterations in the patterns of innervation to the extraocular muscles.

The alterations in mechanics that would result from these age changes would not be expected to produce overall shifts in visual direction of one eye such as that produced by an ophthalmic prism. The effects would be partly dependent on eye position; for example, if the elasticity of the muscles were to change, then the effects would be different for each position of gaze.

To compensate for these types of changes, the adaptive mechanism needs to operate over limited motor fields. Adaptation stimulated by a fusion demand in one particular direction of gaze needs to be confined to a small range of gaze directions and not spread across the whole motor field.

This hypothesis can be tested in the following two ways:

1. *Spread of adaptation* By keeping eye position stable during the adaptation period and then measuring how much oculomotor adaptation has occurred at different eye positions. The hypothesis predicts that adaptation would be maximum when the eyes are in the same position as that at adapting stimulus which the adaptation was given and that it would tail off in other positions to an extent that is proportional to the angular distance from the adapted position.
2. *Adaptation to induced noncomitant deviations* By adapting the oculomotor system to an induced phoria demand whose amplitude varies with eye position. The hypothesis predicts that

DAVID B. HENSON • Department of Optometry, University of Wales, Cardiff CF1 3XF, Great Britain.

the change in phoria from pre- to postadaptation would be proportional to the fusional demand and therefore greatest in the direction with the largest fusional demand and least in the direction with the smallest fusional demand.

SPREAD OF ADAPTATION

The spread of adaptation can be measured by repeating the early experiments of Henson and North (1980) with the following additions to the experimental protocol: (1) ensuring that the position of the eyes within the orbit is kept reasonably constant during the adapting period and (2) including at the end of the adapting period a measure of the extent of adaptation at different orbit positions.

The first problem is relatively straightforward to deal with and simply requires a fairly accurate system for head alignment. The second problem can be solved with a dental bite arrangement that permits rotation of the head in both the vertical and horizontal meridians about a center of rotation that coincides with that of one eye. This arrangement is necessary to ensure that the eye continues to look through the center of the Maddox rod used for the phoria measurements. In this type of experiment it is important to ensure that the oculomotor system maintains its adaptation throughout the measurement period and that no binocular visual experience is obtained during the measurement process. For this reason, phoria-measuring techniques rather than fixation disparity ones are preferred. Measurement of fixation disparity involves the use of binocularly fusible targets that, even when exposed for just a few seconds, can influence the adaptive mechanism.

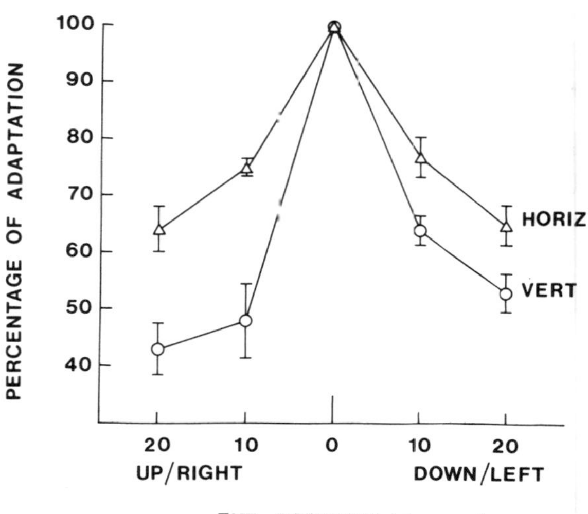

FIGURE 15-2. Spread of adaptation in the vertical and horizontal meridians.

The results of this type of experiment (Figs. 15-1 and 15-2) clearly support the hypothesis that adaptation takes place over limited motor fields. The extent of adaptation is maximum in the visual direction where binocular visual experience was obtained, and it then tails off from this by an amount dependent on the angular distance from the adapted position.

The spread of adaptation is found to be greater in the horizontal meridian than in the vertical one. This may reflect different adapting channels. It can also be seen that the spread of adaptation is large, being reduced to only 60% of the maximum 20° from the adapted position in the horizontal meridian and just over 40% in the vertical meridian.

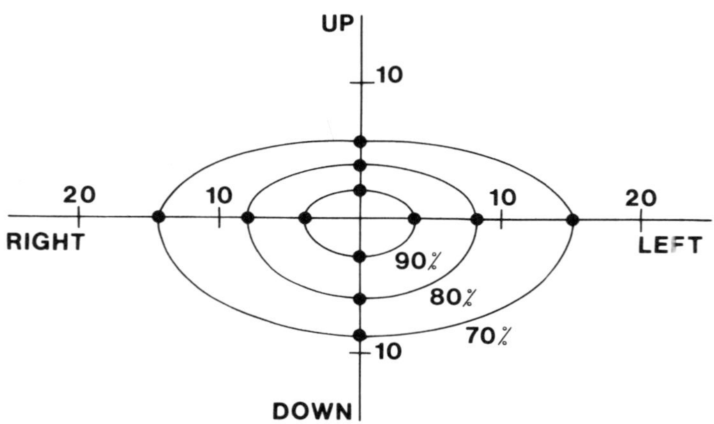

FIGURE 15-1. Spread of adaptation from a single adaptation position that coincides with the center of the diagram.

ADAPTATION TO INDUCED NONCOMITANT DEVIATIONS

To test further the hypothesis that the oculomotor system is capable of adapting over limited motor fields, Henson and Dharamshi (1982) induced a noncomitant deviation by fitting one eye with a soft contact lens and then correcting the induced ametropia with a spectacle lens. Because the contact lens moves with the eye, the combined contact/spectacle lens system induces a prismatic effects whose magnitude is dependent on eye position and the dioptric power of the lens system. The further the subject's line of sight is from the optical center of the spectacle lens, the greater are the prismatic effects and the fusional demand. The effects can be viewed as a magnification of one eye's motor field.

With the same piece of apparatus as that used to measure the spread of adaptation, the phoria amplitude can be measured at a series of different gaze directions. As expected, subjects initially demonstrate a phoria whose amplitude increases as the line of sight moves away from the spectacles' optical center. After a period of adaptation during which the subjects went about their normal office/lab duties, repeat measures of their phorias demonstrated an adaptation whose amplitude was dependent on eye position (see Fig. 15-3).

These data clearly demonstrate that, as hypothesized, the oculomotor system is capable of adapting to induced noncomitant deviations, although the time course of the adaptation process is much slower than that experienced with step displacements such as those produced by an ophthalmic prism. Again, it can be seen that both the rate and the total amount of adaptation are greatest in the vertical meridian.

The data represented in Fig. 15-3 indicate that adaptation in the superior field, particularly at the extreme location (22° above the straight-ahead position), was less than that in the inferior field. This later finding was attributed to the reduced amount of visual experience our subjects obtained in the superior motor field as compared to the inferior one. During normal office/lab duties it would be quite rare for a subject to spend a great deal of time in upward gaze. This explanation was tested further by a modification to the above experiment in which the spectacle lens aperture was reduced with opaque material to allow only an 18° field of view. With this modification the rate of adaptation in areas outside of the 18° motor field was significantly reduced (Sethi and Henson, 1985), indicating that, as expected, the rate of adaptation is dependent on the amount of binocular visual experience gained at each location.

The adaptation shown in Fig. 15-3 is for an induced anisometropia of 3 D. Increasing the amount of anisometropia increases the fusional demands on the oculomotor system. When subjects were fitted with a contact/ spectacle lens system that induced 4.5 D of anisometropia, the rate of adaptation was significantly reduced in both the center and the periphery of the motor field (see Fig. 15-4). This slowing of the adaptation rate does not appear to be solely the result of the amplitude of the fusion demands. At the center of the motor field, where the line of sight is close to the optical center of the spectacle lens, the fusion demand was small, and yet the adaptation rate was notably slow. This effect can be explained by hypothesizing that the oculomotor system is limited not solely by the extent of the fusional demand but also by the rate of change of demand with eye position.

Additional experiments in which subjects were required to adapt to an anisometropia of 3 D for a prolonged period of time (2–3 days) and then adapt to a further 1.50 D of anisometropia (taking the overall degree of anisometropia to 4.50 D) indicate that some improvement in the ability to adapt to this level of anisometropia does occur when the adaptational demand is presented in small steps (Sethi and Henson, 1984). Ogle *et al.* (1967) reported a similar finding using step displacements.

An unwanted side effect of the contact/ spectacle lens system is an induced anisokonia that increases with the dioptric power of the system. The inability of the ocular motor system to adapt, in these experiments, to large amounts of noncomitance may in part be a result of the unwanted side effects of the induced anisokonia.

The type of adaptation reported here is identical to that required of naturally occurring anisometropes when they receive their first spectacle correction. The results of this experiment suggest that, with time, these patients would adapt their oculomotor systems so that their phorias would not vary when they looked through different parts of their spectacle lenses. The work of Ellerbrock and Fry (1942), Ellerbrock (1948), and Allen (1974) indicates that this is indeed the case. The ability of these patients to adapt is dependent on the degree of anisometropia, there being a tendency for the adaptive process to lag behind the fusional demand in high levels of anisometropia (Allen, 1974).

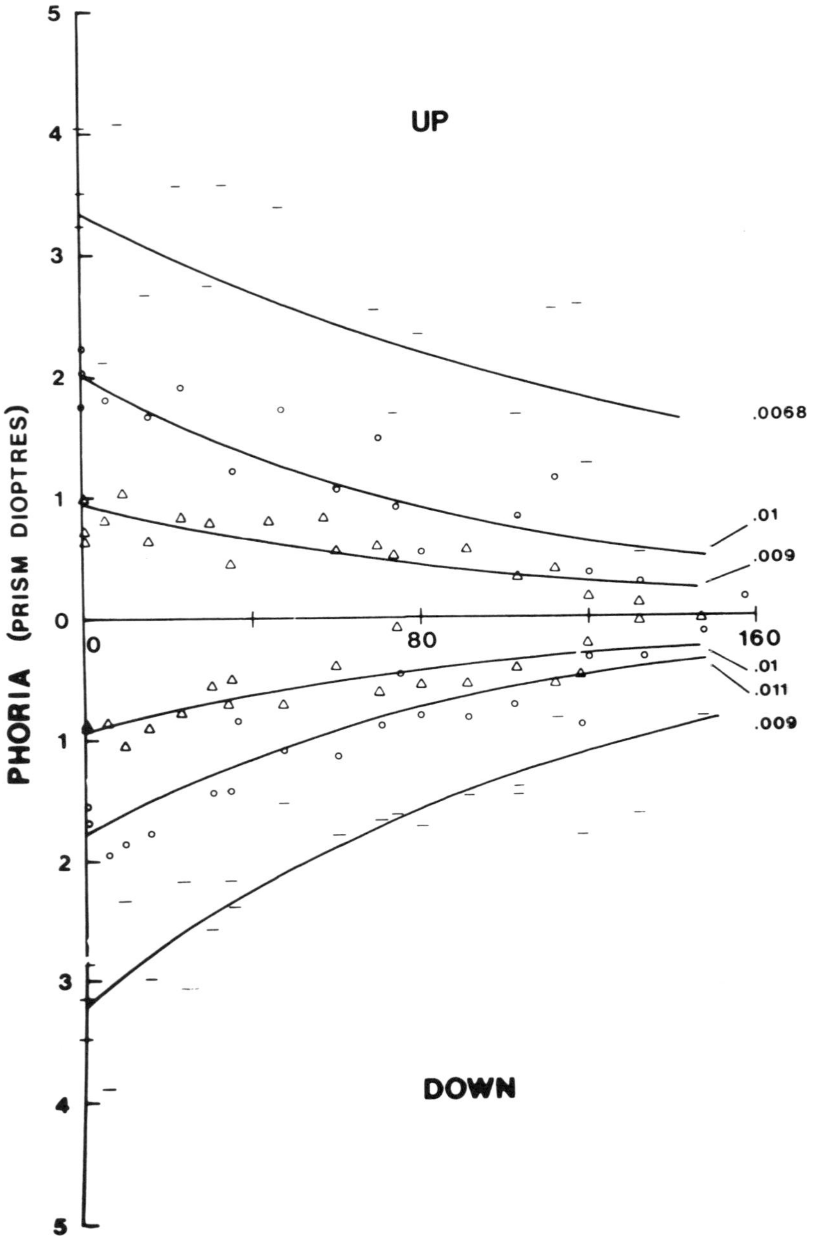

FIGURE 15-3. Noncomitant oculomotor adaptation induced by 3.00 D of anisometropia. Each line in this figure represents the change in phoria for positions of gaze: 7.6°, 14.9°, and 21.8° from the straight-ahead position. The lines drawn through the data are the best-fitting exponential curves of the form $y = P \cdot \exp(-bx)$. The value of b is given at the end of each line.

The difference in the rate of adaptation between step displacements and noncomitant ones is dramatic. Whereas step displacements appear to be corrected for within a period of a few minutes, noncomitant ones produced by inducing an anisometropia take several hours. It is clear from the results of experiments that measure the spread of adaptation that step displacements in one particular direction of gaze produce an adaptation that spreads over a fairly large range of gaze directions. In the case of an induced anisometropia, this spread could hinder the adaptive process by producing an inappropriate adaptation in neighboring gaze directions. As the eyes move about the motor field, there would, in the early stages of adaptation, be a good deal adapting in one direction as a result of spread, followed by adapta-

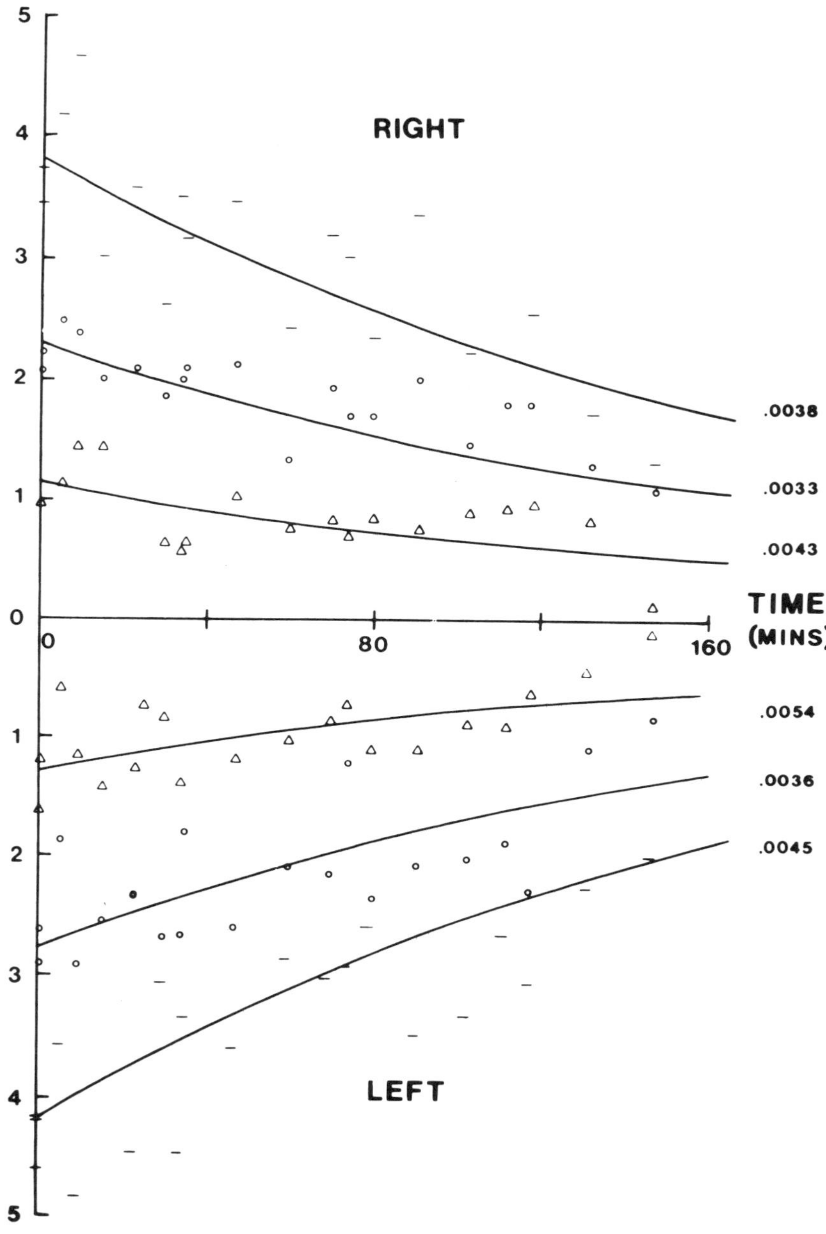

FIGURE 15-3 (cont.)

tion in the other direction as a result of visual experience. What is clear is that the oculomotor system is capable, within limits, of overcoming this problem, presumably by reducing the width of the adapting channels, although this process places an additional burden on the system which reduces the rate of adaptation.

CONCLUSIONS

The initial hypothesis was that the oculomotor system needed to be able to adapt its binocular coordination in order to be able to deal with naturally occurring age changes in orbital mechanics and that this process must be capable of operating over lim-

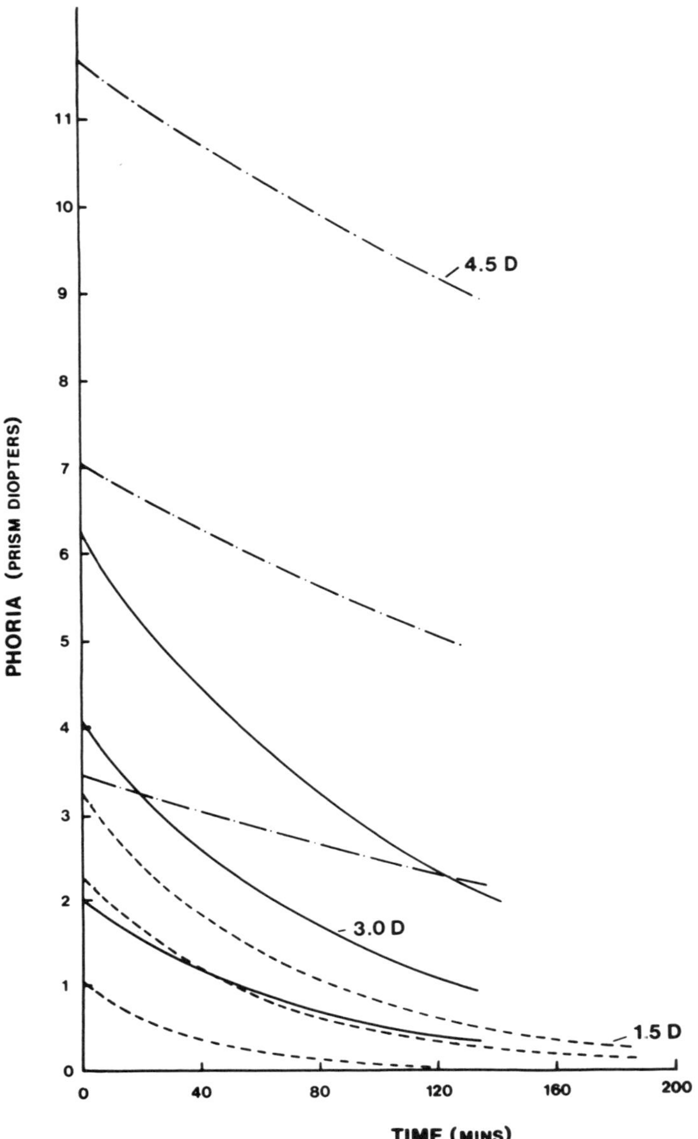

FIGURE 15-4. Noncomitant oculomotor adaptation in the vertical meridian induced by 1.50 D (dashed line), 3.00 D (solid line), and 4.50 D (dash–dot line) of anisometropia. Measurements were taken 7.6°, 14.9°, and 21.8° from the straight-ahead position. The results from up and down positions have been algebraically added.

ited motor fields. It is clear from the results cited that the oculomotor system is capable of such adaptation, although there appear to be limits to the degree of noncomitancy that it can deal with, and these limits vary from one individual to another.

REFERENCES

Allen, D. C., 1974, Vertical prism adaptation in anisometropes, *Am. Acad. Optom. Physiol. Opt.* **51**:252–259.

Carter, D. B., 1965, Fixation disparity and heterophoria following prolonged wearing of prism, *Am. J. Optom. Arch. Am. Acad. Optom.* **42**:141–152.

Ellerbrock, V., 1948, Further study of effects induced by anisometropic corrections, *Am. J. Optom. Arch. Am. Acad. Optom.* **25**:430–437.

Ellerbrock, V., and Fry, G A., 1942, Effects induced by anisometropic corrections. *Am. J. Optom. Arch. Am. Acad. Optom.* **19**:444–459.

Henson, D. B., and Dharamshi, D. G., 1982, Oculomotor adaptation to induced heterophoria and anisometropia, *Invest. Ophthalmol.* **22**:234–240.

Henson, D. B., and North, R., 1980, Adaptation to prism-induced heterophoria, *Am. J. Optom. Physiol. Opt.* **57**:129–137.

Miller, J. E., 1975, Aging changes in extraocular muscles, in: *Basic*

Mechanisms of Ocular Motility and Their Clinical Implications (G. Lennerstrand and P. Bach-Y-Rita, eds.), Pergamon Press, Oxford, 47–61.

North, R. V., and Henson, D. B., 1981, Adaptation to prism-induced heterophoria in subjects with abnormal binocular vision or asthenopia, *Am. J. Optom. Physiol. Opt.* **58**:746–752.

North, R. V., and Henson, D. B., 1982, Effects of orthoptics upon the ability of patients to adapt to prism-induced heterophoria, *Am. J. Optom. Physiol. Opt.* **59**:983–986.

Ogle, K. N., Martens, T. G., and Dyer, J. A., 1967, *Oculomotor Imbalance in Binocular Vision and Fixation Disparity*, Kimpton, London.

Schor, C. M., 1979, The relationship between fusional vergence eye movements and fixation disparity, *Vis. Res* **19**:1359–1367.

Sethi, B., and Henson, D. B., 1984, Adaptive changes with prolonged effect of comitant and incomitant vergence disparities, *Am. J. Optom. Physiol. Opt.* **61**:506–512.

Sethi, B., and Henson, D. B., 1985, Vergence-adaptive change with prism-induced noncomitant disparity, *Am. J. Optom. Physiol. Opt.* **62**:207–216.

Sethi, B., and North, R. V., 1987, Vergence adaptive changes with varying magnitudes of prism-induced disparities and fusional amplitudes, *Am. J. Optom. Physiol. Opt.* **64**:263–268.

Weale, R. A., 1963, *The Aging Eye*, H. K. Lewis, London.

16

Adaptive Regulation of Accommodative Vergence Interactions

CLIFTON SCHOR

INTRODUCTION

The near-visual response consists of concomitant changes in accommodation, vergence, and pupil size. All three of these motor systems share a common organization of phasic–tonic control that characterizes their transient and sustained motor responses. This dual-control organization for accommodation and vergence is indicated by their open-loop decay time constants, which increase with closed-loop stimulus duration (Schor, 1979; Schor *et al.*, 1986a). A phasic–tonic organization of pupil control has also been interpreted from observations of pupillary escape, which occurs with large but not small initial pupil size (Sun and Stark, 1983).

PHASIC–TONIC ORGANIZATION

Traditionally, the phasic components of accommodation and convergence have been referred to as optical reflex accommodation (Fincham, 1951) and fusional (Maddox, 1893) or disparity vergence (Stark, 1979), respectively. The tonic components of accommodation and convergence have been referred to as the resting focus (Leibowitz and Owens, 1975) and phoria (Maddox, 1893), respectively. As indicated by their titles, phasic controllers have much shorter response times than tonic controllers. Phasic components have transient responses that are completed in approximately 1 sec but that decay rapidly within 10–15 sec when the stimulus is removed by opening the feedback loop. For example, opening the accommodative loop with a pinhole pupil, which eliminates feedback of retinal image blur, will allow optical reflex accommodation to relax to the resting focus in 1–10 sec (Schor *et al.*, 1984). Figure 16-1 illustrates both rapid (1 sec) and long-term (30 sec) decay time constants of accommodative aftereffects following removal of an accommodative stimulus by placing the subject in darkness. Similarly, Fig. 16-2 illustrates that opening the vergence feedback loop by occluding one eye after 5 sec of disparity stimulation will eliminate feedback in the form of retinal image disparity and allow disparity vergence to relax to the phoria posture in 5–10 sec (Schor, 1979). In contrast, response times for tonic adjustments following 1 min of disparity stimulation are at least an order of magnitude longer (15–60 sec) than phasic responses. Tonic responses of accommodation are revealed by increase in open-loop decay time constants in excess of 30–60 sec (Schor *et al.*, 1984, 1986a), which may extend to much longer aftereffects following longer stimulus periods (10 min) (Ebenholtz and Zander, 1987).

Tonic Adaptation

The phasic–tonic control of accommodation and convergence has three unique attributes. The first is that their tonic responses are adaptable, which gives each system the capability of long-range adjustments of its set points or resting positions. The tonic components of accommodation and convergence are capable of a marked range of adaptation in response to both short- and long-term demands on the oculomotor near response. Short-term aftereffects of accommodation appear after only several minutes of stimulation with lenses (Schor *et al.*, 1986a). These

CLIFTON SCHOR • School of Optometry, University of California, Berkeley, California 94720.

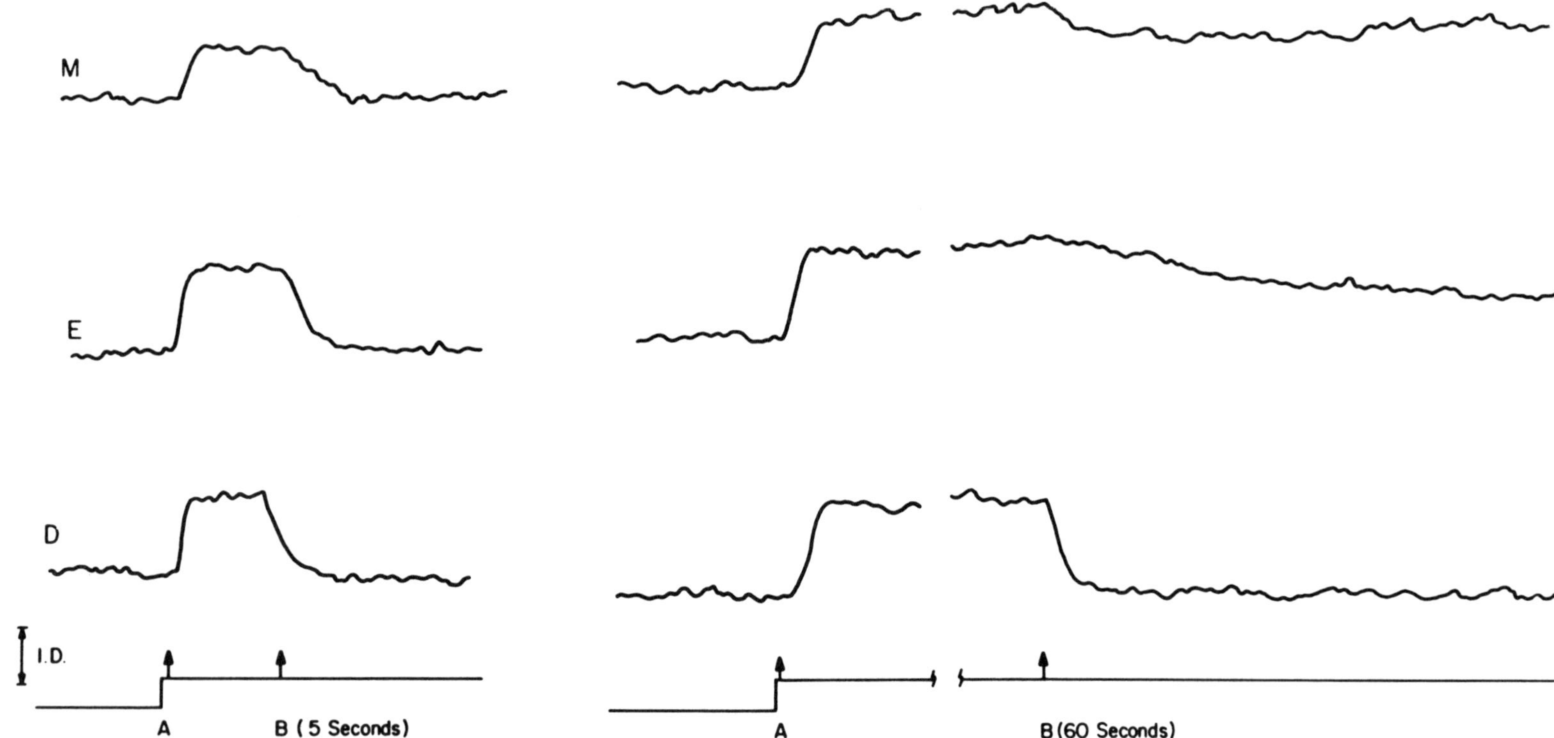

FIGURE 16-1. Different time constants for relaxation of accommodation under three different open-loop conditions [Maxwellian view (M), empty field (E), and darkness (D)] after stimulating accommodation monocularly for either 5 or 60 sec. Arrows marked A and B beneath the recordings indicate when accommodation was stimulated and when the loop was opened, respectively. Incomplete relaxation of accommodation after long-term stimulation demonstrates a tonic aftereffect.

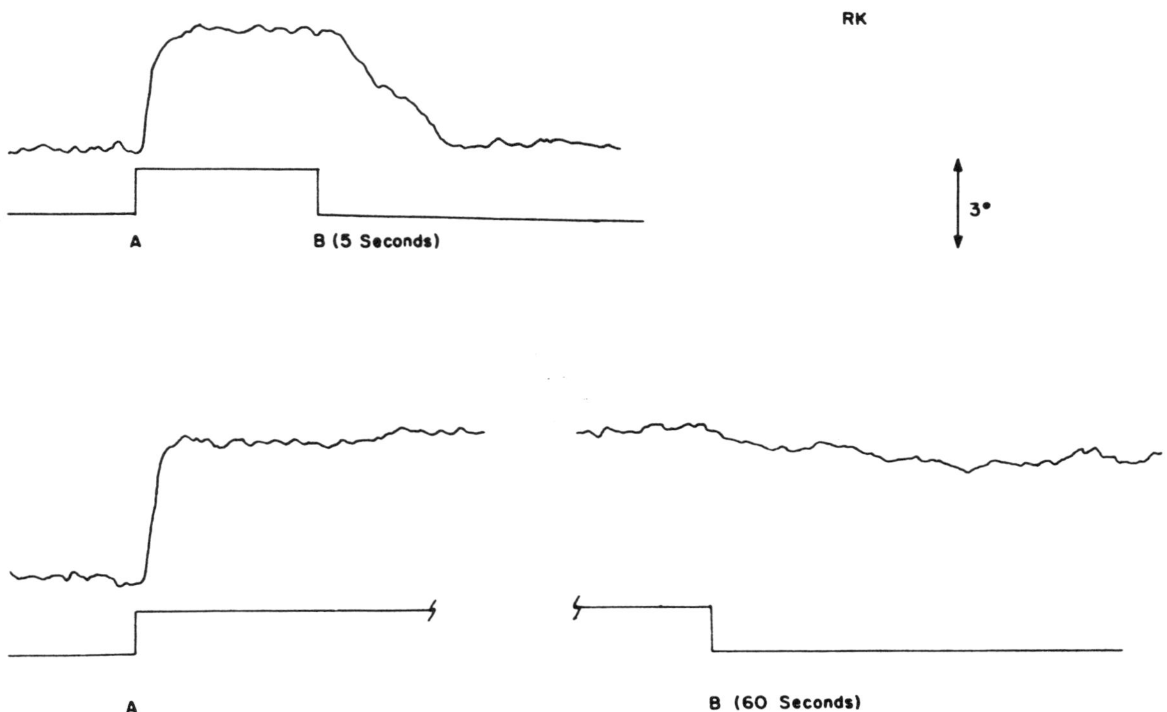

FIGURE 16-2. Disparity-induced vergence, showing time constants for relaxation of fusional vergence after stimulating convergence for 5 sec (upper trace) and 60 sec (lower trace). A, onset of stimulation; B, occlusion of one eye. Incomplete relaxation of convergence after long-term stimulation and convergence indicates slow fusional vergence.

aftereffects of accommodation are most readily observed in photopic levels of illumination (Schor and McLin, 1988). Figure 16-3 illustrates that after stimulating accommodation for 1 min, an aftereffect is present when a pinhole pupil is placed before the open eye, but the aftereffect disappears momentarily when the lights are extinguished and then reappears when the target is again illuminated while the pinhole pupil remains before the open eye. This extinction effect occurs when accommodative aftereffects are disrupted by scotopic light levels, and it illustrates an interaction between the aftereffect and stimulation of the cone system. Aftereffects follow both positive and negative accommodative response from the base-line resting focus of accommodation; however, aftereffects are greater following positive than negative accommodation (Ebenholtz, 1983).

Similar robust vergence aftereffects result from short-term vergence responses to prism (prism adaptation) (Schor, 1979); changes in heterophoria result from wearing vertical (Ellerbrock, 1950) as well as horizontal prism (Ogle and Prangen, 1953). Adaptation occurs more rapidly in response to horizontal than vertical prism (Ellerbrock, 1950), and usually with larger responses to convergent than divergent stimuli (Schor, 1979).

Long-term aftereffects of accommodation and vergence play an important role in tuning the refractive state of the eyes, a process known as emmetropization, and binocular alignment (orthophorization) during development. Much of the emmetropization process of the growing eye is controlled passively by proportion scaling of axial length, corneal curvature, and anterior chamber depth. However, because the newborn eye is typically hyperopic, residual refractive errors could be corrected actively by accommodation, which in turn could stimulate long-term changes in the resting focus of accommodation that would minimize demands on optical reflex accommodation at prominently used viewing distances. A more active role of control is required of vergence adaptation during development because a comparable passive scaling processes does not exist to compensate the resting posture of vergence as the interpupillary distance increases. At birth, an infant's phoria, measured during sleep, is biased in the exophoric direction. However, this phoria is reduced to orthophoria within the

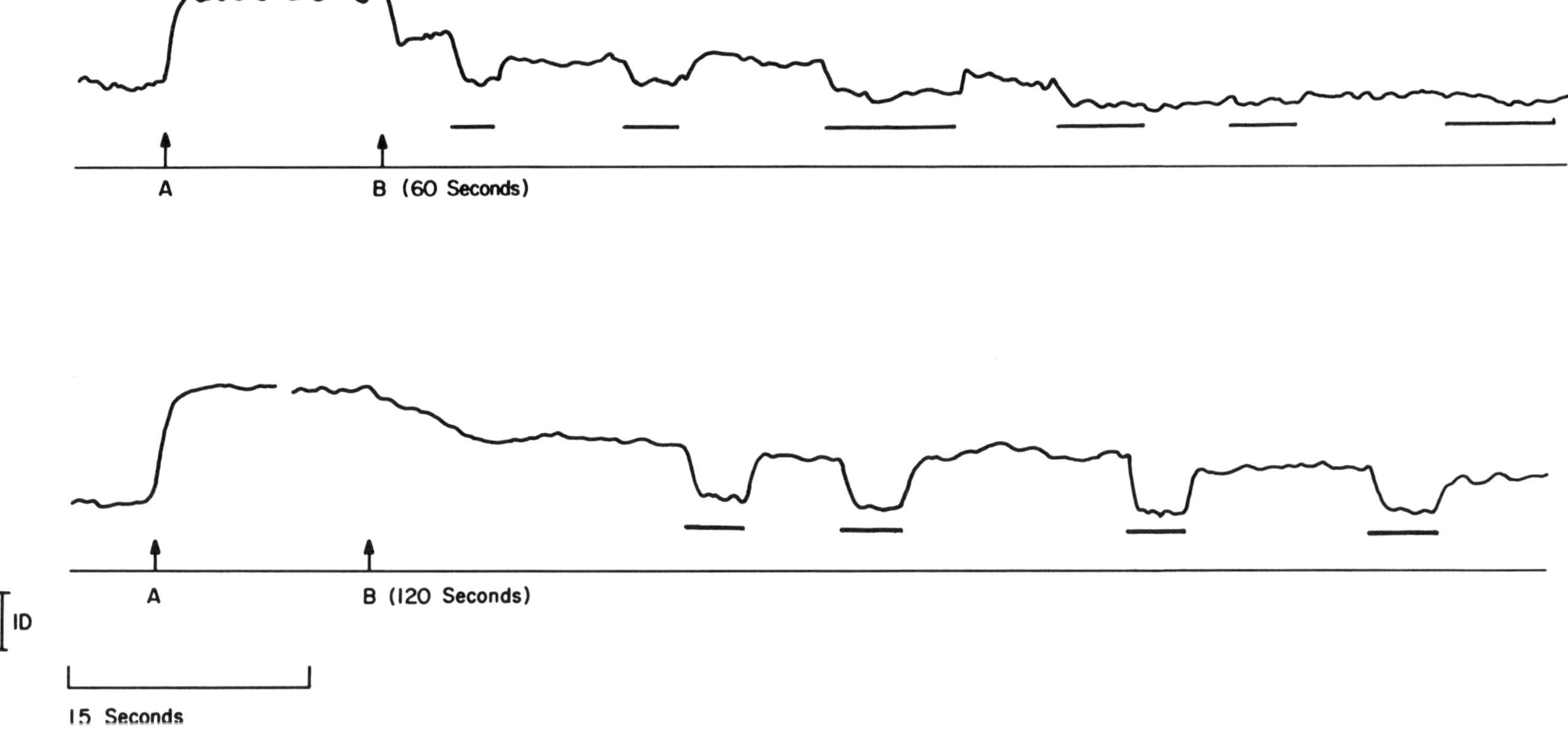

FIGURE 16-3. The decay of a tonic aftereffect of accommodation is shown during open-loop (Maxwellian view) conditions interrupted by brief periods of darkness (underlined segments). Adaptation was for 1 min to a 2-D stimulus (top) or for 2 min to a 3-D stimulus (bottom). During dark periods accommodation went to its resting level, but it returned to the adapted state with the Maxwellian view.

first 6 weeks of development (Rethy, 1969), just as it is in other vertebrate species such as the cat (Sherman, 1972) and owl (Knudsen, 1989).

External and Internal Stimuli

The second unique attribute of phasic–tonic control of accommodation and vergence is that these two controllers operate serially. The phasic component responds to external stimuli present in the retinal image, such as blur, disparity, size, and changing size. The tonic mechanisms adapt in response to effort or innervation by the phasic systems. Thus, blur will not stimulate tonic accommodation, but a phasic accommodative response to blur will result in a change in the resting focus. This phasic–tonic serial organization results in an efficient division of labor. The phasic component provides a brisk response, and the tonic one maintains it. The tonic response is most useful for stimulus conditions that remain static for long periods of time. Because these are components of a negative feedback control system, tonic adaptation removes the stimulus from the phasic control element and allows it to remain idle until a stimulus change occurs. Accordingly, the phasic–tonic organization of components within a negative feedback system results in a stimulus velocity sensitivity for optical reflex accommodation and disparity vergence as well as a stimulus amplitude sensitivity for tonic adaptation of accommodation and vergence (Schor and Kotulak, 1985). Although long-term adaptive processes adjust set points or resting posture of accommodation and vergence for a commonly utilized viewing distance, they are unable to shift these set points rapidly from one distance to another. Phasic mechanisms fulfill this need for small refinements of motor response; however, the effective stimuli have small ranges limited to less than 2 D for accommodation (Fincham, 1951) and less than 2–4° for vergence (Schor *et al.*, 1986b; Erkelens, 1987).

Responses to larger changes in distance are stimulated by proximal or depth cues such as the dynamic cue of looming (McLin *et al.*, 1988) and the static cue of relative size (Bedell and Wick, 1989). The large motor responses of accommodation and vergence to perceived distance are coordinated by the cross-link interactions of accommodative vergence and vergence accommodation, which effectively move the resting position or set point of one motor system (accommodation or vergence) along with the phasic proximal response of the other motor system. Because cross-coupling occurs in both directions between accommodation and vergence, it is not clear which motor system, if not both, leads this interaction. Maddox (1893) emphasized the role of accommodation and accommodative vergence in minimizing the demands on disparity vergence control in the near response. Fincham (1951) took the opposite point of view by suggesting that convergence accommodation served as a coarse adjustment mechanism to bring the retinal image into approximate focus during binocular viewing conditions. However, both directions of interaction are utilized in response to specific proximal cues to distance. Accommodative vergence is the primary mode of interaction in response to kinetic depth cues such as looming (McLin *et al.*, 1988), whereas convergence accommodation dominates the response to static depth cues such as relative size and target elevation (Wick and Bedell, 1989). The main role of these cross-link interactions appears to be to guide the open-loop response of accommodation and convergence in response to large perceived changes in viewing distance. These responses are then further refined by the closed-loop negative feedback systems of optical reflex accommodation and disparity vergence.

Phasic Stimulation of Cross-Coupling Interaction

The third unique attribute of phasic–tonic controllers is that cross-coupling interactions between accommodation and vergence are stimulated by the phasic but not the tonic control elements. This organization is illustrated schematically in the block diagram of Fig. 16-4. The upper and lower loops illustrate the negative feedback control systems that fine-tune the accommodative and vergence responses, respectively. The cross-link interactions of accommodative convergence and convergence accommodation are shown to occur between the phasic and tonic control elements of the two systems. This organization is surmised from the frequency response function of accommodative vergence and vergence accommodation. Figure 16-5 illustrates accommodative convergence, stimulated monocularly by sinusoidal blur oscillation at a low (0.06 Hz) and a high (1.0 Hz) temporal frequency. At the higher temporal frequency, accommodative vergence occurs synergistically with the accommodative response. However, at the low temporal frequency, there is no accommodative vergence response even

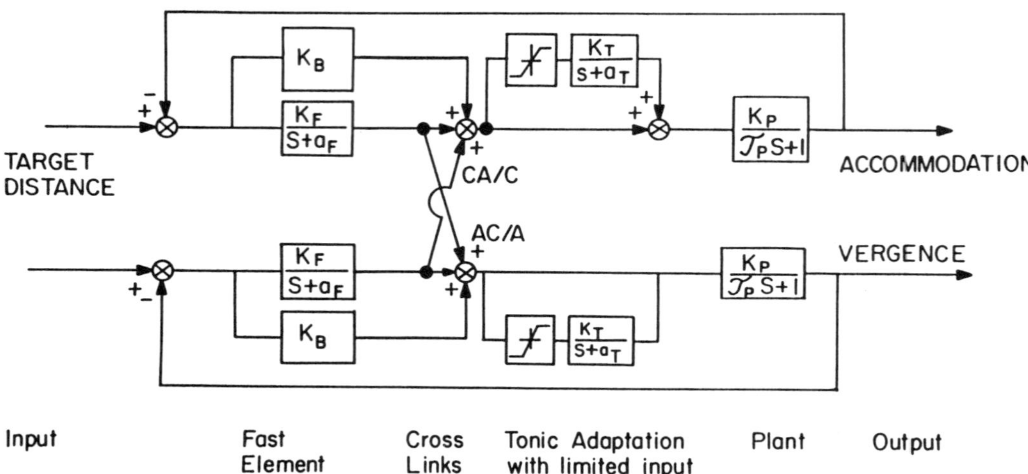

FIGURE 16-4. System model representation of synchenetic interactions between accommodation and vergence. The two motor systems are interconnected at points in their feedforward paths located between phasic and tonic neural integrators.

though the gain of the accommodation response remains high. Similarly, Fig. 16-6 illustrates convergence accommodation, stimulated binocularly by sinusoidal disparity oscillation at a low (0.05 Hz) and a high (1.0 Hz) temporal frequency while pinhole pupils are placed before each eye. As with accommodative vergence, at the high temporal frequency, vergence accommodation occurs synergistically with the disparity vergence response. However, at the low temporal frequency, there is no vergence accommodative response even though the gain of disparity vergence remains high. Figure 16-7 illustrates the gain of accommodative vergence (AC/A ratio) and the gain of vergence accommodation (CA/C ratio) plotted as a function of temporal frequency. The solid lines illustrates the attenuation of these cross-link interactions at both low and high temporal frequencies. The dashed lines represent gains when negative feedback loops are opened. The effects of tonic adaptation on cross-link interactions are eliminated under these conditions, allowing the gain to continue to increase unrestricted by tonic adaptation as temporal frequency is decreased.

The reception or termination of cross-link shown in Fig. 16-4 is proposed to occur at a summing junction that is also located between the phasic and tonic components in the feedforward paths of accommodation and convergence. This organization is indicated by adaptive changes in tonic vergence stimulated monocularly with accommodative vergence (Fig. 16-8) and aftereffects illustrating adaptive changes in the resting focus of accommodation stimulated binocularly with vergence accommodation while pinhole pupils were placed before both eyes (Fig. 16-9).

This organization provides a means by which adaptable tonic components of accommodation and vergence are able to influence the AC/A and CA/C ratios by altering the negative feedback loop for these cross-link interactions and by enhancing the cross-link innervation at the receiving site by tonic adaptation (Schor, 1986). As indicated by the temporal frequency response for cross-link interactions, the gains of the initial phasic responses of accommodative vergence (AC/A ratio) and vergence accommodation (CA/C ratio) are higher during the open-loop response to perceived distance than they are during the subsequent closed-loop fine-tuned responses to blur and disparity, during which tonic adaptation gradually takes over and relieves the phasic innervation of cross-coupling interactions between accommodation and convergence. This frequency-dependent organization increases the velocity of proximal responses by accommodation and vergence to large changes in perceived distance and markedly reduces and minifies the contribution of accommodative vergence and vergence accommodation during steady, fixed viewing conditions.

A MECHANISM FOR MODIFYING THE AC/A AND CA/C RATIOS

The mechanism by which tonic adaptation might interact with the cross-coupling is illustrated by the

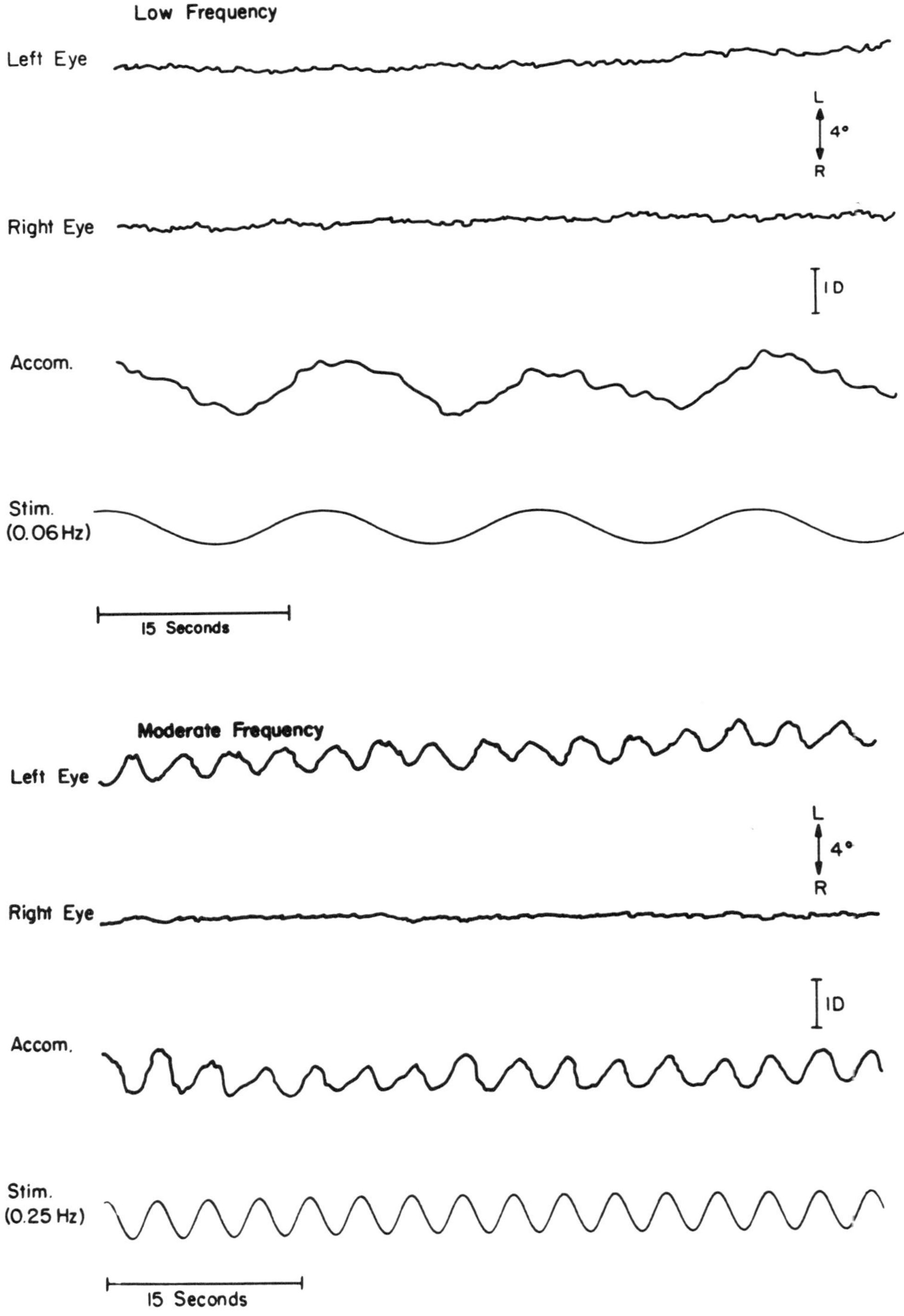

FIGURE 16-5. Accommodation was stimulated in the right eye while the left eye was occluded. Movements of the left eye were stimulated by accommodative vergence. Shown are the accommodative vergence responses to low- (0.06 Hz) and moderately high-frequency (0.25 Hz) sinusoidal blur stimuli presented under "closed-loop" conditions. Accommodative vergence was reduced with the low-temporal-frequency stimulus.

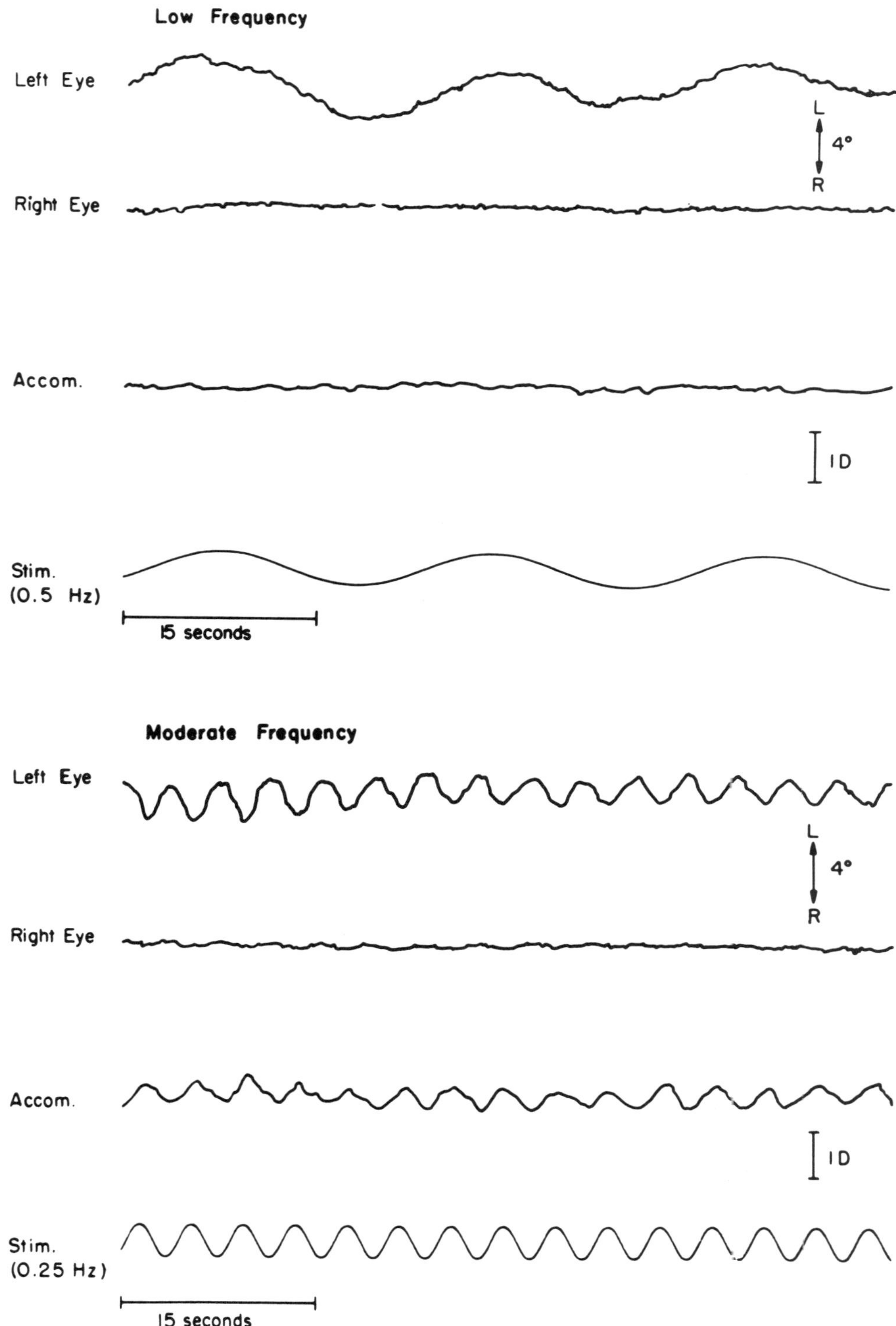

FIGURE 16-6. Disparity vergence was stimulated with an asymmetric disparity presented to the left eye in order to allow accommodation to be monitored in the right eye. The accommodative loop was opened by a 0.75-mm pinhole aperture stop imaged in the eye pupil plane in Maxwellian view. Shown are the vergence accommodation responses to a low (0.05 Hz) and a moderately high (0.25 Hz) frequency of sinusoidal disparity oscillation presented under "closed-loop" conditions. Vergence accommodation was reduced with the low-temporal-frequency stimulus.

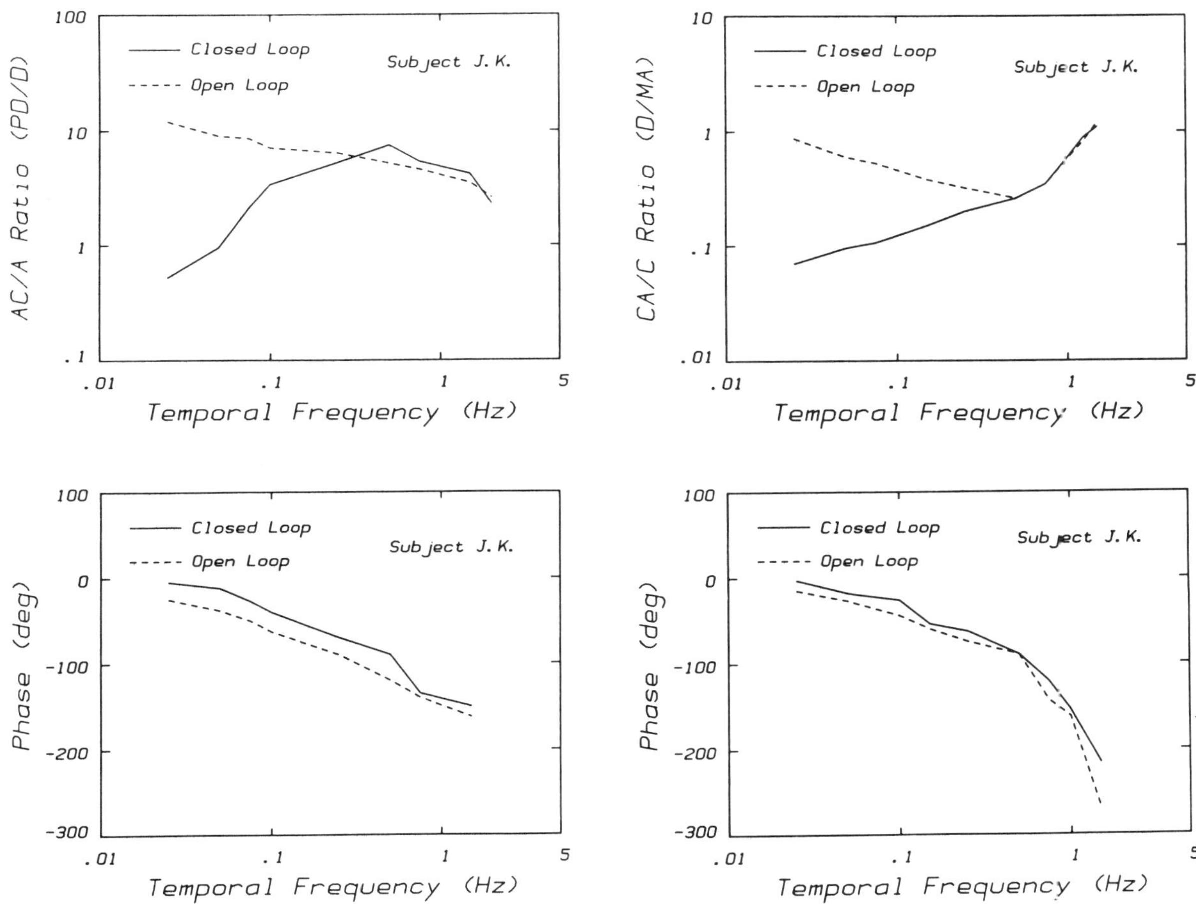

FIGURE 16-7. Gain and phase plots for accommodative vergence and vergence accommodation responses to sinusoidal blur and disparity stimuli presented under open-loop (dashed line) and closed-loop (solid line) conditions. Gain is the response AC/A or CA/C ratio. Phase lag equals the difference between response and stimulus phase. Stimulus amplitude for open- and closed-loop conditions were ±0.5 and ±1.0 D, respectively, for accommodation and ±1 PD and ±10 PD for vergence, respectively.

block diagrams in Fig. 16-10. The upper loop in the figure has the same organization as either feedback loop shown in Fig. 16-4. The cross-link is shown to originate after the site of the phasic controller and before the site of the tonic controller. This loop is reorganized in the lower half of the figure with the cross-link of accommodative vergence or vergence accommodation as the system output. When this is done, the adaptable tonic integrator is in the negative feedback loop of the system, and accordingly any tonic shifts in response level are subtracted from the input to the system. Placing the integrator in the feedback path causes the system output to be equal to the derivative or velocity of the system input. Accordingly, the AC/A ratio is reduced as tonic adaptation to low temporal frequencies increases. When tonic accommodation is the principal component of the accommodative response to low temporal frequencies, there is little if any phasic activity in the feedforward loop to stimulate accommodative vergence. However, when phasic accommodation is the principal component of the accommodative response to higher temporal frequencies, there is more phasic activity in the feedforward loop to stimulate accommodative vergence. Expressed in engineering terms, when the output of an integrator, located in a negative feedback loop, is subtracted from the system input, it yields a phasic or transient stimulus or input to the feed-forward loop.

The effectiveness of regulating the gain of AC and CA by altering the rate of tonic adaptation has been illustrated by fatigue of tonic adaptation (Schor and Tsuetaki, 1987). Fatigue is produced by continuous tracking of repetitive ramp stimuli for accom-

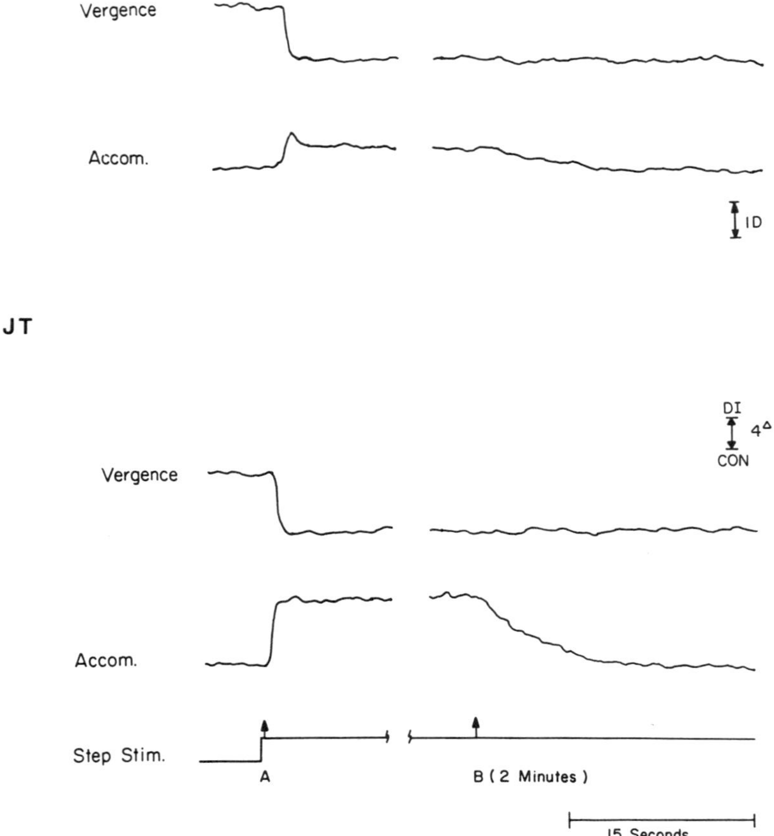

FIGURE 16-8. Aftereffects of accommodation and vergence shown for subject J.T., who had a low CA/C ratio (0.3 D/MA) and a high AC/A ratio (8 PD/D). Aftereffects shown in the upper half of the figure followed 2 min of adaptation to base-out prism while the accommodative loop was opened by pinhole pupils. After 2 min of adaptation (B), the vergence loop was opened by occluding the left eye. The vergence response persisted while accommodation decayed rapidly. Aftereffects in the lower half of the figure followed 2 min of monocular adaptation to a 2-D accommodative stimulus. The accommodative loop was opened at B with a pinhole pupil. As it did following vergence adaptation, the vergence response persisted while the accommodative response decayed rapidly.

modation (2 D at 0.75 Hz) and vergence (5° at 0.5 Hz) (Figs. 16-11 and 16-12). Fatigue occurs rapidly after tracking these stimuli for only 4 min. When produced by accommodative tracking, fatigue results in a reduction in the duration of both accommodative aftereffects and vergence aftereffects (Figs. 16-13 and 16-14). After fatigue reduces aftereffects of accommodation, there is a marked increase of the AC/A ratio (Fig. 16-15). Similarly, after fatigue reduces aftereffects of vergence stimulated by prism, there is a marked increase in the CA/C ratio (Fig. 16-16) (Schor and Tsuetaki, 1987). Clearly, the gain of cross-link interactions becomes elevated by a reduction in the gain of tonic adaptation.

Imbalanced Tonic Adaptation Produces Abnormal AC/A Ratios

This reciprocal relationship between adaptability of tonic components and the gain of cross-link interactions suggests a possible mechanism underlying abnormally high and low AC/A ratios observed in convergence excess and convergence insufficiency anomalies that fall within the Duane–White classification of near-point anomalies (Duane, 1897; Tait, 1951). The abnormal AC/A ratios in these patients can be accounted for by too little or too much tonic adaptation of accommodation.

Measures of the AC/A ratio were categorized as abnormally high if they exceeded 7 PD/D and low

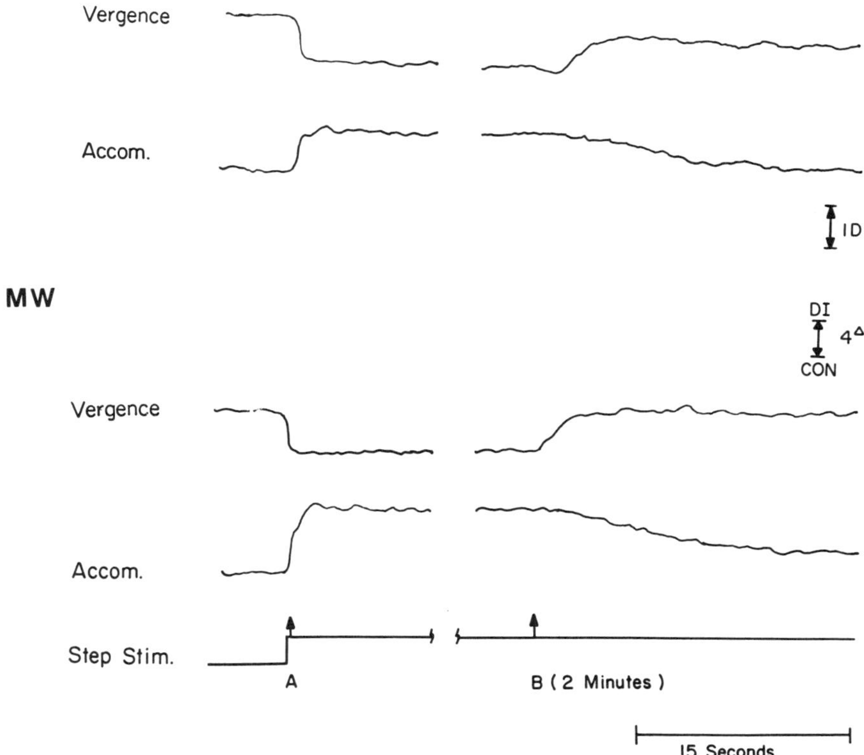

FIGURE 16-9. Aftereffects of accommodation and vergence are shown for subject M.W., who has a low AC/A ratio (2.5 PD/D) and a high CA/C ratio (0.9 D/MA). Aftereffects shown in the upper half of the figure followed 2 min of adaptation to 6 base-out prism while the accommodative loop was opened by pinhole pupils. After 2 min of adaptation (B), the vergence loop was opened by occluding the left eye. The accommodative response persisted while vergence decayed rapidly. Aftereffects, in the lower half of the figure, followed 2 min of monocular adaptation to a 2-D accommodative stimulus. The accommodative loop was opened at B with a pinhole pupil. As it did following vergence adaptation, the accommodative response persisted while the vergence response decayed rapidly.

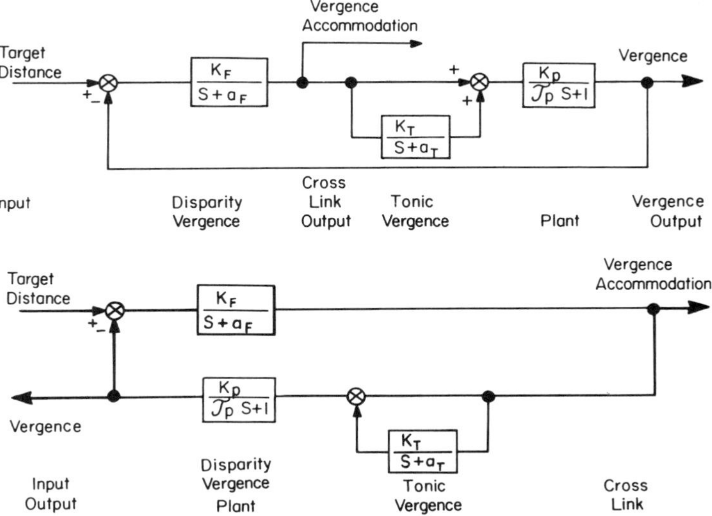

FIGURE 16-10. The influence of adaptable tonic vergence on the phasic response of vergence accommodation is illustrated in the two block diagrams.

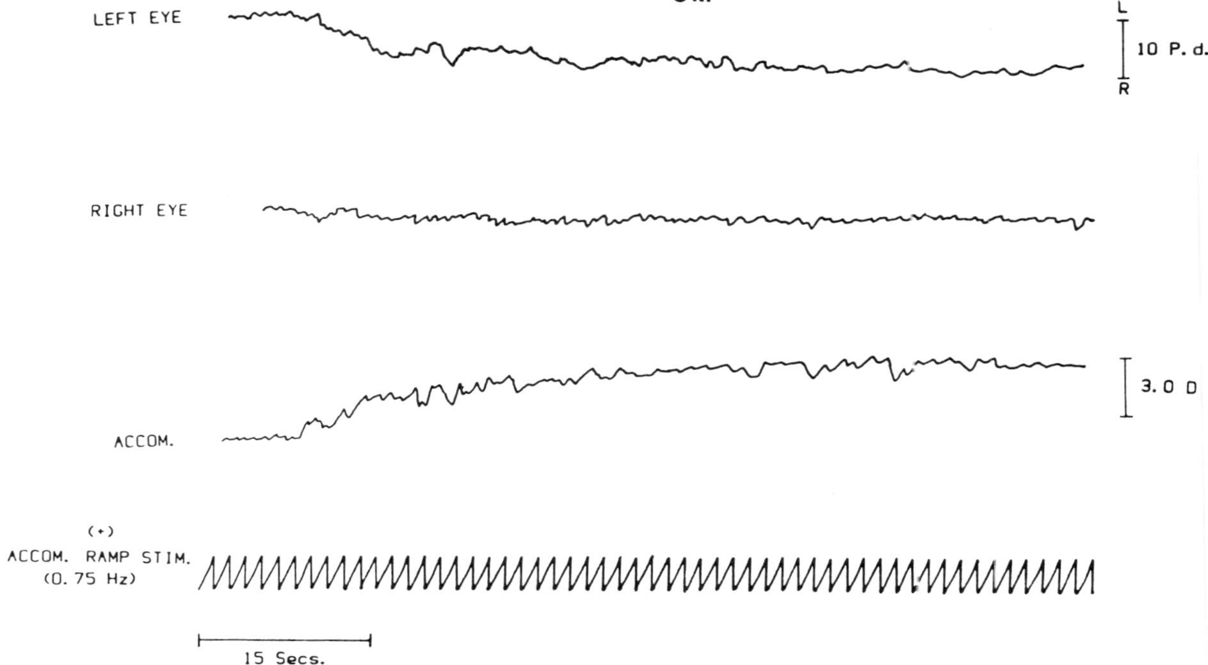

FIGURE 16-11. Accommodative and monocular accommodative vergence tracking responses are illustrated for subject C.M. in response to a 2-D, 0.75-Hz positive accommodative ramp stimulus presented monocularly for 4 min. Occasionally, as illustrated in the recording, the accommodation response level increases during ramp tracking. Vergence also becomes more esophoric because of accommodative vergence.

TABLE 16–1
Gradient and Response AC/A Ratios and Accommodative and Vergence Adaptation After-Effects for Subjects Categorized as Converge Excess or Insufficiency. After-Effects are Expressed as a Ratio of Changes in Open/Closed Loop Amplitudes and Duration in Seconds.

Subject	Clinical category	Gradient AC/A	Response AC/A	Accommodative ratio	Accommodative duration(s)	Vergence ratio	Vergence duration(s)
				Low AC/A (CI)			
LS	CI	2	2.7	0.3	135	0	0
CF	CI	0	0	0	15	0	0
CH	CI	2	0.33	0.23	80	0	0
JC	CI	2	1	0	20	0	0
GC	CI	3	0	0.5	120	0.11	45
MM	CI	0	0	0.8	120	0	0
GF	CI	1	1	0.5	85	0.30	50
RG	CI	3	3.2	0.67	120	0	0
MS	CI	2	0.8	0	20	0.25	60
AS	CI	2	0	0.23	60	0.17	45
Mean		1.7	0.903	0.323	77.5	0.083	20
				High AC/A (CE)			
SB	CE	10	12.8	0.2	55	0.2	40
MC	CE	8	10	0	10	1.0	125 (> 120)
JL	CE	11	13	0	0	0.27	60
LV	CE	12	20	0.17	105	0.73	135
LL	CE	10	11	0	20	0.80	120
Mean		10.2	13.36	0.074	38	0.6	96

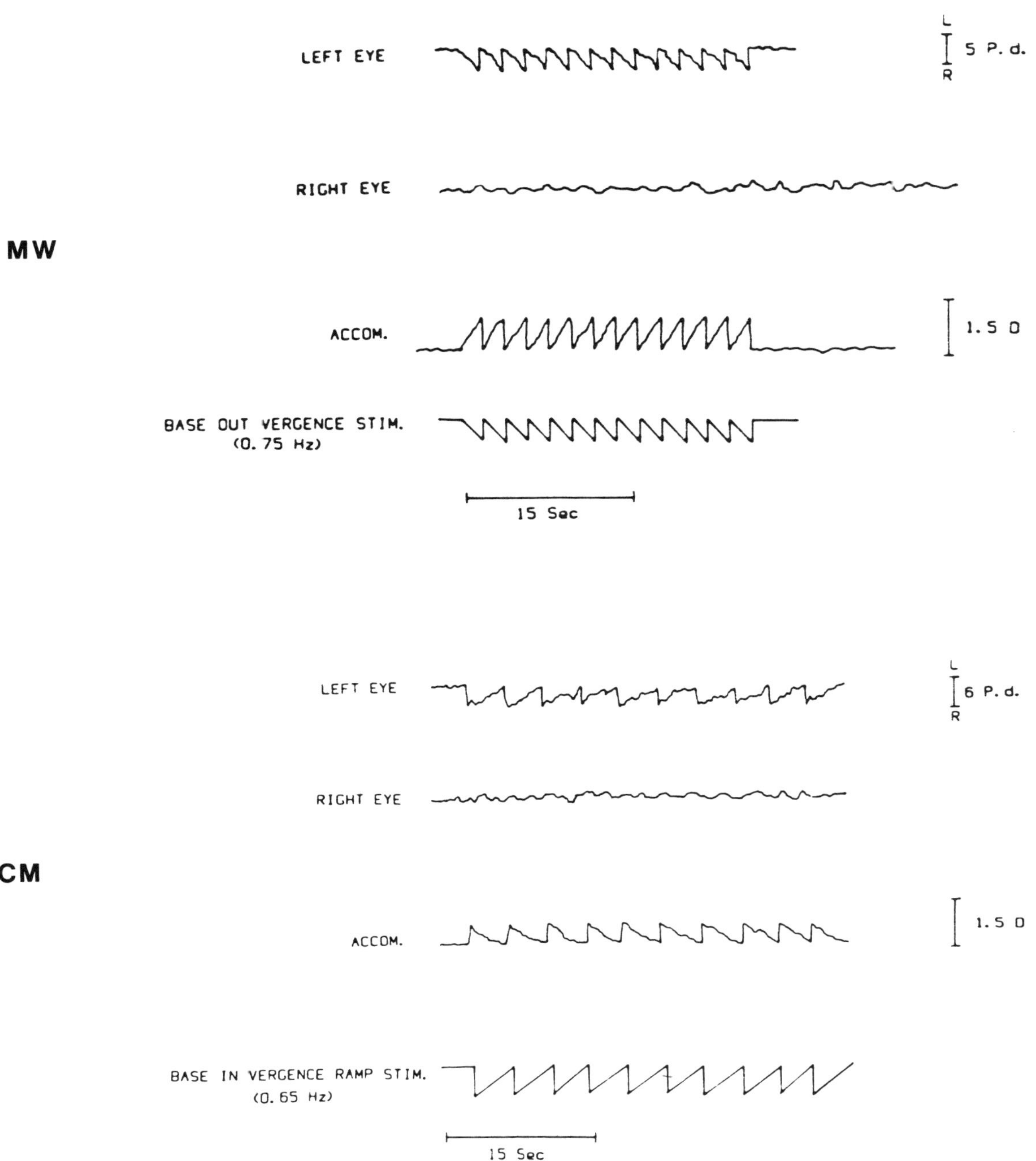

FIGURE 16-12. Disparity vergence and vergence accommodation tracking responses to 5-PD, 0.75-Hz base-out (top) and base-in (bottom) ramp stimuli are shown for subject M.W. As convergence or divergence is stimulated, accommodation makes a concurrent ramp movement because of vergence accommodation.

if they were less than 3 PD/D. Table 16-1 illustrates that aftereffects were nearly five times greater in amplitude and twice as long in duration for the low-AC/A category than the high-AC/A group. Interestingly, vergence aftereffects had the reverse trend for the two groups. Adaptation of vergence was eight times greater in amplitude and lasted five times longer in the high- than low-AC/A categories. This result is consistent with our observation that vergence adaptation can be stimulated by accommodative vergence.

Similarly, vergence aftereffects were five times greater and lasted three times longer in the low- than high-CA/C category. In addition there was significantly more adaptation of accommodation in the high-CA/C category than the low-CA/C category.

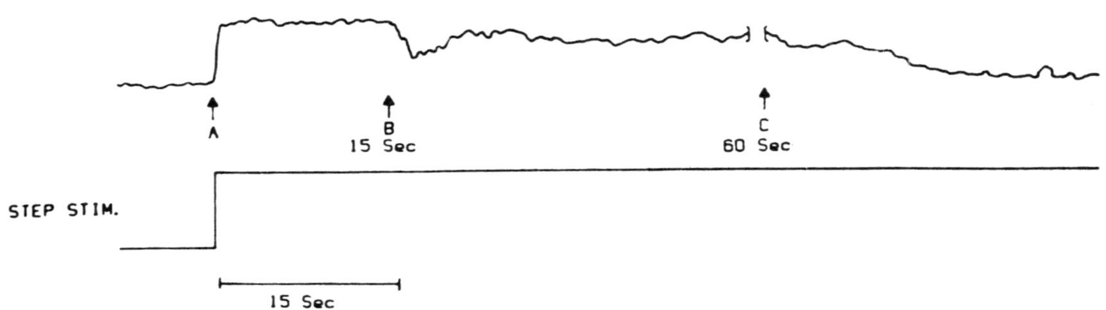

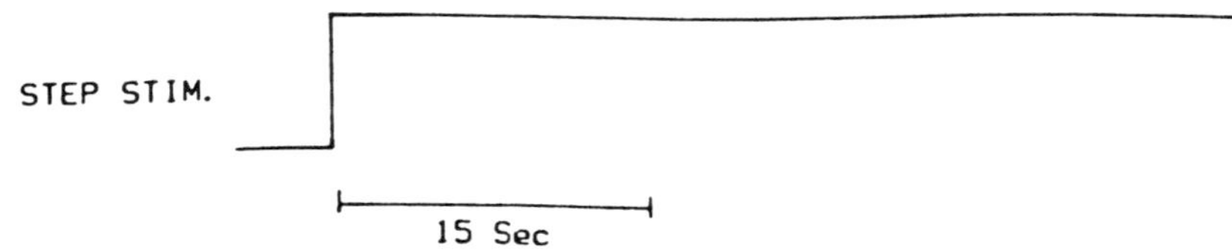

FIGURE 16-13. Tonic accommodative aftereffects are illustrated for subject C.M. after stimulating accommodation monocularly by 2 D for 15 sec (A) and then opening the accommodative loop with a projected pinhole pupil (B). The time discontinuity at C illustrates the initial decay of the accommodative aftereffect 1 min after the accommodative loop was opened. Aftereffects of accommodation are observed prior to (top) but not following (bottom) performance of the accommodative tracking task. Aftereffects lasted for 10 min before the tracking procedure, and they decayed in several seconds following accommodative ramp tracking.

The extremely high and low values for the AC/A ratios are influenced significantly by marked imbalances in adaptation of tonic accommodation and tonic vergence. Excessive adaptation of accommodation and too little adaptation of vergence results in a low AC/A ratio, whereas the opposite extremes result in a high AC/A ratio. Correction of the imbalance could restore a normal AC/A ratio as shown by the fatigue study (Schor and Tsuetaki, 1987).

DEVELOPMENTAL AND AGING FACTORS REQUIRING CALIBRATION OF CROSS-LINK INTERACTIONS

The tonic adaptive regulation of accommodative vergence and vergence accommodation is necessary throughout life to maintain coordinated motor responses to large changes in viewing distance. Continued increase in interpupillary distance after 6 weeks of life requires no further adjustment of the phoria for distance targets; however, some reduction of near exophoria is still required. As suggested by Maddox (1893), differential adjustment of the near and far phoria is most readily accomplished by increasing the proportion of accommodative convergence during early infancy. Indeed, as visual acuity develops rapidly during the first 2 years of life, the eyes' depth of field decreases, allowing the infant to sense a retinal image blur of near objects and to accommodate and converge more fully than when accommodation is unable to sense and respond to blur. A similar facilitation of the accommodative near response occurs by means of convergence accommodation with the development of

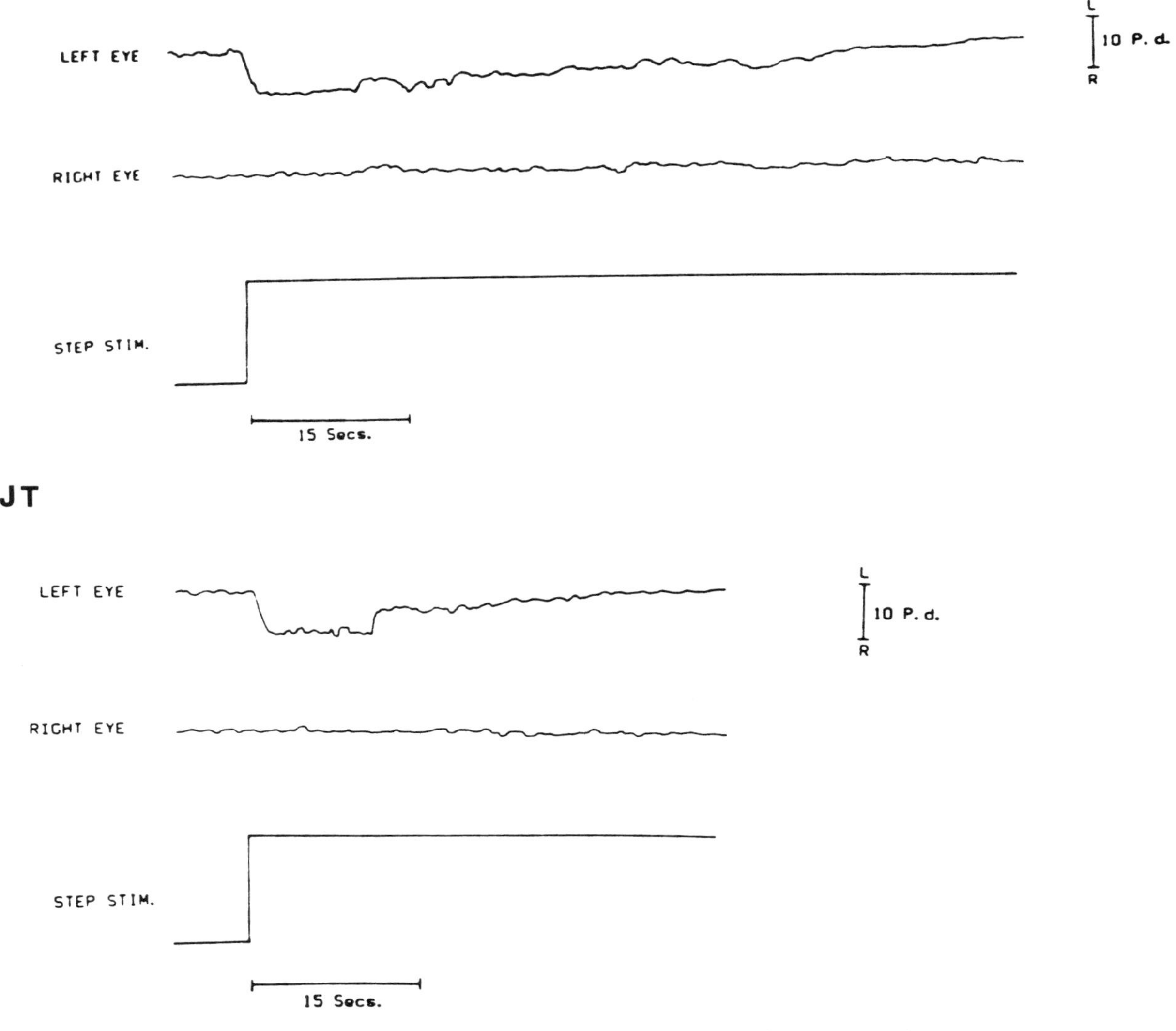

FIGURE 16-14. Tonic aftereffects of vergence are observed in subject J.T. after stimulating disparity vergence with a 10-PD base-out stimulus for 15 sec prior to (top) and following (bottom) 4 min of the disparity vergence ramp tracking task. The vergence loop was opened after 15 sec of disparity stimulation by occlusion of the left eye. Vergence decayed slowly prior to the tracking task (45 sec), and it decayed quickly (within several seconds) following the vergence ramp tracking task.

accurate convergence at 4 months of life (Aslin, 1977). Disparity evokes both convergence and convergence accommodation, which Fincham (1951) believed was a coarse adjustment mechanism for accommodation whose residual errors were corrected by the fine adjustment mechanism, optical reflex accommodation. Convergence accommodation becomes reduced in the early 20s, when the lens begins to sclerose. The tonic adaptation process would only adjust the resting focus for one viewing distance, and this set point would then be moved to new distances by convergence accommodation. As with vergence, this would eliminate the need to develop a different resting focus for all perceived stimulus distances. Increases in interpupillary distance and changes in plasticity of the interocular lens produce new demands on convergence and accommodation responses to near viewing distances. These changes require continued tuning or recalibration of cross-coupling gains of accommodative vergence and vergence accommodation. This recalibration process can be accomplished by altering the response time of adaptable tonic vergence and tonic accommodation (Schor and Tsuetaki, 1987).

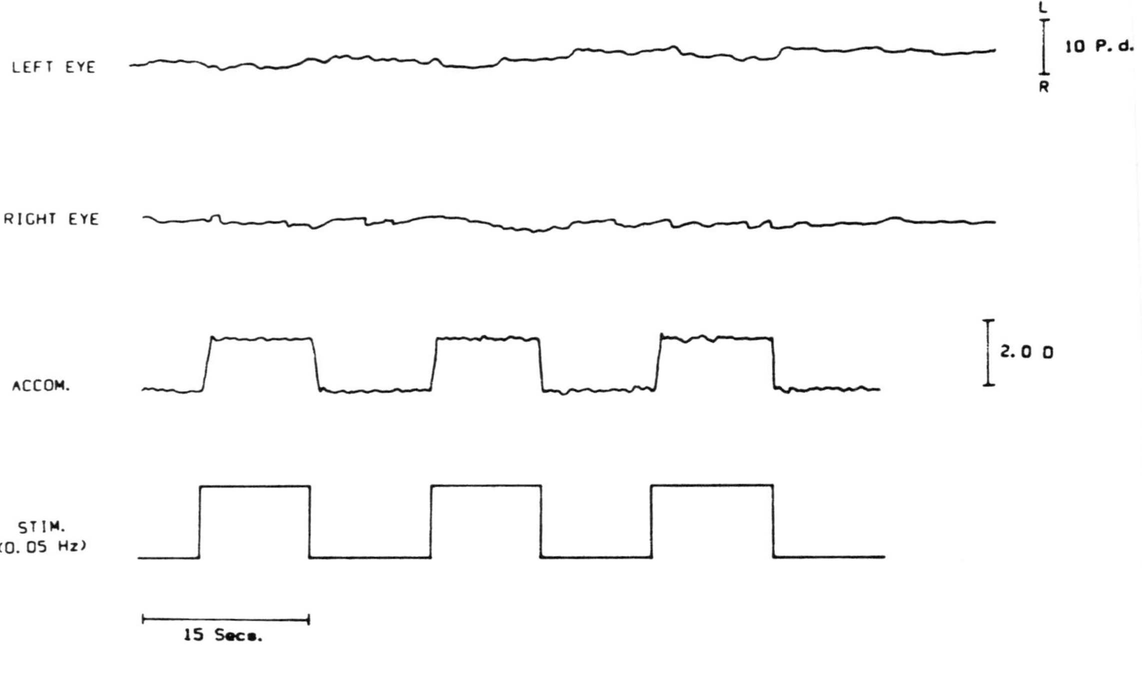

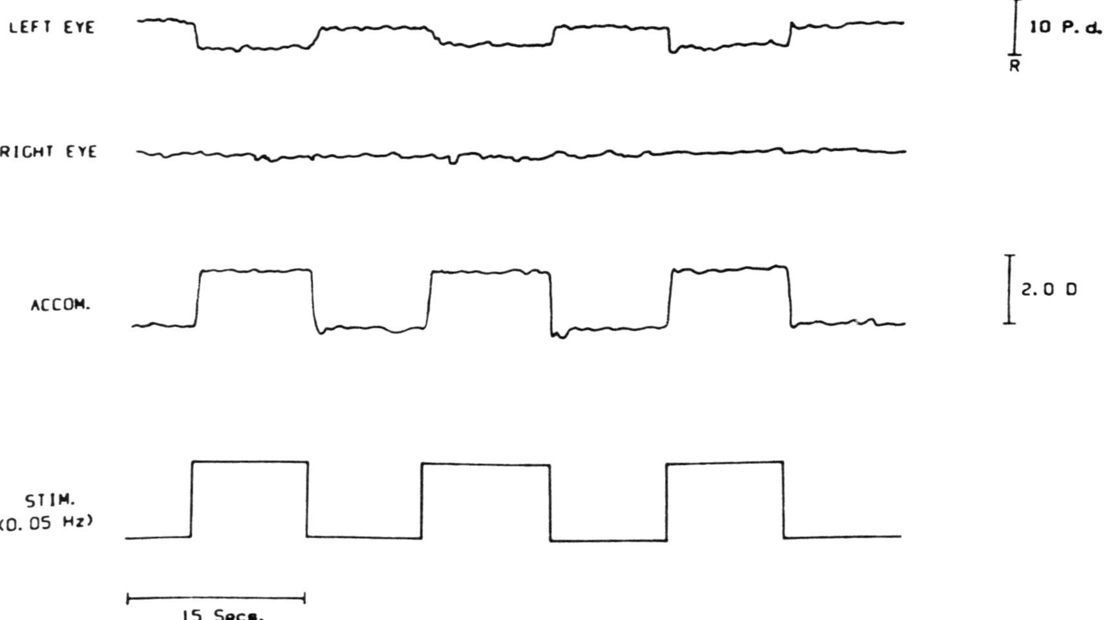

FIGURE 16-15. Accommodative vergence responses to a 2-D, 0.05-Hz monocular accommodative stimulus are illustrated prior to (top) and following (bottom) the accommodation ramp tracking task (2 D at 0.75 Hz for 4 min). The AC/A ratio increased from <2 PD/D to 4 PD/D as a result of the ramp tracking task.

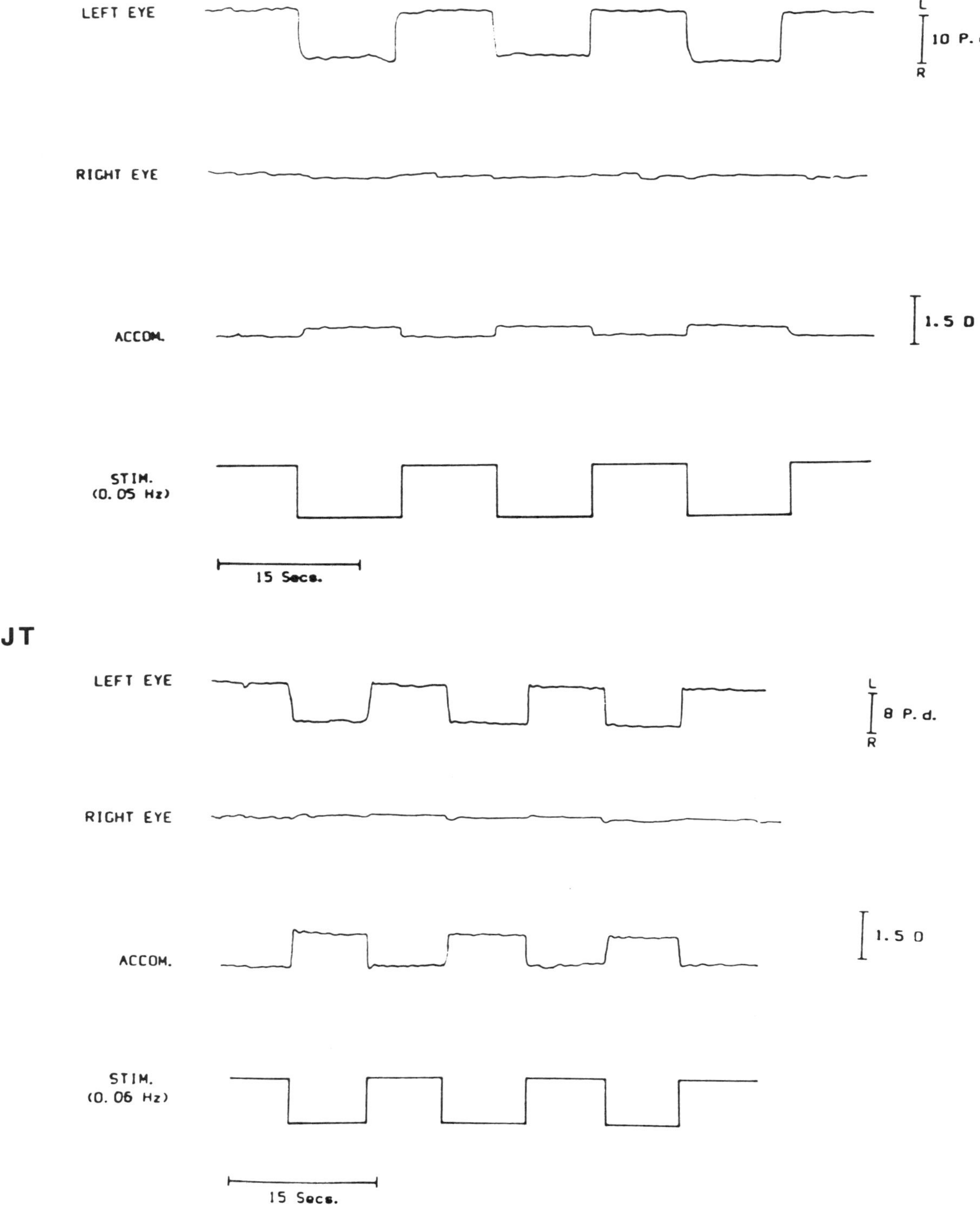

FIGURE 16-16. Vergence accommodation responses to an 8-PD base-out, 0.05-Hz disparity-step stimulus are illustrated for subject J.T. prior to (top) and following (bottom) performance of the vergence ramp tracking task (5 PD BO at 0.75 Hz for 4 min). The CA/C response ratio increased from 0.2 D/MA to 0.6 D/MA as a result of the vergence ramp tracking task.

SUMMARY

The near visual response or near triad consists of a concomitant increase in accommodation, vergence, and pupillary constriction. All three motor systems exhibit phasic and tonic responses as evidenced by their open-loop decay time constants. The phasic response adapts readily to efforts of accommodation and vergence. This adaptation provides a means of fine-tuning the emmetropization and orthophorization process during development. Stimulation of accommodation under monocular viewing conditions or of vergence while wearing a pinhole pupil before each eye also elicits all three motor responses. The cross-coupling between accommodation and vergence provides a means of dynamically adjusting the tonic set points of the two motor systems to a near or far working distance. The accommodative vergence cross-links play the dominant role in the proximal response to dynamic continuous changes in viewing distance while convergence accommodation dominates cross-link interactions during the proximal response to large step changes in viewing distance.

The magnitude (gain) of the changes in vergence stimulated by accommodation (AC/A ratio) and the magnitude of accommodation stimulated by vergence (CA/C ratio) can be modified by adaptation of accommodation and vergence to lenses and prisms, respectively. For example, fatigue of adaptable accommodation causes an increase in the AC/A ratio, and fatigue of vergence adaptation causes an increase of the CA/C ratio. Finally, patients with abnormally high AC/A ratios (convergence excess) demonstrate little if any adaptation of the resting focus of accommodation, whereas patients with abnormally low AC/A ratios (convergence insufficiency) demonstrate robust adaptation of the resting focus of accommodation to lenses. Active tuning of the AC/A and CA/C ratios by tonic adaptation of accommodation and convergence plays an important role during development when there is rapid growth of the interocular separation as well as later in life in response to gradual sclerosis of the lens. The tuning is important to maintain accurate responses of accommodation and vergence to dynamic and static proximal cues to distance.

REFERENCES

Aslin, R., 1977, Development of binocular fixation in human infants, *J. Exp. Child Psychol.* **23:**133–150.

Bedell, H., and Wick, B., 1989, Magnitude and velocity of proximal vergence, *Invest. Ophthalmol. Vis. Sci.* **30:**755–760.

Duane, A., 1897, A new classification of the motor anomalies of the eye based upon physiologic principles, *Am. J. Ophthalmol. Otol.* Part I. October 1896, Part II, January 1897. **6:**84.

Ebenholtz, S. M., 1983, Accommodative hysteresis: Or precursor for induced myopia? *Invest. Ophthalmol. Vis. Sci.* **24:**513–515.

Ebenholtz, S. M., and Zander, P. A. L., 1987, Accommodative hysteresis: Influence on closed loop measures of far point and near point, *Invest. Ophthalmol. Vis. Sci.* **28:**1246–1249.

Ellerbrock, V., 1950, Tonicity induced by fusional movements, *Am. J. Optom. Arch. Am. Acad. Optom.* **27:**8–20.

Erkelens, C. J., 1987, Adaptation of ocular vergence to stimulation with large disparities, *Exp. Brain Res.* **66:**507–516.

Fincham, E. F., 1951, The accommodation reflex and its stimulus, *Br. J. Ophthalmol.* **35:**381–393.

Knudsen, E. I., 1989, Fused binocular vision is required for development of proper eye alignment in barn owls, *Vis. Neurosci.* **2:**35–40.

Leibowitz, H. W., and Owens, D. F., 1975, Night myopia and the intermediate dark focus of accommodation, *J. Opt. Soc. Am.* **65:**1121–1128.

Maddox, E. C., 1893, *The Clinical Use of Prism,* 2nd ed., John Wright and Sons, Bristol, p. 83.

McLin, L., Schor, C. M., and Kruger, P., 1988, Changing size (looming) as a stimulus to accommodation and vergence, *Vis. Res.* **28:**883–893.

Ogle, K. N., and Prangen, A., 1953, Observations of vertical divergences and hyperphorias, *Arch. Ophthalmol.* **49:**313–334.

Rethy, I., 1969, Development of simultaneous fixation from the divergent anatomical eye position in the neonate, *J. Pediatr. Ophthamol.* **6:**92–96.

Schor, C. M., 1979, The relationship between fusional vergence eye movements and fixation disparity, *Vis. Res.* **19:**1359–1367.

Schor, C. M., 1986, Adaptive regulation of accommodation vergence and vergence accommodation. Glen Fry Lecture, *Am. Acad. Optom. Physiol. Opt.* **63:**587–609.

Schor, C. M., and Kotulak, J., 1985, Dynamic interactions between accommodation and convergence are velocity sensitive, *Vis. Res.* **26:**927–942.

Schor, C. M., and McLin, L., 1988, The effect of luminance on accommodative and convergence aftereffects, *Clin. Vis. Sci.* **3:**143–154.

Schor, C. M., and Tsuetaki, T., 1987, Fatigue of accommodation and vergence modifies their mutual interactions, *Invest. Ophthalmol. Vis. Sci.* **28:**1250–1259.

Schor, C. M., Johnson, C., and Post, R., 1984, Adaptation of tonic accommodation, *Ophthalmol. Physiol. Opt.* **4:**133–137.

Schor, C. M., Kotulak, J. C., and Tsuetaki, T., 1986a, Adaptation of tonic accommodation reduces accommodative lag and is masked in darkness, *Invest. Ophthalmol. Vis. Sci.* **27:**820–827.

Schor, C. M., Robertson, K. M., and Wesson, M., 1986b, Disparity vergence dynamics and fixation disparity, *Am. J. Optom. Physiol. Opt.* **63:**611–618.

Sherman, S. M., 1972, Development of interocular alignment in cats, *Brain Res.* **37:**187–192.

Stark, L., Kenyon, R. V., Krishnan, V. V., and Ciuffreda, K. J., 1975, *Disparity Vergence:* A proposed name for a dominant component of binocular vergence eye movements, *Am. J. Optom & Physiol Optics* **57:**606–609.

Sun, F., and Stark, L., 1983, Pupillary escape intensified by large pupillary size, *Vis. Res.* **23:**611–615.

Tait, E. F., 1951, Accommodative convergence, *Am. J. Ophthalmol.* **34:**1093–1107.

Part II

Conclusions

KEVIN O'REGAN

Like all biological systems, the visual system is able to adapt to changes produced by growth, aging, or damage. The subject of Part II, which I organized with Dr. John Findlay from Durham, was to explore the capacity and limitations of the visual system in making such adaptations.

Professor Howard from Toronto began by reviewing and distinguishing long-term and short-term perceptual adaptation. He stressed the importance of classifying them according to whether they occur in a proprioceptive system, an egocentric system—that is retinocentric, head centric, or body centric—or whether they occur in a frame of reference external to the body. Professor Howard showed how some forms of perceptual adaptation can be understood in terms of models in which information is coded by populations of cells, each cell being sensitive to wide, overlapping ranges rather than narrow ranges of visual attributes. Doctor Cornilleau-Peres from Paris described a study she has done with the support of Essilor in which she observed to what extent and at what speed the distortions caused by cylindrical lenses can be perceptually compensated.

Intense work on adaptation in the oculomotor system started in the early 1970s, although as usual Helmholtz had anticipated it. Professor Miles from N.I.H. and Dr. Gauthier from Marseilles were among the first to demonstrate the plasticity of the vestibuloocular reflex, and Professor Miles gave a very good review of this area, which is highly relevant to the problem of eye–head coordination in spectacle wearers. Progressive lenses pose particular problems in this sense, and this question was the purpose of Dr. Gauthier's study. He showed in particular the importance of acuity demands in determining eye–head coordination adaptation. Another important oculomotor subsystem is the saccadic system, where it is now known that the precision of saccades is continuously being recalibrated. Dr. Deubel from Munich showed how this process, which is one necessary aspect of adaptation to spectacle corrections, operates in independent local lobes corresponding to different radial directions.

The area of accommodation and vergence is a classical area of optometry. Adaptations in this system are present in both tonic and phasic components. Dr. Henson from Cardiff, discussing tonic adaptation, showed that adaptation of phoria can occur in both comitant and noncomitant ways. Dr. Schor from Berkeley reviewed the various changes that can occur in the AC/A and CA/C ratios and suggested that many near visual disorders that had been classified on the basis of AC/A ratio might be better described in terms of imbalanced motor adaptive disorders of accommodation and vergence. Finally, Dr. Judge from Oxford reported progress in efforts to find a neurophysiological substrate of control and adaptation in the accommodation and vergence systems.

In conclusion, it is clear that the area of visuomotor adaptation is currently highly active. Judging from the contributions to Part II, our new understanding may soon be of significance to optometrists.

KEVIN O'REGAN • Laboratory of Experimental Psychology, CNRS, EHESS, Université René Descartes Paris V, 75006 Paris, France.

Part IIIA

MECHANISMS FOR SEPARATING SUPERIMPOSED IMAGES

Part IIIA

Introduction

LAWRENCE STARK

Jorge Luis Borges, who died just a few years ago, wrote many fantastic tales, one of which deals with simultaneous vision. In the tale, an obscure writer (presumably Borges himself) is invited by a madman to his cellar in the middle of Argentina, where there is a spot on earth that contains all other places. This spot is called an *Aleph*. He is lying in the cellar, and suddenly the spot appears to him as forewarned:

> In that single gigantic instant I saw millions of acts both delightful and awful; not one of them amazed me more than the fact that all of them occupied the same point in space, without overlapping or transparency. What my eyes beheld was simultaneous, but what I shall now write down will be successive, because language is successive. Nonetheless, I'll try to recollect what I can.
>
> On the back part of the step, toward the right, I saw a small iridescent sphere of almost unbearable brilliance (1), at first I thought it was revolving; then I realized that this movement was an illusion, created by the dizzying world it bounded. The *Aleph*'s diameter was probably little more than an inch, but all space was there, actual and undiminished. Each thing (a mirror's face, let us say) was the universe. I saw the teeming sea; I saw daybreak and nightfall; I saw the multitudes of America; I saw a silvery labyrinth (it was London); I saw, close up, unending eyes watching themselves in me as in a mirror; I saw all the mirrors on the earth, and none of them reflected me; I saw in a backyard of Soler Street the same tiles that 30 years before I'd seen in the entrance of a house in Fray Bentos; I saw bunches of grapes, snow, tobacco, lodes of metal, steam; I saw convex equatorial deserts and each one of their grains of sand. I saw the rotted dust and bones that had one deliciously been Beatriz Viterbo; I saw the circulation of my own dark blood; I saw the coupling of love and the modification of death; I saw the *Aleph* from every point and angle, and in the *Aleph* I saw the earth and in the earth the *Aleph* and in the *Aleph* the earth. I saw my own face and my own bowels; I saw your face; and I felt dizzy and wept, for my eyes had seen that secret and conjectured object whose name is common to all men, but which no man has looked upon—the unimaginable universe.*
>
> —*The Aleph and Other Stories 1933–1969*

Our own vision "sees" only one thing at a time, perhaps. Indeed, the scanpath theory suggests that a single cognitive model dominates perception and controls lower level processes of vision by means of active looking fixations or shifts in attention. The Necker cube illusion is perhaps a demonstration of simultaneous vision reduced to its simplest form. One looks at the inner corners and perceives how they flip back and forth; there are a large number of other illusions that our perceptions produce. Our brain is creating all these illusions, just as the writer was creating the story.

In Part IIIA, we look at the question of superimposed images and simultaneous vision and the way the brain disambiguates these, from two different points of view. From the scientific viewpoint, where many careful scientists have taken objective measurements, various operators in the visual brain are studied, along with how the visual system disambiguates images using clues from accommodation, vergence, motion, stereopsis, and binocular vision. A series of chapters tell us about that aspect. From another point of view, we have conjectural information on how the brain perceives in a top-down fashion. One of the earliest and most capable workers in this field was Alfred Yarbus, the famous Russian eye movement and vision scientist. He died about a year ago without ever having been able to leave Russia because of his status as *persona non grata*.

I would like to dedicate this section to his memory: a beautiful example of his work known to visual scientists around the world is the superimposition of eye movements on the head of Queen Nefertiti.

*Borges, J. L., 1970, The Aleph and Other Stories (N. T. de Giovanni, ed. and transl.), E. P. Dutton, New York.

LAWRENCE STARK • School of Optometry, University of California, Berkeley, California 94720.

17

A Breakdown in Simultaneous Information Processing*

RICHARD F. HAINES

INTRODUCTION

The ability to detect accurately, recognize, and extract useful information from two superimposed sources of visual information at the same time is of both theoretical and practical interest. The scientific literature in such areas as the psychology of attentional processes, cognitive processing, sensory channel capacity, and other fields contains numerous such studies. Among them are the early auditory perception research of Broadbent (1952, 1958) in which he proposed a single-channel filter theory of information processing, Treisman's (1960) modified sequential processing model involving a two-stage filter, and Deutsch and Deutsch's (1963) preattentive parallel processing model. Most past visual research employed simple stimuli (point light sources, simple geometric shapes, etc.). Nevertheless, the effectiveness of information transfer using more complex, real-life stimulus fields demanded further study from the standpoint of helping to design modern instrumentation for commercial turbojet airplane flight. This study combined both basic and applied objectives, as will be seen.

A pilot landing an airplane is concerned with two principal fields of information, his external view through the windshield and the instrument panel. They differ in their location within the pilot's field of view, their angular extent, range of brightness and color, dynamic characteristics, and various other sensory and cognitive dimensions. All the pilot needs to do is to shift his gaze back and forth from one to the other sequentially and keep them both in sufficient optical focus (Weintraub *et al.*, 1985) to be able to extract the required flight guidance and control information they contain. It is my belief that the very process of nodding the head up and down (during this information gathering), of rotating the eyeballs within the bony sockets, and of the image translation across the retina associated with these saccades (each) acts to add significant coded "bits" in the afferent information stream to the central nervous system to signify a change in information source by way of previous associations.

The main purpose of this investigation was to find out how well pilots could accurately and rapidly extract information from either one or both of two superimposed sources of flight guidance and control information.

METHOD

Procedure and Experimental Design

Each pilot was given extensive familiarization in the control of a NASA Ames Research Center fixed-base (no cockpit motion) flight simulator, which was programmed to fly like a Boeing model 272 turbojet airplane. They were given a battery of vision tests (all possessed a Class A medical certificate) and shown a 20-min video tape explaining and demonstrating the various features of the head-up display (HUD) in flight. They were not told about the possibility of runway obstruction but only that

*This chapter is based on an earlier study by E. Fischer, T. A. Price and the author (Fischer *et al.*, 1980) conducted at NASA–Ames Research Center.

RICHARD F. HAINES • Research Institute for Advanced Computer Science, Ames Research Center–NASA, Moffett Field, California 94035.

the purpose of the study ". . . was to test the HUD under various environmental conditions."

Flight training consisted of flying five increasingly difficult approaches without HUD; HUD training followed, using simple maneuvers in clear visibility and calm winds. The difficulty of the flight conditions was continually increased by adding turbulence, crosswind, wind shear, and low visibility (simulated fog). The author sat in the right seat serving as the First Officer during all training and data runs; I operated flaps, landing gear, and verbalized whatever landing-related information the pilot asked for (e.g., altitude and distance to go). A colleague (E. Fischer) was also present to ensure that all experimental conditions were accurately preset and that the pilot fully understood what was required prior to each run. This pretest training averaged about 3 hr, about 2 hr of which were devoted to flying with the HUD.

Each trial began with the simulator's flight being controlled by an autopilot (to ensure uniform initial conditions for all runs and all pilots) in level flight at 1500-ft altitude and 8 miles from the runway. The First Officer kept looking outside; he pressed a button and called out the words "ground in sight" and, later, "runway in sight" at the earliest possible moment during the low-visibility runs. During the no-HUD trials the pilot remained looking down at the instrument panel until he heard the words "ground in sight" and then looked up quickly toward the outside scene. He would look up and down several more times before making the decision to land or go around (Haines *et al.*, 1980). The pilot was instructed to say "decision" when he felt he had enough real-world information to decide whether to land or to go around. The runway obstruction was not mentioned in advance except on two occasions by the First Officer by mistake. During the HUD runs the pilot looked at the HUD symbology and external scene at the same time throughout the entire approach to landing.

Four of the eight pilots were presented the runway obstruction condition with the HUD first, and the other four encountered it with the standard instrument panel first. Each also received a second obstruction trial from 13 to 21 runs later but with the opposite HUD-present or HUD-absent condition to counterbalance the presentation order.

An unobtrusive infrared (Honeywell) oculometer was installed inside the cockpit to monitor the pilot's eye movements during the trials (approx. 1° accuracy). The output from this system was a small white dot, which was accurately superimposed on a videotape record of what the pilot was seeing through the windshield. In addition, another camera was aimed at the pilot's head; its image was inlaid into a corner of the main scene for later analysis.

HUD Symbology Information

The computer-generated HUD symbols were seen at the pilot's normal eye position by reflection off of a semitransparent glass. Both the HUD and the external scene were collimated to apparent optical infinity (Od). A three-sided trapezoidal outline of a runway was seen on the HUD. It always remained in proper registration with the image of the "real" (external scene) runway, as is shown in Fig. 17-1. This photograph was taken from the pilot's eye position at an altitude of 72 feet and airspeed of 131 knots.

Other HUD-presented information included a digital airspeed, altitude, heading, vertical rate, angle of attack, etc. We will not be concerned with the other available information except to say that the symbology was carefully designed to provide all the information a pilot usually requires to make these approaches (Bray, 1980).

Simulator

The scene presented outside the simulator's forward windows was obtained from a full-color, 900:1 scale model with accurate runway, approach lights, surrounding terrain, terminal building, etc. The flight control inputs made by the pilot controlled the movement of a small, medium-resolution color TV camera relative to this model and its runway. It provided a realistic scene, which was displayed on large-raster color monitors located behind large-diameter collimating lenses mounted to the simulator. Electronically generated "fog" was produced by a white raster overlay, which effectively reduced the scene's contrast over a wide range. This fog effect can be seen at the top of Fig. 17-1. The cockpit forward windows each subtended 45° arc in width.

Test Subjects

Eight commercial pilots currently certified to fly Boeing model 727 aircraft from two domestic airlines were tested. All possessed 20:20 distance acuity, normal color and depth perception, and no dysfunctions that may have influenced their performance. They were paid for their services and were also highly motivated because of their interest in

FIGURE 17-1. Simulator pilot's forward visual scene at an altitude of 72 feet and 131 knots with runway obstruction clearly visible.

having an opportunity to learn more about head-up displays. The three First Officers had flown an average of 1113 hr, and the Captains an average of 2350 hr.

RESULTS AND DISCUSSION

During the approaches in low visibility there was a period of about 4–5 sec after break-out before the runway obstruction became visible as an obstacle. During the first 1–2 sec of runway visibility the retinal size of the obstructing airplane's image was still fairly small, allowing the green HUD symbology to partially obscure it. However, flying with the HUD, the pilot was head-up at the first opportunity to perceive the obstacle. In contrast to this, when flying (no HUD) with head-down instruments, most of the pilots glanced up quickly (when the First Officer called out "ground in sight") and then looked back at their instruments again. Some pilots made several more quick glances up and down; each would come head up permanently only after they heard the First Officer say "runway in sight." By this time the obstruction was clearly visible.

Table 17-1 presents the mean response time (sec) and type of response made only during the runway obstruction trials. It should be remembered that the subjects made many other landing approaches under difficult flight conditions both before and after these two particular trials.

Response time represents the time from the first opportunity to see the obstruction to the first positive recognition made by the pilot (most often a verbal exclamation!). For the no-HUD runs, response time represents the time from the final look up to their response [this was considered the available time for perceiving the obstacle, although earlier (brief) opportunities were possible]. Also shown is the type of response made: MA, executed a missed approach or go-around; IMA, intended missed approach when he saw the obstruction and called for a MA, but an experimenter terminated the run before it could be carried out.

As seen in Table 17-1, two pilots never saw the obstructing airplane in front of them at all. Both of these runs were with HUD first; i.e., they were the pilots' first exposure to the obstruction with the HUD present. Pilot D was a First Officer with about 2000 hr in the 727; his performance with both head-down instruments and the HUD was considered good. He flew 21 data runs prior to the first obstacle encounter. The obstruction became increasingly visible about 4 sec after breakout from the fog and cloud layer, yet the pilot gave no indication of seeing it. He was pleased with his approach "setup," as indicated by his comments (altitude given in parentheses): ". . . oh, it looks good (110 ft altitude) . . . the HUD looks good (90 ft) . . ." The experimenter terminated the run at an altitude of about 67 ft* when he said, "Oh, wait a minute! It looked good, the flare bars were coming up . . . then the picture disappeared." The subsequent conversation between the First Officer and the Captain was, "I saw an airplane. Did you see it?" "No." "You didn't see it?" "No, sir."

TABLE 17–1
Response Time and Type of Response to Airplane on the Runway

	First exposure		Second exposure	
Pilot	Response time (sec)	Type of response[a]	Response time (sec)	Type of response[a]
	No HUD		HUD	
A	—	MA, warned	2	MA
B	1	MA	3	MA
C	—	MA, warned	3	MA
G	3	IMA	3	IMA
Mean	2.0		2.75	
	HUD		No HUD	
D	6[b]	Never saw	2	MA
E	5	IMA	2	MA
F	6[b]	Never saw	1	MA
H	5	IMA	1	IMA
Mean	5.5[b]		1.5	

[a]MA, missed approach; IMA, intended missed approach.
[b]These values are not response times, since the pilot never saw the airplane; rather, they denote the available time in which the airplane could have been seen; the mean also includes these values.

Pilot F was a high-flight-time Captain who demonstrated exceptionally good performance both with and without HUD. The runway obstruction run was his seventh data run. He indicated his "Decision (140 ft) . . . to land (110 ft)," and proceeded to do so. The experimenter terminated the run at an altitude of 50 ft.* The pilot was surprised. Captain: "Didn't get to flare on this one." First Officer "No you didn't . . . I was just looking up as it (the picture) disappeared, and I thought I saw something on the runway. Did you see anything?" Captain: "No, I

*The startlingly large apparent size of the runway obstruction at this point may be appreciated by reference to Fig. 17-1, where the separation distance is even greater.

did not." The experimenters suggested that an equipment failure was probably to blame. Both of these pilots saw the obstruction during the second exposure without HUD (13 runs and 21 runs later, respectively) and executed missed approaches. Later, when he was shown the videotape of this run, Pilot D said, "If I didn't see it (the tape), I wouldn't believe it. I honestly didn't see anything on that runway."

On all other trials all of the other pilots did see and react to the obstacle. Mean response time was longer with the HUD than without it (4.13 versus 1.75 sec) and was also longer on the first exposure than on the second (3.75 versus 2.13 sec). Following their first encounter with the runway obstruction, most pilots were on the alert for it during later runs, being more cautious about calling out "decision." Other findings are presented elsewhere (Fisher *et al.*, 1980).

The above finding suggests that the HUD information may restrict or even inhibit perception of information from the external scene when flying a simulator. The longer perception and response times during runs with the HUD may be the result of a combination of causes that can be discussed in three groups: cognitive–psychological factors, visual–optical factors, and operational factors. Of central concern, of course, is the degree to which these results may be generalized to the real world.

Cognitive–Psychological Factors

(1) The landing phase of flight is particularly stressful and calls for concentration of attention by all members of the flight crew. The HUD may well contribute to this focusing or may detract from it. It should be noted, however, that the 2- to 3-sec attentional–response lag without the superimposed HUD information is still highly significant during an actual high-speed approach. Much can happen within 2 to 3 sec time when traveling at 135 knots. A pilot who encountered a real runway obstruction (without HUD) stated, "It may be difficult to believe that I looked and did not see an aircraft coming toward me. . . . I should have seen the other aircraft, and I bear the responsibility for not having seen it. . . ." (NASA, 1978). (2) Attentional distraction produced by other ongoing cockpit activities should also be considered as a contributor to this effect. (3) Since landings occur at the end of a flight, the flight crew are probably more tired, and the mental effort of switching back and forth between simultaneously presented information may be slowed. (4) Because of its highly compelling appearance, HUD information may "capture" the pilot's attention altogether thereby blocking the perception or subsequent processing of the runway obstruction. (5) The runway obstruction was completely unexpected, so it did not "fit" into the pilot's expectancy model, if he had one. Flight training simulators do not present runway obstructions, but they should! (6) An important part of the HUD-presented information were two horizontal, linear "flare bars" that rose from the bottom of the HUD's optical display field somewhat unexpectedly beginning at an altitude of only 50 ft. When they appeared to climb past the HUD's central circular (airplane) symbol, the pilot was supposed to pull back on the control yoke gently to flare the airplane for a smooth landing. Pilot D was probably waiting expectantly for the appearance of these bars and even looking down for them to some degree.

Visual–Optical Factors

(1) There is the possibility that some of the HUD symbology may have overlaid or occluded the image of the airplane early in the approach. One could recognize the airplane after 2–4 sec of runway visibility if one were looking at it and expected it to be there, however. (2) Because most of the symbols occupied the center of the HUD's optical display field, they may have caused the pilot's eye scan to become more constrained than otherwise. Nevertheless, the obstructing airplane's image was in this same region of the visual field. (3) Although these pilots said that the handling characteristics of the simulator were good, the outside scene's visual quality was not as good. The color images were less focused than found in the real world, and the horizontal raster lines gave a less than real impression. Nevertheless, none of the pilots felt that the outside scene's image quality would have inhibited them from seeing the obstruction on the runway.

Operational Factors

(1) Although runway obstructions continue to be an operational problem and "represent a significant safety problem" (NASA, 1978), definite procedures have been established to help cope with them. The problem of runway obstructions must be faced with or without HUD. (2) With or without HUD, this (landing) phase of flight is particularly stressful and requires focused attention. (3) Since the HUD pro-

vides more accurate visual information than does the outside scene (during the final flare and touchdown), pilots may shift their attention to it for primary flight control. That is, the highly precise nature of the HUD information may have led these pilots to ignore some or all of the available outside visual cues. (4) The pilots knew that they were in a ground simulator where the physical consequences of making mistakes is not particularly great. This awareness may have contributed to a lowering of critical attentional focus both inside the cockpit and outside. (5) In real life, the First Officer would probably have called out the presence of the obstruction to the pilot. This was not done here on purpose. This absence of "normal" cockpit procedures probably reinforced the impression that no runway obstruction could occur in the simulator. In the two trials where the First Officer did announce the presence of the obstacle, the Captain had no difficulty executing a timely missed approach and go-around. (6) These pilots said that they felt confident in flying the HUD after only 25–30 approach trials, but it is possible that they had concentrated on flying the HUD and not the entire airplane, i.e., where all aspects of the visual, intellectual, and physical environment must be integrated smoothly and efficiently.

CONCLUSIONS

Although this study has raised some interesting questions concerning the capability of pilots to process visual information simultaneously, one should not jump to the conclusion that these results automatically apply to the real world, at least not without more study. The sample size is small, and the other differences between the simulation environment and the real world make it unwarranted to extrapolate the findings. These results should have theoretical applications as well. Further research is needed that controls for all of the factors discussed above. This will not be an easy task but should be performed anyway because of the criticality of the consequences involved should a pilot encounter a real obstruction with the HUD some day.

SUMMARY

Eight experienced commercial pilots flew a NASA B-727 turbojet airplane simulator under daytime, low-visibility (fog) conditions toward a realistic collimated TV image of a runway. Each made 18 approaches using a head-up display (HUD) that presented all necessary flight guidance and control information (e.g., altitude, airspeed, position, heading, angle of attack) and 13 using conventional instrument-panel-presented information. On two randomly chosen trials they were presented with unexpected, "critical" information in the form of a large jet airplane located directly ahead of them on the runway. Could they see and avoid it in the relatively few seconds available? This would be a useful test of simultaneous processing of visual information located in approximately the same region of the visual field. Four of the pilots flew the obstruction trials with HUD on first, and the other four flew it with the no-HUD condition first. It was found that (1) flight performance was generally better with the HUD, (2) perception time of the runway obstruction was longer with the HUD, perhaps suggesting cognitive interference by the superimposed information, and (3) two pilots flying the HUD approach first never saw the obstruction at all! These results are discussed in terms of possible explanations for the breakdown in simultaneous information processing

REFERENCES

Bray, R. S., 1980, *A HUD Format for Application to Transport Aircraft Approach and Landing*, NASA Technical Memorandum 81199, Washington, D.C.

Broadbent, D. E., 1952, Listening to one of two synchronous messages, *J. Exp. Psychol.* **44:**51–55.

Broadbent, D. E., 1958, *Perception and Communication*, Pergamon Press, London.

Deutsch, J. A., and Deutsch, D., 1963, Attention: Some theoretical considerations, *Psychol. Rev.* **70:**80–90.

Fisher, E., Haines, R. F., and Price, T. A., 1980, *Cognitive Issues in Head-up Displays*, NASA Technical Paper 1711, Washington, D.C.

Haines, R. F., Fisher, E., and Price, T. A., 1980, Head-up transition behavior of pilots with and without head-up display in simulated low-visibility approaches, NASA Technical Paper 1720, Washington, D.C.

NASA, 1978, *Aviation Safety Reporting System: Eighth Quarterly Report*, NASA Technical Memorandum 78540, Washington, D.C.

Treisman, A. M., 1960, Contextual cues in selective listening, *Q. J. Exp. Psychol.* **12:**242–248.

Weintraub, D. M., Haines, R. F., and Randle, R J., 1985, Head-up display (HUD): II: Runway to HUD transitions monitoring eye focus and decision times, in *Proceedings of the Human Factors Society 29th Annual Meeting*, (Santa Monica: Human Factors Society of America), pp. 615–619.

18

Ambiguous Figures: A Paradigm for Separation of Superimposed Images

LAWRENCE STARK

INTRODUCTION

How does the brain separate simultaneously viewed images? A number of preattentive, primitive, hard-wired visual processes are discussed in other chapters in Part IIIA. These include stereo viewing, motion as a disambiguating process, accommodation, and vergence. The residual "classical" case can be defined as two clearly viewed superimposed images not in relative motion with respect to one another. Still, the brain can handle this case. Figure–ground separations and cognitive shifts with different interpretations of an ambiguous figure form the paradigm that here concerns us.

Our approach is to utilize measurements of eye movements to make inferences concerning higher-level visual processes. The scanpath theory or scanpath mechanism for top-down vision (Fig. 18-1) postulates that a motor–sensory cognitive representation in the brain controls the active looking process in a sequential fashion (Noton and Stark, 1971b; Ellis and Stark, 1973; Stark and Ellis, 1981).

A principal aim of this chapter is to present experimental eye movement data showing scanpath sequences used in viewing ambiguous figures. This may be adduced as evidence that higher-level perceptual processes can act to separate superimposed images.

LAWRENCE STARK • School of Optometry, University of California, Berkeley, California 94720.

METHOD

Stimuli have been presented to the subjects in a variety of ways including placing a drawing of a scene or a figure directly in front of the subjects. Slides presented on a backlit translucent screen or material presented on a computer graphics screen or on a video monitor can be used under control of the experimental computer program.

Eye movement measurement hardware utilizes either a television eye movement camera with computer processing of the eye movement pictures or reflecting infrared glasses that record analogue signals related to the direction of eye gaze. The observing and looking subject signals which aspect or interpretation of the ambiguous figure his mental representation consists of at the moment.

An online computer acts as a control and data acquisition system so that pictures can be presented and eye movements recorded under the control of laboratory-generated software programs. These programs, having passed through many generations in our laboratory, are always a compromise between the desire to be as flexible as possible and the need for program discipline so that a feasible and clean structure emerges.

Experimental protocols have been different for different phases of the results presented below and are best discussed with the results themselves. It is important to realize that explicit or implicit instructions to the subjects contribute significantly to the experimental structure. Experiments were carried out at the University of California at Berkeley under the auspices of the Committee for the Protection of Human Subjects.

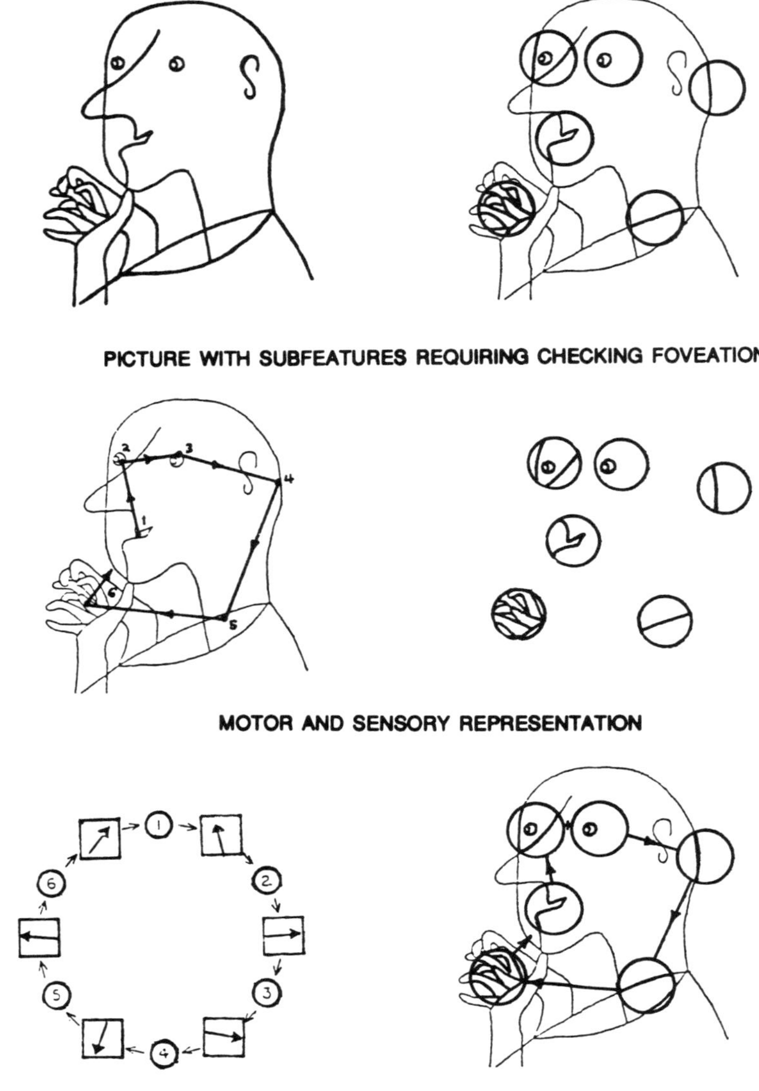

FIGURE 18-1. Scanpath theory. (scanpath mechanism for top-down vision)

RESULTS

Early Evidence for Scanpaths

Studies by Noton and Stark (1971a–c) demonstrated that repetitive sequences of saccades visited the important subfeatures of the pictures presented. These repetitive sequences constitute the definition of the "scanpath." It was further found that an individual had different scanpaths for different pictures and that different individuals had different scanpaths for the same picture. These dynamic and sequential studies of eye movements triggered the scanpath theory and were in retrospect supported by earlier studies by Buswell (1935), Yarbus (1967), and Jeannerod *et al.* (1968) that suggested the occurrence of "cycles" and the continual revisiting of subfeatures by the looking eye movement process. However, these early static studies did not have the dynamic data presentations necessary to put forward a dynamic hypothesis such as the scanpath theory.

The variance of the scanpaths suggests that a higher-level visual recognition process, unique for each individual, underlies the generation of the scanpath. Following our early studies, several researchers took it on themselves to support or disprove the scanpath hypothesis. Locher and Nodine

(1973) assessed the actual recognition performance of his subjects and the "quality" of their scanpath movements. They found that the two measures were often discordant and dismissed the scanpath hypothesis; incidently, Noton and Stark had specifically disclaimed consistent concordance of scanpath quality and recognition capability. At the same time, Locher and Nodine confirmed the basic findings of Noton and Stark, that scanpaths could be easily demonstrated and with approximately the same frequency and variance as claimed. Parker (1978) demonstrated an even higher incidence of regular scanpaths in a recognition task but, in pointing out the importance of peripheral retina in vision, missed the point that an internal cognitive model was likely driving the scanpath sequence.

Ambiguous Figures

When subjects look at ambiguous figures, their mental representation is almost always of one or another of the (usually) two possible representations. Here we have a situation with the same subject looking at the same physical picture. It was exciting then to find that the scanpath pattern changed with changes in the mental cognitive state of the subject (Ellis and Stark, 1978; Ellis and Stark, 1979; Stark and Ellis, 1981).

To the thoughtful visual scientist, it is clear that our vision of the wide world seen in high resolution and color must be an illusion. What we "see" is our current cognitive model; what our retina "sees" is a small foveated region. What our entry-level visual brain "sees" is a sequence of such small regions as the saccades drive our eyeball from hither to yon. In addition, of course, we "see" a low-resolution view of the world in our peripheral retina; the periphery is also most sensitive to flickering and moving targets. Thus, in the ambiguous figure paradigm, what we have set up is an appropriate situation to demonstrate the ability of different cognitive models to drive the saccadic eye movement sequences ("scanpaths") and thus those foveation sequences necessary to check on the particular cognitive model under consideration. In so doing, the experimenter has the opportunity to record the sequences of saccades and fixations and to demonstrate the existence of a scanpath particular to each mental state.

Recent and ongoing studies have collected eye movement recordings from a larger set of ambiguous figures such as the triply ambiguous figure (Fig. 18-2). The Necker cube has been extensively studied and reported elsewhere (Ellis and Stark, 1978), so I concentrate here on presenting more artistic and novel figures. The visual exploration study (Fig. 18-3), carried out by Ms. Maryse Leroy was part of the study of the influence of color and symmetry on visual exploration and scanpath generation. The duck–rabbit ambiguous figure set (Fig. 18-4) was part of a protocol wherein one or another of the limiting cases could be used to train the subject's *a priori* expectations. In this way we could influence the mental representation as perhaps evidenced by an identifying scanpath type when presenting the intermediate case.

Dr. Stephan Brandt has also been refining his visual imagery paradigm. In the essential experiments the subject looks at the stimulus, an irregular checkerboard, for a few moments. Then he is presented with a blank screen with only a faint indication of the relative size and position of the boundary of the checkerboard. The upper panel (Fig. 18-5) represents a result found in approximately 30% of the runs, wherein the eye movement sequence during visual imagery does not clearly resemble the eye movement sequence recorded while viewing. On the other hand, it was usual (more than 50%) to get results showing close concordance of the two scanpaths (bottom panel, Fig. 18-5), the first while actually viewing and the second during visual imagery.

Eye movements while looking at realistic and abstract art might tell us something about the validity of esthetic theories (Zangemeister *et al.*, 1989). We were interested in comparing realistic and abstract paintings viewed by sophisticated and naive subjects. Additionally, Professor Rudolf Groner at the University of Berne, Switzerland, has contributed the "global–local dichotomy" to the development of the scanpath hypothesis. He defined "local" scanpaths as comprised of smaller eye movements interjected within "global" scanpaths composed of larger eye movements (Groner, 1989). Groner's local–global dichotomy, an easy one to measure and quantify, was an apparent behavioral feature of the sequences of eye movements (Fig. 18-6) occurring while viewing works of art. The left-hand histogram of each pair represents the number of saccades in a bin of particular saccadic size range. It can be noted that small saccades are prominent for the realistic figures on the right. The narrower set of histograms represents fixation durations for which no significant findings were observed (Groner *et al.*, 1984).

FIGURE 18-2. Triply ambiguous figure. Upper picture shows old man with moustache and big nose, old woman with gnarled nose and large chin, and young woman in chapeau and fur muff with elegant profile interrupted by eyelash. Left lower figure shows 15 sec of viewing with 39 sequential fixations and multiple switchings signaled and recorded between mental representations. Lower right figure shows four separate occasions when subject signaled that he was "seeing" the old man. Note similar sequence of three successive fixations during each of the four time periods. (From Stark and Ellis, 1981.)

Quantitative Measures

By now, the experienced reader will understand that quantitative assessment of the similarity of one sequence of eye movements to another is a crucial and difficult problem. Indeed, it has been an intermittent focus of our laboratory activities for a number of years. Stark and Ellis (1981) utilized Markov matrices ranging from zero order to second order, focusing, of course, on the classical first-order transition matrix. In order to provide an intuitive understanding of the relationship between the Markov coefficients and the scanpath, we used three model matrices to generate three scanpaths (Fig. 18-7). The scanpaths (lower diagrams) have foveations with random (rectangular) distributions of exact fixation within each half-degree "fixation fovea"; in this way, the number of saccades between foveations can be denoted graphically. The deterministic matrix (left panel) produces a deterministic scanpath with only this residual variation. The random matrix (right panel) generates a confusion of saccades, with all connections being equally likely and no true scanpath evident. The probabilistic matrix (middle panel) may be similar to motor–sensory representations generating usual scanpaths when these are evident in human vision. Further quantitative studies can be carried out with respect to statistical dependency in visual scanning (Ellis and Stark, 1986); an interesting recent suggestion by Rizzo *et al.*, (1990)

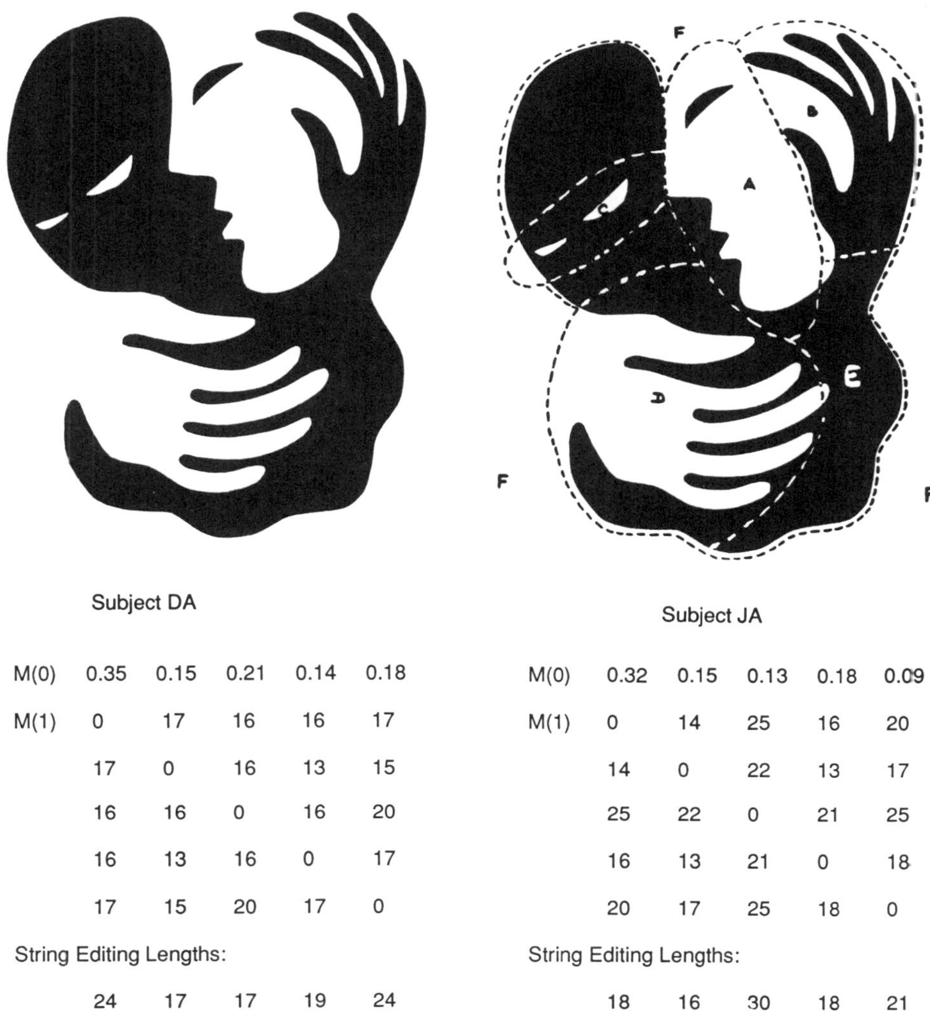

Subject DA

M(0)	0.35	0.15	0.21	0.14	0.18
M(1)	0	17	16	16	17
	17	0	16	13	15
	16	16	0	16	20
	16	13	16	0	17
	17	15	20	17	0
String Editing Lengths:					
	24	17	17	19	24

Subject JA

M(0)	0.32	0.15	0.13	0.18	0.09
M(1)	0	14	25	16	20
	14	0	22	13	17
	25	22	0	21	25
	16	13	21	0	18
	20	17	25	18	0
String Editing Lengths:					
	18	16	30	18	21

FIGURE 18-3. Visual exploration. Upper left picture shows figure–ground alternation; upper right shows five areas into which eye fixations cluster plus one "bin" region (F). Lower numerical arrays represent zero-order [M(0)] and first-order [M(1)] Markov matrices and, as well, string-editing distances for two different subjects viewing this picture (From M. Leroy and L. Stark, work in progress.)

is use of asymmetric λ to score the Markov matrix.

The string-editing measure also appears to be a very useful one. One calculates the minimum cost of insertions, deletions, and transpositions to transform one string, e.g., letters of a misspelled word, into another string, e.g., the correct spelling. Strings of genes on chromosomes are an important application of string-editing distance studies. Scanpaths, sequences of eye saccades between foveations on critical subfeatures of a picture or scene, can also be thought of as strings, and several of our recent research efforts are in this direction (Brandt *et al.*, 1989; Leroy *et al.*, 1989; Hacisalihzade *et al.*, 1989).

DISCUSSION

Philosophy

The possible philosophical positions in metaphysical epistemology are relevant to the scanpath hypothesis. On the one hand, the naive realist might expect a bottom-up, external-scene-driven, nonrepetitive series of fixations (Didday and Arbib, 1972), or possibly even allowing for repetition with control by a simple algorithm. Such an algorithm might assign relative stimulus importance to subfeatures in a scene, with some time dependence to allow already seen features to lose their attraction for a long enough period to allow a fair number of

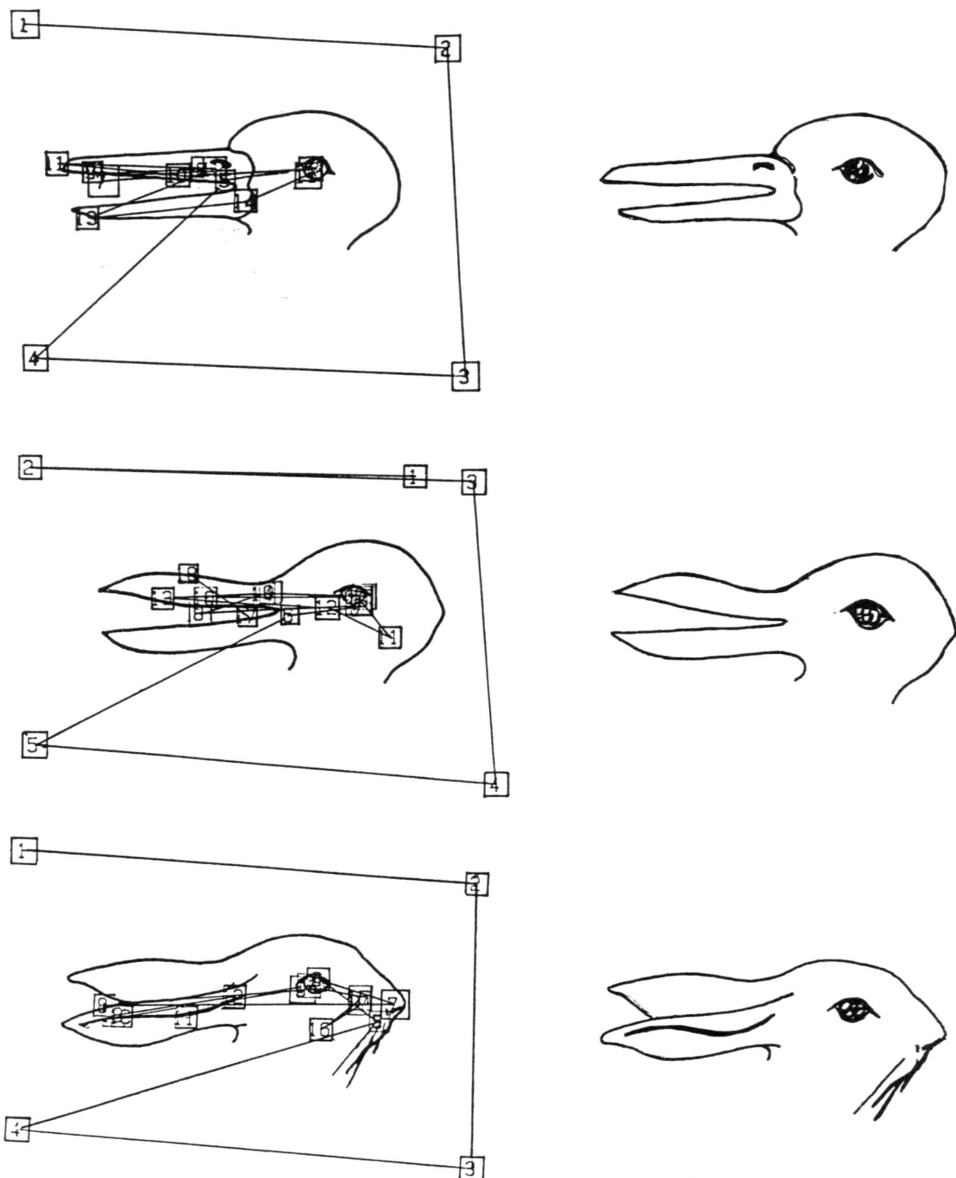

FIGURE 18-4. Ambiguous figure with delimited range. Upper picture is a duck's head, lower picture a rabbit's head, and middle picture the ambiguous junction of the two. On left are shown eye movement sequences including outline calibration eye movements (From S. Brandt and L. Stark, work in progress.)

fixations to occur in a sequence (Prokowski, 1971).

The Platonic world view postulated a full set of ideals inherent at the birth of an individual. Plato might have looked on the scanpath as an obvious consequence of a checking sequence as to which internal ideal matched what was most likely out "there" at any particular time. Indeed, the Kantian view (Fig. 18-8) is not too different but perhaps more elaborated as a result of the impact of modern science on the philosophical intellect; i.e., Newton influenced Kant. Kant's word for perception, *Anschauung,* implies an active looking, searching process driven by the *Vorstellung* or representations in cognition and seeking to add space and time coordinates to sensation, *Empfindung,* blindly driven by the doctrine of specific nerve endings. (I use endings here instead of "energies" to indicate that the filtering is done anatomically rather than functionally, in

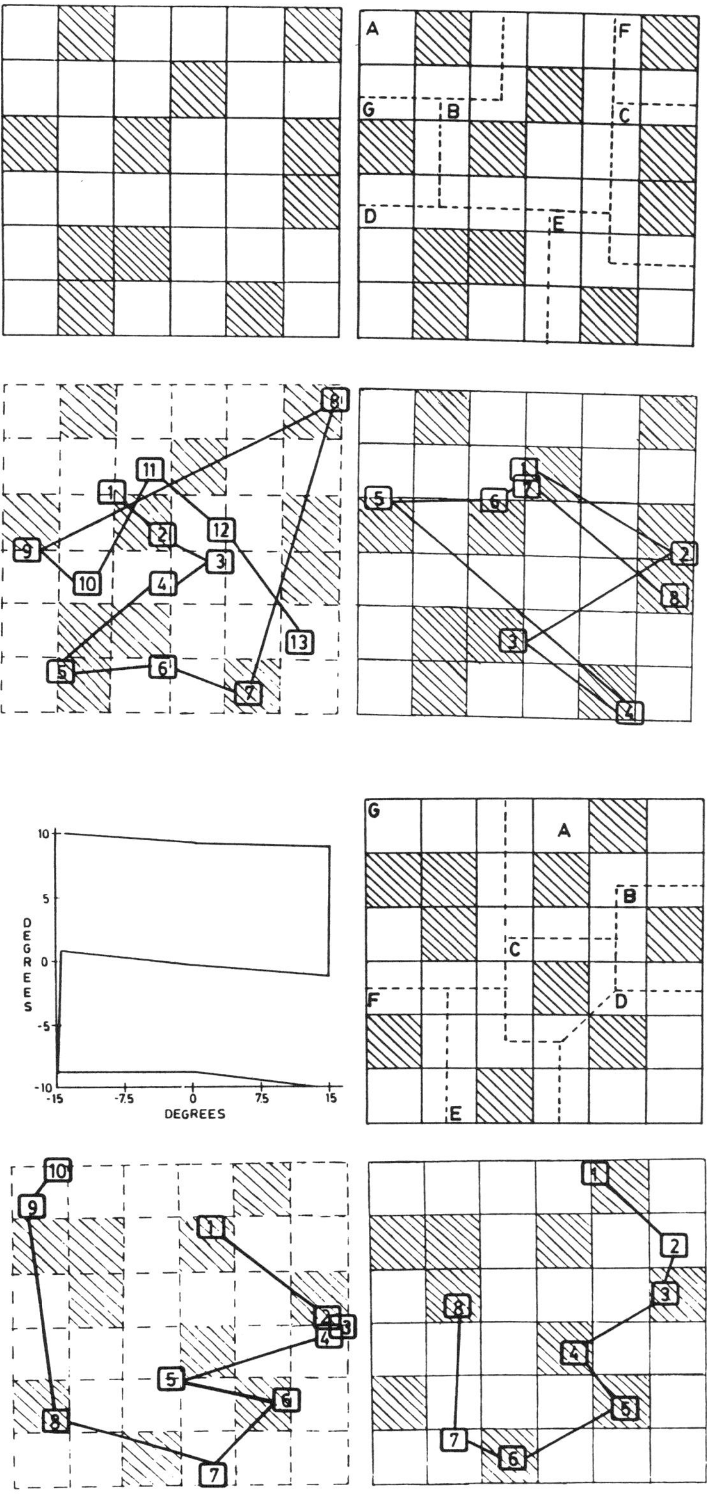

FIGURE 18-5. Visual imagery. Upper panel: Irregular checkerboard (upper left) divided into fixation areas (upper right); eye movement scanpath sequence when viewing the checkerboard (lower right); eye movement scanpath sequence while imagining that particular irregular checkerboard (lower left). Lower panel: Calibration (upper left); another irregular checkerboard divided into eye fixation areas (upper right); scanpath while viewing checkerboard (lower right); scanpath while imagining that particular checkerboard (lower left). (From Brandt *et al.*, 1989.)

FIGURE 18-6. Viewing abstract art. Selection of pictures ranging from realistic (upper right) to figurative (lower right), figurative–abstract (lower left), and abstract (upper left). Histograms show relative excess of small saccades, first "bin" dominant for rightmost "realistic" pictures; this is taken to mean a predominance of local scanpaths. Contrariwise, global scanpaths predominate for "abstract" pictures on left. (From Zangemeister *et al.*, 1989.)

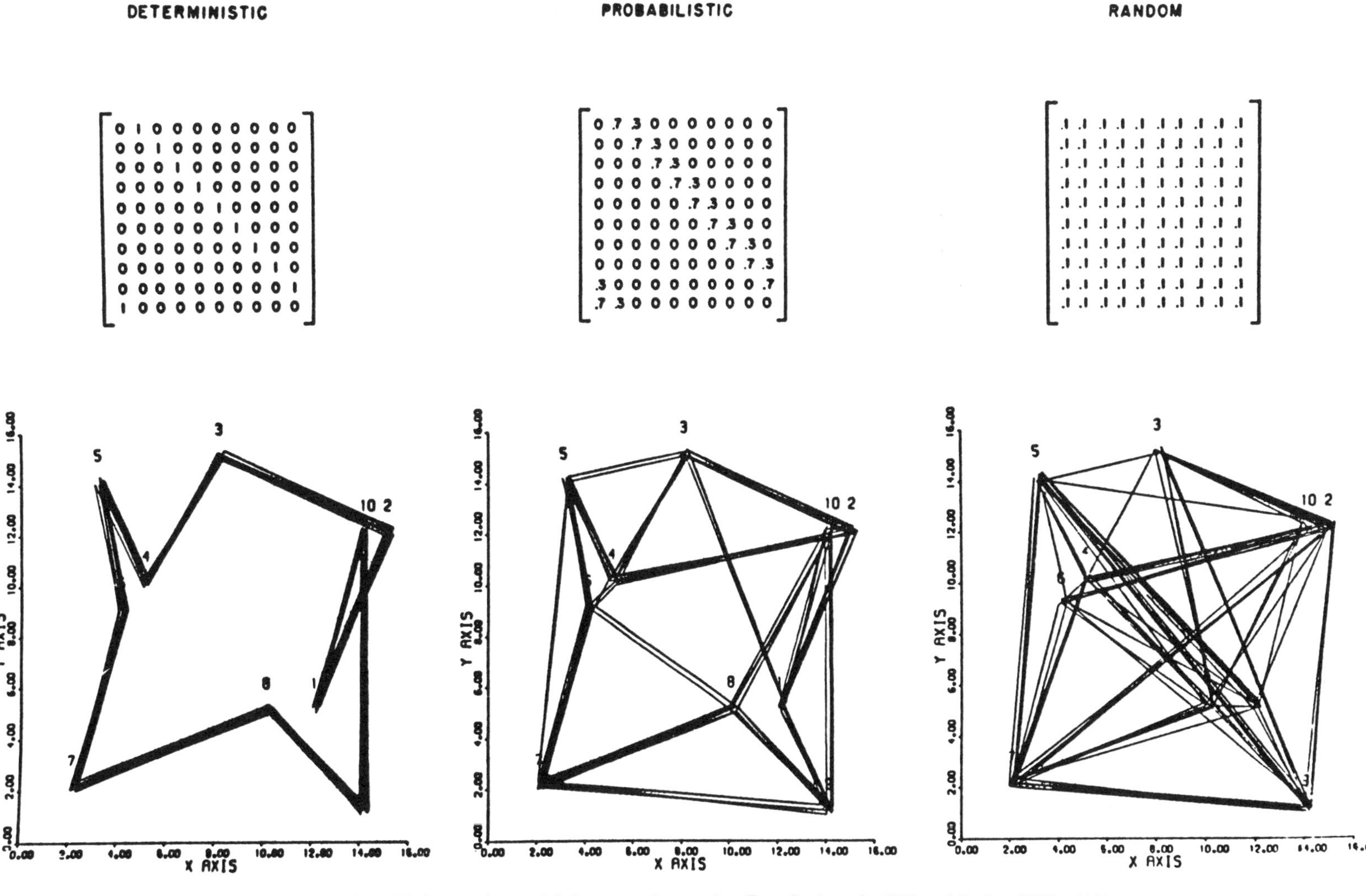

FIGURE 18-7. Model Markov matrices and their generated scanpaths. (From Stark *et al.*, 1979, and Stark and Ellis, 1981.)

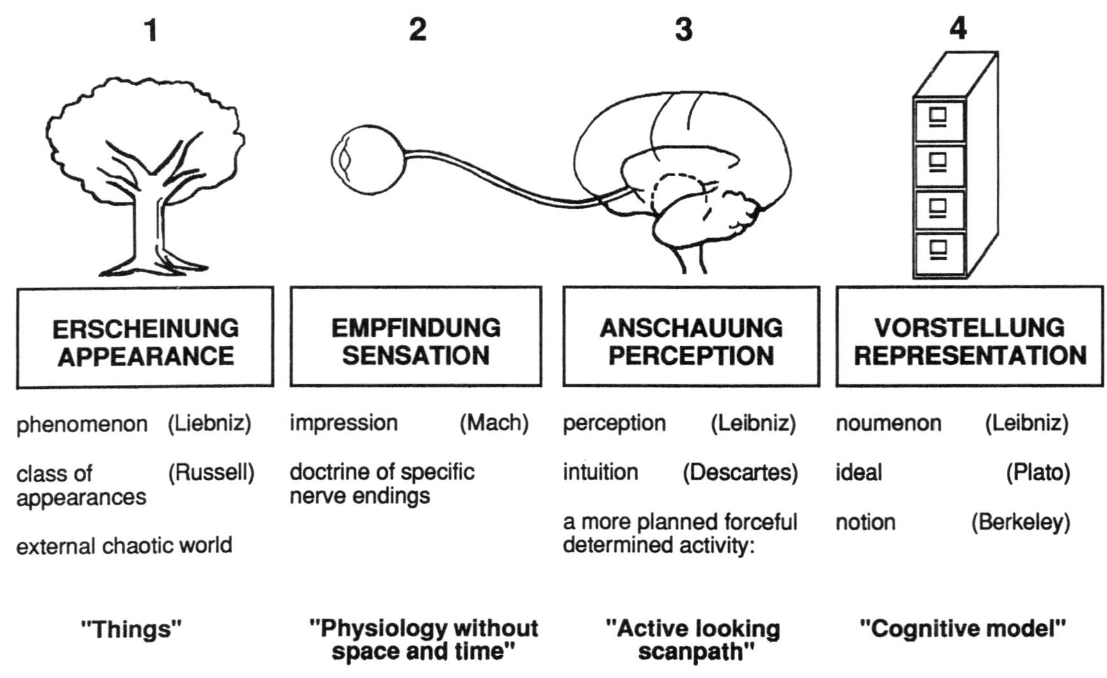

FIGURE 18-8. Kantian definitions for perceptual processes.

my neurological view.) Thus, cognition can come to peace, if not to grips, with appearance or *Erscheinung*.

Other Scanpth Experiments

Other types of picture presentations have been used by cognitive–perceptual researchers. Fragmented figures form an interesting series, and it has been demonstrated that the eye movement patterns are indeed different before and after recognition (Stark and Ellis, 1981). The careful continuous screening of a CDTI (cockpit display of air-traffic information) by pilots carrying out full simulations of restricted-visibility flights has resulted in evidence for higher-order Markov matrix constraints in successive fixations (Stark and Ellis, 1981).

Different tasks might be expected to call forth different cognitive models, and they in turn control different scanpaths. Indeed, the brilliant Russian scientist Alfred Yarbus (1967) demonstrated clear differences in eye movement patterns consequent to different tasks instructed to his observers, results that not only are compatible with the scanpath theory but support it strongly.

Two further preliminary studies might bear reporting here. American Sign Language (ASL) is widely used in deaf families, even by hearing children of deaf parents. A group of ASL-proficient subjects looked more at the signing hand of a picture of Gaudet (the inventor of ASL) than did non-ASL speakers, perhaps a partial promise of support for the Worf hypothesis. Also, a person on psyllicybin, an American Indian psychopharmacological agent, was found to make scanpaths of minute size but with the same shape as larger scanpaths (found the week before and the week after the drug session) that clearly directed her eyes to prominent features of the picture. The subject was aware that her eyes did not travel to the picture subfeatures; she felt that the drug enabled her to avoid this congruence. Perhaps the pictures were too simple to require foveation for recognition.

CONCLUSION

The ambiguous figure paradigm clearly supports the suggestion that higher-level visual processes in perception and cognition complement the preattentive disambiguating visual processes. Although it is difficult to get solid quantitative evidence concerning these higher-level perceptual mechanisms, the measurements of eye movements and their interpretation as scanpaths provide clear documentation of the existence and, indeed, ubiquity of active looking controlled by top-down cognitive models.

SUMMARY

The ambiguous figure paradigm clearly demonstrates switching between mental representations. This is an example of higher-level perceptual–cognitive processes playing a crucial role in separating superimposed images.

Objective eye movement recordings interpreted via the scanpath theory indicate that different mental representations act as top-down cognitive models controlling different active looking scanpaths. More definitive quantitative measures are critically needed; future meticulously refined experiments and clearer philosophical discussions may further our understanding.

ACKNOWLEDGMENTS. I am grateful for discussions with my colleagues J. Allen, S. Brandt, S. Ellis, S. Hacisalihzade, M. Leroy, D. Noton, K. Sherman, W. Zangemeister, and Y. Zeevi and to Essilor/ Varilux (Drs. Obrecht and Tahran) and to NASA–Ames Research Center Cooperative Agreement NCC 2-86 (Dr. Ellis, Technical Monitor) for partial support.

REFERENCES

Brandt, S., Stark, L., Hacisalihzade, S., Allen, J., and Tharp, G., 1989, Experimental evidence for scanpath eye movement during visual imagery, in: *Proceedings 11th IEEE/EMBS, Seattle,*

Buswell, G. T., 1935, How people look at pictures, University of Chicago Press, Chicago.

Didday, R. L., and Arbib, M. A., 1972, *Eye Movements and Visual Perception: A "Two Visual System Model,"* COINS Technical Report 73C-9, University of Massachusetts.

Ellis, S. R., and Stark, L., 1978, Eye movements while viewing Necker cubes, *Perception* **7**:575–581.

Ellis, S. R., and Stark, L., 1979, Reply to Piggins, *Perception* **8**:721–722.

Ellis, S. R., and Stark, L., 1986, Statistical dependency in visual scanning, *Hum. Factors* **28**:421–438.

Groner, R., and Groner, M. T., 1989, Attention and eye movement control: An overview, *European Archives of Psychiatry and Neurological Sciences,* **239**:9–6.

Groner, R., Walder, F., and Groner, N., 1984, Looking at faces: Local and global aspects of scanpaths, in: *Theoretical and Applied Aspects of Eye Movement Research* (A. G. Gale and F. Johnson, eds.), North-Holland, Amsterdam.

Hacisalihzade, S., Allen, J., and Stark, L., 1989, Automatic analysis of eye movement with a PC, in: Procession of the 11th Annual IEEE/EMBS, Seattle 1989.

Jeannerod, M., Gerin, P., and Pernier, J., 1968, Deplacements et fixation du regard dans l'exploration libre d'une scene visuelle, *Vision Res.* **8**:81–97.

Leroy, M., and Stark, L., 1989, Personal Communication.

Locher, P., and Nodine, C., 1973, The influence of visual symmetry on visual scanning patterns, *Percept. Psychophys.* **13**:408–412.

Noton, D., and Stark, L., 1971a, Eye movements and visual perception, *Sci. Am.* **224**:34–43.

Noton, D., and Stark, L., 1971b, Scanpaths in eye movements during pattern perception, *Science* **171**:308–311.

Noton, D., and Stark, L., 1971c, Scanpaths in saccadic eye movements while viewing and recognizing patterns, *Vision Res.* **11**:929–942.

Parker, R. E., 1978, Picture processing during recognition, *J. Exp. Psychol. Hum. Percept. Perform.* **4**:284–293.

Prokowski, F. J., 1971, *A model for human eye movements during viewing of a general, two-dimensional, dynamic display,* unpublished doctoral dissertation, University of Connecticut.

Rizzo, M., Hurtig, R., and Damasio, A. R., 1987, The role of scanpaths in facial recognition and learning, *Annal. of Neurology,* **22**:41–57.

Stark, L., and Ellis, S. R., 1981, Scanpaths revisited: Cognitive models direct active looking, in: *Eye Movements, Cognition and Visual Perception* (Fisher, Monty, and Senders, eds.), Lawrence Erlbaum Associates, Hillsdale, NJ, pp. 193–226.

Stark, L., Ellis, S. R., Inoue, H., Freksa, C. R., and Zeevi, J., 1979, Cognitive models direct scanpath eye movements, in: *XII International Conference on Medical and Biological Engineering, Jerusalem, Israel,*

Yarbus, A. L., 1967, Eye movements and vision, Plenum Press, New York.

Zangemeister, W., Sherman, K., and Stark, L., 1989, Scanpath eye movements while viewing abstract images, in: *Fifth European Conference on Eye Movements* (R. Schmidt and D. Zambarbieri, eds.), Pavia University Press, Pavia, Italy, pp. 261–265.

19

Complementary Cognitive and Motor Image Processing

BRUCE BRIDGEMAN

PHYSIOLOGICAL INTRODUCTION

Everyone shares the irresistible conception that vision is one sense. We experience one coherent visual world and produce visually guided behavior to interact with that world. Extensive laboratory work, however, has shown this introspection to be in error: visual processing has several representations of space, coding different aspects of the information available from vision. The representations operate simultaneously, in parallel, in performing various visual functions.

One approach to the multiple-representations phenomenon is physiological and begins with the observation that over a dozen topographic maps represent the visual world in the cortex (Van Essen *et al.*, 1982). This observation raises a question of the relationship between physiology and function: do all of these maps work together in a single visual representation, or are they functionally distinct? And if they are distinct, how many functional maps are there, and how do they communicate with one another? These questions can be answered with psychophysical techniques, reinforcing the evidence for separate visual representations obtained in physiological studies.

The many visual representations might conceivably support an equal number of distinguishable visual information-processing systems, running quasi-independently in parallel. Physiological evidence, however, indicates that the topographic maps are organized into groups of serially organized representations spreading out from a common starting point on the retinas.

The number of such functionally separable systems is not known, but there is good evidence for at least two modes of visual processing. A primary distinction, with experimental support from a number of directions, is between visual perception on one hand and control of visually guided behavior on the other. The thesis of this chapter is that a common thread from physiological, anatomic, and psychophysical directions distinguishes these two separable functional systems. Other subdivisions may exist, and other kinds of distinctions (such as the magnocellular–parvocellular pathways) may cut across the perception–behavior distinction.

The organizations are parallel and independent in that perception can suffer spatial illusions that are not shared in spatial behavior, and visually guided behavioral orientation can be modified without affecting perception. Psychophysical studies have revealed that subjects are unaware of sizable displacements of the visual world if they occur during saccadic eye movements, implying that information about spatial location is degraded during saccades (Ditchburn, 1955; Wallach and Lewis, 1965; Brune and Lücking, 1969; Mack, 1970; Bridgeman *et al.*, 1975). Yet people do not become disoriented after saccades, implying that spatial information is maintained.

Experimental evidence supports this conclusion. For instance, the eyes can saccade accurately to a target that is flashed (and mislocalized) during an earlier saccade (Hallett and Lightstone, 1976b), and hand–eye coordination remains fairly accurate following saccades (Festinger and Cannon, 1965). How can the loss of perceptual information and the maintenance of visually guided behavior exist side by side?

In an attempt to resolve this paradox, we noted that the two conflicting observations use different response measures. The experiments on saccadic

Bruce Bridgeman • Program in Experimental Psychology, University of California, Santa Cruz, California 95064.

suppression of displacement require a nonspatial verbal report or button press, both symbolic responses. Successful orienting of the eye or hand, in contrast, requires quantitative spatial information with a 1 : 1 correspondence between stimulus position and motor output. This distinction produces a functional, as opposed to anatomic, distinction between the two proposed systems. By definition, then, perceptual tasks with a symbolic output address the cognitive system, whereas isomorphic motor responses address the motor system. The remainder of this chapter examines the validity of the cognitive–motor dichotomy.

HISTORY OF THE SEPARATION OF FUNCTIONS

Interest in the distinction between pathways for spatial and object vision crystallized with a symposium and a series of articles in *Psychologische Forschung* (now *Psychological Research*). In that series, Trevarthen (1968) named the two systems focal and ambient; the focal system was supposed to reside in the geniculostriate pathway and to specialize in pattern recognition. The ambient system, in the superior colliculus and related brainstem structures, handled visually guided behavior. This may be the case in the hamster (Schneider, 1967), but we now realize that both systems have important cortical components in primates (Mishkin *et al.*, 1983). The successor to the ambient system includes occipitoparietal pathways as well as superior colliculus, and the focal system uses an occipitotemporal pathway.

The two visual systems are not equal in size, for relatively little information is required to drive visually guided behaviors. Perception, in contrast, requires sensitivity to fine detail, which requires a large processing capacity for high-spatial-frequency information. For this reason, processing in the focal (cognitive) system is concentrated in the foveal projection, where high-spatial-frequency information is available, although the remainder of the visual field is available to this system as well. Ambient (motor) vision, in contrast, requires a large field of vision, the larger the better, and consequently the fovea plays only a relatively minor role because of its small angular size. Because of the asymmetry in function and the resulting differences in locations of projections, the terms "focal" and "ambient" have been widely misunderstood by others to be related to foveal and peripheral vision, respectively, though the originators of the distinction did not intend this interpretation.

Since the 1960s several workers, using different methods and approaches, have converged on a general distinction between spatial and object vision. They have generated slightly different definitions and a chaotic nomenclature that makes a complete search of this literature difficult: the terms and their principal proponents are given in Table 19-1.

More recently, Post and Leibowitz (1982, 1985) have used a distinction between two types of eye movements, one voluntary and corresponding to the focal system and the other reflexive and corresponding to ambient vision, to interpret several motion illusions. Key to the interpretation is that innervations of the voluntary "pursuit" system are perceived as object motions, whereas reflex "optokinetic" innervations are not perceived. The latter are interpreted indirectly as self-motions.

Thus, induced motion results from the pursuit system, tracking a target, having to counter the effects of a reflexive system that is stimulated by a background frame. Motion of the frame is misattributed to target motion because the extra pursuit effort, required to maintain fixation on the target despite reflex innervation from background tracking, is perceived while the reflex innervations are not. The pursuit–optokinetic distinction maps onto the cognitive–motor distinction reviewed above. As a result, concomitant changes in apparent straight ahead, not predicted by other theories, can be explained (Post and Heckman, 1986). The theory has also been applied to the oculogyral illusion (Post, 1986) and to changes in perceived movement with changes in gain of the VOR.

TABLE 19–1
Nomenclature Used for Components of Vision

Terms		Originator
1. Focal	Ambient	C. Trevarthen
2. Experiential	Action	M. Goodale
3. Cognitive	Motor	B. Bridgeman
4. Cognitive	Sensorimotor	J. Paillard
5. Explicit	Implicit	L. Weiskrantz
6. Object	Spatial	M. Mishkin
7. Overt	Covert	K. Rayner
8. Exocentric	Egocentric	I. Howard

PSYCHOPHYSICAL CHARACTERIZATION OF THE COGNITIVE–MOTOR DISTINCTION

Most of the early work in this field was done in lesioned animals and in neurological patients. Recent work shows that the dissociations are not disconnection syndromes or compensations for deficits induced by lesions but, also exist in normal humans. Both pathways retain a topographic representation of space, but the representations follow different rules, reflecting their differing functions.

In our first experiment on this question (Bridgeman *et al.*, 1979), subjects pointed to the position of a target that had been displaced and then extinguished. Subjects were also asked whether the target had been displaced or not. Pointing accuracy was similar whether the displacement was detected or went undetected because of a simultaneous saccadic eye movement. This implied that quantitative motor control was unaffected by the perceived target position. But it is possible (if a bit strained) to interpret the result in terms of signal detection theory as a higher response criterion for the report of displacement. The first control for this possibility was a two-alternative forced-choice measure of saccadic suppression of displacement. This criterion-free measure showed no information about displacement to be available to the cognitive system under conditions where pointing was affected (Bridgeman and Stark, 1979).

A more rigorous way to separate the two systems is with a double-dissociation paradigm, introducing a signal only into the motor system in one condition and only into the cognitive system in another. We know that induced motion affects the cognitive system, because we experience the effect. But the above experiments implied that the information used for pointing might come from sources unavailable to perception.

We inserted a signal selectively into the cognitive system with stroboscopic induced motion (Bridgeman *et al.*, 1981). A surrounding frame was displaced, creating the illusion that a target had jumped. Target and frame were then extinguished, and the subject pointed open–loop to the last position of the target. Trials where the target had seemed to be on the left were compared with trials where it had seemed to be on the right. Pointing was not significantly different in the two kinds of trials, showing that induced motion did not affect pointing. Information was inserted selectively into the motor system by asking each subject to adjust a real motion of the target, jumped in phase with the frame, until the target seemed stationary. Thus, the cognitive system specified a stable target. Nevertheless, subjects pointed in significantly different directions when the target was extinguished in the left or the right positions. Thus, a double dissociation was obtained: in the first condition apparent target displacement affected only perception; in the second, real displacement affected only motor behavior.

Dissociation of cognitive and motor function has also been demonstrated by giving cognitive and motor systems opposite signals at the same time. A target jumped in the same direction as a frame but not far enough to cancel an induced motion. Immediate saccadic eye movements followed the true direction even though subjects perceived motion in the opposite direction (Wong and Mack, 1981). If a delay in responding was required, however, eye movements followed the perceptual illusion, implying that the motor system has no memory and must rely on information from the cognitive system when the motor map no longer contains the needed information.

A NEW METHOD FOR DISSOCIATING THE SYSTEMS

All of these experiments involve motion or displacement, leaving open the possibility that the dissociations are related in some way to motion systems rather than with representation of visual space *per se*. A new method, however, can test dissociations of cognitive and motor function without motion of the eye or the stimuli at any time during a trial. The dissociation is based on the Roelofs effect (Roelofs, 1935), a tendency to misperceive the position of an edge of a large target presented in an unstructured field. The effect has also been observed as a tendency to perceive the locations of light flashes as closer to the line of sight than their true positions (Mateeff and Gourevich, 1983).

This method takes the Roelofs effect one step further and measures the misperception of target position in the presence of a surrounding frame presented asymmetrically in the field; this is an "induced Roelofs effect" but is called a Roelofs effect below. Positions of targets within the frame are misperceived in the direction opposite the offset of the

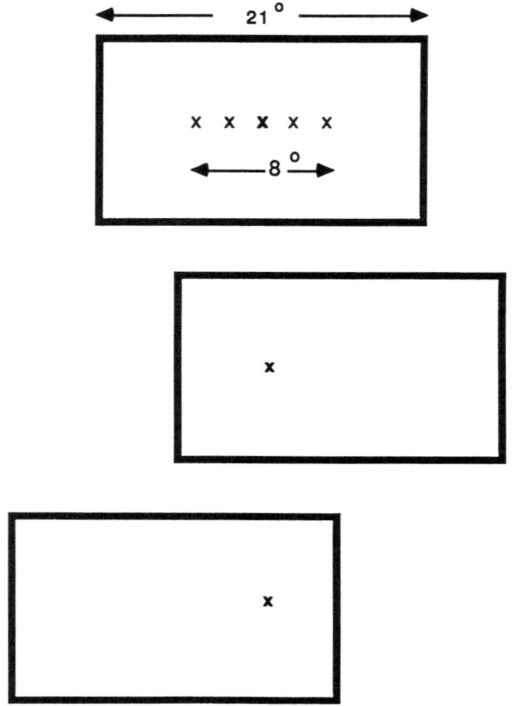

FIGURE 19-1. Stimulus array used in experiments on the Roelofs effect. The frame could be centered (top), offset 5° right (middle), or offset 5° left (bottom). A target appeared in one of the five positions indicated within the top frame. Targets always appeared in the same positions, regardless of frame position, and only one target was visible at a time. The center target (shown in boldface in the top frame) is shown in the frames offset left and right. Only one target and frame were shown in a given trial.

frame. The effect is similar to a stroboscopic induced motion in which only the final positions of the target and frame are presented (Bridgeman and Klassen, 1983). After reviewing some new experiments using this effect, implications for the two-visual-systems theory are discussed.

An Experiment

Subjects sat with stabilized heads before a hemicylindrical screen. A rectangular frame was projected, via a galvanic mirror, either centered on the subject's midline or 5° left or 5° right of center. Inside the frame, an "x" could be projected via a second galvanic mirror (Fig. 19-1).

A pointer with its tip near the screen gave an analogue voltage indicating its position. Perceived target position was recorded on a keyboard. For each trial, one of the five targets and one of the three frames were presented with simultaneous onset, exposed for 1 sec, and simultaneously extinguished. Subjects could not respond until stimulus offset, so that at the time of the response they were looking at a blank field. Thus, the task was a response to an internally stored representation of the stimulus, not a perceptual task.

For judging trials, all subjects showed a Roelofs effect (Fig. 19-2). The mean magnitude of the effect was a difference of 2.0° between judgments with the frame on the left and judgments with the frame on the right. Though small, this effect is reliable; it is present in all subjects under all condi-

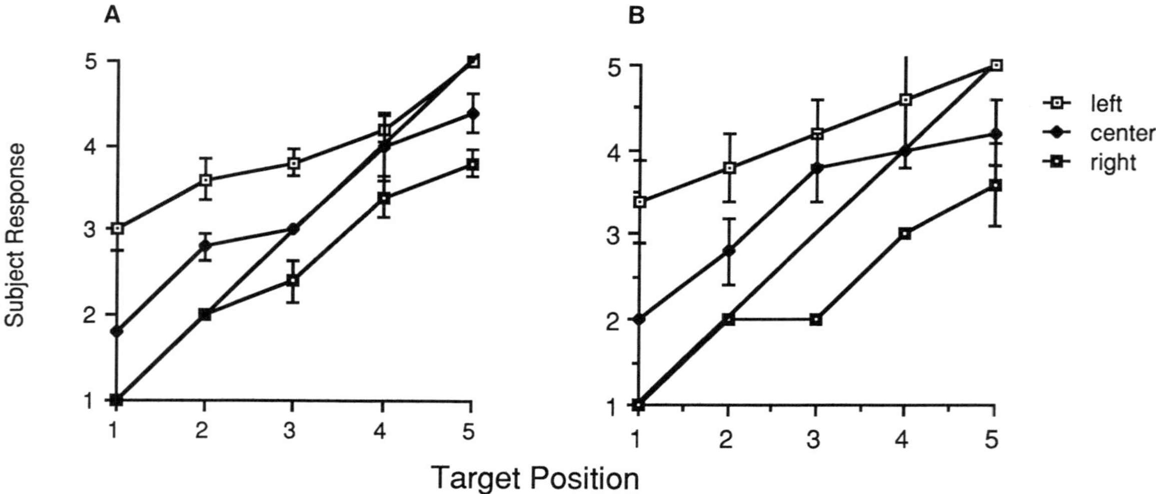

FIGURE 19-2. Judging of target position immediately after stimulus offset. Data illustrated here are from two subjects, though statistical analyses in the text are from all subjects. Lack of error bars indicates that the standard deviation was less than the width of the symbol. A: subject A. B: subject B.

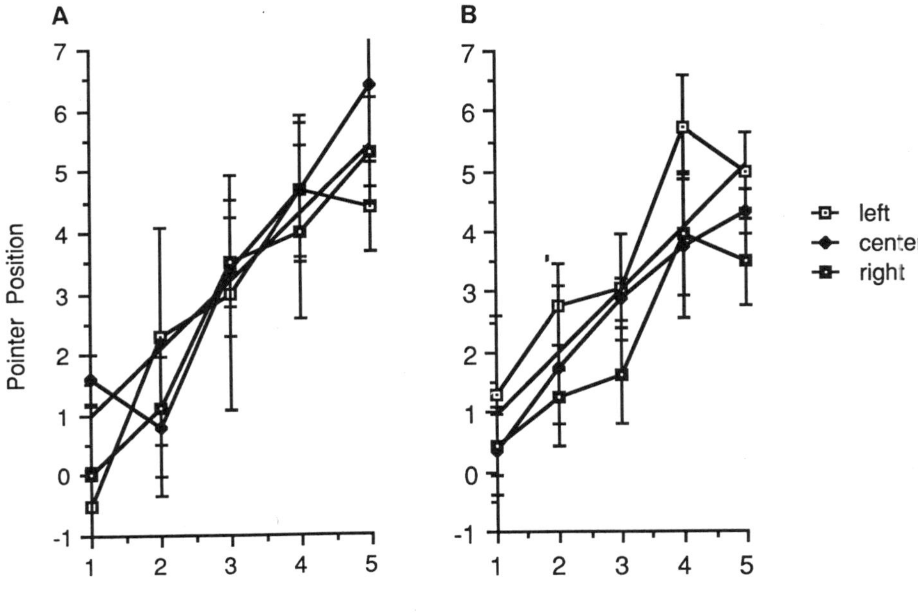

FIGURE 19-3. Pointing to targets under the same perceptual conditions as in Fig. 19-2 in trials randomly intermingled with the judging trials. Subjects A and B correspond to the two subjects in Fig. 19-2. Overlap of the three curves in subject A indicates lack of influence of frame position on pointing behavior. Separation of the curves in subject B indicates a continuing influence of the frame. All subjects showed either the type A or the type B pattern.

tions and is statistically reliable in each one using within-subject statistical analyses.

Pointing trials (Fig. 19-3) yielded a sharp division of the subjects into two groups: five of the ten subjects showed a highly significant Roelofs effect ($P < 0.005$), while the other five showed no sign of an effect ($P > 0.18$). The bimodal distribution (Fig. 19-4) reveals two qualitatively different results; the distribution of significances between subjects is not related to a small, normally distributed effect reaching significance in some subjects and not in others. A given subject showed either a large, robust effect or no sign of influence of the frame. Thus, pointing was qualitatively different from judging for half of the subjects; these subjects showed a Roelofs effect only for judging.

Nine of the ten subjects were also run with a 4-sec delay interposed between display offset and tone. Eight of the nine showed a significant Roelofs effect for the judging task ($P < 0.01$), with a mean difference of 2.12° between pointing when the frame was on the left and when it was on the right.

The major difference between the results in this condition and the no-delay condition was that seven of the nine subjects showed a significant Roelofs effect for the pointing task ($P < 0.05$ for 2 *Ss*, $P < 0.01$ for 7 *Ss*).

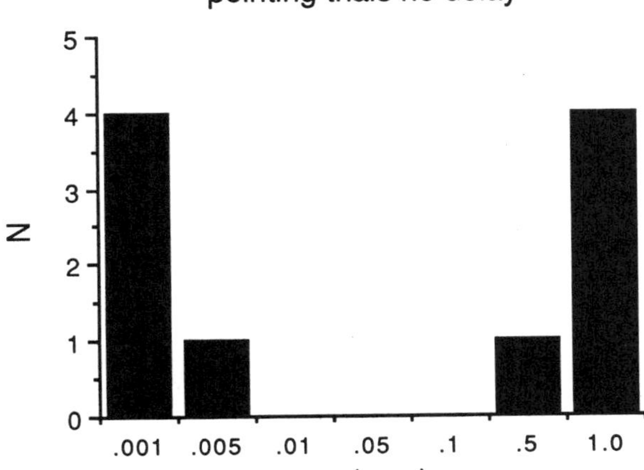

FIGURE 19-4. Statistical significance of within-subjects two-way ANOVA frame (Roelofs) effects. Each bin contains subjects with results at least as significant as the probability indicated on the abcissa but not as significant as the next probability to the left. The bimodal distribution shows two different types of subjects, widely spaced, with no intermediate cases.

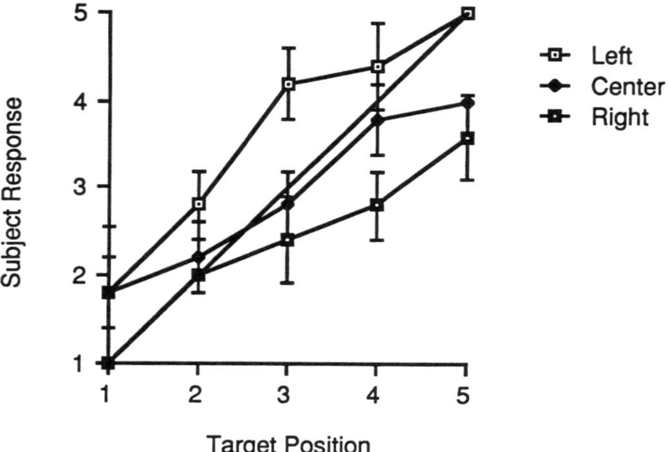

FIGURE 19-5. Judging of target position after an 8-sec delay in subject A. Results are similar to those in Fig. 19-3 (no delay), showing a stable representation of visual positions in the system controlling target position judgments. Display format as in Fig. 19-2.

One of the two remaining subjects showed no significant effect of frame position for either pointing or judging. The other subject whose pointing still showed no effect of the frame was retested with an 8-sec delay between display offset and tone. A Roelofs effect was found both for judging ($P < 0.001$) (Fig. 19-5) and pointing ($P < 0.001$) (Fig. 19-6).

The experiment was repeated with a continuous centimeter-estimation judging measure, so that

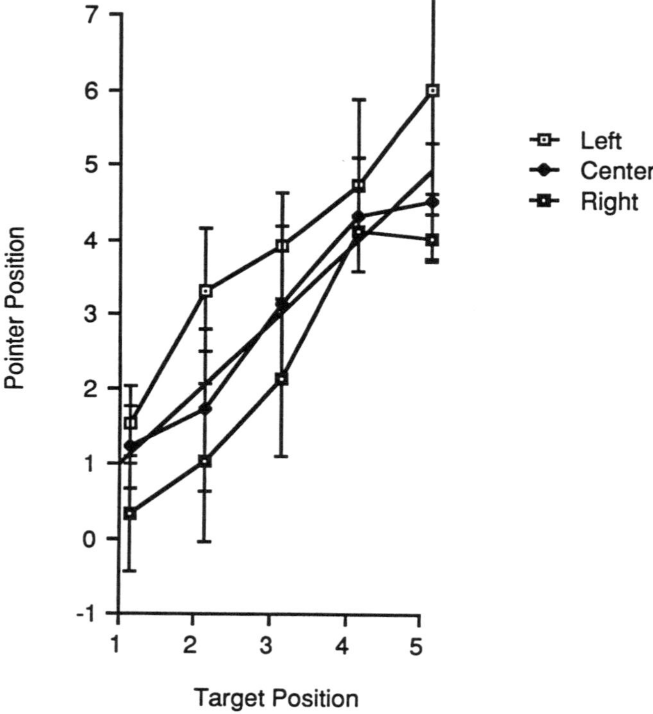

FIGURE 19-6. Pointing to targets after an 8-sec delay in subject A, from trials intermingled with those shown in Fig. 19-5. A Roelofs effect for pointing has appeared, indicating that frame position affects pointing just as it affects judging at this latency, though the effects of a separate spatial representation are still evident: slopes of the lines here are more than 1, in contrast to slopes of less than 1 at all latencies for judging. Display format as in Fig. 19-3.

subjects would not be forced to remain within the 8° range of the target stimuli. Results were similar, though variability in judging was somewhat greater with the centimeter estimation method, resulting in less significant Roelofs effects.

Interpretation of the Results

Interposing a long enough delay before the response forced all subjects to use pointing information that is biased by frame position, even though half of the subjects were not vulnerable to this bias when responding immediately. Differences between pointing and judging to the same target in the same trial block could amount to more than 5°.

These experiments show that perception of a Roelofs effect is robust, being seen by all subjects under all delays. The Roelofs effect in visually guided behavior, though, depends much more strongly on the subjects and conditions. Half of the subjects showed a strong effect of a surrounding frame on pointing behavior, while the other half showed no effect. The bimodality of this distribution suggests that subjects went about the task in two different ways. Since all responses were made in a blank field after the stimuli had been extinguished, the contrast may be related to differing strategies of the subjects; some responded in a motor mode, whereas others switched almost immediately to a cognitive mode, which brought the illusion along with it. The difference between the modes explains the bimodal nature of the distribution in Fig. 19-4, with subjects showing either a large Roelofs effect or none at all.

One need not assume that the two groups of subjects followed different psychological laws, only that they switched from motor to cognitive modes at differing delays after stimulus offset. Discussion in terms of differential accuracy and decay rates of cognitive versus motor responses would redescribe the data but would not explain their source.

Further, all of the subjects showed a Roelofs effect in pointing when a long enough delay was interposed between target presentation and response; a closer titration of delay times would probably show a unique critical delay for obtaining a pointing Roelofs effect in each subject.

The appearance of the Roelofs effect with a delay between stimulus and motor response is reminiscent of the results of Wong and Mack (1981) that saccadic eye movements followed a veridical motion with a short delay but followed a perceived motion in the opposite direction after a long delay. Though the delays used here were longer than those of Wong and Mack, the pattern of results is similar. Thus, it appears that if the motor representation of space possesses a memory for the positions of stimuli no longer present, the memory begins to degrade after no more than a few hundred milliseconds. The duration of this memory and the conditions under which it is degraded are subjects for future research.

Conclusions

In addition to differences in the Roelofs effect, the results have shown a differential decay rate of perceptual responses and pointing responses along with greater variability for pointing than for perceptual measures; how are these to be interpreted? An interpretation that is consistent with cortical neurophysiology as well as with the literature cited above is that the two measures access information from different maps of visual space. The motor map is accessed by a pointing measure that requires a 1 : 1 relationship between stimulus position and behavior; stimulus and response map isomorphically onto one another. The cognitive map, in contrast, requires a categorization in which the relationship between target position and behavior is arbitrary.

Our conclusion is that the normal human possesses two maps of visual space. One of them holds information used in perception: if a subject is asked what he sees, the information in this "cognitive" map is accessed. This map can obtain great sensitivity to small motions or translations of objects in the visual world by using relative motion or position as a cue. The price that the cognitive system pays for gaining this sensitivity is that it loses absolute egocentric calibration of visual space. In calculating dx/dt by differentiation, the constant term (the spatial calibration) drops out.

The other visual map drives visually guided behavior, but its contents are not necessarily available to perception. This map does not have the resolution and sensitivity that the cognitive map has, but it is not required to: a small error in pointing, grasping, or looking is of little consequence. The advantage of this map is its robustness; the "motor" map is not subject to illusions such as induced motion and the Roelofs effect. In this sense it is more robust, but as a result it is less sensitive to small motions or fine-grained spatial relationships. It also has only a short memory, being concerned mainly with the here-and-now correspondence between visual information and motor behavior. If a subject must make motor responses to stimuli no longer present, this system must take its spatial information from the cognitive representation and brings any cognitively based illusions along with it. This is not to say that the sequence cannot be stored in memory and used to improve motor performance in the future; the current egocentric spatial values are lost, however. The relationships of information flow in the two systems are schematized in Fig. 19-7.

Another way of interpreting the relationship between the cognitive and the motor representations of visual space is in terms of the subject's ability to integrate information from the map with other information. The cognitive map's contents can be described and compared with other spatial or nonspatial information, whereas the motor map is generally inaccessible to integration with information from other sources.

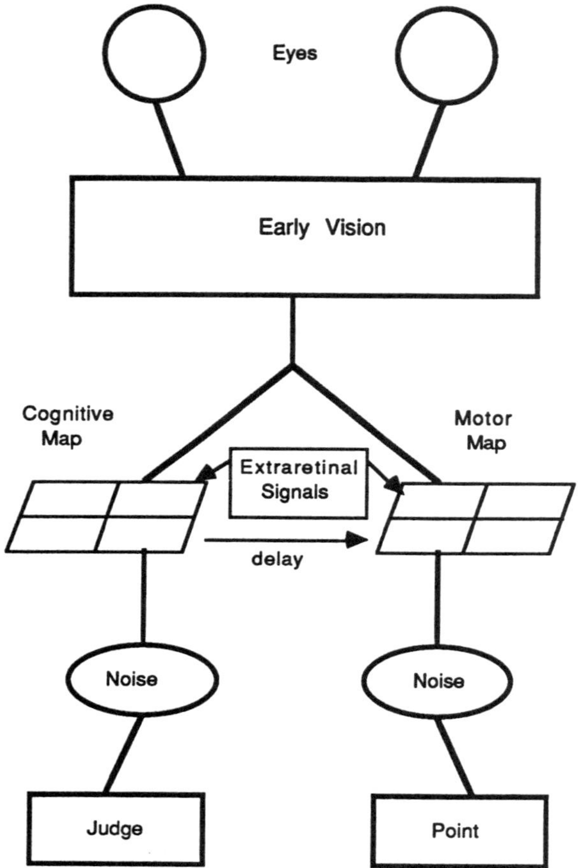

FIGURE 19-7. A proposed information flow scheme for perceptual judgments and visually guided behavior. After a common input stage, spatial information maps into both a cognitive representation (left center) and a motor representation (right center). Extraretinal information does not affect early vision, because receptive fields at the striate cortex and below have retinotopic rather than spatiotopic receptive field organization. The extraretinal information is supplied to the two maps separately because it may affect them differently. If the motor map has no spatial information, it can receive information from the cognitive map. The two maps influence behavior through independent noise sources.

An example of this dichotomy between overt and covert aspects of visual processing is seen in experiments in which a monocularly viewing eye is pressed to separate efference copy from eye position. When the eye is pressed and held, it does not undergo a passive rotation—quite the reverse, it actively resists rotation, for fixation on a target can be maintained while the eye is pressed. As a result, the efferent commands to the eye change, and the efference along with it, while position of the retinal image remains constant. One can easily demonstrate this active resistance by closing one eye and slowly pressing on the other; a fixated object will remain fixated even while apparent motion is seen. Monitoring the movements of the occluded eye provides an objective record of the change in the efference to the eye.

Cognitive components of vision can be measured under these conditions by asking a subject to set a target to appear straight ahead while the eye is pressed. Motor components are measured by open-loop pointing to a target. The cognitive measure is affected by visual context; in a normally illuminated environment, eyepress has little effect on straight-ahead settings, but the same settings correspond to the offset of the efference copy if the judgment is made in darkness with only the target visible. Pointing, in contrast, is determined by the change in efference copy regardless of the illumination conditions, and is always offset by eyepress (Bridgeman and Stark, 1981; Stark and Bridgeman, 1983). Again, the subject is unaware that the cognitive settings and the pointing directions do not correspond.

The difference between cognitive and motor representations in this context is comparable to the distinction between explicit and implicit modes of memory, respectively. The explicit mode is accessible to language and to experiential memory, whereas the implicit mode may hold information that the subject is unaware of or even that is contradictory to the contents of explicit memory (Roediger *et al.*, 1988). Similarly, subjects can hold one position of a stimulus in the cognitive mode and simultaneously hold a different position for the same stimulus in the motor mode.

REMAINING PROBLEMS

The two-visual-systems idea needs further development both in laboratory experimental work and in theory. The theoretical problems must be addressed before empirical work can make further contributions. First, the definitions of the two systems must be clarified and standardized. Some workers have defined the two systems anatomically, some physiologically (in terms of receptive field properties), and still others functionally. The approach taken here is functional, distinguishing the two systems in terms of the task required of an observer. This, I argue, is the primary mode of definition, because anatomic or physiological distinctions make sense only when weighed against behavioral and perceptual criteria. Lesions in monkeys are interpreted with perceptual–motor experiments, and receptive fields are compared to behaviors. The variety of modes of definition has led to some inconsistency and to the possibility that different workers are distinguishing different systems. There may be more than two separable streams of visual information, for example, or different groups may be defining a stream at different points.

It is likely that the two modes are not independent but that limited information exchange occurs even at the most central levels. This is an area that needs further investigation.

SUMMARY

Two distinct modes of visual image processing have been identified in normal humans: first, a cognitive system, serving perception and assessed with perceptual measures; and second, a motor-oriented system, serving visually guided behavior and assessed with open-loop pointing or looking. Experiments in my laboratory and others have identified some of the properties of each system. The cognitive system requires fine-grained, high-spatial-frequency signals for optimal function; because of limitations of retinal sampling, such signals are available only on and near the fovea. Moderate amounts of high-pass filtering have little effect on this system. The motor system, in contrast, requires only lower spatial frequency signals and cannot make use of information at the highest spatial frequencies. This grosser information must be gathered from a large area of the retina, however, ideally including both the fovea and a large region of the retinal periphery. Thus, one can conceive the retinal image as containing two spatially overlapping, complementary images: a high spatial frequency, primarily foveal image used in perception, and a larger, low-spatial-frequency image used to guide visual–motor coordination. Because most of the information in the image is in the high frequencies, the perceptual branch of the system dominates the geniculostriate pathway and has received the most attention from neurophysiologists.

The implication for lens design is that if compromises must be made between image quality and spatial distortion off the center of vision, it is preferable to sacrifice image quality in the high spatial frequencies to preserve spatial relationships.

ACKNOWLEDGMENTS. Partially supported by a Faculty Research Grant from the University of California, Santa Cruz. I thank Jennifer Schroeder and Elizabeth Yeater for assistance with the experiments. The Roelofs effect experiments reviewed here were presented at a NASA symposium in Asilomar, CA, Sept. 1988.

REFERENCES

Bridgeman, B., Hendry, D., and Stark, L., 1975, Failure to detect displacement of the visual world during saccadic eye movements, *Vis. Res.* **15:**719–722.

Bridgeman, B., Kirch, M., and Sperling, A., 1981, Segregation of cognitive and motor aspects of visual function using induced motion, *Percept. Psychophys.* **29:**336–342.

Bridgeman, B., and Klassen, H., 1983, On the origin of stroboscopic induced motion, *Percept. Psychophys.* **34:**149–154.

Bridgeman, B., Lewis, S., Heit, G., and Nagle, M., 1979, Relation between cognitive and motor-oriented systems of visual position perception, *J. Exp. Psychol. Hum. Percept. Perform.* **5:**692–700.

Bridgeman, B., and Stark, L., 1979, Omnidirectional increase in threshold for image shifts during saccadic eye movements, *Percept. Psychophys.* **25:**241–243.

Bridgeman, B., and Stark, L., 1981, Efferent copy and visual direction, *Invest. Ophthalmol. Vis. Sci. Suppl.* **20:**55.

Brune, F., and Lücking, C. H., 1969, Oculomotorik, Bewegungswahrnehmung und Raumkonstanz der Sehdinge, *Nervenarzt* **40:**692–700.

Ditchburn, R., 1955, Eye-movements in relation to retinal action, *Optica Acta* **1:**171–176.

Festinger, L., and Canon, L. K., 1965, Information about spatial location based on knowledge about efference, *Psychol. Rev.* **72:**373–384.

Hallett, P. E., and Lightstone, A. D., 1976, Saccadic eye movements towards stimuli triggered during prior saccades, *Vis. Res.* **16:**99–106.

Mack, A., 1970, An investigation of the relationship between eye and retinal image movement in the perception of movement, *Percept. Psychophys.* **8:**291–298.

Mateeff, S., and Gourevich, A., 1983, Peripheral vision and perceived visual direction, *Biol. Cybernet.* **49:**111–118.

Mishkin, M., Ungerleider, L., and Macko, K., 1983, Object vision and spatial vision: Two cortical pathways, *Trends Neurosci.* **6:**414–417.

Post, R. B., 1986, Induced motion considered as a visually induced oculogyral illusion, *Perception* **15:**131–138.

Post, R. B., and Heckman, T., 1986, Induced motion and apparent straight ahead during prolonged stimulation, *Percept. Psychophys.* **40:**263–270.

Post, R. B., and Leibowitz, H., 1982, The effect of convergence on the vestibulo-ocular reflex and implications for perceived movement, *Vis. Res.* **22:**461–465.

Post, R. B., and Leibowitz, H., 1985, A revised analysis of the role of efference in motion perception, *Perception* **14:**631–643.

Roediger, H., Weldon, M., and Challis, B., 1988, Explaining dissociations between implicit and explicit measures of retention: A processing account, in: *Varieties of Memory and Consciousness: Essays in Honor of Endel Tulving* (H. L. Roediger and F. I. Craik, eds.), 3–41, Lawrence Erlbaum Associates, Hillsdale, NJ.

Roelofs, C., 1935, Optische Localization, *Arch. Augenheilkd.* **109:**395–415.

Schneider, G. E., 1967, Contrasting visuomotor functions of tectum and cortex in golden hamster, *Psychol. Forsch.* **31:**52–68.

Stark, L., and Bridgeman, B., 1983, Role of corollary discharge in space constancy, *Percept. Psychophys.* **34:**371–380.

Trevarthen, C. B., 1968, Two mechanisms of vision in primates, *Psychol. Forsch.* **31:**299–337.

Van Essen, D. C., Newsome, W. T., and Bixby, J. L., 1982, The pattern of interhemispheric connections and its relationship to extrastriate visual areas in the macaque monkey, *J. Neurosci.* **2:**265–283.

Wallach, H., and Lewis, C., 1965, The effect of abnormal displacement of the retinal image during eye movements, *Percept. Psychophys.* **1:**25–29.

Wong, E., and Mack, A., 1981, Saccadic programming and perceived location, *Acta Psychol.* **48:**123–131.

20

The Psychoanatomy of Binocular Single Vision

JEREMY M. WOLFE

INTRODUCTION

Humans, in common with the rest of the vertebrates, have two eyes. In common with a significant subset of the vertebrates, notably predators, the visual fields of those two eyes overlap to a substantial extent. This arrangement offers a number of substantial benefits. Binocular overlap makes possible stereoscopic depth perception (Wheatstone, 1938). Many visual tasks are performed better with two eyes than one (Blake and Fox, 1974; Blake *et al.*, 1981; Jones and Lee, 1981). Two eyes together have a larger field of view than one eye, even in animals with frontal eyes. Finally, two eyes provide insurance against the loss of one eye.

For these advantages of binocular vision, the visual system pays a price. Two eyes, looking at the three-dimensional world from two slightly different vantage points, will receive slightly (or not so slightly) different input. Since the perceiver wishes to perceive only one visual world, the visual system must create *binocular single vision.*

There are in principle two ways to solve the problem of binocular combination. The two retinal images can be combined, or one can be seen while the other is suppressed. The central argument of this chapter is that the human visual system does both. It is suggested that separate pathways generate separate binocular representations in one case by combining the images from the two eyes wherever possible and, in the second, by suppressing one or the other monocular input at each location in the visual field. These two representations are then combined to produce the perceptual experience of binocular single vision.

JEREMY M. WOLFE • Department of Brain and Cognitive Sciences, Massachusetts Institute of Technology, Cambridge, Massachusetts 02139.

THE GEOMETRY OF BINOCULAR VISION

Stereopsis

The tasks of binocular vision are best understood by considering the geometry of two eyes viewing a three-dimensional world from two horizontally displaced vantage points. Figure 20-1 represents a horizontal plane running through the fovea of each eye. The visual axis of one eye can be represented as a line running from the fovea through the lens to some fixated stimulus A. If the eyes are converged on A, the image of A falls on the fovea of each eye (FL and FR, respectively). Each retina is a two-dimensional surface. The two foveas are corresponding points on those surfaces. Obviously, there are other pairs of corresponding points. In Figure 20-1, XL and XR are each the same distance to the right of the fovea. A stimulus, B, some distance to the left of A will cast its images on these corresponding points. The set of all locations that stimulate pairs of corresponding points forms a curved surface known as the *horopter* (Aguilonius, 1613, cited in Boring, 1942). Theoretically, the horopter should be a circle (the Vieth–Muller circle) in the horizontal plane. Reality is a more complicated curved surface (see Boring, 1942; Tyler, 1983). For purposes of this discussion, the important point is that there is a surface, any point on which will stimulate corresponding points in each eye.

The horopter is a thin sheet running through the volume of visual space. It follows that all of the loci in the remainder of that space will cast images on noncorresponding points. Moreover, it follows that stimuli at two different loci can cast images on corresponding points. Thus, stimuli C and D in Fig. 20-1 form images at corresponding points YL and YR.

The noncorresponding points stimulated by items off the horopter bear a geometrically regular

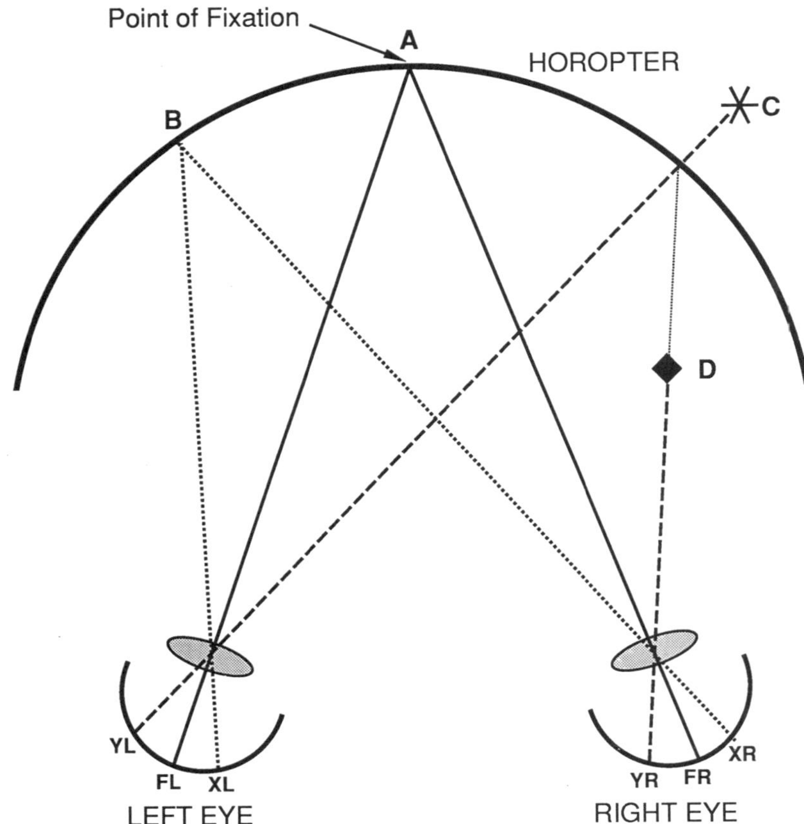

FIGURE 20-1. The set of all corresponding retinal points forms the horopter.

relationship to their distance from the horopter. The greater the distance from the horopter, the greater the horizontal disparity between the stimulated points in the two eyes. This is illustrated in Fig. 20-2. Point A is at a greater depth than point B. The separation of the stimulated retinal loci is greater for A than for B. The direction of the separation gives information about direction in depth; in front of or behind the horopter. Thus, in Fig. 20-2b, point A is in front of the horopter. Its image in the left eye is to the right of the fovea while its image is to the left in the right eye. This is *crossed disparity*. The situation is reversed for point B, behind the horopter, yielding *uncrossed disparity*.

The utility of binocular disparity as a cue to depth was made explicit only in the midnineteenth century by Wheatstone (1938). It is surprising that stereopsis took so long to be formally discovered. The basic concepts were known much earlier. Leonardo da Vinci, though he never built a stereoscope, clearly understood the notion of binocular parallax, noting that no monocular painting could ever perfectly mimic the reality of binocular vision (Boring, 1942). Johannes Kepler in 1604 came even closer to the truth. He gave the first accurate account of the image-forming properties of the eye's optics. This posed a problem. The visual images were formed on the retinas but were seen single out in the world. Kepler proposed that "mental rays" projected the image out into the world along lines emanating from the eyes. These lines from the two eyes would intersect, and the image would be seen at the point of intersection. Though the jargon is somewhat archaic, Kepler's ideas are very similar to later theories. Kepler was not in a good position to pursue the study of binocular depth perception. He was a myope, an exotrope, and prone to bouts of monocular polyplopia (Kaufman, 1974). It is unlikely that he could have used a stereoscope even if he had invented one.

Kepler's theory, couched in more modern terms, allows all horizontal retinal disparities to be fused regardless of size. This is not correct. After the invention of the stereoscope, it became clear that stereopsis was possible only for a range of disparities. Larger disparities simply produced di-

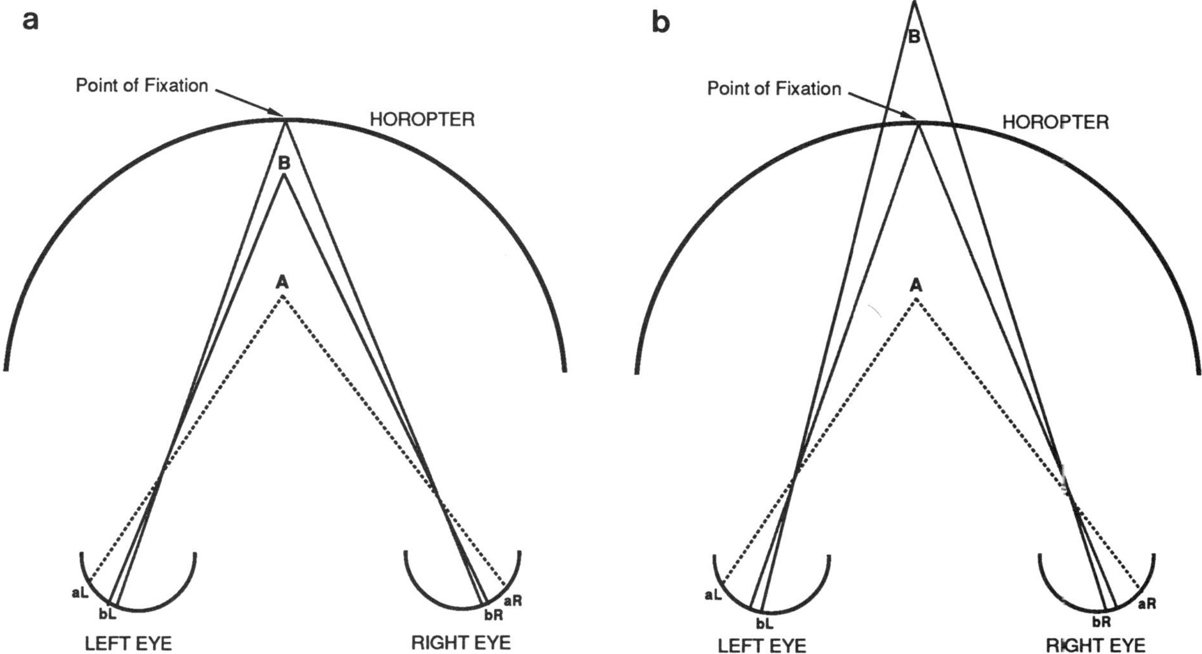

FIGURE 20-2. The relationship between position in depth and retinal disparity. (a) Greater depth yields greater disparity. (b) Stimuli in front of the horopter yield crossed disparity. Stimuli behind the horopter yield uncrossed disparity.

plopia. Panum (1957) modified the notion of corresponding points to account for this fact. He proposed that each retinal point was matched to a range of points in the other eye. A pair of points could be fused if and only if the point in one eye fell within what Panum called the "retinal circle of sensation" defined by the other eye. Projecting this idea into visual space, stereopsis is possible only in a region of space surrounding the horopter. This is known as *Panum's area* (Fig. 20-3). In reality, Panum's area is not as clearly demarcated as it is in Fig. 20-3. For example, it is larger for low-spatial-frequency stimuli than for high (Schor *et al.*, 1984; Schor, 1987).

Suppression and Rivalry

Panum's area defines the spatial limits of stereopsis. Beyond this slab of space surrounding the horopter, we simply lack the neural hardware to support stereopsis. Moreover, the invention of the stereoscope made it easy to study binocular stimuli that did not exist in normal three-dimensional space. Radically different stimuli could be put in each eye. Wheatstone (1838) did this and saw binocular rivalry. When corresponding retinal loci are presented with completely different stimuli, one is perceptually suppressed and the other is dominant. This pattern of dominance and suppression is local. With stimuli subtending more than about 1.0°, the left eye can be dominant in one location and the right in another. It is rule-governed. Bright stimuli tend to dominate over dim stimuli (Kaplan and Metlay, 1964), high contrast over low (Blake, 1977), moving over stationary (Grindley and Townsend, 1965), and so forth. Dominance and suppression vary over time. The usual experience of rivalry is of a continually changing battle between regions of left and right eye dominance. Figure 20-4 shows in a series of static frames an impression of the dynamic nature of rivalry. Interestingly, any visible stimulus is able to suppress any other stimulus, though a weak stimulus will suppress a strong stimulus only briefly (Blake, 1977). For reviews of binocular rivalry, see Walker (1978) or Blake (1989).

THE USES OF BINOCULAR RIVALRY

Since virtually all information from one eye is lost after the point of rivalry suppression,* binocular

*In saying that visual information is "lost" after the point of rivalry suppression, it is important to remember that this does not mean that suppressed stimuli have no impact on the visual system. For example, a suppressed stimulus will still stimulate and fatigue photoreceptors in the eye.

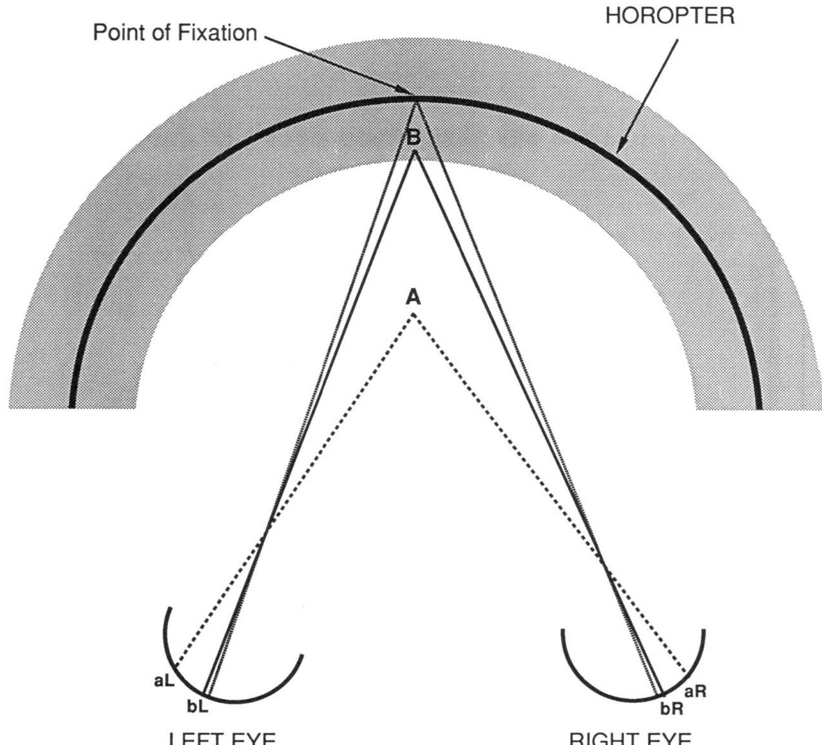

FIGURE 20-3. Panum's area.

rivalry may appear to be a wasteful process. However, it is probably the best solution under the circumstances. The other logical possibility is some form of combination: binocular summation or binocular averaging. The negative consequences of indiscriminate combination are greater than those of suppression. First, combination would create stimuli that do not exist. Given the stimuli in Fig. 20-4, combination would produce a nonexistent grid. Given a vertical branch seen with the left eye and a horizontal branch seen with the right, a monkey without rivalry might reach for a nonexistent intersection. In addition to the danger of false perceptions, superimposition rather than rivalry of disparate images would lead to image degradation. A minimal case would be a black spot on a white background presented to one eye with just the white background presented to the other. Rivalry will yield the perception of a high-contrast black spot on a white ground. Combination, either by summation or averaging, would yield a lower-contrast gray spot on a white ground. It is clear that superimposing two disparate images under normal viewing conditions will produce many cases of this type of contrast reduction.

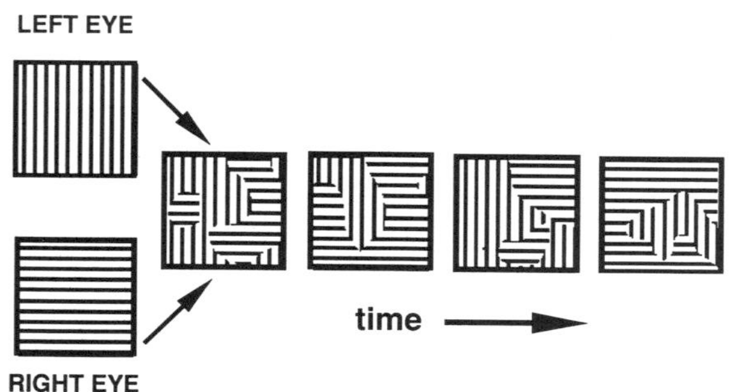

FIGURE 20-4. Binocular rivalry is seen when different stimuli are presented to each eye.

SUPPRESSION THEORIES OF BINOCULAR SINGLE VISION

Suppression theories of binocular vision argue not only for the necessity of rivalry outside of Panum's area but against the necessity or existence of binocular combination within Panum's area. The essence of suppression theory is that only the input to one eye is seen at any given location at any one time. When the two images are identical, the alternation is invisible to the perceiver. Suppression theories date back at least to Porta in the 16th century. Suppression theories were easier to defend prior to the discovery of binocular phenomena such as brightness averaging and, most importantly, stereopsis. Modern suppression theories are forced to deal with these phenomena. Verhoeff (1935, 1959) argued that fusional effects were the result of very rapid alternation. If the inputs from the two eyes alternated rapidly enough, like the flicker of a single light source, the alternation would not be perceived. Verhoeff called this "unification," but it sounds suspiciously like some form of binocular combination. More serious attempts have been made to explain stereopsis within the context of suppression theory (e.g., Hochberg, 1964), but they have not proven to be adequate accounts (see Kaufman, 1974).

Summary

Stereopsis is possible in a region of space surrounding the horopter. Outside of that region and in cases where the two eyes do not view the same stimulus within that region, stereopsis is impossible, and dominance of one eye's input at each location is the rule. Neither a model based entirely on combination nor one based on suppression can be a complete theory of binocular vision. Neither would solve the full range of problems posed by binocular vision, and neither would be adequate to explain the entire set of binocular phenomena.

THE DOMAINS OF STEREOPSIS AND RIVALRY

The binocular visual system requires mechanisms that can combine input from the two eyes (e.g., for stereopsis) and mechanisms that can suppress input from one eye. There are several possible architectures that could realize this requirement. Consider two, here named "stereopsis first" and "parallel pathways." The "stereopsis first" architecture is the more intuitively appealing of the two. It holds that the primary goal of binocular vision is binocular combination and stereopsis. If and only if combination and stereopsis are impossible does the system fall back on rivalry and suppression. In terms of the geometry of visual space, "stereopsis first" argues that combination mechanisms operate within Panum's area while rivalry mechanisms operate outside (or inside, when different stimuli are presented to the two eyes). This two-step process generates a single representation of the binocular visual field having stereoscopic information wherever possible and rivalry/suppression as the default elsewhere (Fig. 20-5b). The clearest exponents of this position are Blake (1989) and O'Shea (1987) (see also Julesz and Tyler, 1976; Blake and Boothroyd, 1985; Blake and O'Shea, 1988).

The "parallel pathways" architecture that will be defended here holds that separate rivalry and stereopsis representations are generated. Rivalry occurs everywhere in the binocular visual field at all times, generating a representation of the visual world that consists of either the left or the right eye's input at each location in the visual field. The stereopsis mechanism operates over a limited range of disparities (roughly Panum's area), generating a representation of the visual world having stereoscopic information in all possible locations and no information in other locations. These two representations are combined to yield binocular single vision (Fig. 20-5a). In the final combination, there may be conditions where the stereopsis representation vetoes the contribution from rivalry (or vice versa, see below). The central point is that both representations are generated.

The two models are similar in their final product, as they must be since the final product is binocular single vision in both cases. They differ in the mechanisms that intervene between the input and that final product. In the next section of this chapter, four lines of evidence will be presented in support of the parallel pathways architecture:

1. Independence. There are separate rivalry and stereopsis pathways through the early stages of visual processing.
2. Coexistence. Rivalry and stereopsis can occur at the same place and time.
3. Silent rivalry. Rivalry occurs, unperceived, when identical stimuli are presented to the two eyes.
4. Double dissociation. It is possible to disrupt rivalry without disrupting stereopsis, and vice versa.

The final section discusses the combination of rivalry and stereoscopic representations into binocular single vision and offers some thoughts about reconciling this theory with a stereopsis first model.

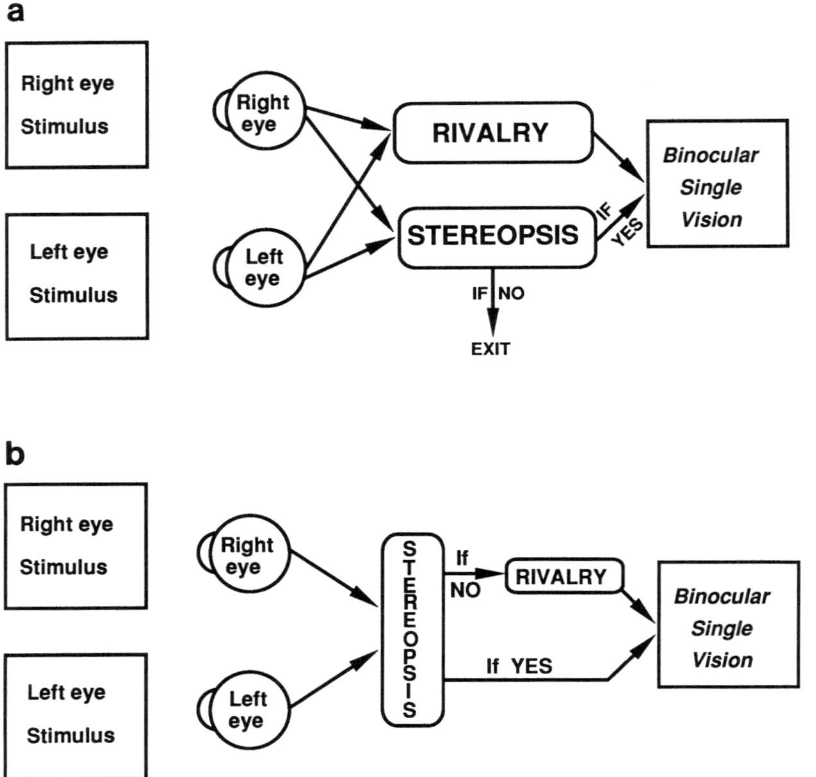

FIGURE 20-5. Models of binocular single vision. This chapter argues for the parallel-pathways model (a). Others argue for a stereopsis-first model (b).

Independence of Stereopsis and Rivalry

THE BORDER OF PANUM'S AREA

If rivalry and suppression occurred only when stereopsis became impossible, we might expect to find a clean transition between stereopsis and rivalry as disparity increases. That is, the border of Panum's area should represent the border between stereopsis and rivalry. However, although stereopsis does fail as disparity gets larger, rivalry does not appear to begin at the point of failure. Ogle (1952) had subjects view a simple three-line stereogram. Two flanking lines were always in the same location in the two eyes. The center line could have a binocular disparity. With small disparities, the center bar was seen in depth, with the amount of depth corresponding to the magnitude of the disparity. As disparity increased, the center bar was no longer fused. However, though four lines were seen, stereopsis was preserved, and depth still corresponded to the magnitude of disparity. At still greater disparity, only the sign of depth was preserved (in front of or behind the horopter). Finally, depth vanished, leaving only the perception of the four bars.

The changes in the appearance of the stereogram are best explained as the sum of independent stereopsis and rivalry mechanisms. The changes in stereopsis with increasing disparity can be explained in the context of a model of stereopsis proposed by Richards (1970, 1971) and given physiological support by Gian Poggio (Poggio and Fischer, 1977; Poggio and Talbot, 1981). Figure 20-6 shows three disparity channels mediating stereopsis: one tuned for "near," crossed disparity; one for "far," uncrossed; and one for zero. Quantitative estimates of depth are determined by comparing the output of two or more channels in a manner similar to that proposed for color (Hurvich and Jameson, 1955; Jameson and Hurvich, 1955) and now more generally for most basic visual features (Braddick *et al.*, 1978). Thus, for crossed disparities in the range of disparities between B and C, two or three channels are stimulated, and quantitative depth can be perceived. Disparities between A and B still stimulate a

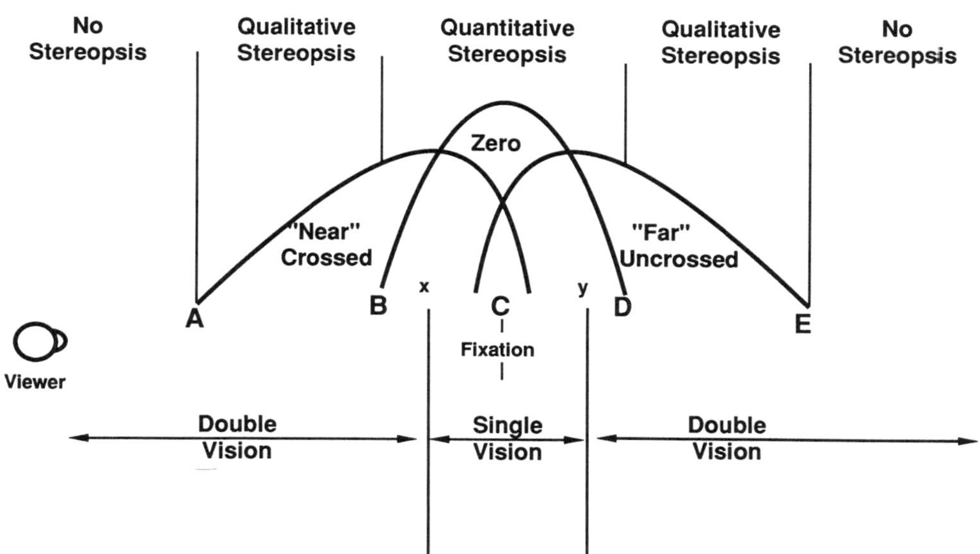

FIGURE 20-6. Stereopsis is mediated by "channels" selectively sensitive to crossed (near), uncrossed (far), and zero disparity. The phenomena of stereopsis are explained by these channels. Diplopia appears to require a separate explanation.

disparity channel and therefore yield an impression of depth. However, because only one channel is stimulated, only the sign and not the magnitude of depth can be computed. Disparities greater than A cannot be seen by the stereopsis mechanism.

Returning to the example of a three-line stereogram, the boundary between single and double vision is unrelated to the limits on stereopsis, casting doubt on a stereopsis-first model that sees rivalry as the default that occurs when stereopsis fails. The transition from single to double vision can be explained by a separate rivalry/suppression mechanism. A dominant monocular stimulus at one location will tend to suppress stimuli in the other eye at that and at neighboring locations (Kaufman, 1963). In Fig. 20-6, when the disparity between the center lines is between X and Y, one line suppresses the other, and a single line is seen. For disparities greater than X or Y, each line is outside of the suppressive reach of the other, and both are seen. By this account, a simple combination of the outputs of stereopsis and rivalry pathways would produce the succession of percepts that accompany increasing disparity. These results, as well as those of Schor (1987) mentioned above, illustrate that the border of Panum's area (defined as the region of single vision and/or stereopsis) is ill defined. If there is a transition from stereopsis to rivalry over that border, it is not a particularly sharp transition.

These same findings can be explained in the context of a stereopsis-first model by assuming that stereopsis with diplopia occurs when low-spatial-frequency mechanisms support large-disparity stereopsis but high-frequency mechanisms do not.

THE PURELY BINOCULAR PROCESS

A second line of evidence in favor of independent mechanisms for rivalry and stereopsis is the existence of a "purely binocular process" early in visual processing. Such a process is activated only by matching or nearly matching binocular stimuli. It could be and probably is the entry to the stereopsis mechanism. Since it is silent in the presence of rivalrous stimuli, it could hardly be part of a rivalry mechanism.

The evidence comes from a series of tilt aftereffect experiments. In the tilt aftereffect (TAE), exposure to one orientation changes the apparent orientation of contours at neighboring orientations (Gibson, 1937; Campbell and Maffei, 1971). Figure 20-7 gives an example. Aftereffects can be used as probes of binocular visual processes (see Blake *et al.*, 1981; Wolfe and Blake, 1985; Wolfe, 1986a, for general accounts). For example, if the adapting stimulus (Fig. 20-7 left) is presented to the right eye and the test stimulus (Fig. 20-7 middle) is presented to the left, the TAE can still be measured (Gibson,

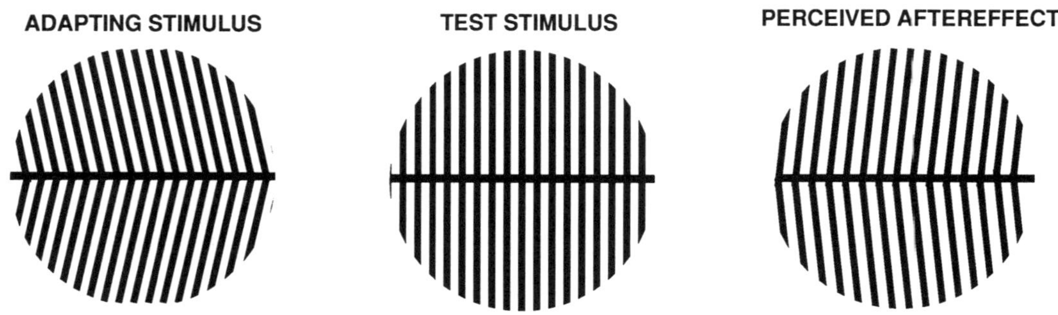

FIGURE 20-7. The tilt aftereffect (TAE). Viewing the first pattern for 30 sec will cause the second pattern to appear to point to the left as in the third pattern. This aftereffect can be used to study binocular pathways in early vision.

1937). Apparently, exposure of the right eye produces effects in a binocular process that can be stimulated by either the right or the left eye. This interocular transfer (IOT) is a general characteristic of aftereffects (e.g., spatial frequency—Blake *et al.*, 1981; motion—Wohlgemuth, 1911; figural aftereffects—Kohler and Wallach, 1944), though there are exceptions (notably the McCollough effect: McCollough, 1965; Over *et al.*, 1973).

Interocular transfer is not complete. For the TAE, the aftereffect in the unadapted eye tends to be about 60–80% of the strength seen in the adapted eye (Fig. 20-8A). This can be explained by assuming the existence of monocular processes that are activated only by adapting and test stimuli presented to one eye. In the example given above, adapting stimuli presented to the right eye affect the binocular *or* process and the right eye's monocular *only* process. Test stimuli presented to the left eye activate the adapted binocular *or* process and an unadapted left eye monocular *only* process. The resulting aftereffect is smaller than the effect seen in the right eye where test stimuli activated only adapted processes (see Blake *et al.*, 1981; Wolfe and Held, 1981; Moulden, 1980, for more on this matter.)

As further confirmation of the existence of monocular *only* processes, the left eye can be adapted to stimuli pointing to the left while the right eye is adapted to stimuli pointing to the right. In this case, roughly equal and opposite TAEs are produced in the two eyes. The effects are small because the contribution of the binocular *or* process has been nullified by the equal and opposite adaptation (see Fig. 20-8B; Wolfe and Held, 1981; Wolfe, 1986a). Similar independent monocular aftereffects have been found for spatial frequency (Sloane and Blake, 1987) and motion (Anstis and Duncan, 1983).

The "purely binocular process" is a second type of binocular process. It responds only to stimulation of the left *and* right eyes. Its existence is also revealed by TAE experiments. Both eyes view the adapting pattern, but not at the same time. After several minutes of alternating adaptation, the aftereffect is measured with the left and right eyes alone and with both eyes together (see Fig. 20-8C). A full-strength TAE is measured with each eye alone. The binocular TAE is significantly reduced in magnitude. This suggests the dilution of the binocular TAE by a process that was not adapted by stimulation of either eye alone but was activated by binocular test patterns. Such a process would be a purely binocular process (Wolfe and Held, 1981). The four processes uncovered by these experiments are included in Fig. 20-9.

A process that becomes active only when matching stimuli are presented to the two eyes would be of obvious use as part of a stereopsis mechanism. Several lines of evidence suggest that the purely binocular process is part of such a mechanism. Notably, cyclopean adapting stimuli do not produce monocular aftereffects. To establish this point, a subject adapts to a random-dot stereogram version of the TAE adapting pattern. This pattern is only visible binocularly. The TAE is measured with normal black-and-white test patterns. When the test pattern is viewed binocularly, a small but reliable TAE is measured (last column of Fig. 20-8D). When the identical test pattern is viewed monocularly, no TAE is measured (Fig. 20-8D: Wolfe and Held, 1982). Looking at Fig. 20-9, these results can be explained if we assume that the cyclopean adapting stimulus activated only the purely binocular process. When both eyes are tested, all four processes are active. A TAE diluted by the presence of three

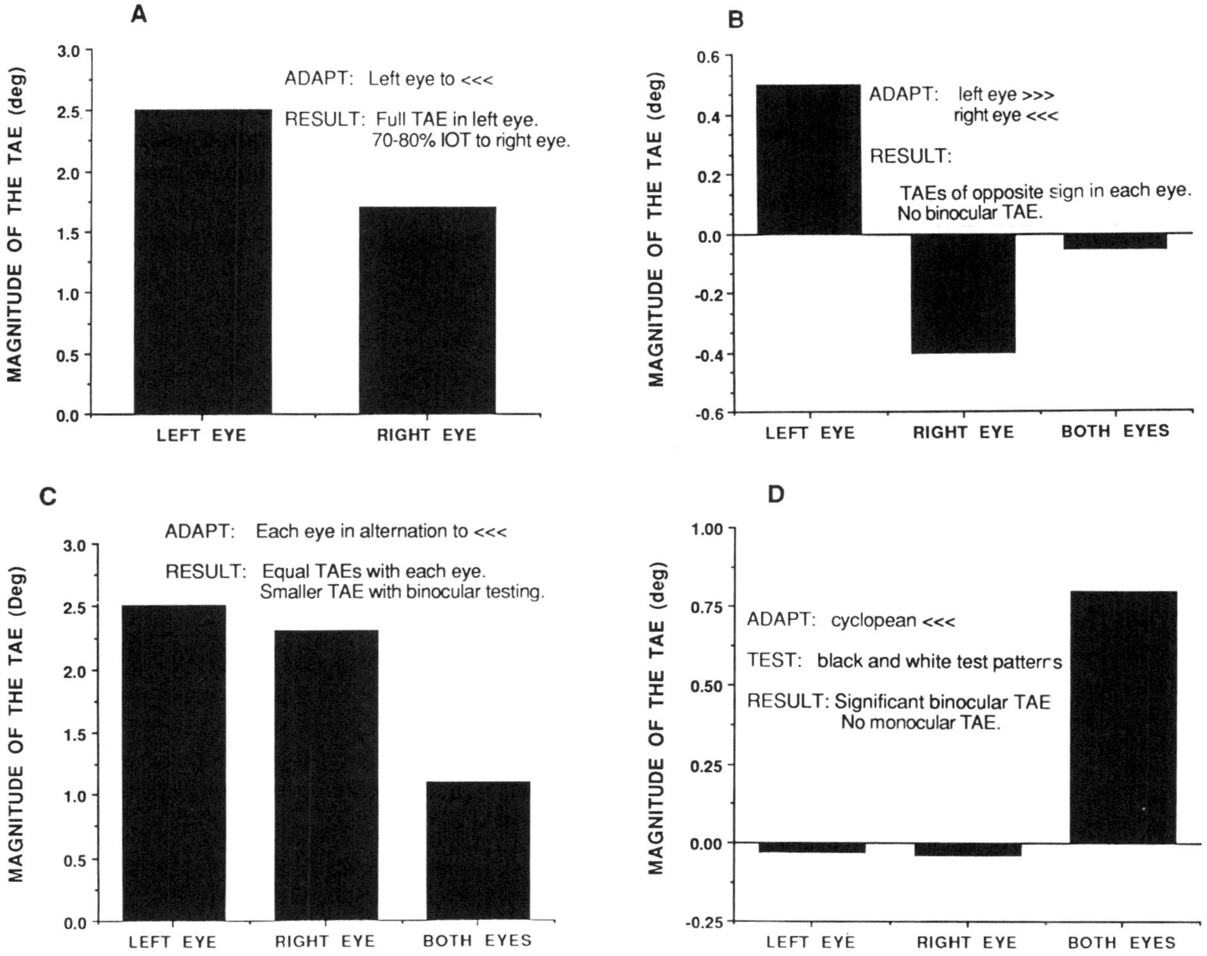

A
MAGNITUDE OF THE TAE (deg)
3.0
2.5
2.0
1.5
1.0
0.5
0.0
ADAPT: Left eye to <<<
RESULT: Full TAE in left eye.
70-80% IOT to right eye.
LEFT EYE
RIGHT EYE
B
MAGNITUDE OF THE TAE (deg)
0.6
0.4
0.2
0.0
-0.2
-0.4
-0.6
ADAPT: left eye >>>
right eye <<<
RESULT:
TAEs of opposite sign in each eye.
No binocular TAE.
LEFT EYE
RIGHT EYE
BOTH EYES
C
MAGNITUDE OF THE TAE (Deg)
3.0
2.5
2.0
1.5
1.0
0.5
0.0
ADAPT: Each eye in alternation to <<<
RESULT: Equal TAEs with each eye.
Smaller TAE with binocular testing.
LEFT EYE
RIGHT EYE
BOTH EYES
D
MAGNITUDE OF THE TAE (deg)
1.00
0.75
0.50
0.25
0.00
-0.25
ADAPT: cyclopean <<<
TEST: black and white test patterns
RESULT: Significant binocular TAE
No monocular TAE.
LEFT EYE
RIGHT EYE
BOTH EYES

FIGURE 20-8. Results of four TAE experiments. Details in text.

Right eye Stimulus
Left eye Stimulus
Right eye
Left eye
RIGHT ONLY
RIGHT OR LEFT
RIGHT AND LEFT
LEFT ONLY
RIVALRY
STEREOPSIS
Binocular Single Vision

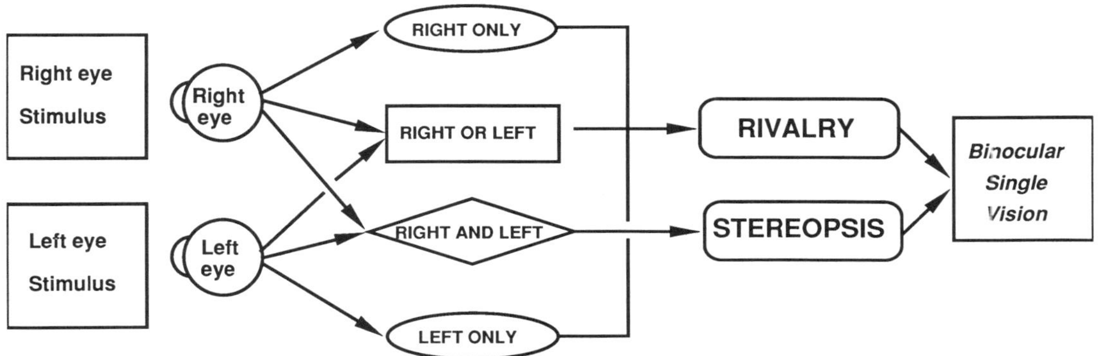

FIGURE 20-9. A more detailed version of the parallel-pathways model showing four types of process at the first stage of binocular interaction: left and right eye *only* monocular processes, an *or* process responding to either left *or* right eye stimuli, and an *and* process responding only to matched binocular stimuli. The *only* and *or* processes give rise to the rivalry pathway. The *and* process gives rise to the stereopsis pathway. The outputs of the two pathways are combined to produce binocular single vision.

unadapted processes is measured. When one eye is tested, the binocular *or* process and the relevant monocular *only* process are active. Neither of these "saw" the adapting stimulus, and, thus, no TAE is measured.

The purely binocular process and the stereopsis mechanism appear to have similar spatial frequency tuning (compare Frisby and Mayhew, 1978a; Wolfe and Held, 1983). Both are more sensitive to monocular blur than to binocular blur (Julesz, 1971; Westheimer and McKee, 1980; Wolfe and Held, 1983). Finally, there is physiological evidence that some of the neurons involved in stereopsis are much more or even exclusively sensitive to binocular stimuli (Poggio and Fischer, 1977; Poggio and Talbot, 1981).

It appears that the monocular processes revealed at this stage of processing are not involved in stereopsis. Recall that adapting each eye to a different orientation produces a different TAE in each eye. Orientational disparity of this sort produces stereoscopic depth if the disparities are physically present in the monocular stimuli. However, orientation differences do not give rise to stereoscopic depth when they are produced in the monocular *only* processes by TAE adaptation (Wolfe, 1986a). The same principle holds for spatial frequency adaptation (Sloane and Blake, 1987).

Rivalry, by contrast, is unlikely to involve the purely binocular process, since that process cannot "see" rivalrous stimuli. Rivalry does not interfere with most forms of adaptation (spatial frequency—Blake and Fox, 1972, 1974; motion—Lehmkuhle and Fox, 1975; tilt—Wade and Wenderoth, 1978; McCollough effect—White *et al.,* 1978). Aftereffects of phase (Blake and Bravo, 1985) and figural aftereffects (Crabus and Stadler, 1973) are the exceptions. Similarly, rivalry does not block IOT (Blake and Overton, 1979; O'Shea and Crassini, 1981). By contrast, adaptation can influence the temporal dynamics of rivalry (Sloane and Blake, 1987). It follows that locus of rivalry must be later in the sequence of visual processing than the locus of adaptation. Therefore, rivalry must involve input from the monocular *only* processes because rivalry requires eye-of-origin information that is lost in the binocular processes.

SUMMARY OF ADAPTATION EXPERIMENTS

Stereopsis presumably occurs at or before the locus of adaptation, since cyclopean stimuli produce aftereffects and those aftereffects can be seen with noncyclopean test patterns (Blakemore and Julesz, 1971; Tyler, 1975; see Fig. 20-8D above). A more complete accounting of the combination of information from these classes of monocular and binocular processes into the pathways for rivalry and stereopsis can be found in Wolfe (1986a).

The adaptation experiments show that there is at least one locus in the visual pathway where separate pathways are involved with rivalry and stereopsis. Note that this independence of the stereopsis and rivalry pathways is not necessarily incompatible with a "stereopsis first" model. Given that stereopsis occurs earlier in processing than rivalry, the presence of stereoscopic information at one spatial location in the purely binocular/stereopsis pathway could act to block rivalry at that location in the other pathway. As the next section shows, the coexistence of rivalry and stereopsis at the same spatial location argues against this hypothesis.

Coexistence of Rivalry and Stereopsis

A series of demonstrations argue that the presence of stereoscopic information at one location does not necessarily preclude the experience of rivalry at that location. The first class of example involves stereoscopic views of standard, three-dimensional objects. Perhaps the clearest example is a simple cube. Washburn and Manning (1934) had subjects binocularly view a cube. Under their conditions, two vertical surfaces were visible: one in the frontal plane and a second side that sloped away from the viewer. The image of the second side would be of quite different size in each eye. Washburn and Manning's subjects could track rivalry between the two monocular images of this side while presumably experiencing stereoscopic depth as part of the perception of the same cube. Asher (1953) produced a series of similar demonstrations. To this class of demonstrations, it may be objected that the rivalry and stereopsis, although contiguous, were not superimposed. That is, one part of the cube (e.g., the front surface) supported stereopsis while another (the side) supported rivalry.

A second class of demonstration shows the coexistence of stereopsis and color rivalry. This is easily seen in anaglyphic presentation of stereograms, where disparate images are presented to each eye by printing one in red and the other in green and placing red and green filters in front of the eyes (e.g., Julesz, 1971). Under these circumstances many observers will see stereoscopic depth unperturbed by the rivalrous battle between red and green that occurs at the same place and time. The border of rivalry may be distinctly seen sweeping across stereoscopic contours. Formal observations have been made by Treisman (1962) using classical stereograms and by Ramachandran and Sriram (1972) using random-dot stereograms. To this class of demonstration two objections may be made. First, there are great differences between individuals in the perception of color rivalry (e.g., Hering, 1920/1964). Second, the parallel pathways might be "color" and "form" rather than stereopsis and rivalry. It has been argued that color and form information follow different physiological pathways (Hubel and Livingstone, 1987) and that they are psychophysically separable in some manner as well (Cavanaugh *et al.*, 1984; Livingstone and Hubel, 1987, 1988; though see Cavanaugh and Favreau, 1985). Moreover, color and form rivalry may be separable (Hastorf and Myro, 1959). Certainly, the binocularity of color processing is different from that of the rest of the visual system (Shattuck and Held, 1974; Over *et al.*, 1973; White *et al.*, 1978).

Rivalry and stereopsis can coexist in space if at least some of the information for each is carried in separate spatial frequency bands. Julesz and Miller (1975) found that band-pass-filtered random-dot stereograms could be seen in the presence of rivalrous noise if the noise was in a separate band. Mitchell's (1969) demonstration of stereopsis from different monocular shapes can be explained in similar terms. When the rivalrous noise and the stereogram occupy the same band, only rivalry is seen. Wolfe (1986a) has produced a different version of this type of demonstration. A random-dot stereogram generates the impression of an oblique, cyclopean grating. Monocular vertical and horizontal gratings are used as rivalry stimuli. The observer sees rivalry between vertical and horizontal gratings at the same time and in the same place as he sees the cyclopean grating.* The cyclopean grating is intact and undisturbed by the ongoing rivalry. As with the Julesz and Miller result, when coexistence fails, only rivalry is seen.

In this final class of demonstration, it is important to note that failures of coexistence are failures of stereopsis and not dominance of stereopsis over rivalry. The failures occur when the monocular stimuli for rivalry interfere with the monocular stimuli for stereopsis. If the stereopsis mechanism cannot match inputs from the two eyes, stereopsis will not be seen. Blake and O'Shea (1988) maintain that this argument fails because the purely binocular process is not stimulated by the monocular stimulus for rivalry. They reason that if stereopsis is mediated by a purely binocular pathway, it should not be disturbed by monocular information. However, the problem in this case is not the intrusion of monocular information into the purely binocular pathway but rather a failure of the purely binocular process to find matching stimuli in the monocular inputs. Consider a fairly extreme example. Suppose the stereoscopic information consists of three dots in each eye, the center dot having a disparity yielding depth. Now suppose 500 dots are added to the left eye's image. Although it may be true that the purely binocular process cannot see the monocular dots and that it could in principle see the binocular ones, the matching process required to determine which dots are binocular will have become nearly impossible, and it is unlikely that stereopsis will be occur. Similarly, in the experiments described here, coexistence fails when the rivalry stimuli simply make the stereo matching problem too difficult, not when stereopsis eliminates rivalry.

Blake and O'Shea (1988) agree that rivalry and stereopsis can occur at the same place and time but note that there is no convincing evidence for rivalry and stereopsis coexisting within the same spatial frequency channel. These data, therefore, could be consistent with a form of the "stereopsis first" model in which stereopsis is done within a channel whenever possible. If stereopsis is impossible, rivalry is produced in that channel. The output of all channels is combined into the final percept, allowing rivalry from one channel and stereopsis from another to coexist.

Silent Rivalry

Given a modified form of the stereopsis-first model, the major issue separating it from the parallel-pathways model is the existence of rivalry during the viewing of identical stimuli. Since there would be no perceptual consequences of an alternation between two identical monocular views, the question is both hard to answer and somewhat arcane. There is some evidence in favor of rivalry under these conditions, though it is necessarily indirect. There are three relevant lines of research: fluctuations in seen position, abnormal fusion experiments, and probe detection experiments.

*Robert O'Shea (personal communication) has produced a similar effect.

FLUCTUATIONS IN SEEN POSITION

If rivalry continues during stereopsis, its output might be expected to influence the apparent position of a target seen in depth. Consider a three-line stereogram with crossed disparity on the center line. The central line should be seen in front of the plane of fixation, but what should be its lateral position? A simple stereopsis-first theory might predict that the lateral position should be exactly between the positions of the monocular lines. A somewhat more sophisticated version would allow for a constant bias toward one monocular position (e.g., Helmholtz, 1909/1924; Hering, 1920/1964; Sheedy and Fry, 1979; Kaufman and Arditi, 1976, offer a dissenting view). An input from the rivalry pathway would be expected to produce a fluctuation in seen position over time. This has been reported by Washburn and Manning (1934), Makous and Sanders (1978), and Tyler (1984), though the matter could do with some more systematic investigation. In particular, it would be more convincing if it could be shown that stimulus manipulations that influence rivalry also influence changes in seen position in stereopsis (Wolfe, 1986). Verhoeff (1933) has shown that increasing luminance in one eye biases the seen position toward that position of the image in that eye, but binocular brightness averaging would probably predict the same result in this case. Notice that the effect of rivalry on seen position will be subtle. Depth without diplopia arises from fairly small disparities. If we assume that the seen position of the line is some average of the information from rivalry and stereopsis pathways, the expected oscillations in seen position caused by rivalry will be very small. Undoubtedly, they would go unnoticed under most circumstances.

In an interesting wrinkle on this point, Rose and Blake (1988) have shown that the seen position of diplopic lines outside of Panum's area seems to be influenced by the stereopsis mechanism. Diplopic lines are seen as closer to a central position.

ABNORMAL FUSION

Normal rivalry does not occur with very brief stimulus presentations, a fact that has been repeatedly discovered over the course of the last century (Hering, 1920/1964; Kaufman, 1963; Bower and Haley, 1964; Goldstein, 1970; Wade, 1973; Anderson *et al.*, 1978; Wolfe, 1983a,b). It appears to take some time to start the rivalry mechanism. A good rule of thumb is that normally rivalrous stimuli will appear abnormally fused if they are presented for less than 150 msec. It also appears to take time to turn off the rivalry mechanism. If presented immediately after the offset of rivalry, a briefly presented pair of normally rivalrous stimulus will appear as a single "snapshot" of rivalry. If a dark ISI of longer than 250 msec intervenes between the offset of rivalry and the onset of briefly presented but normally rivalrous stimuli, those stimuli will be abnormally fused (Wolfe, 1983a). Abnormal fusion is an anomaly of the rivalry pathway and not of stereopsis. It does not disrupt stereopsis (Wolfe, 1986a).

Abnormal fusion can be used as a test for activity in the rivalry pathway. Rivalrous stimuli flashed in isolation appear fused. They appear rivalrous if they follow other stimuli that activate the rivalry pathway (Goldstein, 1970). Wolfe (1986a) found that the effects of preexposure to fused and stereoscopic stimuli on abnormal fusion were identical to the effects of preexposure to rivalry, suggesting that the fused stimuli activated the rivalry mechanism just as successfully as the rivalrous stimuli.

PROBE DETECTION

Suppression makes it harder to detect stimuli presented to the suppressed eye (e.g., Wales and Fox, 1970). If rivalry continues even when the stimuli in the two eyes are identical, then it should be possible to find evidence for this rivalry in probe detection experiments. In general, this is not the case. Using threshold elevation and reaction time methods, numerous studies have found that monocular probes are detected as if both eyes were "dominant" when both eyes view the same stimulus (Fox and Check, 1966, 1968; Blake and Camisa, 1978; Cogan, 1982; Blake and Boothroyd, 1985; O'Shea, 1987). One set of studies found evidence for suppression during stereopsis (Makous and Sanders, 1978), but the bulk of the data suggest that it is as easy to detect monocular probes in either eye during fusion as it is to detect probes in the dominant eye during perceived rivalry.

The probe detection data are among the strongest evidence in favor of a stereopsis-first theory. It is not compatible with a parallel-pathways model unless monocular probes are detected by the purely binocular process.

HOW PURE IS THE PURELY BINOCULAR PROCESS?

How might monocular information get into the purely binocular pathway? One could imagine that matched contours open the logical *and* gate at the entrance to the stereopsis pathway. Once it is open, some monocular stimuli might enter along with the binocular stimuli. For example, imagine a set of black vertical bars presented to each eye. If a white spot is placed on one bar in one eye, it will be continuously visible. One explanation would be that it is seen via the stereopsis pathway. This loosening of the rules of admission to the purely binocular process could account for a number of otherwise problematic phenomena. Color fusion is reported to occur more readily when fused achromatic contours are present (Creed, 1935; Engel, 1958). Further, displays containing matched and unmatched contours are often perceptually ambiguous, giving an impression neither of crisp rivalry nor of complete fusion (e.g., Cogan and Goldstein, 1972; Cogan, 1973). This could be explained if the unmatched contours were producing rivalry in the rivalry pathway and at the same time were being brought into the purely binocular/stereopsis pathway by the pull of the matched contours. Finally, unmatched monocular points can serve as depth cues in some cases (Frisby and Mayhew, 1978b; O'Shea and Blake, 1987; Nakayama and Shimojo, 1988). At the present time, there is no direct evidence for the intrusion of monocular stimuli into the stereopsis pathway. The matter is worth further study. If it did occur, then the failure to find evidence for probe suppression during the viewing of identical stimuli in the two eyes may be caused by detection of monocular probes through the stereopsis pathway.

SUMMARY OF SILENT RIVALRY RESULTS

The existence of silent rivalry between identical stimuli is the critical test between a strict parallel-pathways model and a modified stereopsis-first model that requires only that stereopsis take precedence over rivalry within a spatial frequency channel. Unfortunately, the evidence is equivocal and indirect. Certainly, it must be conceded that the parallel-pathways model can account for the probe detection data only by allowing for some mechanism for the detection of monocular probes by the stereopsis pathway.

Double Dissociation

Double dissociation is a last line of evidence for the independence of rivalry and stereopsis. If manipulation 1 disrupts A but not B whereas manipulation 2 disrupts B but not A, then A and B are said to be doubly dissociated. In neuropsychology, "double dissociation" is taken as evidence for structural independence. Stereopsis and rivalry can be doubly dissociated. As noted above, the rivalry mechanism fails when stimuli are presented for brief periods. Stereopsis does not fail under these conditions (Wolfe, 1986a). Stereopsis, by contrast, can be disrupted by developmental mismatch between the eyes (e.g., strabismus). These conditions may alter rivalry (Smith *et al.*, 1985), but the basic pattern of dominance and suppression appears to continue (Jampolsky, 1955; Burian and von Noorden, 1974; Schor, 1977). Moreover, the suppression of stereoblind constant suppressors appears to be similar to that of normals to the extent that it, too, is abolished when stimuli are briefly flashed (Wolfe, 1986b). Artificial anisometropia, differential blur in the two eyes, also disrupts stereopsis without disrupting rivalry.

Constant Suppression: A Second Mechanism?

There are several anomalous phenomena that do not fit the usual accounts of binocular rivalry. For example, Schor *et al.* (1987) presented matched letters to each eye. For each letter, one eye saw a blurred version while the other saw a focused version. Each eye saw some blurred and some focused letters. The binocular percept was of a full set of focused letters. The blurred letters were continuously suppressed. Blake and Boothroyd (1985) presented vertical gratings to both eyes and a superimposed horizontal grating to one eye. They report seeing a stable plaid percept. The monocular horizontal grating does not appear to rival with the vertical grating in the other eye. They proposed that matched stimuli were removed from possible rivalry. Thus, the vertical gratings fuse and are removed from possible rivalry. The horizontal grating is left with no opposition in the other eye and is seen continuously. Alternatively, it could be that the vertical grating in one eye was continuously suppressed by the plaid in the other. Both of these phenomena could be examples of constant suppression. R. F. Hess (unpublished data,

1989) has provided evidence that neural loci of constant suppression and rivalry suppression are different, at least in amblyopes. Rivalry suppression does not interfere with most forms of adaptation (Blake and Fox, 1974; Blake and Overton, 1979). R. F. Hess (unpublished data, 1989) found that constant suppression does.

Hess was dealing with amblyopes who constantly suppress the weaker eye. It may be that a similar form of constant suppression occurs when a normal eye receives a stimulus that is a subset of the stimulus in the other eye (e.g., blurred). It remains to be proven if this second, earlier site of suppression exists in normal vision. If it does, it can explain a number of phenomena of binocular interaction such as the Blake and Boothroyd and Schor *et al.* demonstrations.

Summary

On the way from two eyes to binocular single vision, the visual system appears to generate a number of psychophysically distinct representations of the binocular visual field. Relatively early in processing there are separate monocular, binocular *or,* and binocular *and* representations. These, in turn, give rise to separate representations of visual information in rivalry and stereopsis pathways. The stereopsis representation contains information whenever and wherever the monocular stimuli fall within Panum's area (defined here as the range of disparities that will support stereopsis). The rivalry representation contains information at all locations in the binocular field.

COMBINING STEREOPSIS AND RIVALRY INFORMATION

The product of the binocular visual system is not a set of two (or more) separate representations. The perceptual product is binocular single vision. Thus, it is important to consider how the intermediate representations interact to give rise to the final perceptual state. Previous accounts (e.g., Wolfe, 1986a) have proposed that the final percept is some sort of weighted average of the contents of the intermediate representations. This must be, at best, an oversimplification. The rich set of phenomena that accompany the combination of information from purely binocular and exclusively monocular sources points to other forms of interaction.

The various cues to three-dimensional space perception may be useful tools in the examination of these interactions because they are vital aspects of the perception of three-dimensional space and because they can be confined to one or sometimes to either of the parallel pathways. Obviously, stereopsis is one cue. Other cues include the various monocularly available depth cues: linear perspective, texture, shading, specular reflection, motion parallax, and so forth. Pictorial depth cue information must be carried in either the stereopsis or rivalry pathways. The information is obviously available to one eye alone, and successful cyclopean presentation of stimuli such as the Necker cube shows that pictorial depth cues can be obtained from purely binocular stimuli (Julesz, 1971).

These roughly independent sources of information must be combined to yield a perception of three-dimensional space. Given the possibility of one cue existing in the rivalry pathway while another is confined to the stereopsis pathway, it can be seen that the interactions between these cues could be used to study interactions between pathways. Bultoff and Mallot (1987) provide a useful list of five of the ways that different sources of information could interact. Each of these can be illustrated with phenomena from the literature on binocular vision and depth perception:

1. Accumulation. Information could accumulate via linear or nonlinear summation or some form of averaging. For example, when depth cues are in agreement at a given location, they can sum to give a richer impression of depth (e.g., Bultoff and Mallot, 1988). In some cases when they disagree, they can average to reduce the apparent depth (e.g., Buckley *et al.,* 1988).
2. Veto. If one source of information is better than another, the lesser source may be suppressed. Pseudoscopic presentation is the classic example of this veto phenomenon. In a pseudoscope, the left and right eye views are reversed. In the absence of other cues to depth, this reverses the perceived depth in the stereogram. However, if pictorial depth cues are available (e.g., as in a standard photograph), reversal of disparity is not adequate to reverse apparent depth. Indeed, depth appears quite normal (Wheatstone, 1852), though it can reverse after adaptation (Shimojo and Nakajima, 1981). In the interaction of the rivalry and stereopsis representations, there may be circumstances that lead to the suppression of

the output of the rivalry pathway by the output of the stereopsis pathway. A particularly plausible occasion for this might be a situation where the contents of the rivalry pathway are a subset of the contents of the stereopsis pathway. This would be an alternative account of the Blake and Boothroyd (1985) demonstration.

3. Cooperation. There can be synergistic relations between sources of information. For example, monocular contours can help in the perception of stereograms (Frisby and Clatworthy, 1974). The cases where rivalry stimuli or isolated points take on depth values may be examples of this cooperation (Frisby and Mayhew, 1978b; Nakayama and Shimojo, 1988).
4. Disambiguation. One source can disambiguate information from another. For example, disparity information can stabilize the ambiguous depth in a Necker cube display.
5. Hierarchy. One source of information can be the input to another process. For example, the cyclopean Necker cube (Julesz, 1971) is evidence that stereopsis information can be the input to depth from linear perspective.

The rich set of possible interactions provides an opportunity to find common ground between parallel-pathways and stereopsis-first models of binocular single vision. The problem with this account of the final combination of stereopsis and rivalry information is that it is far too unconstrained. It allows for a different form of interaction to explain each phenomenon. What is needed is a principled set of rules that govern the combination of these two sources of information. This does not exist at present. There is no obvious reason why it could not exist. Psychophysical experiments have been successful in describing the rules that govern combination of monocular inputs in either rivalry or stereopsis pathways alone. New experiments will need to concentrate on the binocular equivalent of ambiguous stimuli, the cases where the monocular inputs support both binocular combination and perceptible rivalry.

ACKNOWLEDGMENTS. I thank Jeff Schall, Clifton Schor, Robert O'Shea, and Randolph Blake for useful comments and discussion, and Marni Stewart for her help with this paper. We thank IBM for the loan of YODA graphics boards. This research was supported by the National Eye Institute of NIH, the Whitaker Health Sciences Fund, the Educational Foundation of America, and the MIT Class of 1922. Finally, I thank the Essilor Corporation and Lawrence Stark for making it possible for me to attend this conference.

REFERENCES

Anderson, J. D., Bechtoldt, H. P., and Gregory, L. D., 1978, Binocular integration in line rivalry, *Bull. Psychonom. Soc.* **11**:399–402.

Anstis, S. M., and Duncan, K., 1983, Separate motion aftereffects from each eye and from both eyes, *Vision Res.* **23**:161–170.

Asher, H., 1953, Suppression theory of binocular vision, *Br. J. Ophthalmol.* **37**:37–39.

Blake, R., 1977, Threshold conditions for binocular rivalry, *J. Exp. Psychol.* **3**:251–257.

Blake, R., 1989, A neural theory of binocular rivalry, *Psychol. Rev.* **96**:145–167.

Blake, R., and Boothroyd, K., 1985, The precedence of binocular fusion over binocular rivalry, *Percept. Psychophys.* **37**:114–124.

Blake, R., and Bravo, M., 1985, Binocular rivalry suppression interferes with phase adaptation, *Percept. Psychophys.* **38**:277–280.

Blake, R., and Camisa, J., 1978, Is binocular vision always monocular? *Science* **200**:1497–1499.

Blake, R., and Fox, R., 1972, Interocular transfer of adaptation to spatial frequency during retinal ischaemia, *Nature [New Biol]* **240**:76–77.

Blake, R., and Fox, R., 1973, The psychophysical inquiry onto binocular summation, *Percept. Psychophys.* **14**:161–185.

Blake, R., and Fox, R., 1974, Adaptation to invisible gratings and the site of binocular rivalry suppression, *Nature* **249**:488–490.

Blake, R., and O'Shea, R. P., 1988, "Abnormal fusion" of stereopsis and binocular rivalry, *Psychol. Rev.* **95**:151–154.

Blake, R., and Overton, R., 1979, The site of binocular rivalry suppression, *Perception* **8**:143–152.

Blake, R., Overton, R., and Lema-Stern, S., 1981, Interocular transfer of visual aftereffects, *J. Exp. Psychol.* **88**:327–332.

Blakemore, C., and Julesz, B., 1971, Stereoscopic depth aftereffect produced without monocular cues, *Science* **171**:286–288.

Boring, E. G., 1942, *Sensation and Perception in the History of Psychology*, Appleton-Century-Crofts, New York.

Bower, T. G. R., and Haley, L. J., 1964, Temporal effects in binocular vision, *Psychon. Sci.* **1**:409–420.

Braddick, O., Campbell, F. W., and Atkinson, J., 1978, Channels in vision: Basic aspects, in: *Perception: Handbook of Sensory Physiology* (R. Held, W. H. Leibowitz, and H.-L. Teuber, eds.), 3–38, Springer-Verlag, Berlin.

Buckley, D., Frisby, J. P., and Mayhew, J. E. W., 1988, Interaction of texture and stereo cues in the perception of surface slant: Evidence for surface orientation anisotropy in cue integration, *Perception* **17**:A384.

Bulthoff, H. H., and Mallot, H., 1988, Integration of depth modules: Local and global depth measurements, *IOVS ARVO [Suppl.]* **29**:400.

Bulthoff, H. H., and Mallot, H. A., 1988, Integration of depth modules: Stereo and shading, *J. Opt. Soc. Am. [A]* **5**:1749–1758.

Burian, H. M., and Noorden, G. K., 1974, *Binocular Vision and Ocular Motility*, C. V. Mosby, St. Louis.

Campbell, F. W., and Maffei, L., 1971, The tilt aftereffect: A fresh look, *Vision Res.* **11**:833–840.

Cavanaugh, P., and Favreau, O. E., 1985, Color and luminance share a common motion pathway, *Vision Res.* **25**:1595–1601.

Cavanaugh, P., Tyler, C. W., and Favreau, O. E., 1984, Perceived velocity of moving chromatic gratings, *J. Opt. Soc. Am. [A]* **1**:893–899.

Cogan, A. I., 1982, Monocular sensitivity during binocular viewing, *Vision Res.* **22**:1–17.

Cogan, R., 1973, Distributions of durations of perception in the binocular rivalry of contours, *J. Gen. Psychol.* **89**:297–304.

Cogan, R., and Goldstein, A. G., 1972, Reporting fragmentations in the binocular rivalry of contours, *Am. J. Psychol.* **85**:569–584.

Crabus, H., and Stadler, M., 1973, An investigation of the localization of the perceptual process: Figural aftereffects under binocular-rivalry conditions, *Perception,* **2**:67–77.

Creed, R. S., 1935, Observations on binocular fusion and rivalry, *J. Physiol. (Lond.)* **84**:381–392.

Engel, E., 1958, Binocular fusion of dissimilar figures, *Psychol.* **46**:53–57.

Fox, R., and Check, R., 1966, Forced-choice recognition during binocular rivalry, *Psychon. Sci.* **6**:471–472.

Fox, R., and Check, R., 1968, Detection of motion during binocular rivalry suppression, *J. Exp. Psychol.* **78**:388–395.

Frisby, J. P., and Clatworthy, J. L., 1974, Learning to see complex random-dot stereograms, *Perception* **4**:173–178.

Frisby, J. P., and Mayhew, J. E. W., 1978a, Contrast sensitivity function for stereopsis, *Perception* **7**:423–429.

Frisby, J. P., and Mayhew, J. E. W., 1978b, The relationship between apparent depth and disparity in rivalrous-texture stereograms, *Perception* **7**:661–678.

Gibson, J. J., 1937, Adaptation with negative aftereffect, *Psychol. Rev.* **44**:222–244.

Goldstein, A. G., 1970, Binocular fusion and contour suppression, *Percept. Psychophys.* **7**:28–32.

Grindley, G. C., and Townsend, V., 1965, Binocular masking induced by a moving object, *J. Exp. Psychol.* **17**:97–109.

Hastorf, A. H., and Myro, G., 1959, The effect of meaning on binocular rivalry, *Am. J. Psychol.* **72**:393–400.

Helmholtz, H. von, 1924, *Treatise on Physiological Optics* (trans. from 3rd German ed., 1909), The Optical Society of America, New York.

Hering, E., 1920/1964, *Outlines of a Theory of the Light Sense* (L. Hurvich and D. Jameson, trans.), Harvard University Press, Cambridge, MA, p. 250.

Hochberg, J., 1964, Contralateral suppressive fields of binocular combination, *Psychon. Sci.* **1**:157–158.

Hubel, D. H., and Livingstone, M. S., 1987, Segregation of form, color, and stereopsis in primate area 18, *J. Neurosci.* **7**:3378–3415.

Hurvich, L. M., and Jameson, D., 1955, Some quantitative aspects of an opponent-colors theory: II. Brightness, saturation, and hue in normal and dichromatic vision, *J. Ophthalmol. Soc. Am.* **45**:602–616.

Jameson, D., and Hurvich, L. M., 1955, Some quantitative aspects of an opponent-colors theory: I. Chromatic responses and spectral saturation, *J. Ophthalmol. Soc. Am.* **45**:546–552.

Jampolsky, A., 1955, Characteristics of suppression in stabismus, *AMA Arch. Ophthalmol.* **54**:683–696.

Jones, R. K., and Lee, D. N., 1981, Why two eyes are better than one: Two views of binocular vision, *J. Exp. Psychol.* **7**:30–40.

Julesz, B., 1971, *Foundations of Cyclopean Perception,* University of Chicago Press, Chicago.

Julesz, B., and Miller, J. E., 1975, Independent spatial-frequency-tuned channels in binocular fusion and rivalry, *Perception* **4**:125–143.

Julesz, B., and Tyler, C. W., 1976, Neurontropy, an entropy-like measure of neural correlation in binocular fusion and rivalry, *Biol. Cybernet.* **23**:25–32.

Kaplan, I. T., and Metlay, W., 1964, Light intensity and binocular rivalry, *J. Exp. Psychol.* **67**:22–26.

Kaufman, L., 1963, On the spread of suppression and binocular rivalry, *Vision Res.* **3**:401–415.

Kaufman, L., 1974, *Sight and Mind,* Oxford University Press, New York.

Kaufman, L., and Arditi, A., 1976, The fusion illusion, *Vision Res.* **16**:335–343.

Kohler, W., and Wallach, H., 1944, Figural aftereffects: An investigation of visual processes, *Proc. Am. Phil. Soc.* **88**:4.

Lehmkuhle, S. W., and Fox, R., 1975, Effect of binocular rivalry suppression on the motion aftereffect, *Vision Res.* **15**:855–859.

Livingstone, M. S., and Hubel, D. H., 1987, Psychophysical evidence for separate channels for the perception of form, color, movement, and stereopsis, *J. Neurosci.* **7**:3416–3468.

Livingstone, M., and Hubel, D., 1988, Segregation of form, color, movement, and depth: Anatomy, physiology, and perception, *Science* **240**:740–749.

Makous, W., and Sanders, K. R., 1978, Suppressive interactions between fused contours, in: *Visual Psychophysics and Physiology* (J. C. Armington, J. Krauskopf, and B. R. Wooten, eds.), 167–179, Academic Press, New York.

McCollough, C., 1965, Color adaptation of edge-detectors in the human visual system, *Science* **149**:1115–1116.

Mitchell, D. E., 1969, Qualitative depth localization with diplopic images of dissimilar shape, *Vision Res.* **9**:991–994.

Moulden, B., 1980, Aftereffects and the integration of neural activity within a channel, *Phil. Trans. R. Soc. Lond. [Biol.]* **290**:39–55.

Nakayama, K., and Shimojo, S., 1988, Depth, rivalry and subjective contours from unpaired monocular points, *IOVS ARVO [Suppl.]* **29**:21.

Ogle, K. N., 1952, On the limits of stereoscopic vision, *J. Exp. Psychol.* **44**:253–259.

O'Shea, R. P., 1987, Chronometric analysis supports fusion rather than suppression theory of binocular vision, *Vision Res.* **27**:781–791.

O'Shea, R. P., and Blake, R., 1987, Depth without disparity in random-dot stereograms, *Percept. Psychophys.* **42**:205–214.

O'Shea, R. P., and Crassini, B., 1981, Interocular transfer of the motion aftereffect is not reduced by binocular rivalry, *Vision Res.* **21**:801–804.

Over, R., Long, N., and Lovegrove, W., 1973, Absence of binocular interaction between spatial and color attributes of visual stimuli, *Percept. Psychophys.* **13**(3):534–540.

Panum, P. L., 1957, *Physiological Investigations Concerning Vision with 2 Eyes,* Dartmouth Eye Institute, Hanover, NH.

Poggio, G. F., and Fischer, B., 1977, Binocular interaction and depth sensitivity in the striate and prestriate cortex of behaving rhesus monkey, *J. Neurophysiol.* **40**:1392–1405.

Poggio, G. F., and Talbot, W. H., 1981, Mechanisms of static and dynamic stereopsis in foveal cortex of the rhesus monkey, *J. Physiol. (Lond.)* **315**:469–492.

Ramachandran, V. S., and Sriram, S., 1972, Stereopsis generated with Julesz patterns in spite of rivalry imposed by colour filters, *Nature* **237**:347–348.

Richards, W., 1970, Stereopsis and stereoblindness, *Exp. Brain Res.* **10**:380–388.

Richards, W., 1971, Anomalous stereoscopic depth perception, *J. Opt. Soc. Am.* **61**:410–414.

Rose, D., and Blake, R., 1988, Mislocalization of diplopic images, *J. Opt. Soc. Am. [A]* **5**:1512–1521.

Schor, C. M., 1977, Visual stimuli for strabismic suppression, *Perception*, **6**:583–593.

Schor, C. M., 1987, Spatial factors limiting stereopsis and fusion, *Optics News* **13**:14–17.

Schor, C. M., Wood, I., and Ogawa, J., 1984, Binocular sensory fusion is limited by spatial resolution, *Vision Res.* **24**:661–665.

Schor, C. M., Landsman, L., and Erickson, P., 1987, Ocular dominance and the interocular suppression of blur in monovision, *Am. J. Optom. Physiol. Opt.* **64**:723–730.

Shattuck, S., and Held, R., 1974, Color and edge sensitive channels converge on stereo-depth analyzers, *Vision Res.* **14**:309–311.

Sheedy, J. E., and Fry, G. A., 1979, The perceived direction of the binocular image, *Vision Res.* **19**:201–211.

Shimojo, S., and Nakajima, Y., 1981, Adaptation to the reversal of binocular depth cues: Effects of wearing left–right reversing spectacles on stereoscopic depth perception, *Perception* **10**:391–402.

Sloane, M. E., and Blake, R., 1987, Perceptually unequal spatial frequencies do not yield stereoscopic tilt, *Percept. Psychophys.* **42**:569–575.

Smith, E. L., Levi, D. M., Harwerth, R. S., and White, J. M., 1982, Color vision is altered during the suppression phase of binocular rivalry, *Science* **218**:802–804.

Treisman, A., 1962, Binocular rivalry and stereoscopic depth perception, *Q. J. Exp. Psychol.* **14**:23–37.

Tyler, C. W., 1975, Stereoscopic tilt and size aftereffects, *Perception* **4**:187–192.

Tyler, C. W., 1983, Sensory processing of binocular disparity, in: *Vergence Eye Movements* (C. W. Schor and K. J. Ciuffreda, eds.), 199–295, Butterworth, London.

Tyler, C. W., 1984, Sensory processing of binocular disparity, in: *Vergence Eye Movements* (C. W. Schor and K. J. Ciuffreda, eds.), 199–295, Butterworth, London.

Verhoeff, F. H., 1933, Effect on steropsis produced by disparate retinal images of different luminosities, *Arch. Ophthalmol.* **10**:640.

Verhoeff, F. H., 1935, A new theory of binocular vision, *Arch Ophthalmol.* **13**:152–175.

Verhoeff, F. H., 1959, Panum's area and some other prevailing misconceptions concerning binocular vision, *Trans. Am. Ophthalmol. Soc.* **57**:37–47.

Wade, N. J., 1973, Binocular rivalry and binocular fusion of afterimages, *Vision Res.* **13**:999–1000.

Wade, N. J., and Wenderoth, P., 1978, The influence of colour and contour rivalry on the magnitude of the tilt aftereffect, *Vision Res.* **18**:827–835.

Wales, R., and Fox, R., 1970, Increment detection thresholds during binocular rivalry suppression, *Percept. Psychophys.* **8**:90–94.

Walker, P., 1978, Binocular rivalry: Central or peripheral selective processes, *Psych. Bull.* **85**:376–389.

Washburn, M. F., and Manning, P., 1934, Retinal rivalry in free vision of a solid object, *Am. J. Psychol.* **46**:632–633.

Westheimer, G., and McKee, S. P., 1980, Stereoscopic acuity with defocused and spatially filtered retinal images, *J. Opt. Soc. Am.* **70**:772–778.

Wheatstone, C., 1838, Some remarkable and hitherto unobserved phenomena of binocular vision, *Phil. Trans. (Lond.)* **128**:371–394.

Wheatstone, C., 1852, Some remarkable and hitherto unobserved phenomena of binocular vision: Part Two, *Phil. Mag.* **4**:504–523.

White, K. D., Petry, H. M., Riggs, L. A., and Miller, J., 1978, Binocular interactions during establishment of McCollough effects, *Vision Res.* **18**:1201–1215.

Wohlgemuth, A., 1911, On the aftereffect of seen movement, *Br. J. Psychol.* **1**(monograph supplement):1–17.

Wolfe, J. M., 1983a, Afterimages, binocular rivalry, and the false fusion phenomenon, *Perception* **12**:439–445.

Wolfe, J. M., 1983b, Influence of spatial frequency, luminance, and duration on binocular rivalry and abnormal fusion of briefly present, dichoptic stimuli, *Perception* **12**:447–456.

Wolfe, J. M., 1986a, Briefly presented stimuli can disrupt constant suppression and binocular rivalry suppression, *Perception* **15**:413–417.

Wolfe, J. M., 1986b, Stereopsis and binocular rivalry, *Psychol. Rev.* **93**:269–282.

Wolfe, J. M., and Blake, R., 1985, Dissecting the binocular visual system with psychophysical tools, in: *Models of Visual Cortex* (D. Rose and V.G. Dobson, eds.), 192–199, John Wiley & Sons, New York.

Wolfe, J. M., and Held, R., 1981, A purely binocular mechanism in human vision, *Vision Res.* **21**:1755–1759.

Wolfe, J. M., and Held, R., 1982, Binocular adaptation that cannot be measured monocularly, *Perception* **11**:287–295.

Wolfe, J. M., and Held, R., 1983, Shared characteristics of stereopsis and the purely binocular process, *Vision Res.* **23**:217–227.

21

The Role of Binocular Disparity Vergence Eye Movements in Disambiguating Superimposed Retinal Images

KENNETH J. CIUFFREDA, MARK ROSENFIELD, and LAWRENCE STARK

INTRODUCTION

In laboratory investigations of the disparity or fusional vergence mechanism, it is common to present the observer with isolated objects of regard having differing degrees of retinal disparity and to measure the resulting vergence motor response. However, under more naturalistic conditions, the retinal image may be composed from a complex, multilayered object space involving both overlapping and non-overlapping, spatially discrete object planes producing varying degrees of retinal disparity and defocus. Furthermore, these images frequently occupy an extended area of the retina well beyond the central foveal and macular regions. The task of disambiguating or separating these images in an optimal manner, with the requirement for accurate bifoveal fixation, is likely to represent a complex procedure involving multiple sensory and motor inputs.

KENNETH J. CIUFFREDA AND MARK ROSENFIELD • Department of Vision Sciences, State University of New York, State College of Optometry, New York, New York 10010. LAWRENCE STARK • School of Optometry, University of California, Berkeley, California 94720.

EXAMPLES

As an example of the above, consider the situation where an observer gazes out of a window at a complex distant scene such as an urban skyline and suddenly becomes aware of a small object on the window pane that overlaps the primary object of regard within the central visual field (see Fig. 21-1a). This second object would appear to be both blurred and diplopic and could be located at any point between the observer and fixated target. If a decision is made to bifixate this new target, the disparity error is sensed and processed, and accurate vergence upon it should proceed. Furthermore, the magnitude of the vergence response may also provide information concerning the apparent distance of the object of regard (Gogel, 1961; Lie, 1965; von Hofsten, 1976). Let us now imagine that the fixated object is altered to consist of a series of extended targets, such as several trees positioned close to the window, and the observer again becomes aware of this same second object. In this case the secondary target may appear to be clear (providing it lies within the depth of focus) and diplopic, while its retinal image partially overlaps both that of the

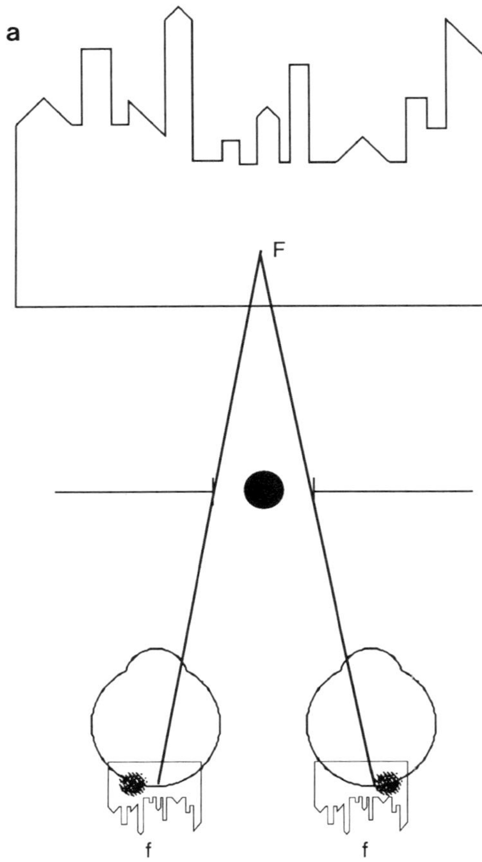

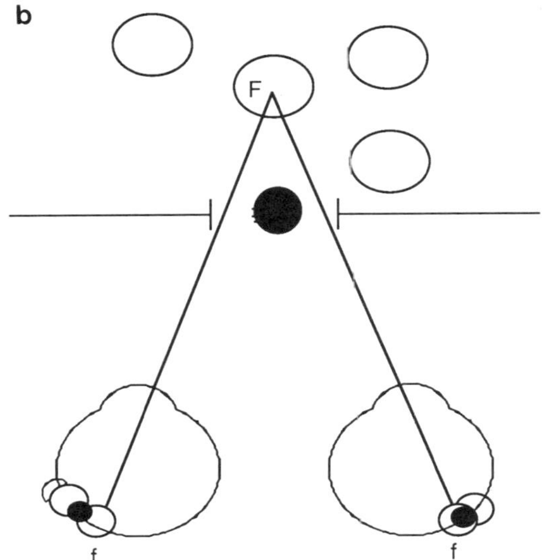

FIGURE 21-1. Complex naturalistic binocular viewing conditions. Subject viewing distant object through a window. Top view. (a) Distance fixation and focus on skyline; near object located on the window. The retinal image from the near object will be defocused and diplopic. (b) Fixation on near tree trunk; nearer object located on the window. The retinal image from the nearer object will be focused (if it lies within the depth of focus) and diplopic. F, bifixation point; f, fovea.

fixated tree and other trees in the field (see Fig. 21-1b).

Another static condition of contemporary interest is that of the "diffractive" contact lens (Walker and Churms, 1987), as shown schematically in Fig. 21-2 (for monocular viewing). Here the observer may have superimposed, full-field, in-focus extended objects at both distance and near. Additionally, overlapping images with varying degrees of defocus may be produced by objects located at intermediate distances. Under binocular viewing conditions, in-focus, disparate images from both distant and near objects may be simultaneously imaged on both retinas. If the retinal disparity is large, then these two images will not be spatially overlapping, as shown in Fig. 21-3. However, if the separation of the two extended objects is reduced, then the degree of retinal disparity will be smaller. Focused, spatially overlapping retinal images will be produced (see Fig. 21-4). Clearly, in the latter example, the process of disambiguation will be more intricate. One can go on to imagine a more complex yet frequently encountered situation in which a scene containing various extended objects over a broad range of viewing distances and with varying degrees of overlap is viewed, resulting in a multitude of differing disparity inputs.

One could also envisage a dynamic situation in which a flock of blackbirds is flying above and toward an observer while an isolated white bird is flying in the opposite direction, i.e., away from the observer and into the flock. If the birds are at a high altitude, e.g., 50 m, then relatively large changes in physical distance would only produce small variations in retinal disparity and defocus. However, at a much nearer distance, e.g., a height of 2 m, the rate of change of these sensory inputs would be much

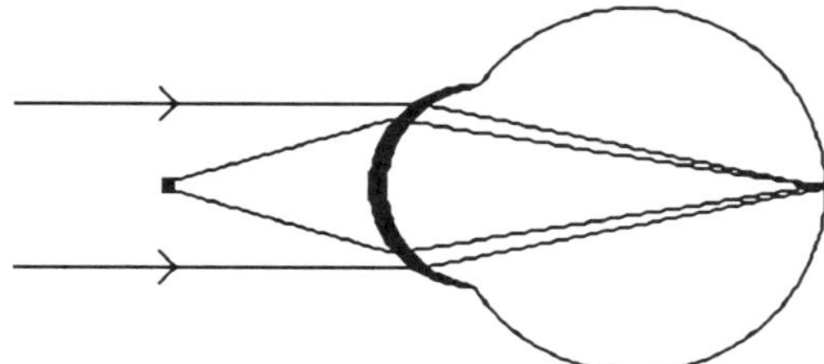

FIGURE 21-2. Schematic view of a diffractive contact lens under monocular viewing. Images from distant and near objects are simultaneously focused on the retina.

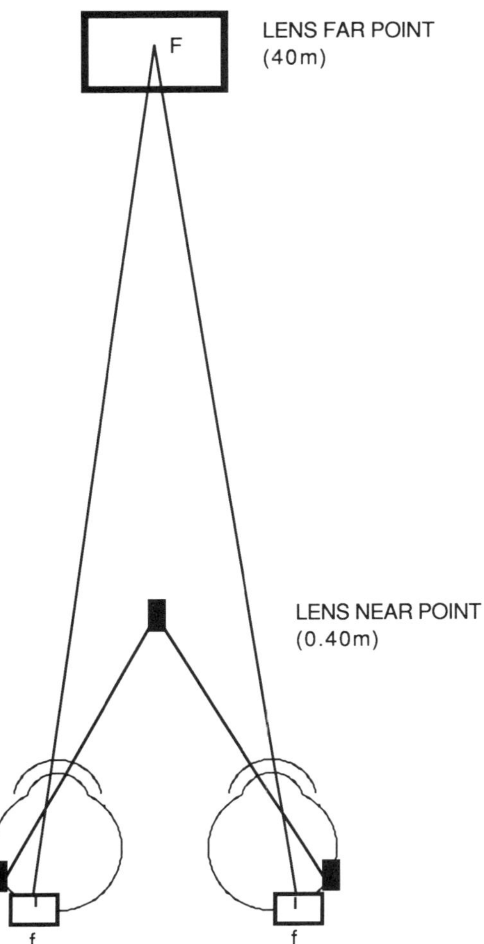

FIGURE 21-3. Distance bifixation with diffractive contact lens. Retinal images from objects at both the lens far point (distance correction) and near point (+2.50 D addition) are in focus but spatially nonoverlapping because of the relatively large disparity.

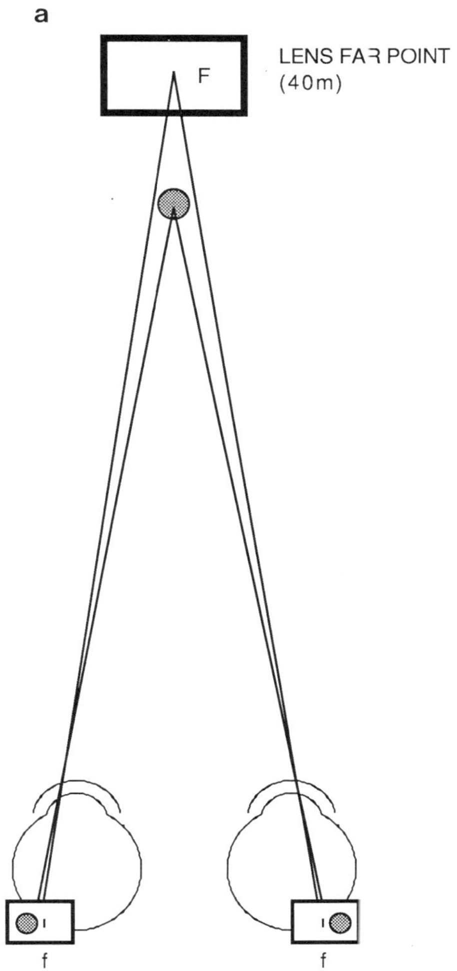

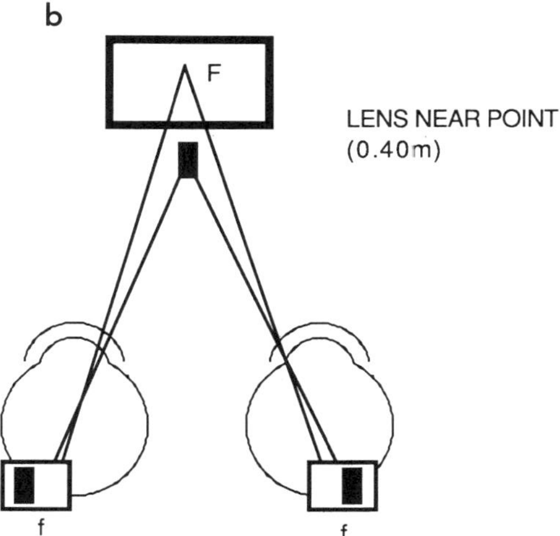

FIGURE 21-4. Distance (a) and near (b) bifixation with diffractive contact lens. Retinal images from two objects located close together are focused (providing both fall within the depth of focus) and spatially overlapping because of the relatively small disparity.

greater and would contribute differentially in the disambiguation and distance localization processes. If the birds approached even closer, e.g., 0.5 m above the ground, the rate of change of these parameters would be yet greater. This is illustrated in Fig. 21-5.

THE VERGENCE RESPONSE

The contribution of individual components of the vergence sensory–motor response, as well as other potential sources of information relating to the perception of apparent distance in the disambiguation process, is considered in Tables 21-1 and 21-2. The classical Maddox components of the aggregate vergence response (Maddox, 1893; Morgan, 1983) are

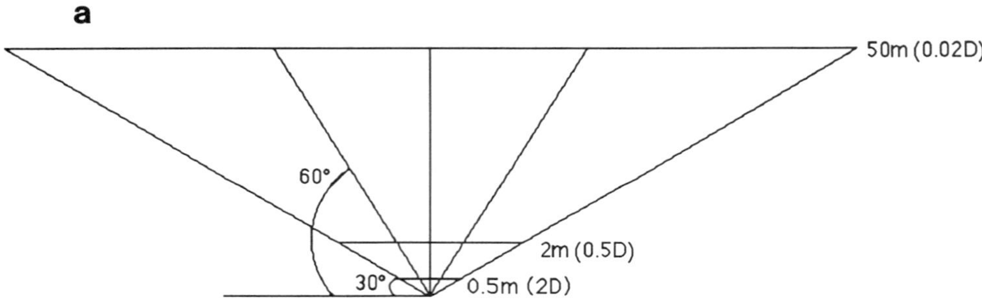

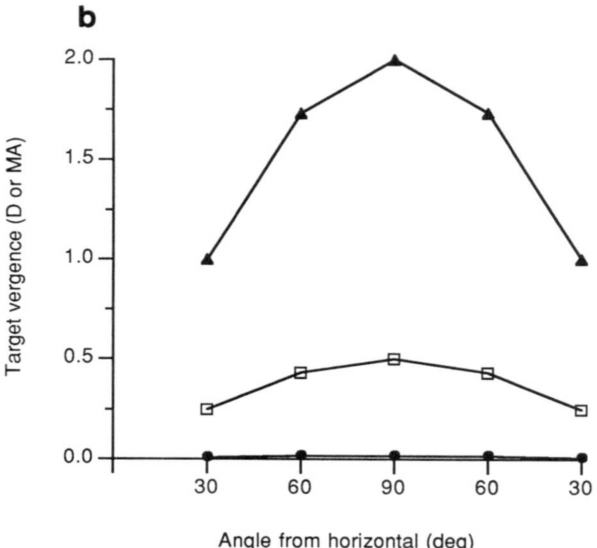

FIGURE 21-5. Changes in target distance, disparity, and defocus for distant (50 m), intermediate (2 m), and near (0.5 m) parallel flight distances from the ground at three different viewing angles (30°, 60°, and 90°). (A) Subject located on ground at point of intersection of lines and tracks lone bird with combined head and eye movements. (B) Plot showing calculated changes in target disparity and defocus for the three different viewing angles and altitudes.

listed in Table 21-1. Assuming the absence of sustained bifixation leading to adaptation of tonic vergence, one may reasonably presume that the level of tonic vergence will remain constant (Hung and Semmlow, 1980; Owens and Leibowitz, 1983; Fisher *et al.*, 1988). The vergence drive resulting from a "psychic awareness of nearness" of the object of regard in space is termed proximal vergence (Hokoda and Ciuffreda, 1983). Recent experiments in our laboratory clearly demonstrate that significant changes in proximally driven accommodation and vergence can only be elicited at distances closer than 3m (Rosenfield *et al.*, 1991). The third component of the vergence response is accommodative

TABLE 21–1
Components of Vergence Response[a] and Their Relative Role in Disambiguating Simultaneously Presented Near or Far Targets

Vergence stimulus	Distance	Near
Tonic	presumed constant	presumed constant
Proximal	Not Significant (Awareness of far)	Significant (Awareness of near)
Accommodative	AR≠0 AC small	AR larger AC larger
Disparity	Small	Large

[a]Maddox (1893).

TABLE 21–2
Other Cues for Distance Perception and Their Estimated Relative Importance in Disambiguating Simultaneously Presented Near or Far Targets

Cues	Distance	Near
Motion	++	++++
Detail/texture	+	++++
Relation to other objects	+	++++
Change in apparent size	+	++++
Stereopsis	+	+++
Size constancy	++	++
Change in vergence level	+	++
Change in accommodative level	+	++

vergence, which is the blur-driven aspect (Ciuffreda and Kenyon, 1983). It was shown in Fig. 21-5b that when one is viewing a far target, e.g., one located at 50 m, large variations in the physical location of the target will result in only small changes in the dioptric accommodative stimulus. Furthermore, the accommodative response at distance is nonlinear, and therefore any changes in accommodation and its synkinetically-driven accommodative vergence resulting from a sizable variation in target location would be negligible. However, at near, for comparable physical changes in target distance, blur will probably be present, resulting in substantial accommodative vergence. Similar arguments can be made with regard to retinal disparity and disparity vergence (Jones, 1983; Tyler, 1983).

In addition to the Maddox components of vergence, other sources of information related to distance perception may influence the aggregate response (Table 21-2; see Ittleson, 1960, and Tyler, 1983, for detailed reviews). The impact of the first six cues listed in Table 21-2 on distance perception would be greater at near than at far, since a given change in the physical distance of a far target would result in less variation in the respective cue than that found when viewing a near target. Changes in vergence level for both far (Gogel, 1961) and near objects (Lie, 1965; von Hofsten, 1976) as well as changes in blur-driven accommodation at near (Fisher and Ciuffreda, 1988) (the effect of changes in blur-driven accommodation at far has not been tested) can also provide information relating to the apparent distance of a target in space, perhaps contributing most to the proximal aspect.

SUMMARY

Disambiguation is a complex procedure resulting from the processing and combination of several sensory and motor inputs. However, it would appear that when the disparity-vergence input is large, information relating to the output of these disparity-induced vergence eye movements probably plays a major role in the disambiguation process. In addition, several other sources of information may exert varying degrees of influence on this complex pattern- and disparity-matching process, especially when a variety of disparity inputs from multiple discrete and extended objects are simultaneously presented.

REFERENCES

Ciuffreda, K. J., and Kenyon, R. V., 1983, Accommodative vergence and accommodation in normals, amblyopes, and strabismics, in: *Vergence Eye Movements: Basic and Clinical Aspects* (C. M. Schor and K. J. Ciuffreda, eds.), Butterworths, Boston, pp. 101–173.

Fisher, S. K., and Ciuffreda, K. J., 1988, Accommodation and apparent distance, *Perception* **17**:609–621

Fisher, S. K., Ciuffreda, K. J., Tannen, B., and Super, P., 1988, Stability of tonic vergence, *Invest. Ophthalmol. Vis. Sci.* **29**:1577–1581.

Gogel, W. C., 1961, Convergence as a cue to absolute distance, *J. Psychol.* **52**:287–301.

Hokoda, S. C., and Ciuffreda, K. J., 1983, Theoretical and clinical importance of proximal vergence and accommodation, in: *Vergence Eye Movements: Basic and Clinical Aspects* (C. M. Schor and K. J. Ciuffreda, eds.), Butterworths, Boston, pp. 75–97.

Hung, G. K., and Semmlow, J. L., 1980, Static behavior of accommodation and vergence: Computer simulation of an interactive dual-feedback system, *IEEE Trans. Biomed. Eng.* **27**:439–447.

Ittleson, W. H., 1960, *Visual Space Perception*, Springer, New York.

Jones, R., 1983, Horizontal disparity vergence, in: *Vergence Eye Movements: Basic and Clinical Aspects* (C. M. Schor and K. J. Ciuffreda, eds.), Butterworths, Boston, pp. 297–316.

Lie, I., 1965, Convergence as a cue to perceived size and distance, *Scand. J. Psychol.* **6**:109–116.

Maddox, E. E., 1893, *The Clinical Use of Prisms and the Decentering of Lenses* (2nd ed.), John Wright and Sons, Bristol, p. 106.

Morgan, M. W., 1983, The Maddox analysis of vergence, in: *Vergence Eye Movements: Basic and Clinical Aspects* (C. M. Schor and K. J. Ciuffreda, eds.), Butterworths, Boston, pp. 15–21.

Owens, D. A., and Leibowitz, H. W., 1983, Perceptual and motor consequences of tonic vergence, in: *Vergence Eye Movements: Basic and Clinical Aspects* (C. M. Schor and K. J. Ciuffreda, eds.), Butterworths, Boston, pp. 25–74.

Rosenfield, M., Ciuffreda, K. J., and Hung, G. K., 1991, The linearity of proximally-induced accommodation and vergence, *Invest. Ophthal. Vis. Sci. (Suppl.)*, **32**:1125.

Tyler, C. W., 1983, Sensory processing of binocular disparity, in: *Vergence Eye Movements: Basic and Clinical Aspects* (C. M. Schor and K. J. Ciuffreda, eds.), Butterworths, Boston, pp. 199–295.

von Hofsten, C., 1976, The role of convergence in visual space perception, *Vision Res.* **16**:193–198.

Walker, P., and Churms, P., 1987, The diffractive bifocal contact lens, *Optician* **194(2 Oct.)**:21–24.

22

Disambiguation of Objects by Stereopsis and Motion Cues

CHRISTOPHER W. TYLER

INTRODUCTION

One of the major tasks faced by organisms in a complex environment is disambiguating the objects in that environment from their backgrounds and from each other. The task is already challenging when there are a few objects in view with multifaceted surfaces covered with multicolored patterns and lit with varying degrees of light and shadow. This might be exemplified by a scene of bathers in multicolored swimwear on the beach. But the disambiguation becomes nearly impossible when the scene is a deeply three-dimensional one of foliage high in the trees of a thick forest. When primates first adopted such an arboreal habitat, they were faced with the task of locomotion through this three-dimensional realm of branches heavily masked by a veil of leaves. The stakes for disambiguating the true distances from tree to tree were high; the price of failure was to drop many meters onto a ground inhabited by carnivorous beasts. Thus, the evolutionary pressure to develop effective depth perception in this random-leaf world was at a high level.

This evolutionary view of the arboreal requirements for complex perception predicts a high level of sophistication of the neural mechanisms devoted to the task. This view gives extra motivation to the study of disambiguating mechanisms in complex scenes, which might otherwise be merely a scientific curiosity. The wealth of special-purpose processing strategies that are being uncovered may further serve as a model for interactive serial–parallel organization for processing of other sensory and cognitive attributes.

CHRISTOPHER W. TYLER • Smith-Kettlewell Eye Research Institute, San Francisco, California 94115.

SEGREGATION OF OBJECTS

Objects in the world are inherently segregated from one another. If they were not, they would necessarily be fused components of the same object. A major component in making sense of the spatial world, therefore, is the ability to perceive such segregation when it is masked in some way. The masking may consist of the following:

1. Masking directly by other objects. This is the situation of occlusion, in which one object is behind another from the perspective of a viewer in three-dimensional space. Although the objects are segregated in space, one is hidden behind the other in the two-dimensional representation projecting to each (or only one) eye. In this case, the reconstruction problem becomes one of determining the nature of an object for which there is no direct information; this can only be solved by inference from the probable shapes of objects in the world.
2. Masking by the camouflage of the surface color features. Here the problem is the confusion between the surface features of the object and its physical boundaries. The problem is compounded if there are surface features in common between the object and its background. A typical case where common features occur is with lighting filtered by, for example, the foliage of the trees between the light source and the object. Another case is that of camouflage developed in an animal's evolution to match the animal's usual background features. However camouflage arises, it is important to have disambiguating mechanisms available to avoid being fooled by the similarity of surface features. There are sev-

eral sources of disambiguating information; relative motion and binocular disparity for two-eyed observers are the two that form the focus of the present discussion.
3. Masking by the inherent limitations of visual processing. This is a category that is not often included in the discussion of object masking. It is nonetheless important, and not just for a biological image processor. Many inherent limitations are also encountered in three-dimensional image-processing machines, because they are inherent in the problem of three-dimensional reconstruction from two-dimensional images. Examples of such limitations are the objects in a field of view being too close or too far away relative to the convergence axis of the eyes (or, for motion, the observer moving too rapidly) to give rise to a depth signal. Alternatively, the objects to be disambiguated may be too close together (or the observer moving too slowly) to be segregated by small differences in depth signal.

FIGURE 22-1. Schematic of the three categories of perceptual segregation—local, regional, and global segregation.

CATEGORIES OF PERCEPTUAL SEGREGATION

There are at least three categories of perceptual segregation available for the task of disambiguation: local, regional, and global segregation (Fig. 22-1).

Local segregation is the lateral separation of images that would otherwise be fused together in the two-dimensional image. For the two-dimensional image alone, the visible separation of images is a question of two-point resolution or visual acuity and has a typical value of about 0.5–1′ in the fovea. Beyond this separation threshold, isolated images are easily segregated into their individual identities.

If the spatial separation is produced by a binocular disparity, then the two images to be segregated are being viewed dichoptically—one to each eye. In this case small separations that would be clearly visible monocularly give rise to fusion in the dichoptic situation. It is not until much larger separations (or binocular disparity) are reached that perceptual segregation (or diplopia) is obtained on the basis of local information (Ogle, 1950).

The case of motion is quite analogous. For small separations in space and time (or small movements), the different spatial positions are coded as belonging to the same object moving, and hence there is no perceptual segregation, although the movements of the object may be well above the position threshold for two-point resolution over time. For large changes in position, apparent movement breaks down, and two separate objects are seen. This is a failure of the motion system to unify the object over time, in much the same way that the binocular fusion system fails to fuse object with large disparities, giving rise to segregation of the two images involved.

Regional segregation is the segregation of objects into separate lateral groups or clusters on the basis of some feature in common between the members of the cluster but different between clusters. In the case of stereopsis, the differentiating feature would be a binocular disparity difference between the elements of one cluster and those of another cluster; the clusters would then appear at different depths from each other. In the case of motion, the differentiating feature would be the direction of motion of the elements of a cluster. In this case, whether the clusters appeared at different depths would depend on a further level of interpretation of the motion information.

Global segregation is used to describe segregation along the dimension of the segregating cue, but globally in the sense that it occurs everywhere in the image space. Thus, for depth cues from binocular disparity, global segregation corresponds to segregation of the image into two transparent depth planes, so that one can look through one plane to the other, everywhere in the image. For motion, global segregation gives a sense of two arrays of objects moving over one another, again with a perception of transparency and perhaps a separation in depth.

LOCAL STEREOSCOPIC SEGREGATION LIMITS

Local segregation based on disparity cues amounts to a fusion/diplopia task (Panum, 1858). The ability to perform such segregation depends profoundly on the properties of the retinal images of the objects being viewed. Data for such segregation were obtained (Schor *et al.*, 1989) with narrow-band line stimuli (as shown in Fig. 22-2) with luminance profiles constructed from second-order Cauchy functions (which are similar to difference-of-Gaussian functions). The profile of the bars had a narrow central bar with a wider negative surround, so that only cells with narrow receptive fields in the brain could respond to this stimulus. The fusion limit was measured for these bars presented dichoptically with a binocular disparity.

To give direct impression of the fusion limits obtained with these stimuli, the limiting separation is depicted for a wide bar width (0.5 cycle/° width at half-height) in Fig. 22-2A. When the disparity was in the horizontal direction, the bars appeared to be fused up to a disparity of 30′. Beyond this point diplopia occurred, and two bars could be seen. When the stimulus configuration was changed so that the bars were now horizontal and the disparity was in the vertical direction, the maximum separation for a fused percept was only 10′. Note that, even for these very blurred stimuli, the limits to keep fusion are much less than the 1° width of the stimuli. This implies that the images in the two eyes need to be matched with at least this precision to avoid diplopia.

Figure 22-2B shows the binocular fusion limits for narrow Cauchy bars (16 cycle/° width at half height). The limits became much narrower, so the scale was changed here to minutes instead of degrees. The fusion range for horizontal disparities decreased to 6′, while for vertical disparities it decreased to 3′. Note that for these narrow bars, the binocular fusion limits were now much wider than the center width of the stimuli themselves.

The striking feature of these data is that local segregation is evidently not limited by a particular aspect of the stimulus profiles. Segregation did not occur at a point such as the separation where the monocular profiles cross at the 50% height or the similar Rayleigh limit for bimodal segregation, for example. The segregation point varies by a factor of 20 from about half the mean bar width for the 0.5 cycle/° vertical disparity condition to about 10 bar widths for the 16 cycle/° horizontal disparity condition. This must mean that the segregation limit is imposed by the structure of the neural processing of disparity in the cortex, rather than by the process of

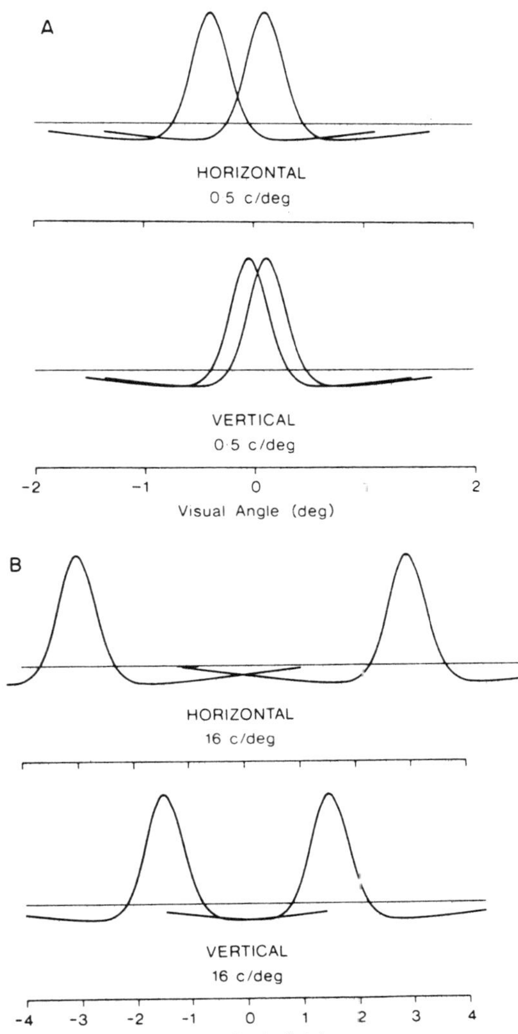

FIGURE 22-2. Depiction of the limiting cases of local segregation by binocular disparity in terms of the stimulus profile positions for the left and right eyes. (A) Segregation limits for 0.5 cycle/° Cauchy bars for horizontal (upper) and vertical (lower) binocular disparities. (B) Segregation limits for 16 cycle/° Cauchy bars for horizontal (upper) and vertical (lower) binocular disparities, with reduced scale on abscissa.

binocular combination of the two monocular images. It may represent the range of disparities to which there are cells tuned in each condition.

LOCAL MOTION SEGREGATION LIMITS

What about the corresponding local limit for motion segregation? Figure 22-3 shows thresholds for a bright bar alternating between two positions with no dark interstimulus interval (ISI). The lower graph shows the limit for detecting motion as temporal frequency is increased. The upper graph shows the upper limit beyond which motion breaks down and two segregated bars are seen. This limit falls precipitously with temporal frequency on a slope of −2. At 2 Hz, the upper limit for coherent motion was about 1°. By 10 Hz, the motion segregation limit had fallen to only about 1′, a sixtyfold decrease for a fivefold increase in frequency. This means that we are far more susceptible to spatial digitization errors in rapidly moving images than in slower ones.

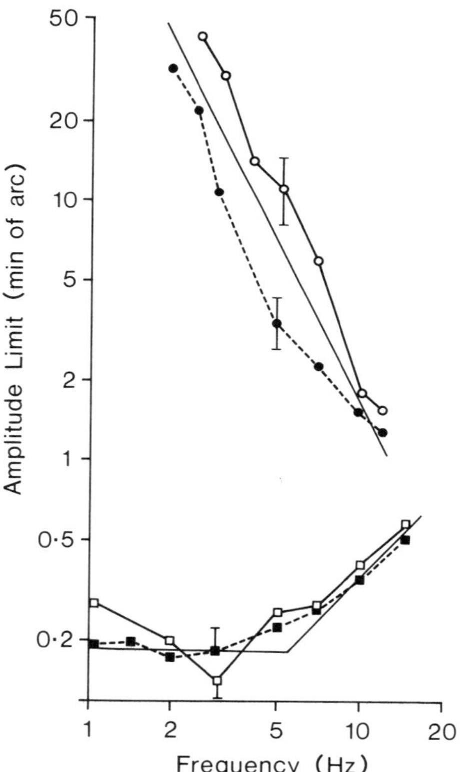

FIGURE 22-3. Amplitude limits for local segregation on the basis of temporal alternation for two observers (filled and open symbols). Upper curves, amplitude segregation limit below which stimuli appear unitary with no segregation; lower curves, detection limit for apparent motion, for comparison.

The steepness of this decline shows that motion segregation is limited not just by a spatiotemporal reciprocity principle but by an extra stage of integration, as though stimulus acceleration was the key parameter governing the behavior. This result is actually a version of Korte's fourth law of apparent motion and implies a specific organization of the motion process that has never been explained, despite decades of motion modeling.

REGIONAL MOTION SEGREGATION LIMITS

As a precursor to the study of regional segregation, one can measure the ability to segregate areas separated by lines in an image. This amounts to asking: How well can we see the wiggle in wiggly lines? The wiggle (or sinusoidal curvature) can be either coarse (low spatial frequency) or fine (high spatial frequency). It turns out that we see intermediate wiggles best; peak sensitivity is found at 2 wiggles/°, with rapid falloff at both high and low spatial frequencies (Fig. 22-4).

The properties of regional segregation are very different. As a reminder, Fig. 22-5A shows the stimulus used to measure regional segregation of a random-dot field on the basis of a motion profile. The spatial profile of the motion is sinusoidal, so this stimulus forms a motion grating consisting of horizontal strips moving in opposite directions.

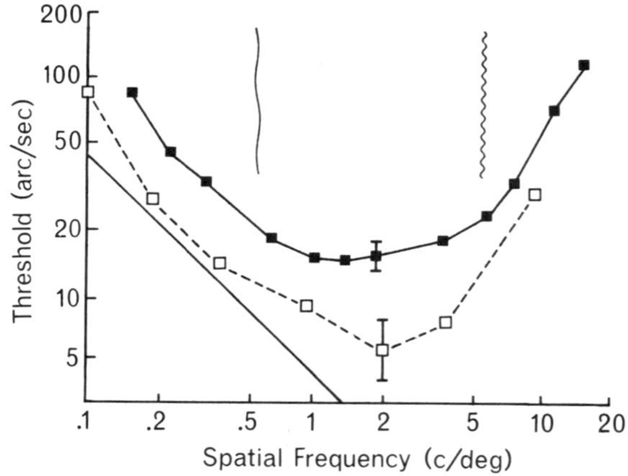

FIGURE 22-4. Threshold for shape of regional segregation in line contours for two observers (filled and open symbols). Contour was a sinusoidal spatial wiggle set to the threshold at which it appeared just not straight, as a function of spatial wiggle frequency. Note that wiggle detection degrades progressively below about 2 cycle/° corresponding to a fixed orientation limit (parallel to oblique line with −1 slope).

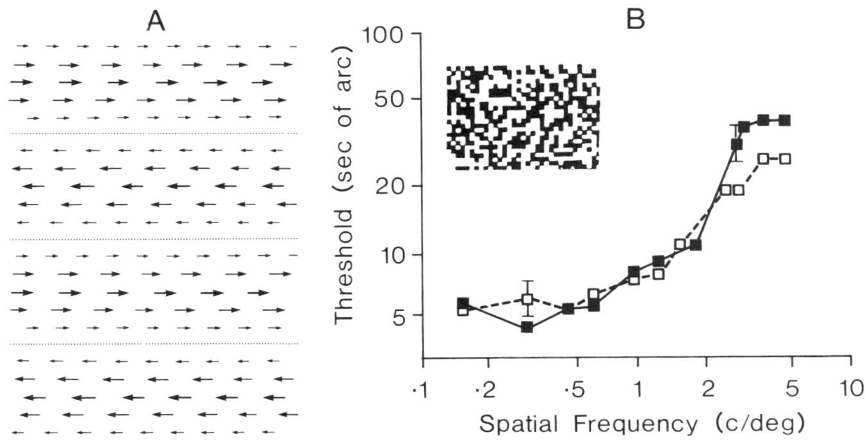

FIGURE 22-5. Regional segregation by motion information for two observers (filled and open symbols). (A) Random-dot motion grating formed by horizontal strips moving in opposite directions. (B) Detectability of segregation of the motion-defined strips as a function of their spatial frequency. Note that detectability at low spatial frequencies is much greater than for contour segregation (Fig. 22-4).

The detectability of motion in this stimulus depends on the spatial frequency of the regional segregation. At high spatial frequency (narrow strips), the motion is hard to detect (just as for the wiggly line), but at low spatial frequencies (broad strips), it becomes much easier (Fig. 22-5B). By 0.1 cycles/° (5 °/strip) the motion threshold is 5″. This is very different from the wiggly line case, where the threshold was about 5′. Thus, regional segregation on the basis of motion allows a far better performance than line wiggle at low spatial frequencies. Thus, contour extraction is not the only process operating in form analysis. Regional segregation on the basis of motion cues provides a spatial segregation of form that allows a far more secure estimate of the region boundaries for large patches than does the contour information.

REGIONAL STEREOSCOPIC SEGREGATION LIMITS

The analogous case for regional segregation in stereopsis is spatial segregation on the basis of disparity differences between contiguous regions. The case corresponding to the motion grating of Fig. 22-5A is the random-dot stereograting reproduced in Fig. 22-6. (Because the grating is invisible to monocular view, this is also known as a cyclopean stereograting.) The dot positions are modulated in disparity so as to produce a horizontal depth corrugation with a sinusoidal profile. Figure 22-6 is an actual stereogram, so that if it is viewed with crossed eyes or with a stereoscope so as to superimpose the left and right eye images in binocular view, the depth profile will be perceived in space.

The depth corrugation of Fig. 22-6 is not constant over the image, but varies in two ways. The spatial frequency increases upwards, and the disparity modulation amplitude increases toward the left, so as to form a contrast- and frequency-swept stereograting. This allows one to view one's own limits of stereograting perception. The higher is the spatial corrugation frequency, the lower is the modulation depth for which it can be processed. The depth corrugation limit therefore runs from the lower left to the upper right in the stereogram of Fig. 22-6. This kind of stimulus was used to obtained the limits of region segregation by disparity differences. Figure 22-7 depicts these limits in terms of the spatial frequency of vertical corrugation into horizontal strips. The lower part of the figure shows the stereograting threshold, with a peak sensitivity at about 0.4 c/°. The right-handed portion of the curve represents the frequency resolution limit at about 4 c/°, and decreasing range of segregation as disparity amplitude is increased (the disparity-scaling effect) is shown in the upper portion of the figure.

No disparity modulation can produce sensations of regional segregation outside these limits, which therefore constitute an important template for the construction of stereograms. This principle is frequently ignored in the presentation of stereograms in such fields as biochemistry, where ster-

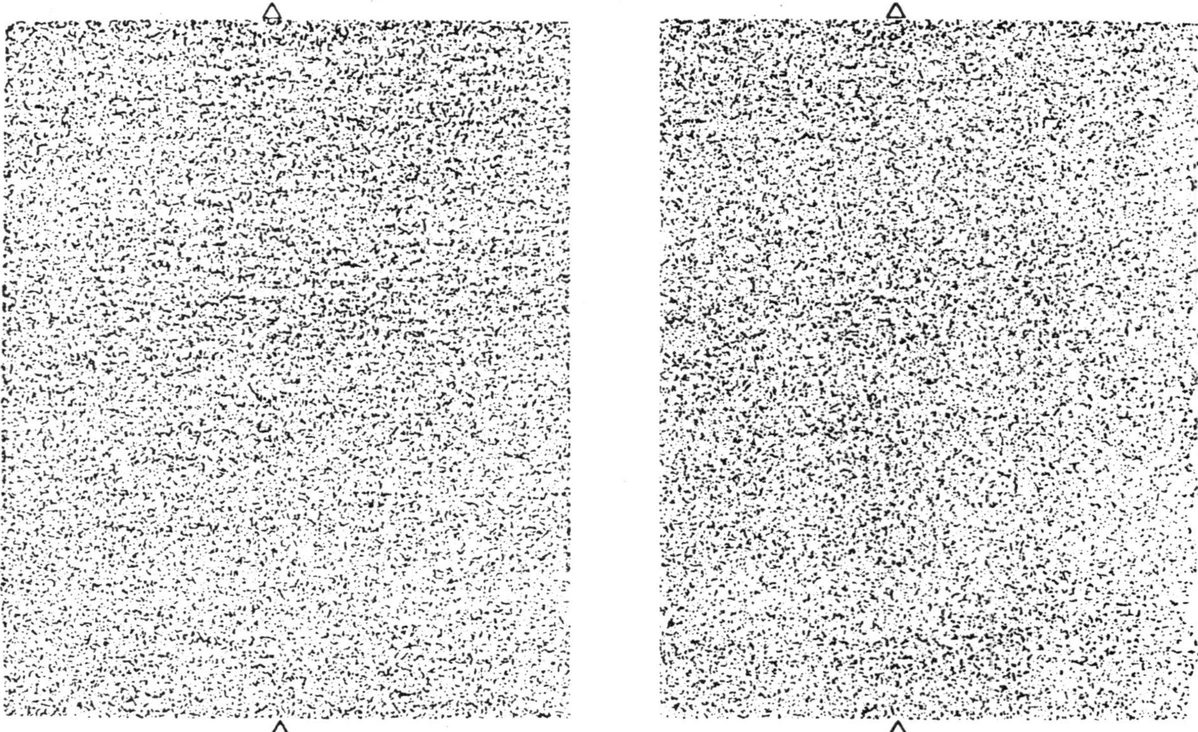

FIGURE 22-6. Demonstration of regional segregation on the basis of binocular disparity cues. This random-dot stereogram may be viewed by free fusion if the eyes are crossed such that the pairs of triangles are superimposed. Both cyclopean spatial corrugation frequency and amplitude are varied across the figure, defining the perceptual boundary of stereoscopic segregation.

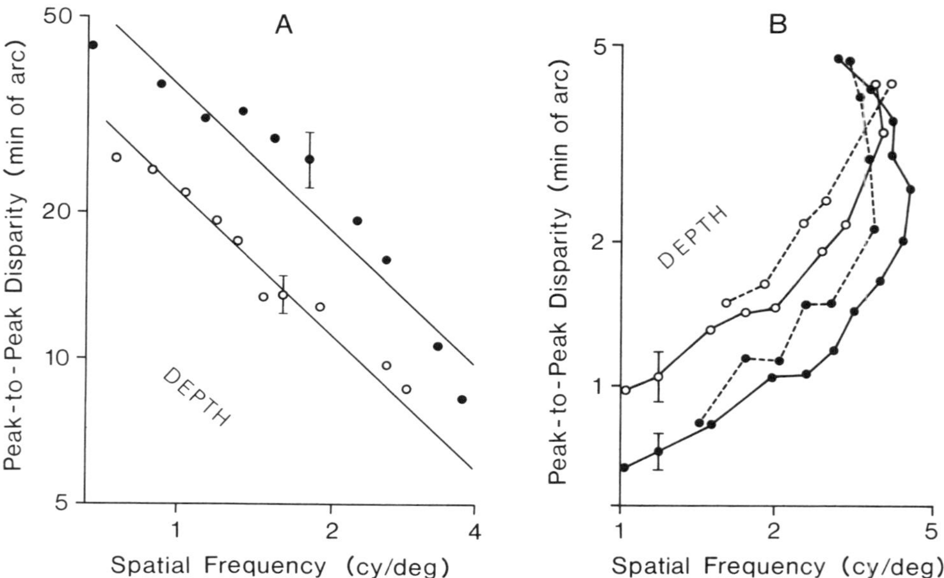

FIGURE 22-7. Measured limits of stereoscopic segregation from the stimulus of Fig. 22-6 for two observers (open and filled symbols). Upper curves show the upper limits of spatial segregation by cyclopean disparity cues, which decrease in inverse proportion to the corrugation frequency. Lower curves show the lower limits of disparity segregation, below which the stimulus appears uniformly flat. Segregation is only perceived between these two limits.

eograms are used for the depiction of large organic molecules. In many cases of fine detail, stereopsis fails because the spatial changes in disparity are too dense for the cortex to process.

A possible reason for this failure lies in the tremendous savings in cortical computation achieved by limiting disparity processing to large-scale spatial regions of disparity difference. The savings has been estimated to be a factor of at least 1000 from a factor of 10 loss in spatial disparity segregation, since the fine detail requires progressively more computations than does the coarse information (Tyler, 1983).

GLOBAL STEREOSCOPIC SEGREGATION LIMITS

What about global segregation into multiple depth planes, rather than planar maps? The focus is on the

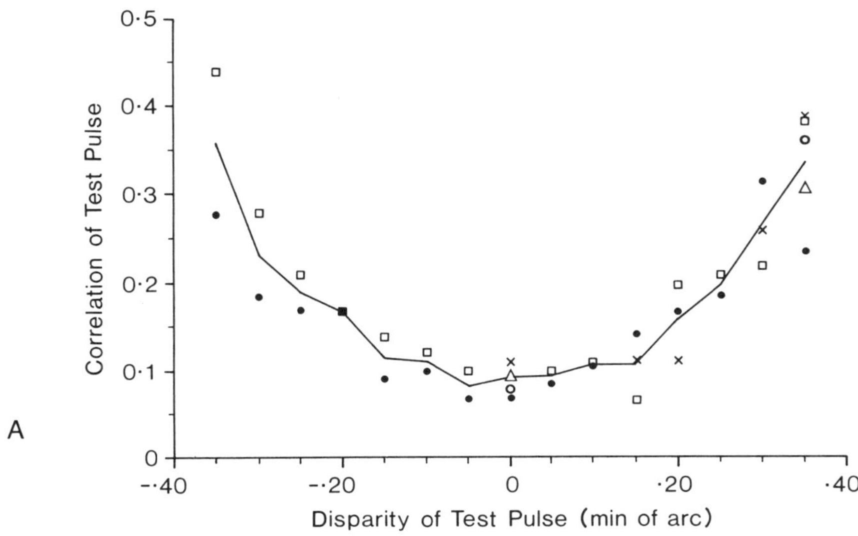

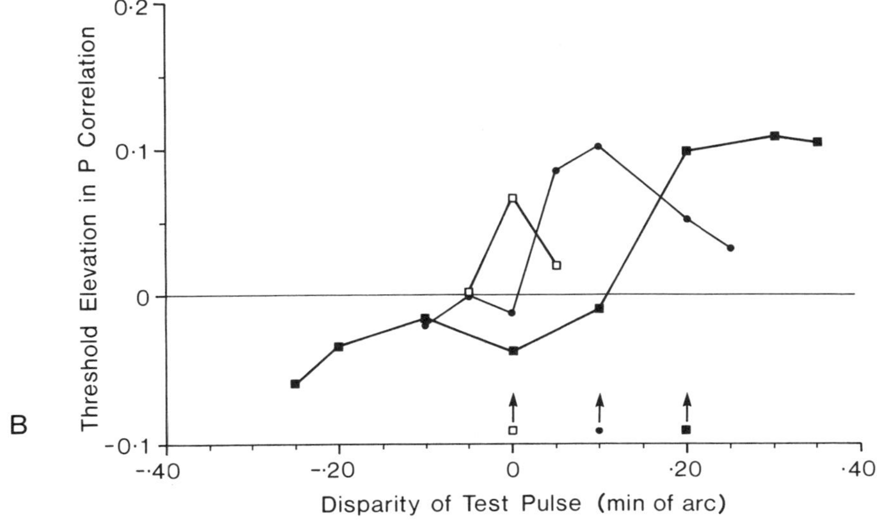

FIGURE 22-8. (A) Detectability of a plane of partially correlated dots in a three-dimensional field of dynamic noise that is binocularly uncorrelated between the two eyes. Thresholds were measured over five sessions for one observer and are expressed in terms of the degree of correlation of the detection plane. (B) Threshold elevation from the base detection level of A by adapting to a plane of 100% correlation at the disparity indicated by the arrows. Note the narrow tuning of the disparity channels revealed by this adaptation near the horopter and the broader tuning for adaptation planes away from the horopter.

stereo case here, since that is what I have mostly worked on. The disparity limits for seeing two transparent planes in a field of random dots vary from about 2′ to 10′, depending on the observer. What is the mechanism underlying this ability for global segregation? First we need to know how one sees a single plane at different depths. Working with Scott Stevenson and Cliff Schor, we have measured the detectability of a single correlated depth plane in a blizzard of dynamic random dots (Stevenson *et al.*, 1990). The correlation is specified such that at 100% correlation all the dots are presented at the disparity of the plane, whereas at 0% correlation the dots are placed at random so that there is no information about the existence of the plane.

Figure 22-8A shows detectability over a range from 40′ crossed disparity (in front of the fixation point) to 40′ behind. Detectability decreases symmetrically on either side of the horopter, with a flat bottom of best detectability of about ±15′ in the center. Detection is measured in terms of the number of correlated dots in the specified depth plane. This stimulus, which was presented for 200 msec only, could be detected for correlation of only 10% of the dots, the others all forming the dynamic three-dimensional blizzard of uncorrelated dots.

How, then, is it possible to see more than one depth plane? We have evidence that there are multiple neural channels tuned to different disparities for this stimulus. We adapted to a fully correlated plane at one disparity for 1 min. We then tested the detectability at different disparities around the adapted disparity. Before each test stimulus, we readapted for 500 msec to the adapting stimulus, with a 500-msec gap between them. The results are shown in Fig. 22-8B in terms of threshold elevation above the base-line detectability. For each adapting disparity, thresholds are elevated in a range around it, but show signs of facilitation, or improvement, for disparities far away. The tuning of the channels is narrow (less than 2′ of disparity) at the fixation distance but substantially broader at greater disparities. These channels presumably correspond to disparity-selective mechanisms, which can process separate depth planes in a global segregation task.

GLOBAL MOTION SEGREGATION LIMITS

Segregation on the basis of motion can similarly exhibit a global variety with very similar properties to those of the stereoscopic limits. This similarity, together with those of the local and regional segregations, suggests that the limits are determined by computational constraints rather than physiological limitations, since motion and stereoscopic cues require very different neural substrates for their processing.

However, the two cues can interact, as shown by situations where motion can form the basis for global segregation by stereoscopic information. This occurs in the random-dot Pulfrich stereophenomenon, where the dot motion in either eye alone is random and unsegregated. With a delay between the two eyes, however, the motions are segregated into planes of different velocities, each at a different stereoscopic depth (Tyler, 1974, 1977). The stereoscopic motion percept is depicted in Fig. 22-9, showing motion in one direction in front of the fixation point and in the other direction behind it. The velocity in each case increases progressively with the degree of depth from the fixation point. This example of stereo/motion segregation suggests that the cortex contains processes specialized for the analysis of stereoscopic motion as a combined cue for the disambiguation of superposed form information.

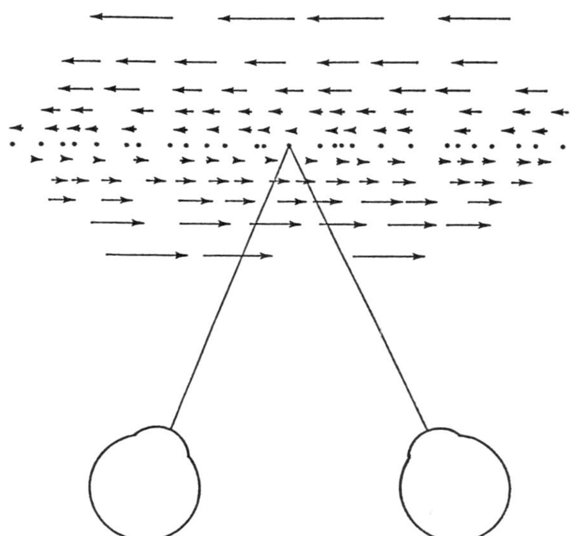

FIGURE 22-9. Depiction of the stereoscopic motion percept of multiple planes moving at different velocities in both directions, which is obtained simply from viewing dynamic visual noise with an interocular delay. Shown in bird's eye view, as if looking down on the observer from above.

A NEW DEMONSTRATION OF SIMULTANEOUS MONOCULAR AND STEREOSCOPIC VISION

As a final example of disambiguation, Fig. 22-10 provides a new demonstration in which one can see simultaneously two stereoscopic depth planes and two separate monocular "sunburst" patterns, one for each eye. To make the demonstration, Fig. 22-10 should be xeroxed onto a piece of transparent acetate. The transparency should then be held over the printed image at a distance of about 1 cm. When viewed binocularly from a distance of about 50 cm, both planes are then visible stereoscopically. An impression of the two simultaneous monocular patterns is given in Fig. 22-11. These four simultaneous percepts overcome both the expected rivalry between monocular patterns that are different in the two eyes and also the rivalry or inhibition between dichoptic and stereoscopic percepts. How can this stimulus generate this degree of unimpeded simultaneity of different visual impressions?

The key to facilitation of perceptual simultaneity is to generate the stimulus from a pattern of separated dots so that the visual system has the opportunity to construct the percept of a transparent plane with another plane behind it. The more distant plane can be perceived as being seen through the front plane by means of the spaces between the dots. Thus, the basic stimulus consists of two planes of

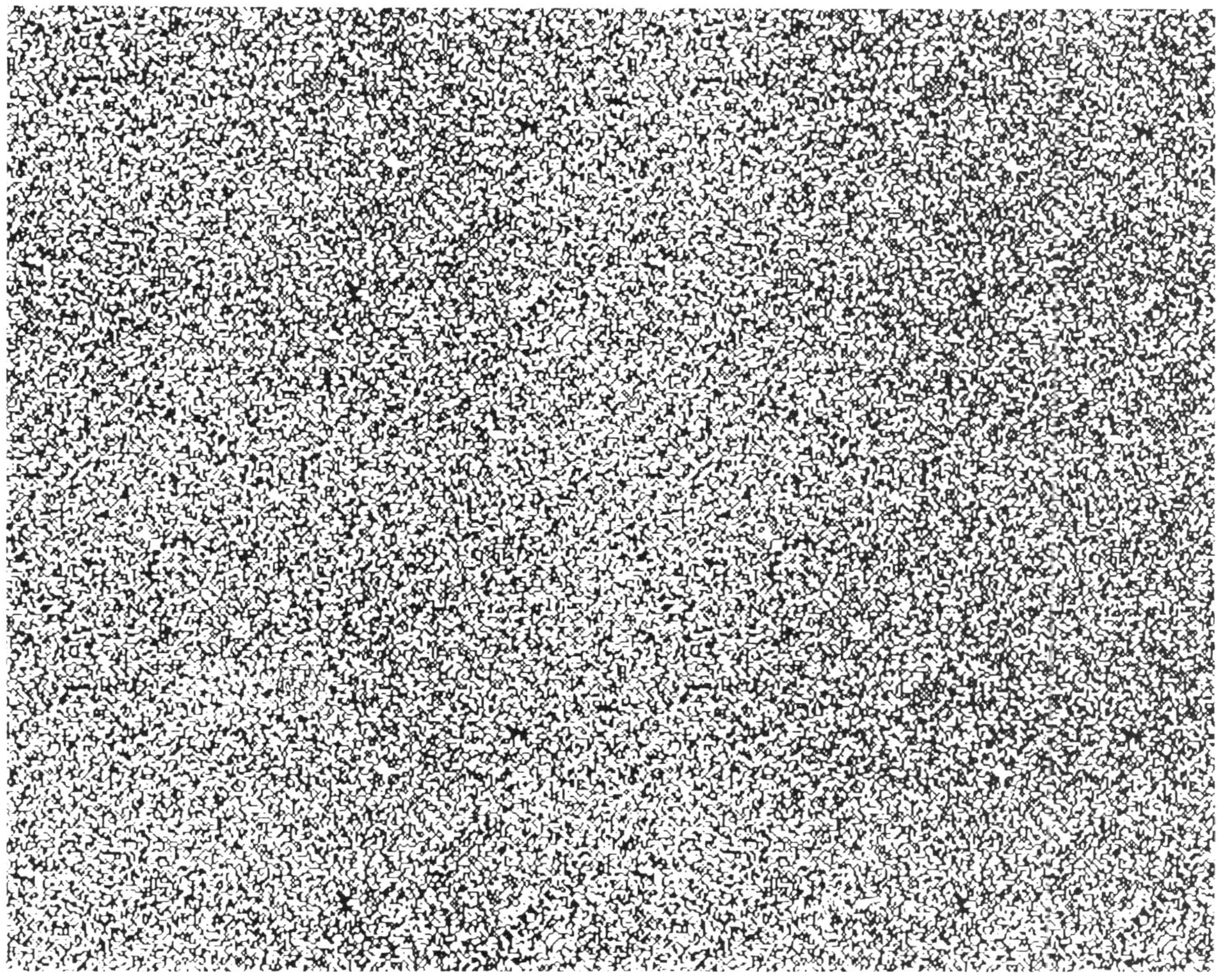

FIGURE 22-10. Demonstration in which four simultaneous percepts can be obtained from a random-dot figure. This demonstration requires the presence of a real disparity, which must be obtained by xeroxing the figure onto transparent acetate, which is then held above the original at a height of 1 cm and viewed binocularly. In addition to the two depth planes, two monocular "sunburst" patterns are visible in this display, particularly when the two planes are moved relative to each other.

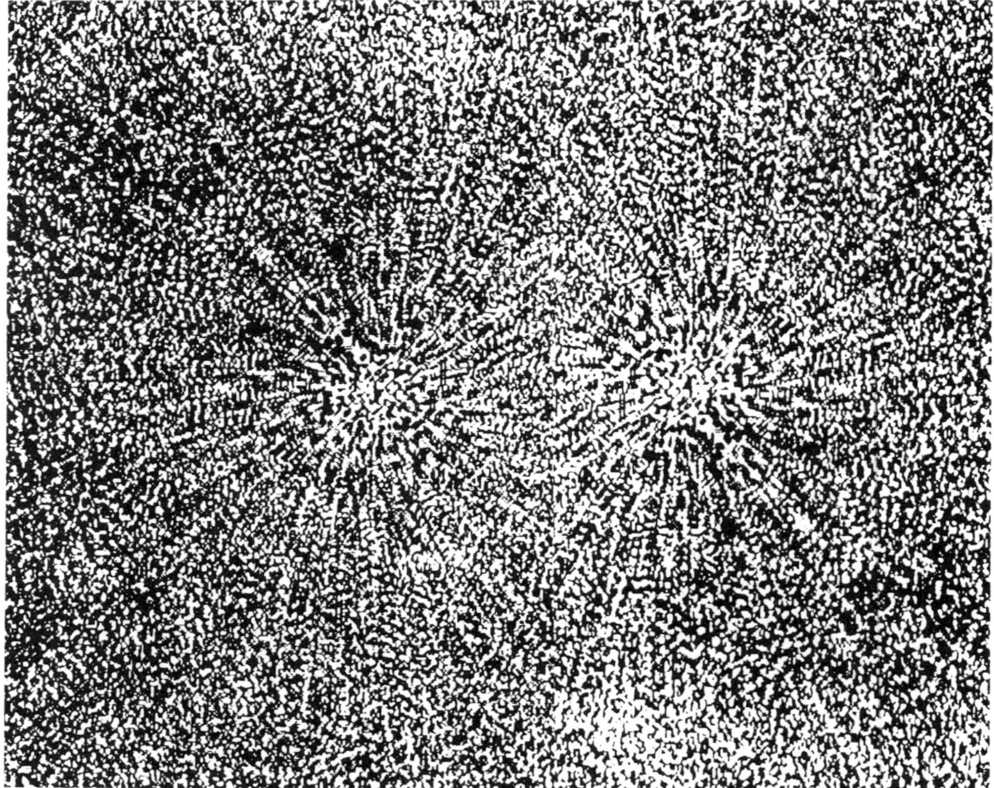

FIGURE 22-11. Depiction of the monocular "sunbursts" that can be seen in the demonstration of Fig. 22-10. These are obtained from the summation of one random-dot image with another, which is increased in size by a small percentage.

dots, one behind the other, as in the original multiple-depth planes of Julesz (1971). The new feature of the present demonstration is that the two planes contain identical sets of dots. They therefore constitute a depth-separated version of the self-additive random-dot patterns of Glass (1964).

However, separating the two identical planes in depth introduces a global expansion of the retinal projection of the front plane relative to the rear plane. When viewed monocularly, this creates the impression of a structured texture in the form of a "sunburst," with streaks emanating from a central point. This sunburst pattern was reported by Glass (1964) for the transformation of a global expansion before addition of the two replicated sets of random dots.

Note that the central region of the expansion pattern has an area that is lighter than the surrounding parts. This is because the expansion is close to zero here, and so the rear dots are obscured by the front dots, resulting in a texture that has the same 50% density as each single plane alone. However, when the expansion causes a significant shift of one plane relative to the other, the dot relationship becomes random. This means that the combined dot density becomes 75% because the black dots in front can obscure white spaces behind, but the transparent spaces in front do not allow black dots behind to be seen. The light area is thus half the brightness of the surround because the latter consists of only 25% white dots.

The nature of the global expansion is such that as the relative positions of the front and rear planes shift, so does the sunburst pattern with its light central region. The effect of monocular viewing is to see the center at the position of alignment with the viewing eye, which is the line from the eye perpendicular to the plane of the images. The same is true for the other eye, which is of course at a different point in space specified by the interpupillary distance (in primary position). This means that in binocular view, each eye is presented with the information of a monocular sunburst pattern and that the centers of the two sunbursts are separated by the interpupillary distance.

The remarkable thing about the demonstration is that, under the right conditions, the predicted summation pattern of two separated sunbursts is what is perceived. There is little or no dichoptic rivalry between the two patterns. This is presumably because they are made up of random dots with very similar elements—one set at least can project identically in the two eyes if vergence is at the distance of one of the two planes, although the other one will then project with some binocular disparity. Nevertheless, the perceived sunburst textures are combined in the two eyes. This stimulus therefore seems to be an example of a cyclopean texture stimulus in the sense that the dichoptic texture is projected up to the cortex in a way that avoids activating the rivalry mechanism and allows the two monocular views be combined into a summed binocular percept with no impedance.

SUMMARY

In many stimuli, two-dimensional pictorial cues do not carry sufficient information to segregate the objects in the scene. The brain must use other cues such as motion and binocular disparity to perform the disambiguation task. Three types of segregation can be identified—local, regional, and global. In local segregation, the motion or disparity cue is simply associated with some individual pictorial elements and no others. For the regional type, the segregated elements form coherent regions with similar properties that have identifiable boundary edges between them. Global segregation consists of the separation of transparent objects overlying one another so that the segregated elements are spatially intermixed and can only be associated by segregation into separate depth planes.

In terms of the binocular disparity cue, the lower bound for local segregation is the binocular fusion limit for disparity, and the upper bound is the edge of the retina. For regional segregation, the lower bound is the random-dot stereoacuity, and the upper bound is the upper disparity limit, both of which vary with the size of the segregated regions. For global segregation, the lower bound is the diastereoscopic boundary for separation of transparent depth planes, and the upper disparity limit for global segregation remains to be determined.

ACKNOWLEDGMENTS. Supported by NIH grants EY 7890, EY 6883 and RR 5981 and Smith–Kettlewell Eye Research Institute grant #3-77.

REFERENCES

Glass, L., 1964, Moire effect from random dots, *Nature* **223**:578–580.

Julesz, B., 1971, *Foundations of Cyclopean Perception*, University of Chicago Press, Chicago.

Ogle, K. N., 1950, *Researches in Binocular Vision*, W. B. Saunders, Philadelphia.

Panum, P. L., 1858, *Physiologische Untersuchungen über das Sehen mit zwei Augen*, Schwerssche Buchhandling, Kiel.

Schor, C. M., Heckmann, T., and Tyler, C. W., 1989, Binocular fusion limits are independent of contrast, luminance gradient and component phases, *Vision Res.* **29**:821–836.

Stevenson, S. B., Cormack, L. K., Schor, C. M., and Tyler, C. W., 1990, Disparity-tuned channels in human stereopsis, *Invest. Ophthal. Vis. Sci. Suppl.* **31**:95.

Tyler, C. W., 1974, Stereopsis in dynamic visual noise, *Nature* **250**:781–782.

Tyler, C. W., 1977, Stereomovement from interocular delay in dynamic visual noise: A random spatial disparity hypothesis, *Am. J. Opt.* **54**:374–386.

Tyler, C. W., 1983, Sensory processing of binocular disparity, in: *Vergence Eye Movements: Basic and Clinical Aspects* (C. M. Schor and K. J. Ciuffreda, eds.), 199–295, Butterworths, London.

Part IIIB

ACCOMMODATION AND ITS ROLE IN SEPARATING SUPERIMPOSED IMAGES

Part IIIB

Introduction

LAWRENCE STARK

INTRODUCTION

Accommodation, the topic of this Part, is of interest for two reasons: first, for its role in accomplishing the separation of superimposed images; and second, as a vital ocular–motor mechanism in the service of clear vision and whose loss defines presbyopia.

ACCOMMODATION AS A "PRIMITIVE" IN SEPARATING SUPERIMPOSED IMAGES

In Part IIIA, a number of other primitive, preattentive, hard-wired visual processes were described that aided in separation of superimposed images. These were contrasted with cognitive, higher-level visual processes that could also aid in disambiguating superimposed or ambiguous images. Accommodation can certainly be considered as another preattentive mechanism. When we see two superimposed images that result from objects at different distances, we can accommodate onto one of these images and so clear the blur to enable high-resolution vision. The other image, now highly blurred, has only a residual action as a veiling contrast superimposed on the focused and sharply contrasted image.

Stigmatoscopy is an example of such a two-image situation. One highly detailed and attended-to image serves as the stimulus to accommodation. The other superimposed image is a stigma, a small spot of light, that may be initially blurred. The subject can manually adjust the optical distance of the stigma and so minimize its blur. Two of the chapters in this section use the stigmatoscopy method.

In Chapter 26, Vandepol and Stark employed stigmatoscopy to estimate the sole contribution of accommodation to the clinical measure of the amplitude of accommodation that inherently includes the variable depth-of-focus contributed by the pupillary aperture. Their use of stigmatoscopy to define the progress of presbyopia documents its early course during youth and young adulthood, whereas the clinical impression, influenced by adding the depth of focus to true accommodative amplitude, places the onset of presbyopia in late middle life.

Chapter 23 by Semmlow *et al.* used stigmatoscopy to confirm earlier results by Fincham (1955) and Eskridge (1962) on the effect of neuromuscular pharmacological agents on the amplitude of accommodation. By means of a model for nonlinear lenticular elasticity, the experiments were reinterpreted as support for the lenticular theory of accommodation.

Perhaps it is appropriate to mention a few control studies that we have done to define limitations and strengths of stigmatoscopy. When compared to the standard haploscope, measurements with stigmatoscopy agree if the rapid pace permitted the subject in the computerized stigmatoscope is slowed to agree with the rate on the noncomputerized haploscope (Matsunaga *et al.* 1988). This is because a dynamic hysteresis permits the two limbs of the stigmatoscopy curve to diverge somewhat. Further, no difference could be observed with different colored targets; perhaps the method is too insensitive to document the effects of chromatic aberration, or perhaps human sensory mechanisms have preadapted.

ACCOMMODATION AS A VITAL OCULAR–MOTOR MECHANISM

One of the pioneers in studying the physiology of accommodation is Professor Fergus Campbell of the University of Cambridge, Great Britain. In the 1950s, Campbell, with two colleagues, John Robson and Gerald Westheimer, now professors at Cambridge and Berkeley, respectively, defined the 2-Hz

LAWRENCE STARK • School of Optometry, University of California, Berkeley, California 94720.

oscillation in accommodation that I now propose should be named the "Campbell oscillation."

Campbell and colleagues showed that this oscillation reduced drastically under "open-loop" conditions, thus implying that some feedback property of closed-loop accommodation was responsible for its generation. In an early control study on accommodation (Stark *et al.*, 1965), Campbell's experiments on the "Campbell oscillation" were compared with new experiments documenting a saturating nonlinearity. These latter experiments had also been carried out in Campbell's laboratory by Stark. A describing function analysis led to a control model that indicated that the accommodation system was stable for large-amplitude signals but became unstable for low-amplitude signals; this predicted the Campbell oscillation experimental findings. Recently, Miege and Denieul (1988) have confirmed Campbell's experiments on Campbell oscillation. Additional experiments have suggested that the Campbell oscillation has no visual functional role in accommodation. In particular, it does not remove the even-error behavioral characteristics of accommodation sensing and control. Its presence is merely a consequence of the feedback control properties of accommodation.

Professor Kenneth Ciuffreda of the SUNY State College of Optometry presents an interesting chapter on discrimination of blur. Of special import is the inclusion of clinically impaired amblyopes in his study, an important clinical application.

Professor Clifton Schor, of my own School of Optometry at the University of California at Berkeley, presents a detailed and careful study on subjects with monocular contact lens corrections or "monovision." He suggests that an interocular suppression of the anisometropic blur was the psychophysical and physiological mechanism underlying its clinical success in adapted monovision wearers, who were also able to increase their binocular depth of focus.

A further contribution by Professor John Semmlow of the Department of Biomedical Engineering, Rutgers University, describes the use of the new magnetic resonance imaging (MRI) technique to visualize the accommodative mechanism in a living human subject (himself). These studies may also lead to exciting clinical applications and definitions of presbyopic development.

I am happy to also include mention here of the Essilor Prize paper by Michael W. Neider, Kathryn Crawford, Paul L. Kaufman, and Laszlo Z. Bito. This paper was largely a beautiful and exciting film of experiments done on the active, behaving, accommodative mechanism in monkeys. Careful attention was paid to the camera optics in these iridectomized animal experiments, and the pictures were of the highest quality and beautifully documented the dynamic effects of ciliary muscle contraction. In young monkeys, the extralenticular accommodative mechanism allowed the lens to round up. In older monkeys, the radial sheath of the ciliary ligments or zonule of Zinn crumpled up as ciliary muscle contraction released tension on the axial zonule. This demonstrates that the extralenticular mechanism is still functioning. The lenticular apparatus was unable to take up this slack by rounding up and so accommodating. This was a beautiful demonstration of the Hess–Gullstrand lenticular theory of presbyopia based on the Helmholtzian "dual, indirect, active mechanism" of accommodation.

Two chapters using stigmatoscopy were mentioned above, since they represent use of the ability of the accommodative system to separate superimposed images, employing one to drive accommodation and the other to enable perceptual clearing of blur as an endpoint for the estimation of clear vision distance. They also contribute to our understanding of the central role of loss of accommodation in presbyopia. The early age at which accommodation reduces amplitude, preclinical presbyopia, makes it difficult to accept muscle weakness as a primary mechanism. The nonlinear model of lens viscoelasticity enables interpretation of the pharmacological experiments also in accordance with the Hess–Gullstrand lenticular theory of presbyopia.

SUMMARY

In summary, accommodation is a complex neurological control system with capacity to utilize many clues to distance; it is aided by vergence accommodation. Blur is the basic stimulus to accommodation; spatial and temporal derivatives of blur are especially effective. A switching process controls the multimodal accommodation system with its long delays and nonlinear operators; indeed, low-amplitude feedback "Campbell oscillations" are ubiquitously present. The muscular and biomechanical bases of accommodation are well summarized in the "dual, indirect, active" mechanism proposed by Helmholtz. Finally, we must not forget the continual growth of the lens throughout life; this late developmental feature involves the mechanics of the lens and leads to presbyopia.

Varying accommodation is one means of separating superimposed images. Stigmatoscopy and monovision depend on the ability of humans to use selective attention to focus on one of several superimposed images.

REFERENCES

Eskridge, J. B., 1984, Review of ciliary muscle effort in presbyopia, *J. Ophthalmic Physiol. Opt.* **61:**133–138.

Fincham, E. F., and Watson, J., 1957, The reciprocal actions of accommodation and convergence, *J. Physiol.* **137:**488–508.

Miege, C., and Denieul, P., 1988, Mean response and oscillations of accommodation for various stimulus vergences in relation to accommodation feedback control, *Ophthalmic Physiol. Opt.* **8:**165–171.

Matsunaga, K., Nguyen, A., Sun, F., van de Pol, C., and Stark, L., 1988, Static accomodations response: Stigmatoscopic method to measure development of presbyopia, in: *Proceedings IEEE International Congress Systems, Man Cybernetics, IEEE,* Beijing, pp. 1103–1106.

23

Magnetic Resonance Imaging of the Presbyopic Eye

JOHN L. SEMMLOW, SUSAN A. MENDITTO, and REUBEN S. MEZRICH

INTRODUCTION

Magnetic resonance imaging (MRI) is a noninvasive process for imaging internal structures in the body. High-resolution imaging of the ciliary muscle and lens can be utilized to verify the predictions from the computer simulation model of presbyopia described by Semmlow *et al.* (Chapter 24). Data for this model were obtained from experiments using topically applied drugs to diminish or enhance the effectiveness of the ciliary muscle. Although the model demonstrates that the presbyopic mechanism is primarily lenticular, and that ciliary muscle change is not fundamental to this process, it does not rule out some change in ciliary muscle response as presbyopia advances. However, the relationship between ciliary muscle and lens response during presbyopia cannot be unambiguously determined by analysis of external behavior alone, such as that observed in the drug studies. High-resolution MRI could provide a direct measurement of ciliary muscle response at known levels of lens accommodation response.

The maximum resolution available from standard diagnostic MRI is approximately 0.33 mm, which occurs with an 8-cm field of view (FOV) corresponding to a 256 × 256 pixel matrix size. Fisher (1986) reports that the ciliary muscle ring contracts roughly 0.8 mm in radius during maximum accommodative response. Thus, using standard resolution, the information on ciliary muscle contractile state will be minimal. To address this problem, we have modified the software to achieve a 4-cm FOV corresponding to a theoretical resolution of 0.156 mm. With this modification we can expect to resolve approximately five distinct levels of muscle contraction. However, to achieve this resolution a hardware modification was also required to improve the signal-to-noise ratio (SNR) of the detected image. Specifically, a small surface coil and an appropriate tuning circuit were designed to be placed close to the eye. Both the small size and proximity of this receiving coil lead to reduced extraneous signals and improved SNR.

METHOD AND RESULTS

Images are acquired on a 1.5-Tesla General Electric Signa imager using a T_1-weighted single-echo multislice spin-echo technique with a 256 × 256 pixel matrix. Modification of the pulse-programming language software to increase the amplitude and the time duration of both the frequency-encoding gradient and the phase-encoding gradient resulted in a 4-cm FOV with the corresponding 0.156-mm resolution.

Reducing the FOV by a factor of 2 decreases the SNR by a factor of four since pixel size has been reduced by that amount and four times less signal is available. One way to restore the SNR to its prior level is to increase the number of averages by a factor of 16, since the SNR increases as the square root of the number of averages. This is undesirable because it would also increase the scan time by a factor of 16.

JOHN L. SEMMLOW • Department of Biomedical Engineering, Rutgers University, and Department of Surgery (Bioengineering), UMDNJ, Robert Wood Johnson Medical School, Piscataway, New Jersey 08855. SUSAN A. MENDITTO • Department of Biomedical Engineering, Rutgers University, Piscataway, New Jersey 08855. REUBEN S. MEZRICH • Department of Biomedical Engineering, Rutgers University, and Laurie Imaging Center, New Brunswick Allied Hospitals, New Brunswick, New Jersey 08901.

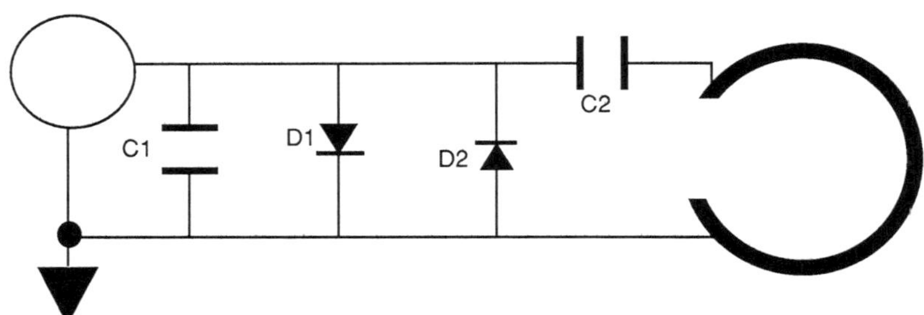

FIGURE 23-1. Schematic drawing of receiving coil and tuning and matching network. Capacitors C_1 and C_2 ensure that the coil resonates at 64 MHz and is matched to the 50-Ω cable. Diodes D_1 and D_2 detune the coil during RF transmission.

We chose instead to modify the hardware by designing a small radio-frequency (RF) surface coil that receives signals primarily from superficial structures within a relatively small volume. Thus, noise originating from deeper structures or surrounding tissues is not detected. Essentially, we are taking advantage of the fact that the structures of interest, the ciliary muscle and lens, lie close to the surface. Our coil consists of a single turn, approximately 2 cm in diameter, and can be placed quite close to the cornea. In order for the coil to receive the maximum amount of signal from the sample, it must be resonant at the same frequency as the hydrogen nuclei (which is 64 MHz for a 1.5-Tesla magnetic field). Similarly, for maximum signal transfer the impedance of the coil must match the cable impedance (50 Ω).

The surface coil and corresponding tuning and matching network are represented schematically in Fig. 23-1. The capacitors C_1 and C_2 accomplish both the matching and tuning, whereas the diodes D_1 and D_2 serve to detune the coil. This detuning provides a safety feature that places the receiving coil off resonance while the imager is transmitting high-level RF energy. Consequently, the coil carries minimal current during this transmit cycle, and both surface heating and image distortion are prevented.

Results will be obtained from subjects with varying degrees of presbyopia. During imaging, lens stimulation will be controlled using visual targets positioned at the subject's near point and at optical infinity (20 ft). The subjects will view a flashing target (to overcome the stabilized image phenomenon, which results in fading of the image) through a mirror. A schematic drawing of the stimulus apparatus is provided in Fig. 23-2.

A representative image obtained with the new surface coil is shown in Fig. 23-3 for a 46-year-old presbyope.

DISCUSSION AND CONCLUSION

The 0.156-mm resolution provides a measure of ciliary muscle contraction for only around five discrete contractile levels. Future work involves ex-

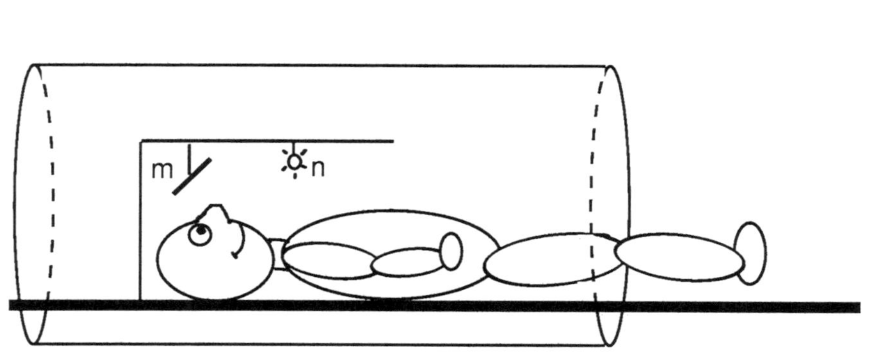

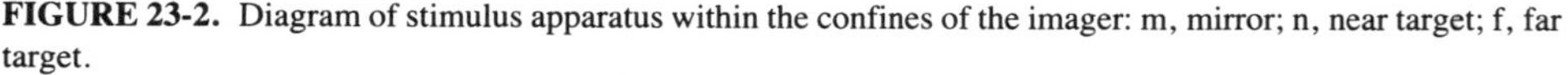

FIGURE 23-2. Diagram of stimulus apparatus within the confines of the imager: m, mirror; n, near target; f, far target.

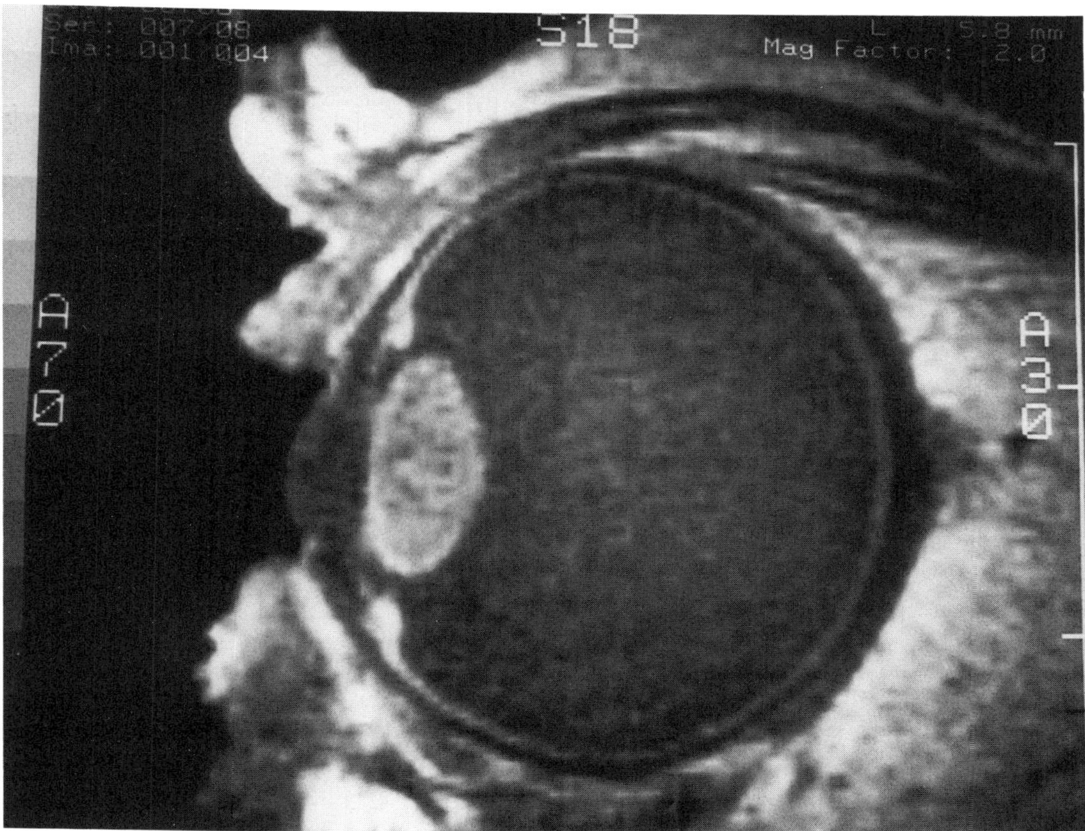

FIGURE 23-3. Representative magnetic resonance image of a presbyopic eye.

tending this discrimination by further increasing image resolution.

The design of smaller, more sensitive receiving coils could produce high enough SNR to make high-quality images with a 3-cm or even 2-cm FOV both possible and feasible. However, additional software modifications and optimization of imaging parameters will be necessary to achieve this increased resolution. In addition, a number of image-processing techniques such as noise reduction and contrast enhancement can be employed to improve image details.

Once adequate image resolution has been achieved, experiments will acquire images at several levels of accommodative response for both drugged and undrugged conditions in subjects with varying degrees of presbyopia.

SUMMARY

The relationship between ciliary muscle contraction and lens accommodation in presbyopia has been a longstanding yet unresolved question. By measuring the contractile state of the ciliary muscle using high-resolution magnetic resonance images of the eye, changes in the relationship between ciliary muscle contraction and lens accommodation response that occur with advancing presbyopia could be definitively determined. However, the resolution of standard diagnostic magnetic resonance imaging systems is somewhat marginal for measuring ciliary muscle contractile state. Modifications are described to both hardware and software components of a commercial magnetic resonance imaging system leading to improved resolution. A representative image of a presbyopic eye is presented, and future work is discussed.

REFERENCE

Fischer, R. F., 1986, The ciliary body in accommodation, *Trans. Ophthalmol. Soc. U.K.* **105:**208–219.

24

The Relationship between Ciliary Muscle Contraction and Accommodative Response in the Presbyopic Eye

JOHN L. SEMMLOW, LAWRENCE STARK, CORINA VANDEPOL, and AN NGUYEN

INTRODUCTION

Donders (1864) described presbyopia as a farsightedness with age that was unrelated to the hyperopia caused by corneal refractive error. Donders, like Hess (1901), was in agreement with the Helmholtz theory of accommodation, which has been described as a "dual" (there are lenticular and extralenticular components), "indirect" (the ciliary muscle acts through the suspensory ligaments), and "active" (contraction of the ciliary muscle produces increased accommodation) theory (Stark, 1987). Hess was able to demonstrate several consequences of the Helmholtz theory such as the lack of tension in the suspensory ligaments when the lens was fully accommodated. He also demonstrated the lack of vitreous support during accommodation when he observed a vitreous bulge produced by an accident and found no change during accommodation. Hess supported the lenticular theory of presbyopia, that is, that changes in the lens substance and capsule alone result in an inability of the lens to respond to extralenticular forces.

JOHN L. SEMMLOW • Department of Surgery (Bioengineering), UMDNJ Robert Wood Johnson Medical School, and Department of Biomedical Engineering, Rutgers University, Piscataway, New Jersey 08855. LAWRENCE STARK, CORINA VANDEPOL, AND AN NGUYEN • School of Optometry, University of California, Berkeley, California 94720.

Under the lenticular theory ascribed to Hess (1901) and Gullstrand (1908), the strength of the ciliary muscle (or other extralenticular processes) should not be diminished in the region where the lens is still capable of fairly normal responses (the "manifest" region). Above this level a "latent" region should exist where the muscle, though still capable of contracting, cannot produce a change in accommodation because of limitations in the mechanical properties of the lens (Fischer, 1969a,b, 1973; Alpern, 1969).

Since the responsiveness, or effectiveness, of the accommodative motor processes (both lenticular and extralenticular) are normal in the nonpresbyopic or "manifest" response region, variables that involve responsiveness, such as the AC/A stimulus ratio, will be unchanged in this region. Conversely, responsiveness is negligible within the presbyopic or "latent" region so that within this region the AC/A response ratio would become very large. This implies that the AC/A response relationship will be highly nonlinear if measured over a range that includes both manifest and latent regions. The more commonly measured AC/A stimulus ratio will also increase in the presbyopic region provided the subject continues to make an accommodative effort. Since maximum accommodation would occur well before maximum ciliary muscle contraction, the lenticular theory also implies that neither an increase nor a decrease in contractile ability would alter the maximum accommodative response.

Though it is commonly held that changes in the mechanical properties of the lens are an underlying cause of presbyopia, there is evidence that extralenticular processes play a significant role. Duane (1909, 1925) compared the loss of accommodation in presbyopia to that seen following the administration of dilute drops of atropine. As with atropine, both the overall amplitude of accommodation and the amount of accommodative response to a given stimulus reduce proportionally. Careful studies by Fincham (1955) and later by Eskridge (1972, 1984) showed that if the maximum ciliary muscle effort is augmented or diminished (for example, through the use of a parasympathomimetic or parasympatholytic drug), maximum accommodation will change accordingly. In the Eskridge experiments, both increases and decreases in maximum accommodation were produced. This finding implies that ciliary muscle weakness at least contributes to the loss of accommodation.

Under the extralenticular theory ascribed to Duane (1909) and Fincham (1937, 1955), the relationship between ciliary muscle contraction and lens shape change is diminished throughout its range of activity. This implies that both the AC/A stimulus and response ratios will be larger in presbyopes than in normals over both regions. Also implied is a more linear AC/A response relationship than expected under the lenticular theory. The same will be true of the AC/A stimulus ratio, but only if the presbyope maintains an accommodative effort in the region where accommodation is still possible.

Perhaps the strongest argument against the extralenticular theory is found in evidence that indicates that the ciliary muscle does not weaken with age. Several researchers (Donders, 1864; Duane, 1909; Marg, 1987; Sun *et al.*, 1987) have shown that loss of accommodation begins at an early age, and Sun *et al.* (1987) have pointed out that this would be unusual for a disorder caused by muscle weakness. In addition, Fisher (1977) has determined that the strength of the ciliary muscle actually increases with age.

Additional support for the lenticular theory is found in impedance cyclography experiments, which provide a qualitative measure of ciliary muscle activity. Saladin and Stark (1975) showed ciliary activity in the absence of an accommodative response, indicating a true latent region as required by the lenticular theory. Similarly, Swegmark (1969) showed that the amount of ciliary muscle activity required for a given accommodative response does not change with age, as implied by the extralenticular theory.

Though the two theories imply experimentally verifiable differences, neither enjoys unambiguous support. Results from studies that track changes in the AC/A stimulus ratio with age have been mixed: some show little change or even a gradual decrease with age (Eames, 1933; Alpern, 1950; Morgan and Peters, 1951; Tait, 1951; Davis and Jobe, 1957), but others (Breinin and Chin, 1973), including two longitudinal studies (Fry, 1959; Eskridge, 1983), have shown a marked increase in stimulus AC/A with age. Note that the lenticular theory predicts an unchanged stimulus AC/A only if the accommodative response is limited to the manifest or nonpresbyopic region, a restriction that is difficult to maintain unless accommodation is objectively measured. If the stimulus is sufficient to drive the lens into the latent or presbyopic region, both theories predict that the stimulus AC/A will increase with age. This dependence on the response could account for some of the conflicting reports.

There are two main objectives for the work presented here. First, we report the results of several experiments that reexamine and confirm the drug experiments of Fincham and Eskridge, extending the methodology over a larger stimulus range and including statistical confidence limits. The second objective is to reconcile the apparent conflicts in past findings as mentioned above by combining our experimental results with a quantitative model of the lenticular and extralenticular processes. Results based on computer simulations of this model show that the drug study data, previously taken as evidence for the extralenticular theory, are compatible with a lenticular theory when the highly nonlinear characteristics of the presbyopic lens are taken into account.

METHODS

The experimental protocol requires the presentation of a reproducible stimulus to the accommodative system and the measurement of the resultant response under both drugged and undrugged conditions. To insure an accurate, repeatable stimulus, disparity vergence stimulation was used to induce accommodative responses through convergence accommodation. This stimulus is a more effective and reliable stimulus to accommodation than blur, particularly in presbyopes. Subjects were positioned in a headrest, 24 cm from the disparity targets, which consisted of two vertical lines viewed separately in each eye. While the right eye target was fixed, the left eye target could be displaced horizontally under computer control. The range of this displacement provided disparity stimulation between 0 and 14 meter angles (MA).

Accommodation responses were monitored using a stigmatoscope. While the disparity targets were fused, the subject adjusted a small, point-like target for minimum blur. Because the small target is viewed through a Badel optometer, its position corresponds to the accommodative response of the eye. Once the correct target position has been achieved, a switch controlled by the subject signals the computer to measure target position.

Both stimulus generation and response measurement were under computer control. After each measurement the computer recorded the response and adjusted the disparity stimulus to the next level. Measurements were made from 0 to 10 MA (0 to 14 MA for subject K.C.) in increments of 1 MA. Each set of measurements was repeated eight to ten times, and the responses to each stimulus level were averaged. Because of the automated stimulus presentation and response measurement, an entire experiment could be completed in about 12 min.

Each experiment was performed on the normal eye and under a variety of drugged conditions. Enhanced ciliary muscle output was achieved by monocular instillation of eserine (1 drop, 0.5%), and experiments were performed at 30 min, 60 min, and 130 min following the application. Monocular instillation of mydriacyl (2 drops, 1%) was used to reduce ciliary muscle effectiveness. Experiments were performed at 80 and 120 min following drug application. Drug applications were under the direction of a licensed physician.

All experiments were performed on four volunteer subjects (C.V., K.C., A.H., and J.S.); however, only two subjects (C.V. and K.C.) had sufficient accommodative range to provide useful data for the model. These subjects were early presbyopes but otherwise had good binocular vision with substantial fusional range. Model simulations were done on an HP 9000 (series 550) with full graphics capability.

RESULTS

Experiments using topically applied drugs to enhance or diminish the effectiveness of the ciliary muscle were performed in several subjects. The accommodative responses to disparity stimulation under various drugged states, along with the undrugged, baseline condition, are shown in Fig. 24-1. This figure shows the combined results from five separate drug experiments and a baseline experiment, and each data point shows the average and standard error of eight to ten trials. The lower curves were obtained after topical application of mydriacyl, and the upper curves were obtained after topical application of eserine. Superimposed on these experimental results are the results of computer simulations of the model described below. These data clearly confirm the findings of Fincham and Eskridge, extending the range of the input stimulus and adding confidence limits in the form of standard errors.

MODEL ANALYSIS

The data of Fig. 24-1 have been analyzed with the aid of a quantitative model of the accommodative motor apparatus. The overall structure of this model representation is shown in Fig. 24-2. The first element represents the ciliary muscle as well as the neural processes that drive it. This element is modeled as a third-order polynomial of the input stimulus, disparity vergence. This polynomial relationship, which is somewhat subject dependent, is suggested by the responses seen at the lowest accommodation levels (square data points in Fig. 24-1), where, presumably, the lens responds more or less linearly to ciliary contractions. (This assumption is supported by the fact that all the response curves are fairly linear where they traverse this lower response region.)

The application of topical drugs is assumed to proportionally enhance or diminish ciliary muscle output but not to change the basic form of the muscle's response function. Hence, the drug effects are modeled by a single gain parameter, the "pharmacological gain," which multiplies the output of the first element. The resultant signal, l_m, represents the effective contractile response of the ciliary muscle.

The third element represents the lens. It receives input from the ciliary muscle in the form of a

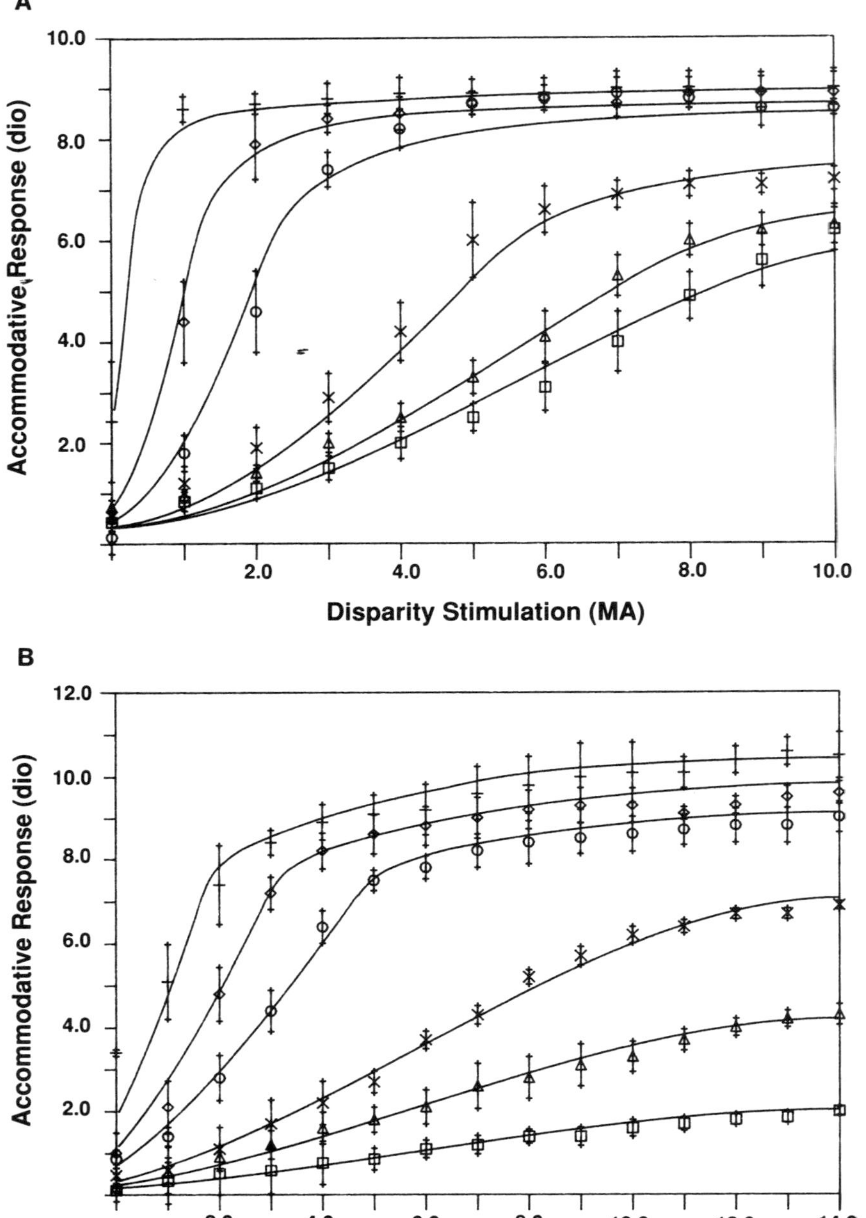

FIGURE 24-1. Accommodative responses to disparity vergence stimulation for a variety of drugged conditions along with the undrugged, baseline response: A, subject C.V; B, subject K.C. Conditions (from lower to upper data points): (A) mydriacyl (1%), 80 min, 120 min; baseline; eserine (0.5%), 130 min, 60 min, 30 min; (B) mydriacyl (1%), 10 min, 60 min, 105 min; baseline; eserine (0.5%), 75 min, 40 min. The solid lines are the results of model simulation with various gain settings for the effectiveness of the ciliary muscle. Pharmocological gain (from lower to upper curves) normalized to the baseline responses: (A) 0.41, 0.62, 1.0, 4.2, 11.0, 55.0; (B) 0.27, 0.54, 1.0, 2.7, 4.4, 7.8.

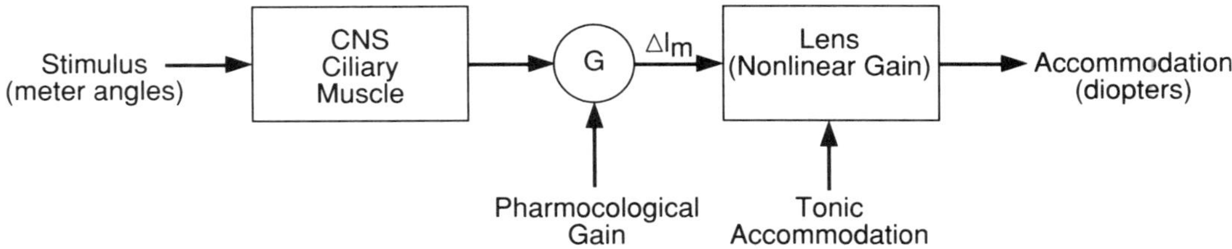

FIGURE 24-2. Organization of the three-element accommodation model.

contractile response and produces an output in diopters of accommodation. This element is nonlinear and is represented by a mathematical function that quantifies the responsiveness of the lens (defined as the ratio of the change in dioptric power to the change in ciliary muscle length) as a function of the accommodative output. Typical examples of this function are shown in Fig. 24-3A,B. These functions have been empirically determined to provide a good fit to the data from the two subjects. Note that these response functions represent highly nonlinear processes that have a constant input/output ratio (of 1.0) up to about 6.0 or 7.0 D (depending on the subject), above which they decrease in a smooth but precipitous fashion.

Simulation results agree quite well with the data. Fig. 24-1 shows that all six experimental responses are accurately predicted (within the standard error of the measurement) by the simple three-element model. We emphasize that the entire set of response curves is constructed simply by changing a single parameter, the pharmacological gain.

The model has also been applied to the data of Eskridge (1972, 1984). Figure 24-4 shows that the model provides a good match to these data as well. (Unfortunately, Eskridge does not provide information on the standard error of his measurements.) The lens response function derived from the Eskridge data is shown in Fig. 24-5 and follows the same pattern as the other two response functions.

One experiment frequently cited in support of the lenticular theory is the impedance cyclography experiment of Saladin and Stark (1975). Using the change in electrical impedance of the eye as a measure of ciliary muscle contraction, they showed that ciliary muscle contraction occurs in the "latent" region, where no further accommodative response is found. Thus, a plot of impedance change (representing ciliary muscle contraction) versus accommodative response shows a sharp vertical transition at the point where accommodative response is maximum (Fig. 24-6, x points). Using the data from one of our subjects (C.V.) as an example, the model predicts very similar behavior. As shown in Fig. 24-6 (solid line), a plot of ciliary muscle response, as predicted by the model, against model accommodative response shows the same vertical transition. (Note that the two curves should not be expected to be the same, since different subjects are involved and the impedance cyclography measurement is subject to substantial experimental error). Thus, the model predicts this important experimental finding.

DISCUSSION

The primary use of the model is as an analytical tool, and no effort was made to develop homeomorphic relationships between model mechanisms and specific physiological processes. As an analytical tool, the model allows us to abstract more information from the data-rich drug experiments. That is, model structure is determined not only from the data points of a single experiment but from the manner in which these curves vary under drugged conditions.

The simplicity of the model adds to its strength. As with any nonlinear process, it is impossible to prove that the model is unique (that is, that this model is the only one capable of representing the data); however, it is the simplest configuration that can represent all the experimental findings. To strengthen the model predictions, it would be beneficial if the model could be "calibrated" by actual measurement of the lens–muscle relationship. Measuring the response of the ciliary muscle *in vivo* is difficult but may be possible using the new technique of magnetic resonance imaging (MRI). A related project, briefly described elsewhere (Chapter 23, this volume), uses MRI to obtain high-resolution images of the muscle and lens. Future experiments will acquire images at several levels of accommodative response for both drugged and undrugged conditions. An approximate measure of ciliary mus-

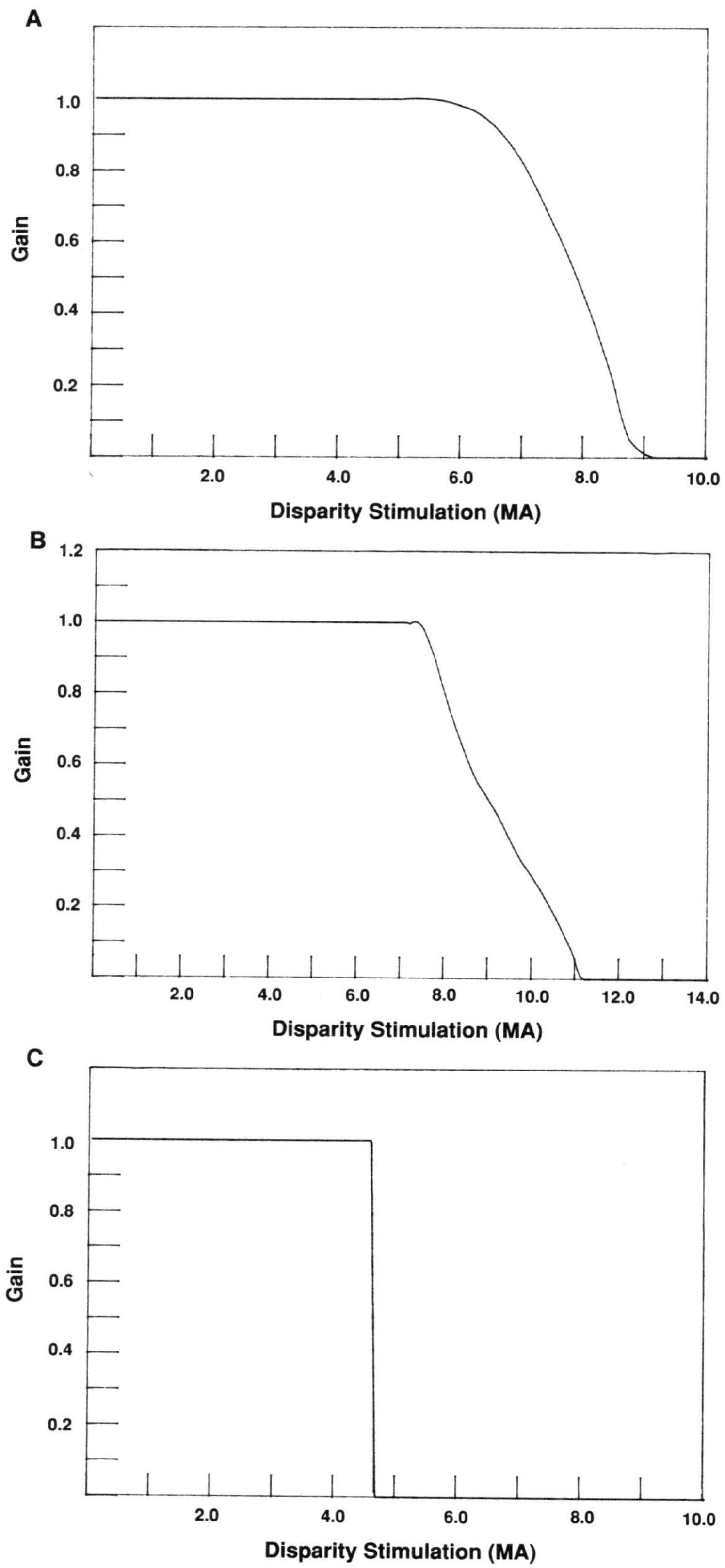

FIGURE 24-3. Gain function representing lens responsiveness in our early presbyopes: (A) subject C.V.; (B) subject K.C. (C) Theoretical response function implicitedly assumed by the lenticular theory.

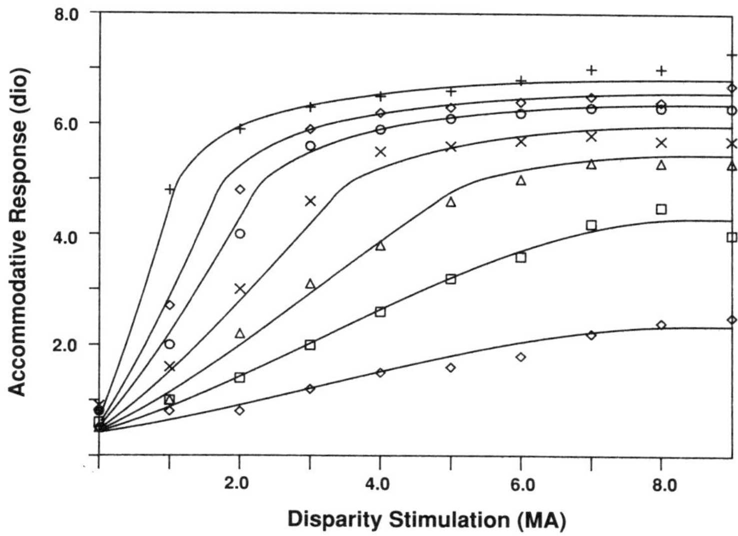

FIGURE 24-4. Data from the drug experiments of Eskridge (data taken from Figs. 1 and 2 of Eskridge, 1984). Solid lines represent the output of the model with various gain settings. Conditions (from lower to upper data points): homatropine (5%) 8 hr, 19 hr, 22 hr; baseline; eserine (0.5%), 180 min; 90 min; 30 min. Pharmacological gain (from lower to upper curves) normalized to the baseline level: 0.21, 0.43, 0.67, 1.0, 1.6, 2.2, 3.7.

cle response at several accommodative response levels can then be obtained directly from the muscle image.

In our model, presbyopia is represented as a smooth, yet precipitous decrease in lens responsiveness at the higher levels of accommodation (Figs. 24-3 and 24-5). Except for the two upper curves in Fig. 24-1, representing the strongest influence of eserine, the maximum accommodative response level corresponds with maximum ciliary muscle output, and no true latent region exists, in conflict with one implication of the lenticular theory. For these subjects, this applies to the normal, undrugged response as well. Yet, the extralenticular theory makes no provision for the highly nonlinear response characteristics of the presbyopic lens (Figs. 24-3 and 24-5). Although not providing the "hard" saturation commonly assumed in the lenticular theory (Fig. 24-3C), these nonlinear lens characteristics do provide very rapid and deep attenuation of lens responsiveness at the higher accommodative levels. This highly nonlinear feature

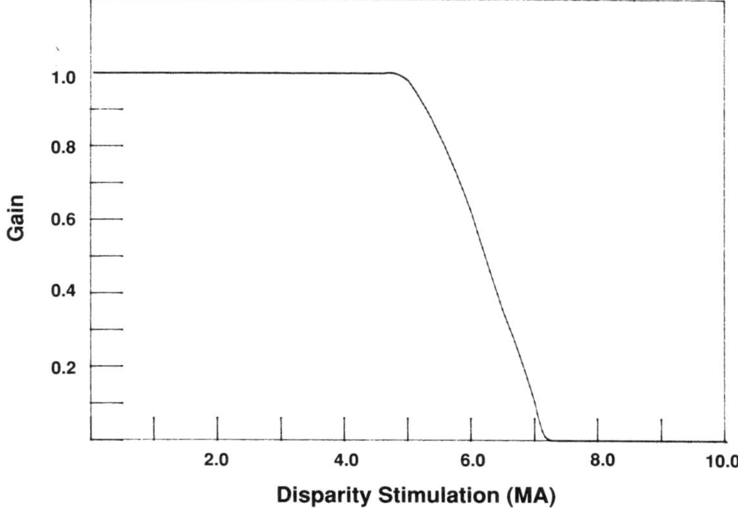

FIGURE 24-5. Nonlinear gain function representing lens responsiveness of Eskridge's subject.

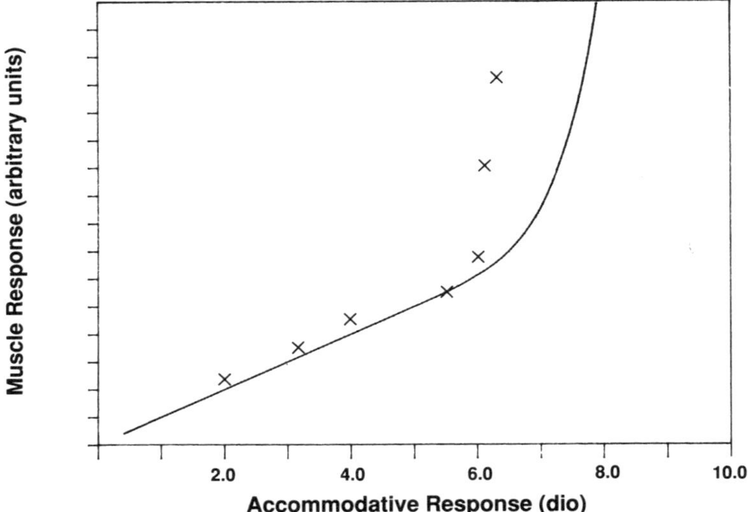

FIGURE 24-6. Model simulation of the Saladin and Stark (1975) impedance cyclography experiment. The points are experimental results showing impedance change (representing ciliary muscle contraction) as a function of accommodative response (data taken from the 22-year-old subject of their Fig. 5). The solid line is the model prediction of the results of this experiment if it were performed on subject C.V.

gives the normal accommodative response curve the appearance of a latent region but allows the large drug-induced changes in muscle efficiency to produce increased maximal accommodation. Indeed, at the highest accommodative level induced by eserine (two upper curves, Fig. 24-1), the responsiveness of the lens has diminished to zero, and a true latent region exists. It is this highly nonlinear process in the model that reconciles the apparently conflicting data described previously.

The fact that all response curves could be developed simply by applying a multiplying pharmacological gain to the basic muscle response is a particularly important finding of the modeling analysis. This indicates that the effects of presbyopia can be accurately represented by lens alteration alone—the ciliary muscle need not change. Though this finding does not rule out possible changes in the ciliary muscle as presbyopia advances, it does demonstrate that muscle changes are not fundamental to the mechanism of presbyopia, supporting the lenticular theory.

In a parallel modeling study, Sun and Stark have examined the excellent studies of Fincham, which track changes in the convergence accommodation to vergence stimulus ratio (CA/C) with advancing presbyopia. Though originally interpreted to support the extralenticular theory, the Sun and Stark biomechanical model has shown that these results are also compatable with the lenticular theory.

In summary, the experimental data in conjunction with the model analysis show that ciliary muscle changes are not fundamental to the presbyopic mechanism, in contradiction with the extralenticular theory. Though presbyopia is principally a lenticular process, a true latent region, usually assumed under the lenticular theory, does not exist, at least in the normal response of early presbyopes. When the highly nonlinear response characteristic of the presbyopic lens mechanism is taken into consideration, the experimental findings of Fincham and Eskridge are reconciled with other evidence in support of a purely lenticular cause for the presbyopic deficit.

SUMMARY

Presbyopia, an inevitable consequence of growing older, is generally believed to be caused, at least in part, by a decrease in the ability of the lens to change its shape. Two theories have evolved from a long-standing debate, theories that differ with regard to ciliary muscle involvement and that predict different relationships between ciliary muscle contraction and lens shape changes. Under the lenticular theory, developed from early contributions by Hess and Gullstrand, muscle changes are not an important consideration in presbyopia. The muscle's ability to modify the lens is essentially normal in the region where the lens is still capable of response (the "manifest" region). Above this level a "latent" region exists where the muscle, though still capable of contracting, cannot produce a change in accommodation because of lenticular limitations. In the extralenticular theory, initially developed by Duane and Fincham, ciliary muscle changes play a role. In addition to a decreased maximal response, muscle weakness reduces the response to a given stimulus over the entire accommodative range. An important implication of this theory is that maximum accommodative response is associated with maximum muscle effort, and if additional contraction were possible, a higher accommodative response could be achieved.

Experiments using topically applied drugs to enhance or diminish the effectiveness of the ciliary muscle in early presbyopes show the appropriate change in maximum response, supporting the extralenticular theory. However, an analysis of the data using a computer-simulated model of the muscle–lens relationship indicates that the effectiveness of the ciliary muscle is unchanged in the manifest region, in contradiction to the implications of the extralenticular theory. Model simulations provide a resolution of these contradictions showing that the effects of presbyopia can be accurately represented by a highly nonlinear change in lens responsiveness alone; changes in the ciliary muscle need not occur.

REFERENCES

Alpern, M., 1950, The zone of clear single vision at the upper levels of accommodation and convergence, *Am. J. Optom. Arch. Am. Acad. Optom.* **27**:491–513.

Alpern, M., 1969, The nature of presbyopia, in: *The Eye*, 2nd ed., Vol. 3 (H. Davson, ed.), pp. 236–241, Academic Press, New York.

Breinin, G. M., and Chin, N. B., 1973, Accommodation, convergence, and aging, *Doc. Ophthalmol.* **34**:109–121.

Davis, C. J., and Jobe, F. W., 1957, Further studies on the A.C.A. ratio as measured on the orthorateur, *Am. J. Optom. Arch. Am. Acad. Optom.* **34**:16–25.

Donders, F. C., 1864, *On the Anomalies of Accommodation and Refraction of the Eye*, New Sydenham Society, London; reprinted 1972 by Milford House, London.

Duane, A., 1909, The accommodation and Donder's curve and the need of revising our ideas regarding them, *J.A.M.A.* **52**:1992–1996.

Duane, A., 1925, Are the current theories of accommodation correct? *Am. J. Ophthalmol.* **8**:196–202.

Eames, T. H., 1933, Physiologic exophoria in relation to age, *Arch. Ophthalmol.* **9**:104–105.

Eskridge, J. B., 1972, Ciliary muscle effort in accommodation, *Am. J. Optom. Arch. Am. Acad. Optom.* **49**:632–635.

Eskridge, J. B., 1983, The A.C.A. ratio and age—a longitudinal study, *Am. J. Optom. Physiol. Opt.* **60**:911–913.

Eskridge, J. B., 1984, Review of ciliary muscle effort in presbyopia, *Am. J. Optom. Physiol. Opt.* **61**:133–138.

Fincham, E. F., 1937, *The Mechanism of Accommodation. British Ophthalmology Monograph Supplement VIII*, George Pulman and Sons, London.

Fincham, E. F., 1955, The proportion of the ciliary muscular force required for accommodation, *J. Physiol. (Lond.)* **128**:99–112.

Fisher, R. F., 1969a, Elastic constants of the human lens capsule, *J. Physiol. (Lond.)* **201**:1–19.

Fisher, R. F., 1969b, The significance of the shape of the lens and capsular energy changes in accommodation, *J. Physiol. (Lond.)* **201**:21–47.

Fisher, R. F., 1973, Presbyopia and the changes in age in the human cristallene lens, *J. Physiol. (Lond.)* **288**:335–358.

Fisher, R. F., 1977, The force of contraction of the human ciliary muscle during accommodation, *J. Physiol. (Lond.)* **270**:51–74.

Fry, G. A., 1959, The effect of age on the A.C.A. ratio, *Am. J. Optom. Arch. Am. Acad. Optom.* **36**:299–303.

Gullstrand, A., 1908, Die optische Abbildung in heterogenen Medien die Dioptrik der Kristallinse des Menschen, *K. Sven. Vetenskapsakad. Handl.* **43**:1–28.

Hess, C., 1901, Arbeiten aus dem Gebiete der Accommodationslehre. VI. Die relative Accommodation, *Albrecht von Graefes Arch. Ophthalmol.* **52**:143–174.

Marg, E., 1987, Presbyopia in man revisited, in: *Presbyopia* (L. Stark and G. Obrecht, eds.), Professional Press Books, New York, pp. 2247–2249.

Morgan, M. W., and Peters, H. B., 1951, Accommodative convergence in presbyopia, *Am. J. Optom. Arch. Am. Acad. Optom.* **28**:3–10.

Saladin, J. J., and Stark, L., 1975, Presbyopia: New evidence from impedance cyclography supporting the Hess–Gullstrand theory, *Vision Res.* **15**:537–541.

Stark, L., 1987, Presbyopia in light of accommodation, in: *Presbyopia* (L. Stark and G. Obrecht, eds.), Professional Press Books, New York, pp. 265–274.

Sun, F., Stark, L., Vasudevan, L., Wong, J., Nguyen, A., and Mueller, E., 1987, Static and dynamic changes in accommodation with age, in: *Presbyopia* (L. Stark and G. Obrecht, eds.), Professional Press Books, New York, pp. 258–263.

Swegmark, G., 1959, Studies with impedance cyclography on human ocular accommodation at different ages, *Acta Ophthalmol.* **47**:1186–1206.

Tait, E. F., 1951, Accommodative convergence, *Am. J. Ophthalmol.* **34**:1093–1107.

25

Discrimination of Blur in Normal and Amblyopic Eyes

KENNETH J. CIUFFREDA

INTRODUCTION

The ability to respond and accommodate appropriately to a defocused and perceptually blurred target requires one to be able to discriminate changes in overall contrast and edge sharpness (Fry, 1955; Fujii *et al.*, 1970; Hamerly and Dvorak, 1981; Watt and Morgan, 1983; Ciuffreda and Rumpf, 1985; Ciuffreda *et al.*, 1987) as well as other attributes (such as chromatic aberration, spherical aberration, etc.) of the retinal image (Fry, 1955; Campbell and Westheimer, 1959; Fujii *et al.*, 1970). Inability to do so would adversely affect accommodative accuracy and quality of the retinal image (Fry, 1955; Fujii *et al.*, 1970; Ciuffreda and Rumpf, 1985). This, in turn, could impair one's visual resolution and visual efficiency, perhaps even leading to general visual discomfort. Such results might be predicted to be found, and indeed exaggerated, in amblyopic eyes in which overall reduced sensitivity is the norm (Schapero, 1971; Ciuffreda *et al.*, 1991). The two experiments described here were performed to test discrimination of blur in normal and amblyopic eyes and relate these findings to their steady-state accommodative performance.

EXPERIMENT 1: DEPTH OF FOCUS

There are a variety of ways to measure the depth of focus of the human eye (Table 25-1). We selected the "detection of blur" method (Ciuffreda *et al.*, 1984), as it appears to be the most widely used and direct approach.

KENNETH J. CIUFFREDA • Department of Vision Sciences, State University of New York, State College of Optometry, New York, New York 10010.

Method

This measure represents the total dioptric range over which a target can be displaced and yet be perceived as in focus by the subject without a change in accommodative state. A noncyclopegic procedure was developed to obtain an estimate of the depth of focus (Fig. 25-1). The subject's head was placed within a headrest/chinrest assembly in a darkened room. The nontested eye was patched, and the refractive correction was placed in the spectacle plane of the viewing eye. The accommodative stimulus consisted of an illuminated near-point Snellen chart placed along the line of sight at a distance of 50 cm. A horizontal luminous aperture ($5 \times 0.3°$) within a Badal system was reflected into the eye by a half-silvered mirror. Thus, the subject saw this luminous slit superimposed on the reduced Snellen chart. The subject was instructed to fixate and maintain in sharp focus at all times a letter in the smallest Snellen row that could be resolved. The luminous slit was slowly moved (method of limits) in the Badal system by the experimenter. The subject was instructed to indicate when the borders of the luminous slit, lying adjacent to the fixated letter, first appeared slightly defocused. Thus, a zone of clarity was mapped out. Four measurements were taken for both increasing and decreasing luminous target dioptric level. The mean and standard deviation were calculated for each end of the range and referenced to the corneal plane. This range, converted to diopters, represented an estimate of the total depth of focus. Values obtained in normal eyes tested using our procedure were within the range reported by Campbell (1957), who used artificial pupils and experienced subjects without cycloplegia.

Results and Discussion

The results for subjects in the various diagnostic groups are summarized in Table 25-2. The depth of

TABLE 25–1
Techniques to Measure Depth of Focus on the Human Eye[a]

1. Detection of blur
2. Reduction of contrast sensitivity
3. Reduction of visual acuity
4. Reduction of vernier acuity
5. Detection of the direction of speckle motion in a laser optometer
6. Photodetector measurements of the line-spread function

[a]Legge *et al.* (1987).

TABLE 25–2
Mean Depth of Focus (D)[a]

Normals	LE[b]	RE
	0.50	0.38
Amblyopes	AE	DE
	0.92	0.52
Former amblyopes	FAE	DE
	0.46	0.48
Strabismics without amblyopia	NDE	DE
	0.78	0.70

[a]Summarized from Ciuffreda *et al.* (1984).
[b]Symbols: LE, left eye; RE, right eye; AE, amblyopic eye; DE, dominant eye; FAE, former amblyopic eye; NDE, nondominant eye.

focus was increased in nine of the 11 amblyopic eyes tested. It was about twice as large in the amblyopic eye as in the fellow dominant eye. In contrast, the depth of focus values were essentially equivalent in each eye for the other three diagnostic groups tested.

This finding of increased depth of focus in amblyopic eyes is consistent with the empirical results with regard to their accommodative stimulus–response function, wherein the accommodative error was consistently increased by 0.25 to 0.75 D or so in the amblyopic eye as compared either to the fellow dominant eye or to visually normal eyes (Ciuffreda *et al.*, 1984) (Fig. 25-2). This finding is also consistent with the contrast discrimination results presented in Experiment 2. Finally, increased depth of focus (reduction of visual acuity criterion) has also been found in patients with unspecified disease-related low vision loss (Legge *et al.*, 1987) (Fig. 25-3); however, they, unlike us, found a progressive increase in depth of focus as visual acuity worsened.

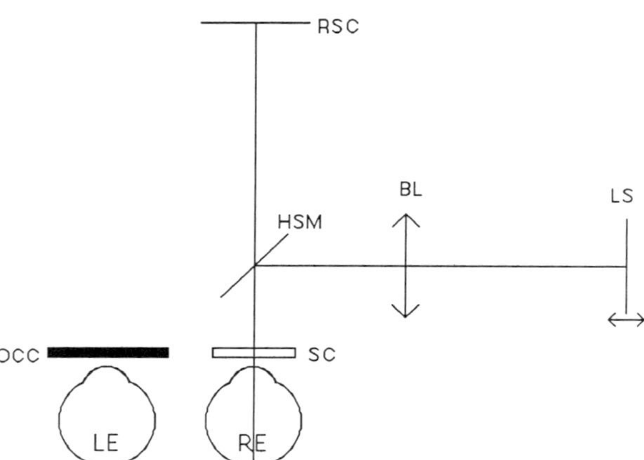

FIGURE 25-1. Top-view schematic diagram of depth of focus apparatus, not drawn to scale. Symbols: LE, left eye; RE, right eye; O, occluder; SC, spectacle correction; HSM, half-silvered mirror; RSC, reduced Snellen chart; BL, Badal lens; LS, luminous slit.

The result suggests reduced neural sensitivity to retinal defocus, presumably related to early abnormal visual experience, i.e., constant suppression in strabismus and monocular form or contrast deprivation in uncorrected anisometropia. In addition, increased oculomotor drift in the amblyopic eye (Ciuffreda *et al.*, 1980), causing smearing of the luminous slit retinal image, and the eccentric fixation of 1° or so commonly found in amblyopic eyes (Brock and Givner, 1952), causing impairment of blur detection with the slightly less sensitive near peripheral retina (Ronchi and Molesini, 1975), can contribute to the finding of increased depth of focus in these eyes.

EXPERIMENT 2: CONTRAST DISCRIMINATION

Method

A successive two-alternative forced-choice procedure (Ciuffreda and Fisher, 1987) was used to measure the contrast necessary to determine that two suprathreshold sinusoidal gratings differed in contrast (Fig. 25-4). The procedure incorporated double-interleaved staircases with a standard grating of 25% contrast and initial comparison gratings of 10% for the ascending series and 40% for the descending series. On a given trial, the randomly ordered standard and comparison gratings were each presented for 150 msec, separated by a 150-msec interval during which only the empty, yellowish-green oscilloscope screen remained visible. These brief presentation times were used to prevent occurrence of increased thresholds in the amblyopic eye because of smearing of the retinal image related to increased ocular drift (Ciuffreda *et al.*, 1980), abnormally rapid fading of the retinal image from enhanced reti-

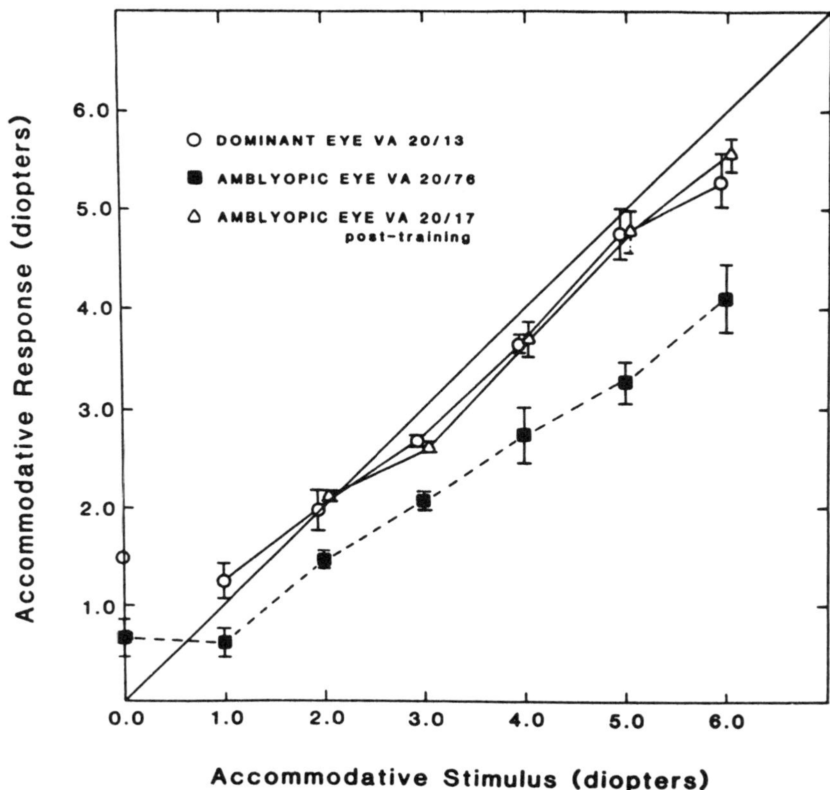

FIGURE 25-2. Accommodative stimulus–response curves in a 12-year-old patient with strabismic amblyopia before and after intensive orthoptic therapy. After therapy, monocular accommodative responses were similar in each eye. In dominant eye ○, VA = 20/13, ACG = 21, and slope = 0.88; in amblyopic eye before therapy ■, VA = 20/76, ACG = 7.8, and slope = 0.68; in amblyopic eye after therapy △, VA = 20/17, ACG = 18, and slope = 0.87. Mean ± 1 SD plotted (Ciuffreda *et al.*, 1984).

nal adaptation (Lawwill, 1968; Ciuffreda *et al.*, 1979), and defocus effects as a result of impaired ability to sustain accommodation (Ciuffreda and Kenyon, 1983). The procedure continued until 100 trials were completed for each staircase. The number of trials was more than sufficient for convergence of the staircases on the 0.707 point, which is considered to represent the contrast difference threshold (Wetherill and Leavitt, 1965).

Viewing conditions were controlled by a circular aperture providing a 3.7° field of view that only permitted the subject to see the glass oscilloscope face and not adjacent portions of the field (Fig. 25-5). The yellowish-green homogeneous aperture surround (20°) was adjusted to have approximately the same space-average luminance (10 cd/m^2) and chromatic composition as the gratings. Given the optical distance of the aperture from the eye (9.5 D), its border did not constitute a significant accommodative stimulus because of the resulting combination of moderate eccentricity (~2°) and large defocus (~7 D) on the retina (Phillips, 1974). Moreover, the aperture vergence exceeded the maximum accommodative amplitude of all subjects. The spatial frequency of the gratings was held constant within a given test session at either 0.5, 2, 4, or 8 cycle/°.

At the beginning of each session, the subject spent 3 min sitting quietly in the dark to allow for completion of transient accommodative changes (Phillips, 1974; Krumholz *et al.*, 1986), after which six measurements of the tonic or resting level of accommodation of the eye to be used in the contrast discrimination task were made with a Hartinger optometer (Hartinger, 1951; Duke-Elder and Abrams, 1970). The oscilloscope screen was then set at an optical distance equivalent to the subject's average monocular tonic accommodative level, thereby establishing conjugacy between the retinal plane and the test gratings without the need for any stimulus-driven (i.e., "blur") accommodation. Such a maneuver prevented the occurrence of elevated thresholds from abnormally large blur-driven accommodative error in the amblyopic eye (Ciuffreda *et al.*, 1984).

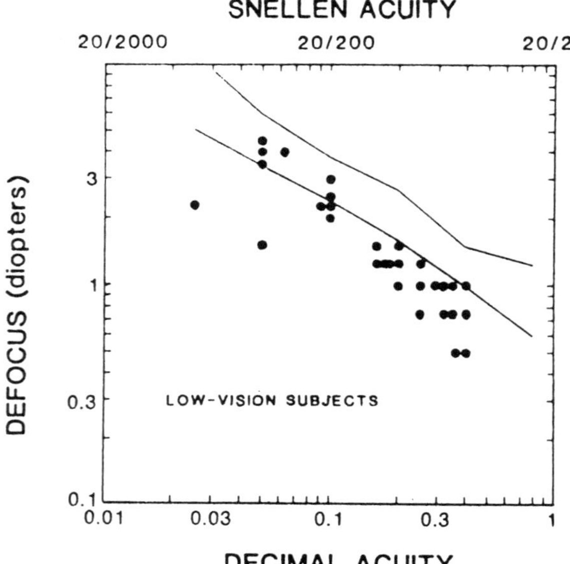

FIGURE 25-3. Tolerance to defocus in low vision. Each point refers to one low-vision eye. The positive-lens defocus required to degrade acuity by one line on a chart (0.1 log unit in character size) is plotted as a function of the subject's acuity. The solid lines are based on predictions from our modulation-transfer data and spatial-frequency filtering considerations (Legge *et al.*, 1987).

The necessary grating vergence level was achieved through an appropriate combination of the dioptric power of a lens placed 10 cm from the eye and the distance of the oscilloscope from the subject, taking into account the magnification of the lens–target combination and the grating spatial frequency. The subject's nontested eye was fully patched, and the refractive correction was placed in the spectacle plane of the viewing eye. On completion of the monocular contrast discrimination testing, the entire procedure was repeated using the formerly occluded eye.

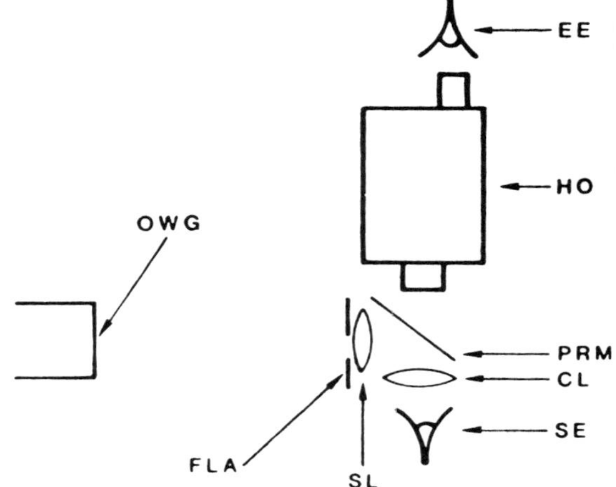

FIGURE 25-5. Schematic diagram of experimental apparatus, top view. Symbols: HO, Hartinger optometer; EE, experimenter's eye; SE, subject's eye; OWG, oscilloscope with gratings; FLA, field-limiting aperture; SL, stimulating lens; CL, correcting lens for ametropia; PRM, partially reflecting mirror. Not drawn to scale (Ciuffreda and Fisher, 1987).

Results

The group results for the amblyopes are presented in Fig. 25-6. There was a significant effect of eye, with the amblyopic eye performing more poorly than the fellow dominant eye. This was true for each spatial frequency. There was also a significant effect of spatial frequency, indicating poorer contrast discrimination ability as spatial frequency increased. There was no interaction, and thus there was no effect of spatial frequency on the magnitude of the interocular contrast discrimination differences.

The group results for the normals and for the strabismics without amblyopia are also presented in Fig. 25-6. In the normals, there was a significant

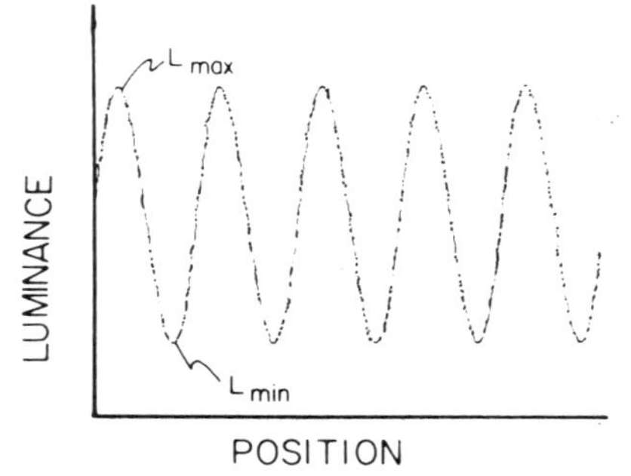

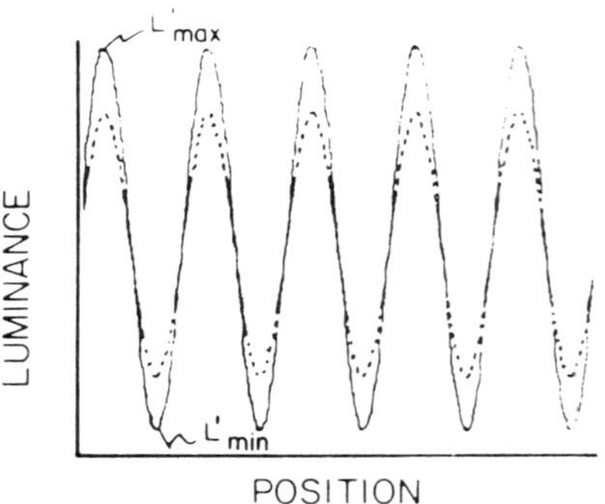

FIGURE 25-4. Stimuli for contrast discrimination task. Left: The luminance profile of a sine-wave grating with the contrast set to the background level. The maximum and minimum luminances are indicated by the arrows. Contrast is defined as: $[L_{\max} - L_{\min}]/[L_{\max} + L_{\min}]$. Right: The luminance profile of a sine-wave grating with a contrast increment added to the background contrast. The background contrast is indicated by the broken line, and the background plus increment by the solid line (Stephens and Banks, 1987).

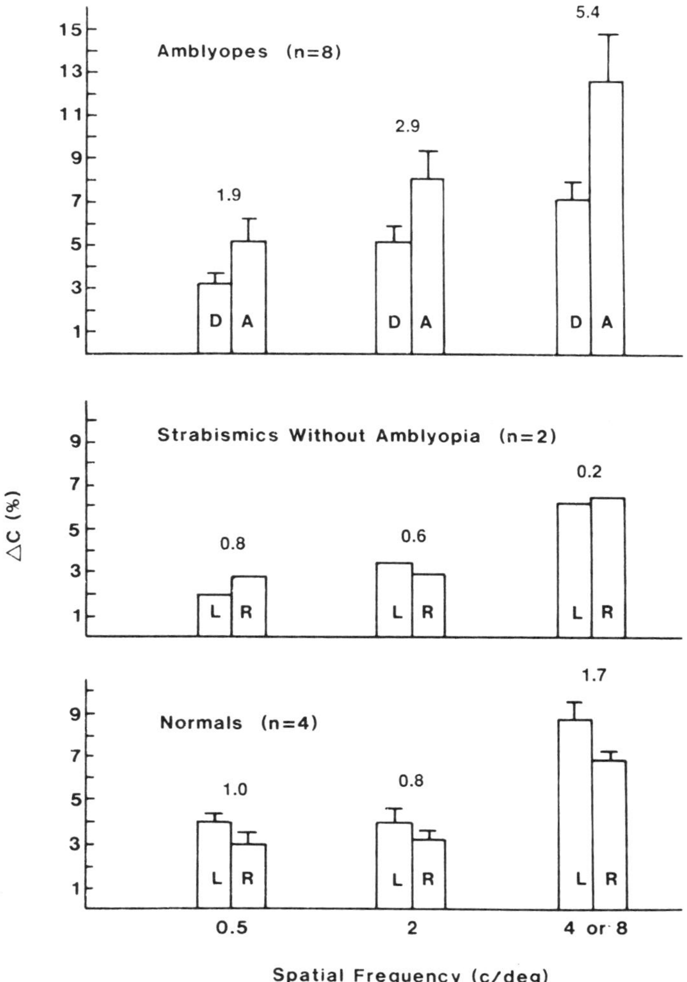

FIGURE 25-6. Contrast discrimination (ΔC) as a function of spatial frequency in each eye for each diagnostic group. Plotted is $\bar{x} \pm 1$ S.E.M. Numbers above each set of bars represent the interocular difference in contrast discrimination in percent. Symbols: A, amblyopic eye; D, dominant eye of the amblyope; L and R, left and right eyes, respectively, of visually normal individuals (Ciuffreda and Fisher, 1987).

effect of eye. Although there was a trend for better performance in the right eye for all spatial frequencies, only the interocular difference at the lowest spatial frequency was significant. There was also a significant effect of spatial frequency, indicating a difference in contrast discrimination ability related to spatial frequency. There was no interaction, and thus there was no effect of spatial frequency on the magnitude of the interocular contrast discrimination differences. The results in the strabismics without amblyopia were clearly within normal limits. There was little interocular difference for any spatial frequency, and contrast discrimination became progressively poorer with increased spatial frequency.

Interocular contrast discrimination was also compared between the amblyopes and normals. There was a significant effect, collapsed across spatial frequency, with this interocular difference being larger in the amblyopes.

Discussion

The results clearly show impairment of contrast discrimination ability in amblyopic eyes, thus confirming and extending the recent results of others (Hess *et al.*, 1983; Bradley and Ohzawa, 1986). Interocular differences in the amblyopes were about 2 to 3.5 times greater than in the normals (for the respective spatial frequencies). The deficit was present at each of the spatial frequencies tested. Such anomalous contrast discrimination in the amblyopic eyes is presumably related to the abnormal visual experience, perhaps involving visual cortical cells sensitive to contrast (Campbell and Kulikowski, 1972), during the early years of life. This includes abnormal binocular interactions as a result of constant strabismic suppression (Burian and von Noorden, 1974) and/or mild monocular form/contrast deprivation from uncorrected anisometropia (Bradley and Freeman, 1981). Apparently, the intermittent suppression experienced in the strabismics without amblyopia was not sufficient to affect contrast discrimination adversely; the periods of normal visual experience were apparently sufficient to maintain a normal sensory status.

The abnormal contrast discrimination ability in amblyopic eyes might contribute to the increased steady-state accommodative error typically found in these subjects, as shown earlier (Ciuffreda *et al.*, 1984). Such a deficit would result in impaired ability to perceive or discriminate small fluctuations in retinal image contrast. If the amblyopic eye requires several times more change in contrast than normal to detect a just noticeable difference, a larger steady-state accommodative error would be allowed to be present without any obvious adverse perceptual consequences such as blur or reduced apparent target contrast. Thus, the accommodative response for a given stimulus would be reduced as compared to normal observers or to the fellow dominant eye of the amblyope. In terms of the accommodation model parameters affected, presence of poor contrast discrimination ability leading to reduced accommodative responses would be manifested as reduced accommodative controller gain and increased depth of focus, as has been found experimentally and modeled with computer simulations (Hung *et al.*, 1983; Ciuffreda *et al.*, 1984) and as presented in Experiment 1.

Lastly, two other points deserve mention. First, could the impaired contrast discrimination in the amblyopic eye result from an immature, almost infant-like neural mechanism for such tasks? Probably not. Contrast discrimination in human infants is about one log unit worse than that in adults, whereas in the present investigation the difference was only about 0.2 to 0.3 log units (Fig. 25-7) (Stephens and Banks, 1987). Second, could the discrimination deficit be caused by the presence of eccentric fixation in the amblyopic eye? Again, probably not. Contrast discrimination at moderate to high suprathreshold levels is relatively independent of retinal eccentricity (Fig. 25-8) (Legge and Kersten, 1987). Thus,

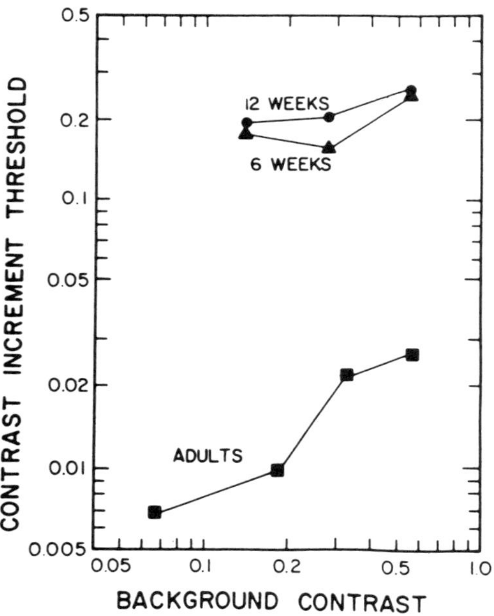

FIGURE 25-7. Infants' and adults' contrast discrimination functions (Stephens and Banks, 1987).

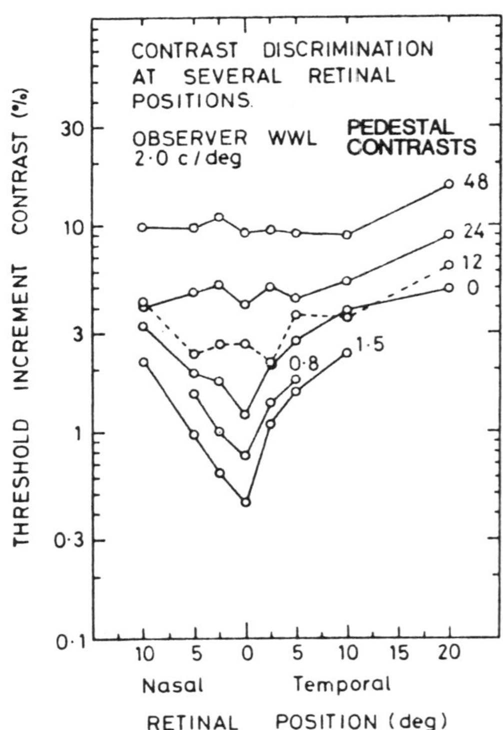

FIGURE 25-8. Threshold-increment contrast plotted as a function of retinal position for several pedestal contrasts. The data are for one observer (W.W.L.). Notice that the curves are much flatter for high-contrast pedestals than for near-threshold pedestals or for detection (a pedestal of zero) (Legge and Kersten, 1987).

once again, early abnormal visual experience resulting in a neurosensory-based moderate depression in sensitivity to contrast changes is probably the basic underlying cause.

ACKNOWLEDGMENT. Supported in part by NIH grant EY03541.

REFERENCES

Bradley, A., and Freeman, R., 1981, Contrast sensitivity in anisometropic amblyopia, *Invest. Ophthalmol. Vis. Sci.* **21**:467–476.

Bradley, A., and Ohzawa, I., 1986, A comparison of contrast detection and discrimination, *Vision Res.* **26**:991–997.

Brock, F. W., and Givner, I., 1952, Fixation anomalies in amblyopia, *Arch. Ophthalmol.* **47**:775–786.

Burian, H. M., and von Noorden, G. K., 1974, *Binocular Vision and Ocular Motility,* C. V. Mosby, St. Louis.

Campbell, F. W., 1957, The depth of field of the human eye, *Optica Acta* **4**:157–164.

Campbell, F. W., and Westheimer, G., 1959, Factors influencing accommodation responses of the human eye, *J. Opt. Soc. Am.* **49**:568–571.

Campbell, F. W., and Kulikowski, J. J., 1972, The visual evoked potential as a function of contrast of a grating pattern, *J. Physiol. (Lond.)* **222**:345–356.

Ciuffreda, K. J., and Fisher, S. K., 1987, Impairment of contrast discrimination in amblyopic eyes, *Ophthalmic. Physiol. Opt.* **7**:461–467.

Ciuffreda, K. J., and Kenyon, R. V., 1983, Accommodative vergence and accommodation in normals, amblyopes and strabismics, in: *Vergence Eye Movements: Basic and Clinical Aspects* (C. M. Schor and K. J. Ciuffreda, eds.), Butterworth, Boston, pp. 101–173.

Ciuffreda, K. J., and Rumpf, D., 1985, Contrast and accommodation in amblyopia, *Vision Res.* **25**:1445–1457.

Ciuffreda, K. J., Kenyon, R. V., and Stark, L., 1979, Saccadic intrusions in strabismus, *Arch. Ophthalmol.* **97**:1673–1679.

Ciuffreda, K. J., Kenyon, R. V., and Stark, L., 1980, Increased drift in amblyopic eyes, *Br. J. Ophthalmol.* **64**:7–14.

Ciuffreda, K. J., Hokoda, S. C., Hung, G. K., and Semmlow, J. L., 1984, Accommodative stimulus/response function in human amblyopia, *Doc. Ophthalmol.* **56**:303–326.

Ciuffreda, K. J., Dul, M., and Fisher, S. K., 1987, Higher-order spatial frequency contribution to accommodative accuracy in normal and amblyopic observers, *Clin. Vis. Sci.* **1**:219–229.

Ciuffreda, K. J., Levi, D. L., and Selenow, A., 1991, *Amblyopia: Basic and Clinical Aspects,* Butterworth, Boston.

Duke-Elder, S., and Abrams, D., 1970, *System of Ophthalmology,* Vol. 5, *Ophthalmic Optics and Refraction,* C. V. Mosby, St. Louis.

Fincham, E. F., 1951, The accommodation reflex and its stimulus, *Br. J. Ophthalmol.* **35**:381–393.

Fry, G. A., 1955, *Blur of the Retinal Image,* Ohio State University Press, Columbus.

Fujii, K., Kondo, K., and Kasai, T., 1970, An analysis of the human eye accommodation system, *Osaka Univ. Tech. Rep.* **20**:221–236.

Hamerly, J. R., and Dvorak, C. A., 1981, Detection and discrimination of blur in edges and lines, *J. Opt. Soc. Am.* **71**:448–452.

Hartinger, H., 1951, Uber ein neues Refraktometer, in: *57th Meeting Deutsch Optisch Gasellschaft,* Munich, pp. 105–108.

Hess, R. F., Bradley, A., and Piotrowski, L., 1983, Contrast coding in amblyopia I. Differences in the neural basis of human amblyopia, *Proc. R. Soc. Lond. [Biol.]* **217**:309–330.

Hung, G. K., Ciuffreda, K. J., Semmlow, J. L., and Hokoda, S. C., 1983, Model of static accommodative behavior in human amblyopia, *IEEE Trans. Biomed. Eng.* **30**:665–672.

Krumholz, D. M., Fox, R. S., and Ciuffreda, K. J., 1986, Short-term changes in tonic accommodation, *Invest. Ophthalmol. Vis. Sci.* **27**:555–557.

Lawwill, T., 1968, Local adaptation in functional amblyopia, *Am. J. Ophthalmol.* **65**:903–906.

Legge, G. E., and Kersten, D., 1987, Contrast discrimination in peripheral vision, *J. Opt. Soc. Am.* **4**:1594–1598.

Legge, G. E., Mullen, K. T., Woo, G. C., and Campbell, F. W., 1987, Tolerance to visual defocus, *J. Opt. Soc. Am.* **4**:851–863.

Phillips, S. R., 1974, *Ocular Neurological Control Systems: Accommodation and the Near Response Triad,* Ph.D. dissertation, University of California, Berkeley.

Ronchi, L., and Molesini, G., 1975, Depth of focus in peripheral vision, *Ophthalmic. Res.* **7**:152–157.

Schapero, M., 1971, *Amblyopia,* Chilton, New York.

Stephens, B. R., and Banks, M. S., 1987, Contrast discrimination in human infants, *J. Exp. Psychol. Hum. Percept. Perform.* **13**:558–565.

Watt, R. J., and Morgan, M. J., 1983, The recognition and representation of edge blur: Evidence for spatial primitives in human vision, *Vision Res.* **23**:1465–1477.

Wetherill, G. B., and Levitt, H., 1965, Sequential estimation of points on a psychometric function, *Br. J. Math. Stat. Psychol.* **18**:1–10.

26

Preclinical Presbyopic Development

CORINA VANDEPOL and LAWRENCE STARK

INTRODUCTION

Determination of the amplitude of accommodation using subjective methods is the most common clinical practice, and for a great number of years since Donders, subjective methods have been used in the laboratory as well. Although objective methods are available, they often require the employment of special procedures such as the use of infrared light, ultrasound, or photography (see Appendix).

Both range and amplitude of accommodation are based on the positions of the far point (the farthest point for which the eye can form a sharp image on the retina) and the near point (the nearest point for which a sharp image can be formed on the retina) of accommodation (Grosvenor, 1982; Cline *et al.*, 1980). The clinical "pushup" method used to determine the amplitude of accommodation generally only measures the near point of accommodation. The far point is assumed to be at a distance of infinity or zero diopters for the fully corrected eye. There have, however, been investigators who have described means of determining the far point as well. Duane (1912), for instance, used a +3.00-D lens to bring the far point within a measurable range. The advantage of determining both the far and near points is clear for the assessment of accommodative range.

Presbyopia, the loss of accommodative amplitude with age, has been studied by numerous investigators starting with Donders in 1864 (Kaufman, 1894; Duane, 1912; Hofstetter, 1944, 1965; Allen, 1956; Hamasaki *et al.*, 1956; Swegmark, 1969; Saladin and Stark, 1975; Matsunaga *et al.*, 1988; Sun *et al.*, 1988). Most studies have employed some form of clinical pushup methodology (although lenses and stigmatoscopy have also been used) to determine the amplitude of accommodation for various age ranges. Once this information is plotted, a formula describing the development of presbyopia may be obtained.

In *stigmatoscopy* the plane conjugate to the retina is determined by the best focus or minimum blur of the stigma, a point target of light. The point of minimum blur is determined by adjustment of the distance of the stigma such that a definite blur is noted at both limits of the depth of focus, and the center point with minimum blur is used as the endpoint. In this way, the pupil aperture size and the depth of focus do not affect the endpoint (Bannon *et al.*, 1950).

In the present chapter we examine the difference in the methodologies of the clinical pushup, Badal pushup, and stigmatoscopic methods of determining accommodative amplitude. By using these comparisons, results of previous investigators can be reconciled and used in concert to develop our understanding of the early progression of presbyopia.

METHODS

Stigmatoscopy

In stigmatoscopy, the subject's attention is directed toward an accommodative target (AT) that has been designed with high-spatial-frequency borders and significant contour to lock the subject's accommodation onto the target (Fig. 26-1). It is lit from behind by a yellow light-emiting diode (LED) with a diffuser filter. While maintaining best focus on AT, the subject brings the stigma (which appears superimposed on the AT) to the point of minimum blur by

CORINA VANDEPOL AND LAWRENCE STARK • School of Optometry, University of California, Berkeley, California 94720.

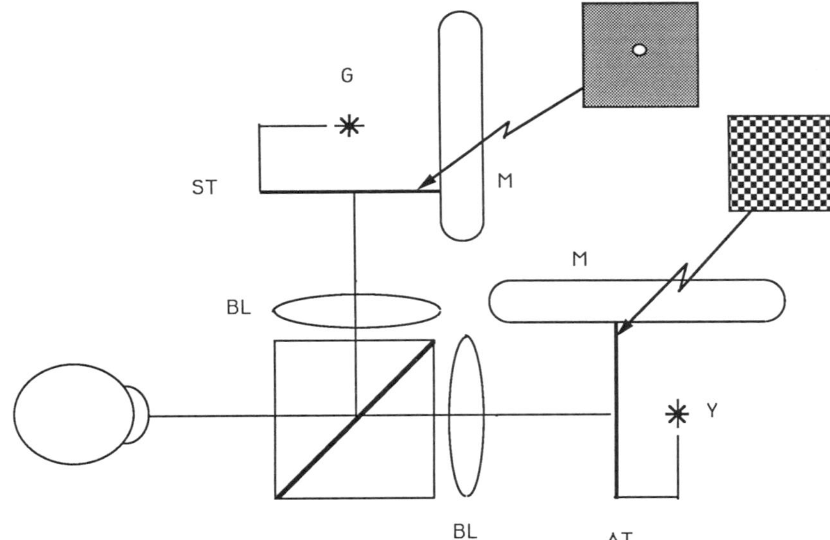

FIGURE 26-1. Stigmatoscope. Beam splitter (actually a collector), represented by cube with heavy diagonal, joins the complex and compelling accommodative stimulus (AT) illuminated by yellow LED (light-emitting diode) (Y) and the stigma point target (ST) illuminated by green LED (G). The Badal lenses (BL) provide appropriate optics for each path; two motors (M) provide for computer control of both optical distances. Of course, the subject (eye in diagram) controls a potentiometer that instructs the computer to position the stigma.

means of a potentiometer (Phillips and Stark, 1977). Focusing of the stigma is accomplished by bracketing about the point of minimum blur using the center of this range as the endpoint. For precision the subject is directed to regard the edges of the pinhole. In control experiments several experienced subjects were directed to use a stigma endpoint criterion of smallest stigma or brightest stigma or sharpest contrast at edge of stigma. Each of these alternative criteria produced results that were not significantly disparate.

The stigma consists of a pinhole 0.75 mm in diameter lit from behind by a green LED and diffuser. The subject's attention is not directed completely away from the AT while the stigma is being focused; instead, the stigma becomes a simultaneously viewed point of regard that is adjusted to match the accommodative posture of the subject. The stigma works as a measure of accommodative posture because it can be ignored as a stimulus to accommodation (especially in the presence of a richly detailed target). However, the subject can evidently still direct his/her attention toward the psychophysical perception of the stigma to adjust it to a point of minimum blur.

Badal Lens System

Although not necessary for stigmatoscopy, the use of the Badal lens system allows greater precision in the determination of higher accommodative levels (Badal, 1876). This is because of the compression of real distances into optical distances; the result is a linear relationship of centimeters of target motion behind the Badal lens to diopters of optical distance. In our setup two Badal lenses are used, one each for the accommodative target and the stigma. Of course, the Badal lens reduces secondary clues such as increase in size and brightness both of the target and of the stigma as each comes optically closer.

Protocol

The subject is instructed to concentrate on keeping the accommodative target clear and then, while maintaining his/her accommodation at that level, to adjust the pinhole of light (stigma) to a point of minimum blur. The accommodative stimuli are presented in a bidirectional mode starting from a far point and progressing in 1-D steps (except for the 9.0-D to 9.5-D interval) toward the subject to complete a run. The stimulus is then backed off in 1-D steps from the maximum setting toward the minimum setting for the next run. The overall range tested was from 2.0 D beyond infinity to 9.5 D of near accommodative stimulus. Each stimulus was presented ten times in this manner; the computerized protocol yielded 130 points in 10 min! The actual speed of each measurement session was controlled by the subject.

Clinical Method with Monocular Pushup

An isolated 20/30 letter was presented to the subject under monocular conditions. As the letter was brought closer to the subject he/she was instructed to report when it first appeared "blurred." The target was then backed off until it was first "clear" again. The distance from the spectacle plane to this point was measured in diopters.

Badal Lens System with Monocular Pushup

The subject viewed the same accommodative target as used for the stigmatoscopic measure. The target was moved nearer to the subject until he/she reported first "blur," then backed off until just "clear" (the same instructions as the clinical method). The dioptric value of this near point was recorded.

Subjects

Thirty-one optometry students, knowledgeable in subjective eye tests and measurements, 22 to 26 years of age served as subjects for this experiment. Participation in the experiment was achieved as part of a demonstration lab of a course in ocular motility that included presentation of the mechanism of accommodation. Clinical information obtained on each subject included refraction, binocular status (e.g., presence or absence of phorias or tropias), date of birth, and sex. A total of 47 students participated in the study; eight subjects whose standard deviation for any point exceeded 1 D were excluded from the study, and records of eight additional subjects who fell outside the 22 to 26 age range were not utilized.

RESULTS

Computerized Protocol

In the computerized stigmatoscopic method, each accommodative stimulus point is presented to the subject ten times. The responses are recorded by the computer, and the average and standard deviation for each point are computed (Fig. 26-2). The accommodation curve produced by the computer accurately depicts the accommodative limits, range, and precision of the subject.

As explained in the *Methods* section, each subject's accommodative amplitude was measured using the computerized stigmatoscopic system, the Badal pushup method, and the clinical pushup method. The stigmatoscopic method also provided information on the range of accommodation of the subject. The subject's clinical data and measurement results were tabulated individually (Table 26-1) and by age group and test type (Table 26-2).

The accommodative amplitude or range mean value for each of the three methods was obtained (Fig. 26-3). The clinical pushup method yielded an average accommodative amplitude of 10.8 D; the Badal pushup method gave an average amplitude of 8.0 D; finally, the stigmatoscopic method showed an average range of accommodation of 5.8 D.

DISCUSSION

Reconcile

One advantage of our using several methods to measure accommodative amplitudes is that we can now reconcile the various studies in the literature that have used disparate methods. A collection of such results using mainly clinical pushup or stigmatoscopy methods (Donders, 1864; Duane, 1912; Hamasaki *et al.*, 1956; Sun *et al.*, 1988) can be compared with our results (Fig. 26-4). With our results from Fig. 26-3 plotted on the same graph, it can be seen that our pushup method results correlated with results from the literature; both document significantly higher measurements of the amplitude of accommodation than that measured with a stigmatoscope. Our results for age 24 ± 2 years show the difference between the stigmatoscopic range of accommodation and the clinical pushup near point of accommodation to be 5.0 D for subjects aged 22 to 26, a large and significant difference. (The mean difference between the stigma results and the Badal near point of accommodation was 2.2 D, and the difference between the Badal NPA and the Clinical NPA was 2.8 D.)

For older subjects these differences apparently reduce, although since our studies do not directly bear on this minor effect we have elected to keep our formulas simple. Hamasaki *et al.* (1956) noted that the mean difference between the values for stigmatoscopy and the pushup methods was 1.75 D for subjects aged 42 through 60 years. They also suggested that the primary difference was caused by the depth of focus of the eye. In the study by Sun *et al.* (1988), which also compared stigmatoscopy with clinical pushup, the mean difference was determined

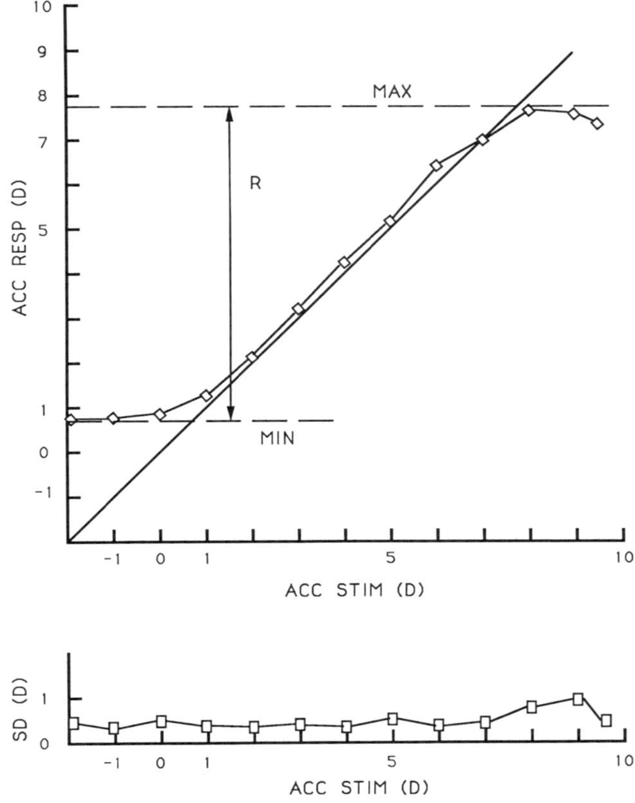

FIGURE 26-2. Accommodative response measured by stigmatoscopy. Accommodative response (ordinate) and stimulus (abscissa) measured in diopters (D). Ten measurements (averaged) at each of 13 stimulus values from −2 to +9.5 D. Range (R = 7.0 D; see vertical arrow) extends from minimum value (MIN = 0.7 D) to maximal value (MAX = 7.7 D) indicated by horizontal dashed lines.

to be 2.0 D. Sun's study involved a fairly large age range, but most subjects still fell within the 40- to 60-year range. Both researchers used the maximum accommodation level measured by the stigmatoscope as their value, whereas our results include the far point measurements resulting in a range of accommodation value. We should also note that the age range of our study was restricted to subjects within the pre-presbyopic population of ages 22 to 26.

Formulas

Duane (1912) and Morgan (1968) contributed "recession of near point" formulas that differ somewhat: Duane (using lenses and clinical pushup) proposed that average amplitude = 18.5 − (0.3) age; Morgan (using clinical pushup) proposed that mean accommodation = 16.0 − (0.25) age. The formula with the larger slope represents zero accommodation amplitude at 61.7 years of age; the formula with a lower slope at 64 years of age. They are equal at 50 years of age.

The formula of Hamasaki *et al.* (1956) (based on stigmatoscopy) is that log amplitude = 3.767 − (0.088) age. It is based on a population of adults aged 42 to 60 and is nonlinear since it focuses on the region where accommodation approaches zero and also where there is a reduction of the differences in amplitudes obtained from different methods.

Our own formula (based on stigmatoscopy) is that average accommodation = 12 − 0.25 (age). It was developed using our point for 24-year-olds and the Hamasaki *et al.* (1956) value for age 44.

It is important to note that all results, and especially our results with younger subjects and with the stigmatoscopy method, emphasize the early development of preclinical presbyopia. The underlying lens changes that proceed through the late teens, 20s, and 30s can not be caused by a senile degenerative change (Saladin and Stark, 1975; Stark, 1988).

TABLE 26–1
Individual Accommodation Test Results and Subject Characteristics

Subject	Female/ male	Age	R.E.	Phoria @ N	Stigmatoscopic measurements Max	Min	Range	Badal NPA	Clinical NPA
10	F	22	–	Esp 4	8.50	0.00	8.50	8.00	11.00
11	M	22	–	Ortho	5.50	0.50	5.00	7.75	12.50
12	M	22	–	Exo 3	8.50	1.00	7.50	9.00	10.00
13	F	22	–	Exo 1	6.00	1.25	4.75	8.50	11.00
14	F	22	–	Exo 4	5.00	1.00	4.00	5.00	9.00
15	F	22	–	Exo 4	6.50	1.00	5.50	8.00	11.00
16	F	23	–	Eso 2	6.75	1.00	5.75	7.50	16.00
17	F	23	–	Exo 2	7.75	0.75	7.00	9.00	11.00
18	M	23	+	Eso 2	6.50	0.50	6.00	7.75	10.00
19	M	23	–	Exo 3	7.50	0.50	7.00	8.25	10.00
20	F	23	–	Exo 4	6.00	0.75	5.25	8.00	10.00
21	M	23	–	Exo 4	6.75	0.75	6.00	7.50	10.00
22	F	23	–	Eso 2	6.75	1.00	5.75	7.50	12.00
23	F	23	–	Eso 2	7.00	1.50	5.50	8.75	12.00
24	M	24	0	Exo 4	5.25	2.25	3.00	7.75	10.00
25	M	24	–	Eso 1	6.00	0.75	5.25	7.00	10.00
26	F	24	–	Exo 5	7.25	0.75	6.50	8.50	10.00
27	M	24	–	Ortho	8.00	1.50	6.50	8.00	10.00
28	F	24	+	Exo 4	7.25	1.50	5.75	8.00	10.00
29	F	24	–	Exo 8	9.25	2.75	6.50	9.50	13.00
30	M	24	0	Exo 2	6.25	1.25	5.00	7.00	11.00
31	M	24	–	Ortho	6.50	1.50	5.00	9.50	12.50
32	M	24	–	Ortho	6.25	0.75	5.50	7.50	10.00
33	F	25	0	Exo 3	6.75	1.00	5.75	7.75	10.00
34	F	25	–	Exo 4	7.25	0.50	6.75	8.50	10.00
35	M	25	–	Ortho	7.50	1.25	6.25	8.75	10.00
36	M	25	–	Ortho	7.25	1.50	5.75	9.00	9.50
37	M	25	0	Exo 5	6.50	0.75	5.75	8.50	12.00
38	F	26	–	Exo 17	7.50	0.75	6.75	9.00	12.50
39	F	26	–	Eso 1	5.25	0.75	4.50	6.50	7.25
40	M	26	0	Exo 2	7.00	0.75	6.25	7.00	11.00

TABLE 26–2
Accommodation Test Results by Age and Summarized

Age	*n*	Stigmatoscope Average	S.D.	Badal Average	S.D.	Clinical Average	S.D.
22	6	5.9	1.7	7.7	1.4	10.8	1.2
23	8	6.0	0.7	8.0	0.6	11.4	2.1
24	9	5.4	1.1	8.1	0.9	10.7	1.2
25	5	6.1	0.5	8.5	0.5	10.3	1.0
26	3	5.8	1.2	7.5	1.3	10.3	2.7
Summary							
24 ± 2	31	Max 6.8	1.0	8.0	0.9	10.8	1.5
		Min 1.0	0.5				
		Range 5.8	1.1				

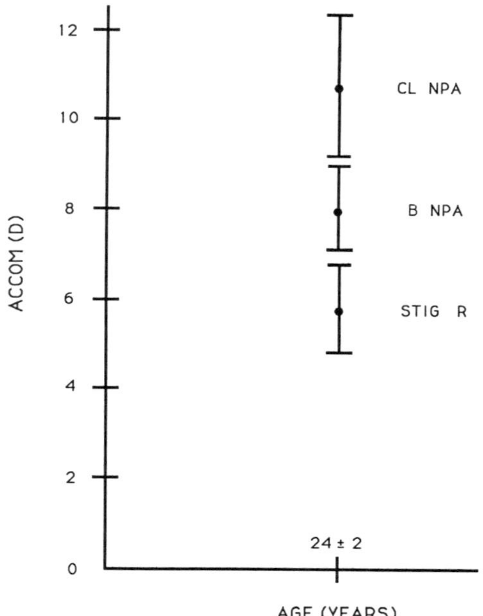

FIGURE 26-3. Three accommodation amplitudes for 24-year-old subjects. The 31 subjects were 24 ± 2 years in age (see Table 26-2). The standard deviation for the clinical near point of accommodation (CL NPA) is greater than that for the Badal pushup method for NPA (B NPA) and that for the range of accommodation measured by stigmatoscope (STIG R). Note that experimental values barely overlap because of the significant hiding of losses of accommodation by B NPA and especially by CL NPA.

Differences in Methods

Factors leading to differences among the three points in Fig. 26-3 are important to consider. They are summarized in Table 26-3.

CLINICAL MONOCULAR PUSHUP

The target used consists of an isolated 20/30 letter. As the target is brought in closer to the subject, several factors improve visual perception. Visual acuity is improved with increased surround brightness and increased target contrast, although these may provide only small increments since the target is already quite contrasted in a bright surround. Further, as the target enlarges in visual angle, it certainly becomes more robust to blur. The endpoint is determined by the subject's estimate of the blur of the target, therefore requiring a judgment on the part of the subject. Additionally, there is a strong proximal effect because of the dynamic nature of the test. The subject might jump across the pupil depth of focus and use the "active" limb of accommodation to bring the near point of accommodation nearer (Fig. 26-5). The clinical method does not include a measurement of the far point since it is assumed to be at infinity based on the accuracy of the subject's refractive correction.

TABLE 26–3
Comparison of Features of the Three Methods

	Stigma	Badal Pushup	Clinical NPA
Minimum blur as endpoint (multiple adjustment)	+	0	0
Blur threshold as endpoint (lazy limb) (aperture)	0	+	+
Pupil depth of focus	0	0	+
Measures far point	+	+	0
Change in angular size of target (spat. freq. and VA)	0	0	+
Change in brightness of target	0	0	+
Target type	Geometric	Geometric	Letter
Proximal effect			
Static	+	+	0
Dynamic	0	+	+
Optical aperture (we believe it cannot have an effect even in Badal)	+	+	0
Near point	6.8	8.0	10.8
Amplitude	5.8	8.0	10.8

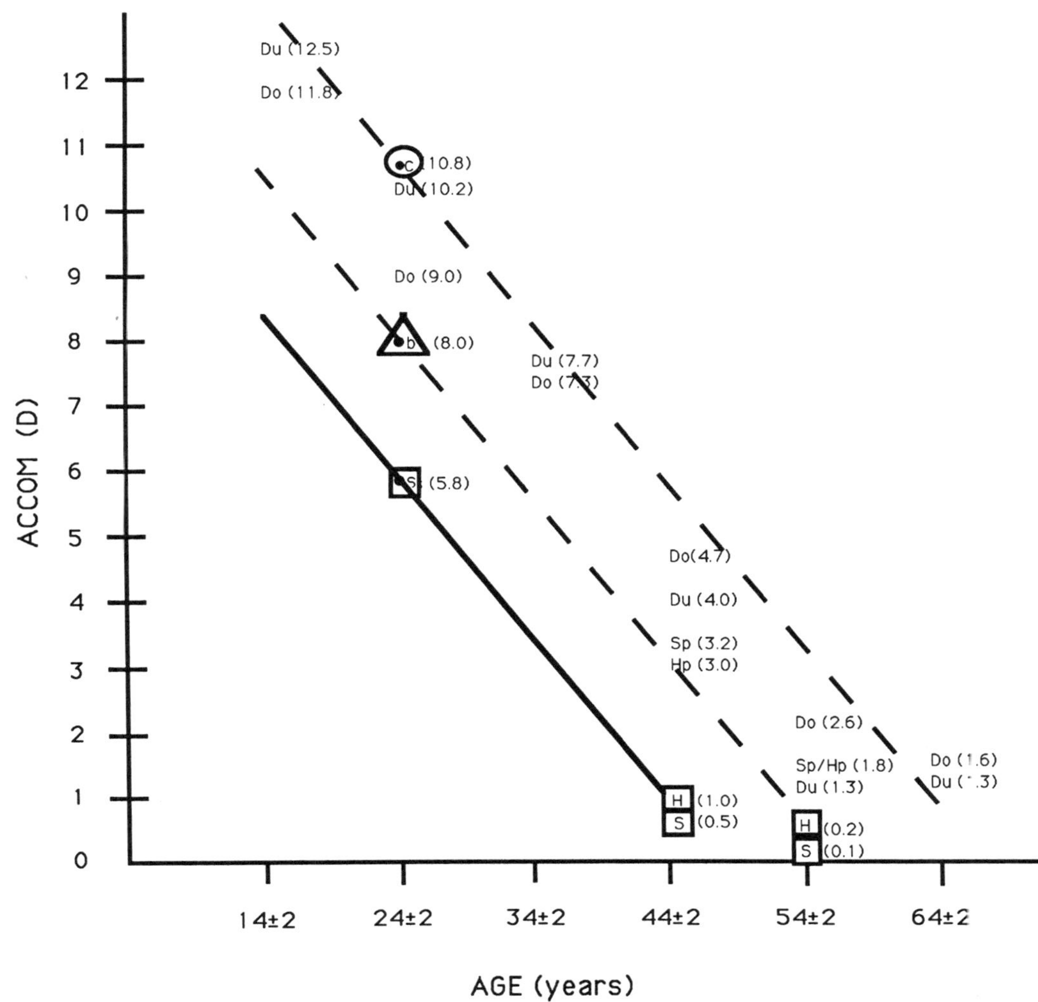

FIGURE 26-4. Preclinical loss of accommodation. Our points for accommodative amplitude for 24-year-olds determined by clinical NPA (open circle), Badal NPA (open triangle), and stigmatoscopic range (open square). These are plotted with values derived from experimental data points of Donders (Do), Duane (Du), Hamasaki (Hp), and Sun (Sp) for clinical pushup accommodation and points determined by Hamasaki (H square) and Sun (S square) using the stigmatoscopic method. Solid line represents the formula: Average accommodation = 12 − (0.25) age, as derived from our stigmatoscopic data for age 24 ± 2 and Hamasaki's stigmatoscopic data for age 44 ± 2. The dashed lines are set parallel to this line and shifted by amounts appropriate for our 24-year-old data points; the extensions of these lines are speculative (see text).

PUSHUP BEHIND BADAL WITH TARGET

With the arrangement of our measurement device this test has to be done monocularly. Again, the endpoint is determined by the subject's judgment of target blur. The adjustment of the target is not to a point of minimum blur. As in the case of the clinical pushup, the target is brought closer to the subject until first blurred, then reversed until the target is just clear (a point of acceptable blur). The endpoint is dependent on the pupillary depth of focus and might nest on the "lazy" limb of accommodation (Fig. 26-5). The dynamic proximal effect plays a role in this test as well, since the subject knows the target is being brought in closer; in this case the endpoint might rest on the active limb. Two advantages over the clinical near point of accommodation measurement are the ease with which a far point can be measured and the absence of changes in angular size or brightness of the target that could provide extraneous stimuli to accommodation.

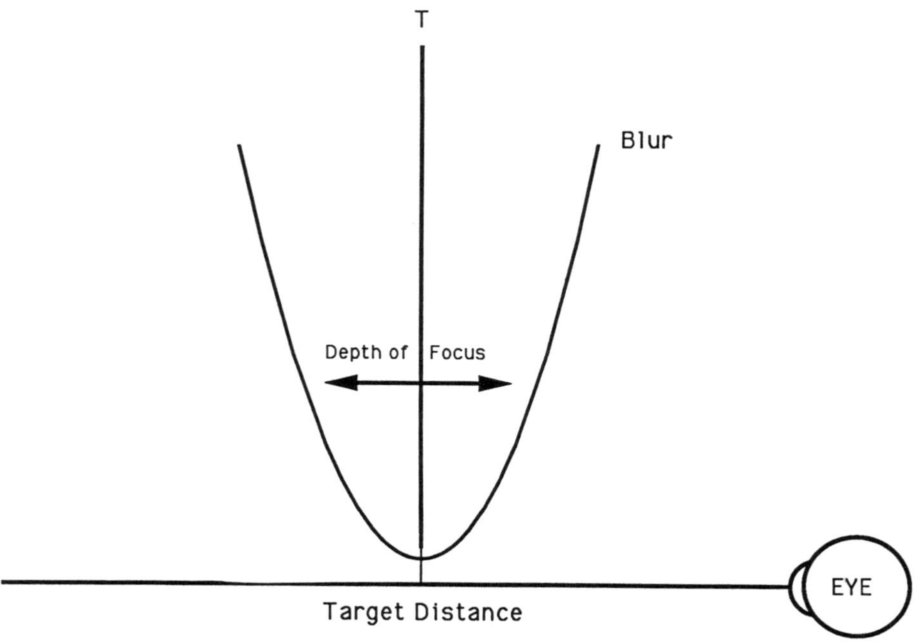

FIGURE 26-5. Magnitude of blur. As clear vision distance (plane in target world conjugate to retina) separates from target plane (T), the amount of blur increases on either side of T approximately as the parabola marked BLUR. With a finite depth of focus (as a function of pupil size and other factors such as visual task demanded of the subject and quality or contrast of the image), the clear vision distance has a permitted range of values. Ordinarily, the accommodative mechanism could rest on the lazy limb of the parabola (clear vision distance beyond target distance). With stigmatoscopic method, the point of minimum blur can be ascertained. (If the parabolic curve were in diopters blur, it would not be symmetrical on spatial distance abscissa.)

STIGMA AND BADAL

This method eliminates the pupillary depth of focus since the subject chooses the center of the depth of focus as endpoint. This point is the setting that corresponds to minimum blur. Instead of assuming infinity to be the far point, stigmatoscopy measures the far point. Again, because of the optics of the Badal lens system there is no change in angular size or brightness of the target as it is presented to the subject at different stimulus settings. There is, however, a static proximal effect caused by the arrangement of the measurement apparatus.

SUMMARY

Stigmatoscopy would not be too difficult to move from the laboratory to the clinical setting. Advantages of this might be to obtain more precise measurements of the preclinical loss of accommodation. Also, exploration of factors that might lead to amelioration of presbyopia could be put on a firm foundation. An example of this latter point is in the study shown in Fig. 26-6. Here examples of "Olympic" accommodation might lead to some structured exercising of accommodation to delay the clinical appearance of presbyopia.

Although clinical methods show presbyopia to reach clinical presence after the age of 40 years, stigmatoscopy studies clearly indicate that in terms of the accommodation contribution by itself, the development of presbyopia occurs during early adulthood. Although the clinical measurement ignores the early appearance of preclinical loss of accommodation, it is well to remember that it is, of course, the measurement that correlates with the patient's perception of the particular extent of his presbyopia, since it measures performance under real-life conditions where the pupil contributes to the amplitude of accommodation.

CONCLUSIONS

Clinical measures of the amplitude of accommodation inherently include the variable depth of focus contributed by the pupillary aperture. Stigmatoscopy, wherein the subject adjusts the optical dis-

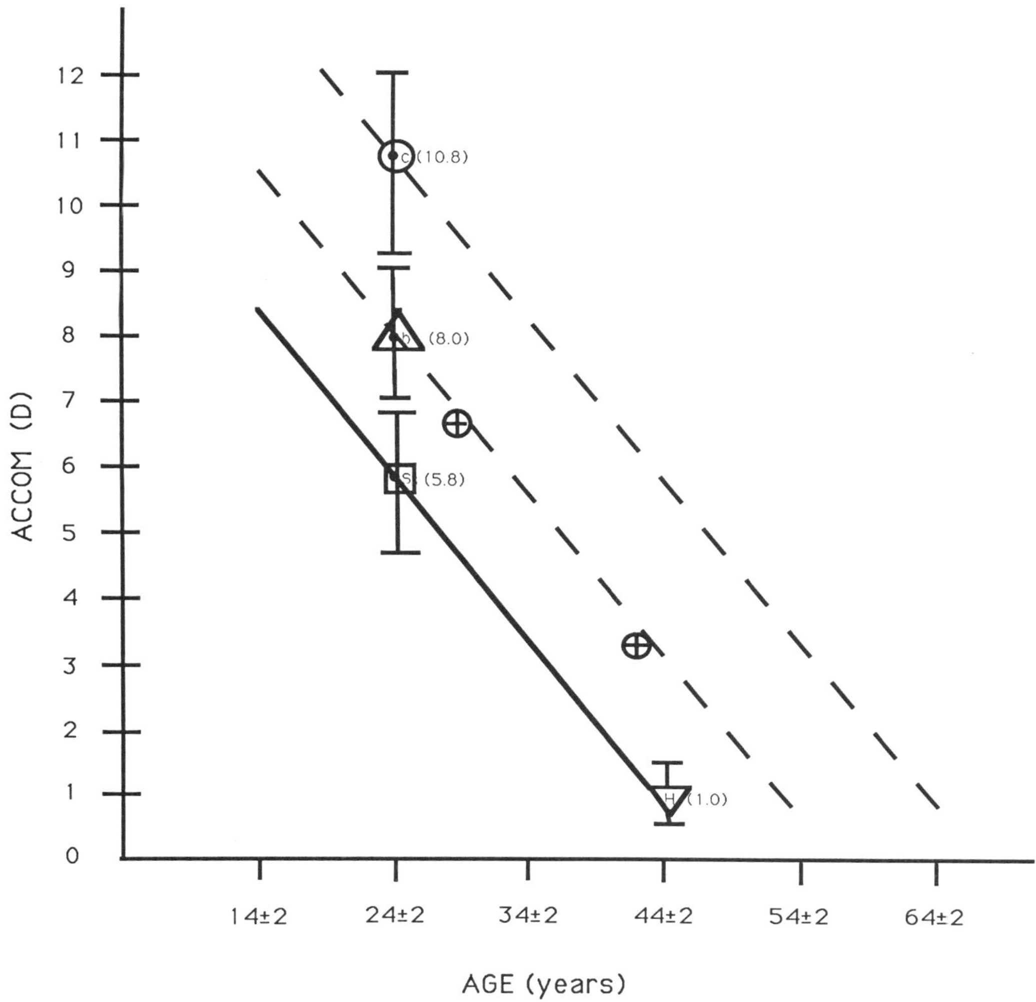

FIGURE 26-6. Olympic accommodation. Two additional subjects (circles with inscribed cross) showed much greater accommodation than expected. One was a 29-year-old moderate exophore (triangle), the other a 42-year-old intermittent high exotrope (other symbols as in Figs. 26-3 and 26-4). It was speculated that continued convergence accommodation, utilized by such subjects, might exercise the rheology of the lens and so delay the mechanical changes producing presbyopia.

tance of a point target for minimum blur, locates the plane conjugate to the retina and thus the accommodative setting at the midpoint of the depth of focus. We compare stigmatoscopy with clinical measures for a large group of cooperative subjects and relate our findings to earlier work going back to Donders (1864), who introduced quantitative methods into the study of presbyopia more than a century ago. Use of stigmatoscopy to define the progress of presbyopia documents its early course during youth and young adulthood, whereas the clinical impression, influenced by adding the depth of focus to the true accommodative amplitude, places the onset of presbyopia in late middle life.

APPENDIX: OBJECTIVE DYNAMIC METHODS OF MEASURING ACCOMMODATION

Dynamic measurements of accommodation have a history associated with infrared, direct-recording optometers. Three retinal image methods have been employed: the lensometer principle, which measures the retinal spread function of a point or line target (Miege and Denieul, 1988), the retinoscope principle, which measures direction of travel of a moving image (Safir, 1970; Kruger, 1979), and the Scheiner principle, which converts blur into prismatic movement (Campbell and Robson, 1959; Allen and Carter, 1960; Lovasik, 1983; Cornsweet, 1970). These optical methods have been improved with doubling techniques, infrared lighting and photocell techniques, and nulling methods to remove large-amplitude nonlinearity.

Another class of instruments rely on physical changes in the lens. Included here are ultrasound, x-ray imaging, general photographic techniques, and the third Purkinje method (O'Neill and Stark, 1968; Phillips and Stark, 1977). This latter tracks motion of the third Purkinje image, a reflection from the lens anterior surface and also from isoindicial surfaces interior to but close to the front surface of the lens. In spite of difficulties in calibration, an advantage of the third Purkinje image method is that a good deal of light is reflected from the lens front surface as compared to the small amount of light reflected from the retina with the three optical methods.

An indirect but important method is impedance cyclography, which estimates ciliary muscle force by measuring a decreased impedance secondary to "laking" of blood beneath the sclera and outside of the muscle as it contracts.

An exciting new method employs MRI, magnetic resonance imagery, a nondestructive technique suitable for use on normal healthy individuals (Chapter 23, this volume).

ACKNOWLEDGMENT. We want to express our appreciation for helpful discussions to our colleagues Dr. An H. Nguyen and Professor Fuchuan Sun and also for support to the Essilor/Varilux and NASA/Ames Research Center Cooperative Agreement NCC-286, (Dr. Stephen Ellis, Technical Monitor).

REFERENCES

Allen, M. J., 1956, The influence of age on the speed of accommodation, *Am. J. Opt. Arch. Am. Acad. Opt.* **33**:201–208.

Allen, M. J., and Carter, J. H., 1960, An infrared optometer to study the accommodation mechanism, *Am. J. Optom.* **37**:503–407.

Badal, D., 1876, A method yielding at one time and in a single operation measures of ocular refraction and visual acuity, *Ann. Ocul.*

Bannon, R. E., Cooley, F. H., Fisher, H. M., and Textor, R. T., 1950, The stigmatoscopy method of determining the binocular refractive status, *Am. J. Opt. Arch. Am. Acad. Opt.* **27**:371–384.

Campbell, F. W., and Robson, 1959, A high-speed infrared optometer, *J. Optom. Soc. America,* **49**:268–272.

Cline, D., Hofstetter, H. W., and Griffen, J. R., 1980, *Dictionary of Visual Science,* 3rd ed., Chilton, Radnor, PA.

Cornsweet, T. N., and Crane, H. D., 1970, Servo-controlled infrared optometer, *J. Optom. Soc. Am.* **60**:548–554.

Donders, F. C., 1864, *On the Anomalies of Accommodation and Refraction of the Eye,* New Sydenham Society, London.

Duane, A., 1912, Normal values of the accommodation at all ages, *Trans. Sect. Ophthalmol. A.M.A.* 365–391.

Grosvenor, T., 1982, *Primary Care Optometry,* The Professional Press, Chicago, pp. 190–191.

Hamasaki, D., Ong, J., and Marg, E., 1956, The amplitude of accommodation in presbyopia, *Am. J. Opt. Arch. Am. Acad. Opt.* **33**:3–13.

Hofstetter, H. W., 1944, A comparison of Duane's and Donder's tables of the amplitude of accommodation, *Am. J. Optom. Arch. Am. Acad. Optom.* **21**:345–363.

Hofstetter, H. W., 1965, A longitudinal study of amplitude changes in presbyopia, *Am. J. Opt.* **42**:3–8.

Kaufmann, J., 1894, *Die absolute und relative Akkommodations-Breite in den verschiedenen Lebensaltern,* Inaugural Dissertation, Georg-Augusts University, Gottingen.

Kruger, P. B., 1979, Infrared recording retinoscope for monitoring accommodation, *Am. J. Optom.* **56**:116–123.

Lovasik, J. V., 1983, A simple continuously recording infrared optometer, *Am. J. Optom. Physiol. Optometrics,* **60**:80–87.

Matsunaga, K., Nguyen, A., Sun, F., van de Pol, C., and Stark, L., 1988, Static accommodation response: Stigmatoscopic method to measure development of presbyopia, *Proceedings IEEE International Congress Systems, Man Cybernetics,* IEEE, Bejing.

Miege, C., and Denieul, P., 1988, Mean response and oscillations of accommodation for various stimulus vergences in relation to accommodation feedback control, *Ophthal. Physiol. Opt.* **8**:165–171.

Morgan, M. W., 1968, Accommodation and vergence, *Am. J. Opt.* **45**:417–454.

O'Neill, W. D., and Stark, L., 1968, Triple function ocular monitor, *J. Optom. Soc. Am.* **58**:570–573.

Phillips, S., and Stark, L., 1977, Blur: A sufficient accommodative stimulus, *Doc. Ophthalmol.* **43**:65–89.

Saladin, J. J., and Stark, L., 1975, Presbyopia: New evidence from impedance cyclography supporting the Hess–Gullstrand theory, *Vis. Res.* **15**:537–541.

Saladin, J. J., Usui, S., and Stark, L., 1974, Impedance cyclography as an indicator of ciliary muscle contraction, *Am. J. Opt.* **51**:613–625.

Stark, L., 1988, Presbyopia in light of accommodation, *Am. J. Opt. Phys. Opt.* **65**:407–416.

Sun, F., Stark, L., Wong, J., Nguyen, A., Lakshminarayanan, V., and Mueller, E., 1988, Changes in accommodation with age: Static and dynamic, *Am. J. Opt. Phys. Opt.* **65**:492–498.

Swegmark, G., 1969, Studies with impedance cyclography on human ocular accommodation at different ages, *Acta Ophthalmol.* **47**:1186–1206.

27

Ocular Dominance, Accommodation, and the Interocular Suppression of Blur in Monovision

CLIFTON SCHOR and PAUL ERICKSON

INTRODUCTION

Traditionally, large amounts of anisometropia are considered an obstacle for normal binocular vision, and clinicians strive to equate the visual acuity in the two eyes with optimal refractive correction. Anisometropia could depress binocular measures of visual acuity if the blurred image of the ametropic eye reduced the contrast of the eye with less ametropia. Anisometropic blur also produces a marked elevation of the stereoscopic depth threshold, which is considerably greater than the threshold elevation produced by equal blurring of the two ocular images (Lit, 1968; Peters, 1969; Ong and Burley, 1972; Levy and Glick, 1974; Goodwin and Romano, 1985; Lovasik and Szymkiw, 1985; Schor and Heckmann, 1989). Finally, anisometropia presents an ambiguous stimulus to accommodation whenever the test target is proximal to the unequal far points of the two eyes.

In spite of these contraindications, one form of contact lens prescription for presbyopes intentionally introduces anisometropia. This correction, which is referred to as monovision, prescribes one eye with a single-vision far point correction and the other eye with a correction for the near working distance. This prescription has the advantage of the absence of a division between bifocal zones, which, when present in alternating-vision bifocal contact lens corrections, must move synchronously in the two eyes in order to maintain the binocular registration of corresponding optical zones in the two eyes. When present in simultaneous-vision bifocal contact lens correction, the retinal image produced in both eyes is degraded. However, the disadvantages described above for anisometropia do apply to the monovision correction. There are the potential reductions of visual acuity, stereo acuity, and accommodative accuracy in prepresbyopes. Surprisingly, these potential problems either do not materialize or are partially overcome by a unique form of interocular suppression that occurs regionally between corresponding retinal points (Schor *et al.*, 1987).

Spatial Characteristics of Interocular Suppression of Blur

Figure 27-1 is a stereogram that illustrates the regional suppression of anisometropic blur. The two halves of the stereo pair contain a complementary mixture of blurred and in-focus optotypes. Under monocular conditions, each stereo half-image is composed of both blurred and in-focus images. However, under binocular viewing of the stereo pair, neither eye perceives a blurred image because the blurred components of the images are suppressed by the clear ones. It is readily apparent that the entire field of one eye is not suppressed but rather that there is interocular suppression between corresponding regions of the two retinal images. Fusion of anisometropic blur is possible because both clear and blurred images contain common low-

CLIFTON SCHOR AND PAUL ERICKSON • School of Optometry, University of California, Berkeley, California 94720.

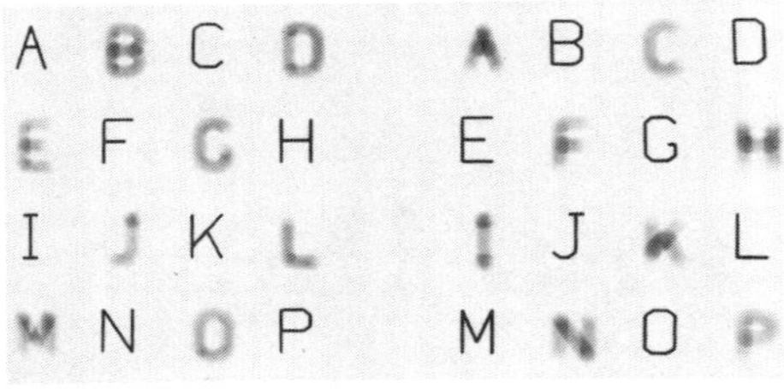

FIGURE 27-1. The upper pair of letter arrays from a stereogram containing a complementary mixture of clear and blurred optotypes. Below is the appearance of the fused stereogram. Blurred images are suppressed by an interocular process.

frequency detail even though the blurred image lacks high-frequency detail. This interocular suppression of blur was first described quantitatively by Fahle (1982), who observed while fusing a blurred and in-focus grating of the same orientation that during fusion the in-focus image was visible for longer periods of time than was the blurred image. The following experiments investigate the effect of target contrast and size on this interocular suppression of anisometropic blur.

Contrast Thresholds for Anisometropic Blur

Target contrast in naturally occurring scenes differs markedly from photopic to scotopic viewing conditions. Under photopic conditions, there are both bright (increment) and dark (decrement) contours, whereas in scotopic conditions only light contours are visible. The physical contrast between bright objects and their background is also much greater under scotopic than photopic conditions. In addition, the perceived size of bright, high-contrast objects on a dark background is exaggerated in comparison to the reverse contrast condition (dark on bright). This perceived increase in size of bright objects was described by Helmholtz (1962) as irradiance. He noted that the perceived size of blur circles would also be increased by this perceptual phenomenon and that the blur would also appear proportionally greater with small than with large bright targets. Finally, he observed that imperfections in the eyes' optics would produced irregularities in the blur circles that would be more noticeable with small than large targets.

Monovision patients report that anisometropic blur is suppressed effectively under photopic viewing conditions but that anisometropic blur is not suppressed binocularly under mesopic and scotopic luminance conditions such as while driving a car at night (McLendon *et al.*, 1968). In addition, the binocular contrast sensitivity function tested with a monovision correction is at least as good as the monocular contrast sensitivity function for the eye receiving a clear image (Loshin *et al.*, 1982), presumably as a result of interocular suppression. Based on these and our own pilot observations, we investigated the effect of target contrast and size on the interocular suppression of blur produced by anisometropia. We also compared the magnitude of blur suppression of the left and right eye after wearing a monovision correction for a 1-day period and observed a form of eye dominance for interocular suppression of blur that emerged during the initial adaptation response to a monovision correction.

METHODS

Contrast Increment Threshold for Dichoptic Blur

A luminance increment test spot was produced by back-illuminating a circular aperture ranging in diameter from 0.75 mm to 6 mm with a 60-watt incandescent bulb. The apertures were drilled in a thin 6 × 6 inch white card. The luminance of the background surrounding the aperture was controlled by front illumination with a variable source to produce changes in the increment contrast of the test spot (Fig. 27-2). The luminance of the test spot was kept constant at 620 cd/m^2 while the surround luminance varied from 0.5 cd/m^2 to 620 cd/m^2. Subjects viewed the front surface of the card at a distance of 1 m while wearing a +1.00 collimating lens over both eyes to eliminate accommodation and with an additional near lens (+1.25 D, +1.75 D, or +2.50 D) worn before one eye to produce anisometropic blur of that eye. Subjects increased the luminance of the background with a rheostat control until the clear image seen by the eye with only the +1.00 lens began to suppress the blurred image perceived by the other eye. Five settings were made for combinations of each aperture size of the test spot with the amount

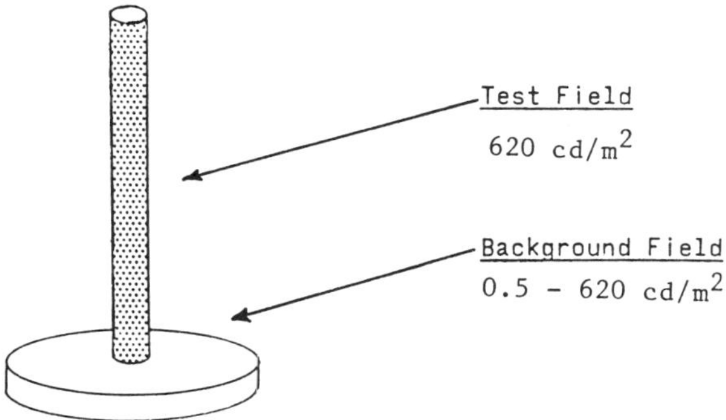

FIGURE 27-2. A small spot of fixed luminance (620 cd/m^2) and variable diameter (0.75 to 6 mm) is centered on a background of variable luminance (0.5 to 620 cd/m^2).

of additional positive lens worn before the left and right eye. Test spot size was varied randomly for a given amount of anisometropic blur. Once all test targets were presented, the amount of anisometropia was changed, and the spots were presented, the amount of anisometropia was changed, and the spots were presented again in another randomized order. The sequence of presenting the three anisometropic additions was also randomized. The examiner preset the rheostat knob to an arbitrary low luminance between each threshold setting by the subject. Thresholds were calculated as a Michaelson contrast of test increment over surround luminance using the formula

$$(\text{test} - \text{background})/\text{background} \text{ or } (\text{test}/\text{background}) - 1$$

Subjects were instructed to increase background luminance until the clear image first became intermittently visible. At lower luminance levels the blurred image was always visible. At the 1-m test distance, test spots subtended visual angles of 3′, 6′, 9′, 12′, and 15′ of arc. These angles approximate the angles subtended by car headlamps viewed in the opposite lane at distances ranging from 500 to 100 ft.

Before beginning the experiment, subjects viewed the spot on a dark background, first with the in-focus eye and then with the out-of-focus eye, to demonstrate the appearance of the stimulus in its two possible states. Then the subject viewed the same target binocularly with the anisometropic lens addition in order to demonstrate the unsuppressed blurred image on the dark background. While viewing the blurred test spot monocularly, subjects were also shown that it always appeared blurred over the whole range of background luminance levels. After this, the subjects practiced three times, wearing the anisometropic lens addition, raising the background luminance until the binocularly viewed test spot appeared clear.

Adaptation of Suppression to Anisometropic Blue

Adaptability to monocular blur was investigated by having two subjects, who were not monovision patients, wear a +2.00-D lens in addition to their refractive correction over the left eye and in a second test over the right eye for a period of 1 day. Contrast thresholds for blur suppression were measured both before and after adaptation while subjects were tested with a +1.75-D lens worn monocularly over the adapted or nonadapted eye.

RESULTS

Preadapted Interocular Suppression of Blur

The maximum contrast at which interocular suppression of blur occurred is plotted for the two spectacles wearers in Figs. 27-3 and 27-4 as a function of test spot diameter. The magnitude of interocular suppression of blur is interpreted as being proportional to the maximum luminance contrast at which the in-focus image was perceived. Thus, strength of suppression increased with the height of the plotted data. The blurred image was always seen at contrasts greater than this upper limit. Each point represents the mean of five thresholds, and the bar represents the standard deviation about the mean. Separate plots are shown for data taken with the blurring test lens worn over the right and left eye. Each figure illustrates curves that represent results with 1.25-D, 1.75-D, and 2.5-D anisometropic blurs. Comparison of contrast curves for different lens additions illustrates that interocular suppression of blur was greater for small than large anisometropic lens additions. Contrast thresholds for detecting blur were at least an order of magnitude greater with the 1.25-D lens addition than with the 2.50-D lens addition. In all cases, the interocular suppression of blur increased dramatically with test spot size. Over the range of spot diameters, the thresholds for both subjects increased by as much as three orders of magnitude. In this preadapted state, the contrast thresholds for detecting blur were equal

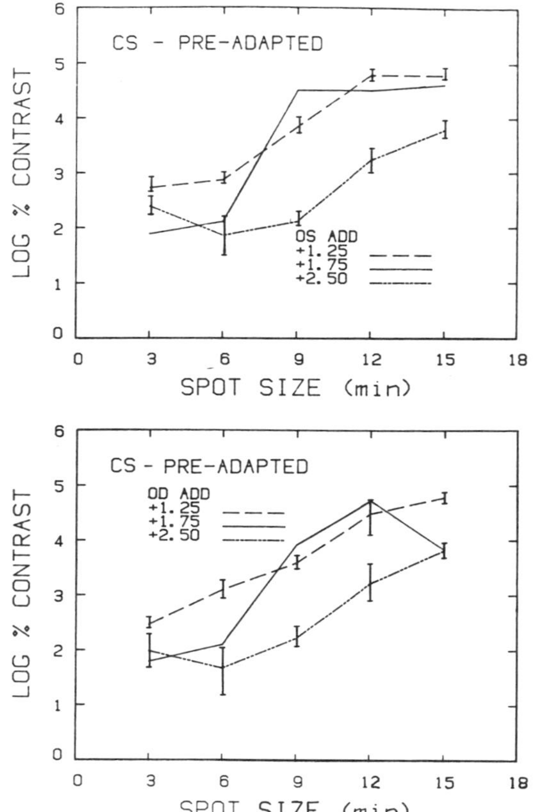

FIGURE 27-3. Before adaptation, the threshold contrast at which anisometropic blur is suppressed is plotted for a presbyopic observer as a function of test spot diameter. Suppression of anisometropic blur of the left and right eye (1.25, 1.75, and 2.50 D) is shown in the upper and lower halves of the figure, respectively. Standard deviations about the mean are shown for each datum point.

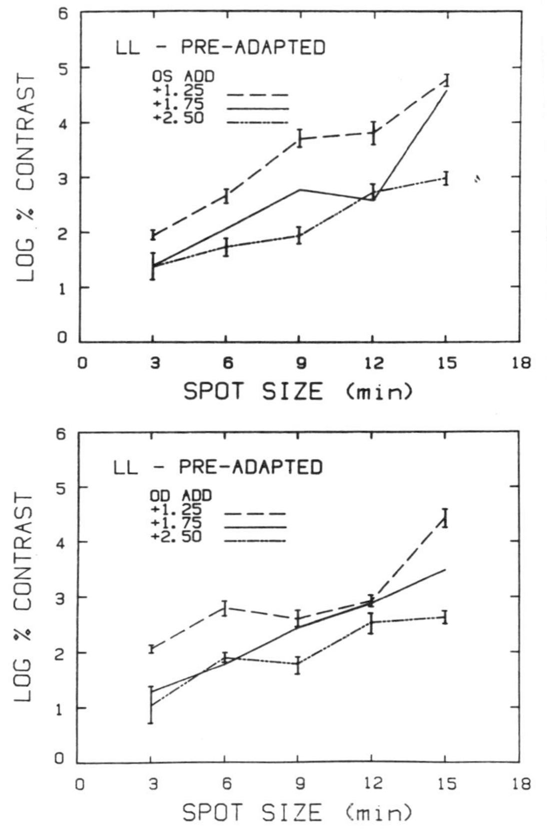

FIGURE 27-4. Before adaptation, the threshold contrast at which anisometropic blur is suppressed is plotted for a prepresbyopic observer as a function of test spot diameter. Suppression of anisometropic blur of the left and right eye (1.25, 1.75, and 2.50 D) is shown in the upper and lower halves of the figure, respectively.

when test lens additions were worn before the left or right eye.

Adaptation of Interocular Suppression to Anisometropic Blur

An anisometropic refractive error was produced for the same two subjects by placing a +2.00-D lens before their non-preferred (left) sighting eye. After the subjects adapted to the anisometropia for 1 day, the lens was removed, and interocular suppression of blur was investigated with the same procedure as used in the previous experiment. A +1.75-D lens was first placed before the left eye that had worn the +2.00-D lens during adaptation, and the contrast threshold for detecting anisometropic blur was measured as a function of spot size. The experiment was then repeated with the +1.75-D lens placed before the contralateral (right) eye. Two weeks later the adaptation procedure was repeated for 1 day in the same subjects with the +2.00-D lens worn over the preferred (right) sighting eye.

Figure 27-5 illustrates the contrast thresholds for detection of anisometropic blur produced with the +1.75-D lens over the right (OD) and left (OS) eyes following 1 day of adaptation to a +2.00-D lens placed over the subject's left eye. Contrast thresholds are plotted in log numbers. Log numbers ranging from 1 to 5 correspond to background luminances of 563, 320, 56, 6, and 0.6 cd/m^2, respectively. These backgrounds produce a range of contrasts with the 620 cd/m^2 test spot extending from 10% to 100,000%, respectively.

Both subjects had higher contrast thresholds (greater interocular suppression of blur) when the blurring test lens was placed over the left than over the right eye. This difference between the two eyes is in marked contrast to the preadapted data in which

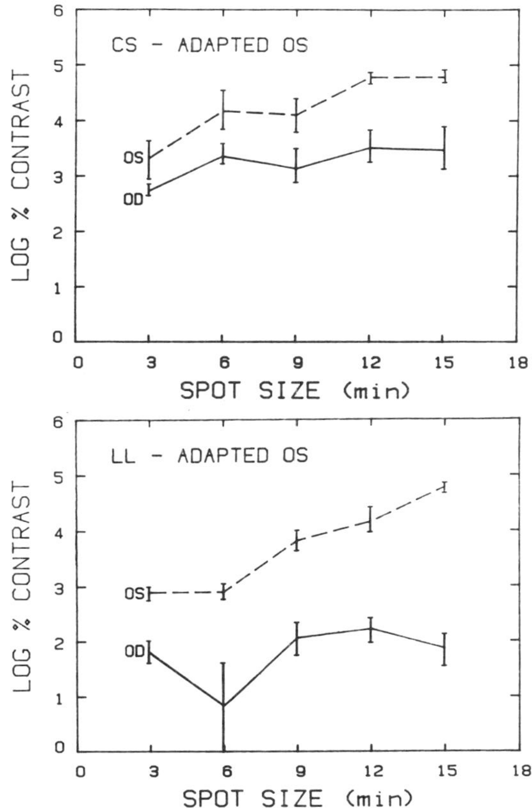

FIGURE 27-5. After adaptation for 1 day to a +2.00-D lens placed before the left eye. The figure illustrates contrast thresholds at which anisometropic blur of +1.75 D of the left (OS) and right (OD) eye is suppressed. Thresholds are plotted for a presbyopic observer (top) and a prepresbyopic observer (bottom) as a function of test spot diameter.

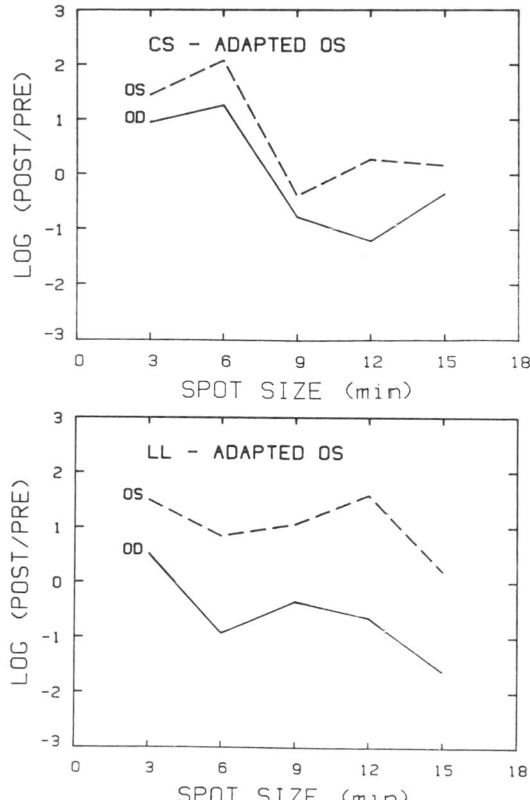

FIGURE 27-6. The change in contrast threshold from the preadapted state after adaptation for 1 day to a +2.00-D lens before the left eye is plotted as a ratio of post-/preadapted contrast threshold as a function of spot size for a presbyopic observer (top) and a prepresbyopic observer (bottom). Clearly, adaptation resulted in an increased suppression by the right eye of small blurred spots seen by the left eye as well as decreased suppression by the left eye of large blurred spots seen by the right eye.

the contrast thresholds for suppressing blur of the right and left eyes were equal. This change in ocular dominance following adaptation is quantified in Fig. 27-6. Here the ratio of post-/preadapted contrast threshold is plotted as a function of test spot diameter on a log-linear scale. Increases and decreases in interocular suppression following adaptation are shown by data plotted above and below the zero level, respectively. The figure clearly illustrates an increase of interocular suppression by the right eye of blurred small spots seen by the left eye and a decrease of interocular suppression by the left eye of blurred large spots seen by the right eye. These figures demonstrate marked changes in ocular dominance for interocular suppression of nearly two orders of magnitude.

Figure 27-7 illustrates contrast thresholds for detecting anisometropic blur produced with a +1.75-D lens over the right and left eye following adaptation for 1 day to a +2.00-D lens placed over the subject's right eye. The results for the presbyopic spectacle wearer C.S. (Fig. 27-7, top) were similar to those following adaptation to the +2.00-D lens worn over the his left eye (Fig. 27-5, top). That is, contrast thresholds for interocular suppression of blur were higher when the test blurring lens (+1.75) was placed over the left than over the right eye. The ratios of post-/preadapted contrast thresholds shown in Fig. 27-8 (top) confirm that interocular suppression was increased from the preadapted level for both eyes when tested with small spot sizes and that the interocular suppression was approximately the same as preadapted levels for both eyes when tested with the large spot sizes. Interestingly, interocular suppression of blur was greater for the right than the left eye whether the

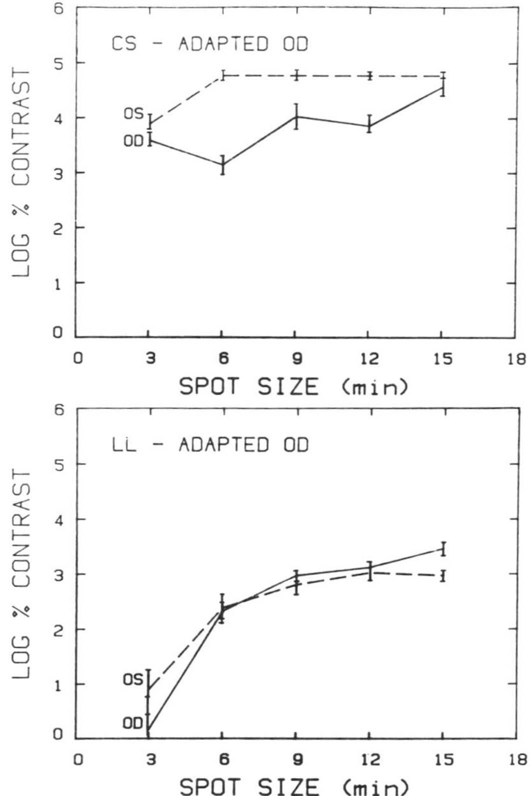

FIGURE 27-7. After adaptation for 1 day to a + 2.00-D lens placed before the right eye, the upper figure illustrates contrast thresholds at which anisometropic blur of +1.75 D of the left (OS) and right (OD) eye is suppressed. Thresholds are plotted for a presbyopic observer (top) and a prepresbyopic observer (bottom) as a function of test spot diameter.

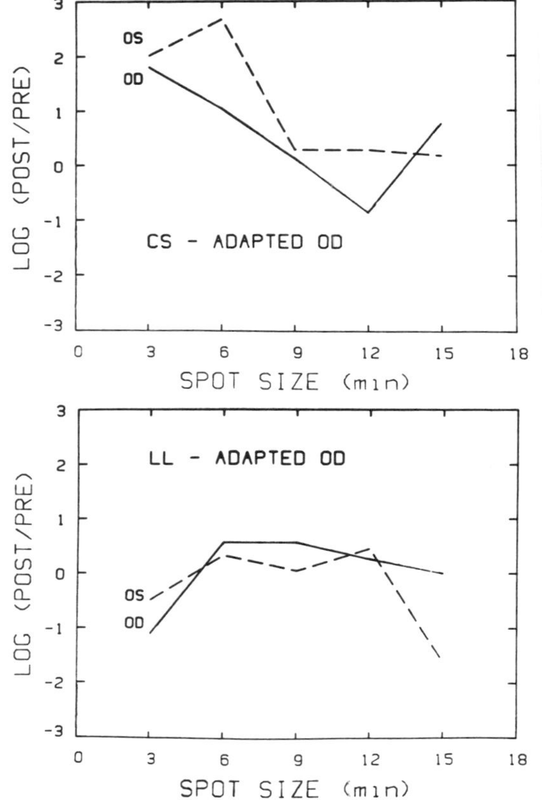

FIGURE 27-8. The change in contrast threshold from the preadapted state after adaptation for 1 day to a +2.00-D lens before the right eye is plotted as a ratio of post-/preadapted contrast threshold as a function of spot size. In the presbyopic observer (top), adaptation resulted in an increased suppression by the right eye of small blurred spots seen by the left eye, as well as decreased suppression by the left eye of large blurred spots seen by the right eye. For the prepresbyopic observer (bottom), interocular suppression of blur was unchanged from the preadapted state, unlike the effects shown in Fig. 27-5 (bottom).

adapting lens was worn before the right or left eye (Figs. 27-5, top, and 27-7, top).

The postadaptation results for the prepresbyopic spectacle lens wearer (L.L.) were very different when the +2.00-D adapting lens was worn over the dominant (right) sighting eye. Postadapted contrast thresholds for detection of anisometropic blur were similar when the test blurring lens (+1.75 D) was placed before the left or right eye (Fig. 27-7 bottom). These contrast thresholds differed little from the preadapted thresholds, as shown by the ratios of post/pre contrast thresholds for blur detection, plotted in the bottom half of Fig. 27-8. Interestingly, postadaptation interocular suppression may have been reduced for both eyes when tested with small spot targets and for the right eye when tested with a large blurred test spot before the left eye. Clearly there was a marked difference in this subject's adaptation response to a +2.00-D lens worn before the right eye and the left eye. That these adaptation aftereffects are not simply a result of practice with the test condition rather than the adaptation condition was demonstrated by the restoration of the initial or baseline ocular dominance pattern for interocular suppression anisometropic blur after subjects C.S. and L.L. had readapted to equal clarity of the two ocular images. Furthermore, the reliability of these aftereffects, as well as the effects of target size and degree of anisometropia, are demonstrated by the clear separation of the standard deviations plotted about the means of thresholds for these various test conditions.

Accommodation Responses to Anisometropic Blur

For monovision adds less than 2 D, there is typically a range of target distances through which small amounts of accommodation can produce a clear image in either eye (Erickson, 1988). Thus, in early presbyopia, clear vision with monovision requires successful coordination of accommodation and blur suppression. Ideally, the monovision wearer should see clearly throughout the monocular range of clear vision of both eyes. That is, the binocular clear range should be continuous and equal to the sum of the monocular clear ranges. Each monovision range of clear vision depends on the amplitude of accommodation and a complex interaction of factors in the visual system (Campbell, 1957; Ogle and Schwartz, 1959; Campbell and Gregory, 1960; Woodhouse, 1975) and the target (Crawford, 1936; Campbell, 1957; Ogle and Schwartz, 1959; Tucker and Charman, 1975; Atchison *et al.*, 1979) that determine the depth of focus. To determine the effectiveness of the visual system in coordinating these functions, we (Schor and Erickson, 1988) studied dynamic patterns of accommodation and blur suppression in subjects wearing monovision lenses. The tests simulated changing target distances both with accommodation-controlled and allowed to act freely.

METHODS

Binocular Depth of Focus

The depth of focus was measured during binocular viewing conditions using the lenticular-septum stereoscope (Brewster, 1856) illustrated schematically in Fig. 27-9. The stereoscope consisted of two +2.00-D lenses placed in the spectacle plane before each eye in rearrangement of a simple optometer and an additional +1.5D monovisions before the left "near" eye. Each eye view two independent targets through 3-mm artificial pupils. The depth of focus for this pupil size is typically ±0.3 D for high-resolution targets (Campbell, 1957; Ogle and Schwartz, 1959).

The variation in image vergence that occurs without noticeable blur is classically defined as the depth of focus. This definition implies that the optical system focusing the image remains static. In the case of the eye, this implies that the state of accommodation is held constant. This can be ac-

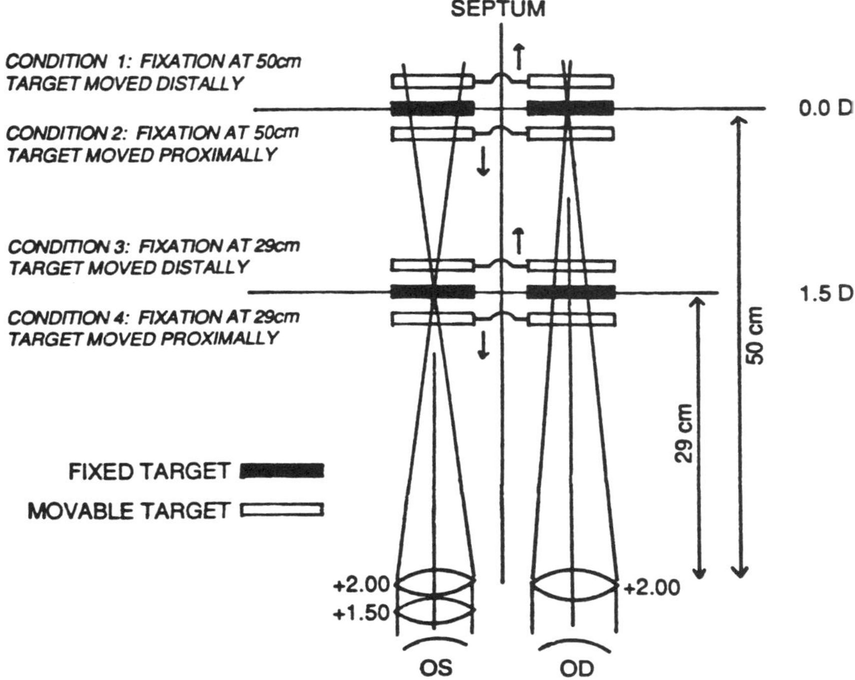

FIGURE 27-9. Schematic overhead view of the lenticular septum stereoscope used to measure depth of focus. Distance ranges were measured relative to the 50-cm target distance; near ranges relative to the 29-cm target.

complished rigorously with cycloplegia. Since our goal was to study suppression under natural conditions, we chose to control accommodation subjectively. Thus, small amounts of accommodation could conceivably occur without causing noticeable blur in the fixed target.

Each eye's target was shielded from the other eye by a septum. Fusion targets were transparent slides that consisted of three spots separated horizontally by 1.5° (Fig. 27-10). The central fixation spot was surround by two fused vertical lines, one above and one below, presented to both eyes. As illustrated in Fig. 27-10, the upper vertical line was presented at either 50 cm or 29 cm from the eyes. These fixed distances corresponded to dioptric vergences of zero and −1.45 D, thus placing a target at a point conjugate to the far point of one of the two eyes wearing the anisometropic correction. The lower set of fused vertical lines was on another set of slides that were on a movable trolley. This second set of slides could be moved together either proximally or distally from the distance of the set of stationary slides. The opaque lines and spots appeared dark on the transparent background luminance of 35 cd/m^2. The separation between the two eyes' targets was set to equal the interpupillary distance. The eyes were aligned parallel to one another with prism to avoid parallax artifacts during target motion. Because the eyes' visual axes were parallel, targets remained superimposed and fused over the entire range of movable and fixed target separations.

The depth of focus in the presence of 1.5-D anisometropia was determined at each of the two

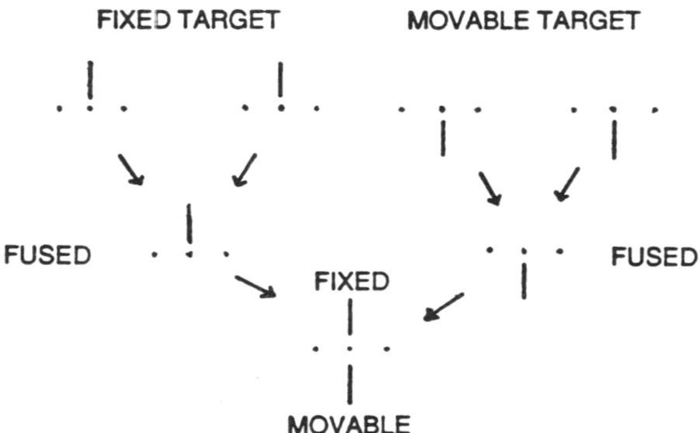

FIGURE 27-10. Target viewed by subjects during measurements of depth of focus. The right-hand components of each "fixed" and "movable" target (upper row) were seen by the right, distance-corrected eye. The left-hand components were seen by the left, near-corrected eye. These components were respectively fused (middle row) to form fixed (upper) and movable (lower) vertical lines of the fused target (bottom row).

focal distances in question. When the three dots that were common to the two sets of transparent slides were fused and viewed through the 1.5-D anisometropia correction, the fused image appeared clear in either the far (50 cm) or near (29 cm) fixed test distance. This clear vision response was the result of interocular suppression of the monocular blurred image (Schor and Erickson, 1988) of either the far eye when fixating at the near distance or of the near eye when fixating at the far target distance. Subjects were instructed to fixate and fuse the upper stationary vertical line while simultaneously moving the lower fused vertical line, first proximally and then distally, until blur was detected in the lower fused image. The monocular depth of focus of each eye was determined with the same procedure and apparatus with one eye occluded. Monocular depths of focus were determined about the 50-cm setting for the right "far" eye and about the 29-cm setting for the left "near" eye. Each subject was given three practice trials to become familiar with the apparatus. Three measures were taken and averaged for each of the eight test conditions (near and far limits for right eye, left eye, and both eyes fixating at near and far distances).

Table 27-1 describes the characteristics of the six subjects who participated in the study (Study s). Ages ranged from 27 to 50 years. Ocular dominance was determined with a sighting task in which subjects viewed a distant target through an approximately circular opening formed by the fingers of the outstretched hands. The task was repeated five times. Dominance was classified as right or left if at least four of the trials resulted in sighting with the same eye. Amplitudes of accommodation were measured by Donder's pushup method. All subjects wore full refractive corrections during the experiment, and none were experienced monovision wearers.

Accommodative Response to Anisometropic Blur

Dynamic responses of accommodation were measured from the right eye with an SRI objective infrared optometer based on the Scheiner principle (Randle, 1970) operating from a retinally reflected infrared beam. The instrument is sensitive to dioptric fluctuations of 0.1 D and has a frequency response of 400 Hz. During measures of accommodation, pupils were dilated with two drops of 2.5% phenylephrine hydrochloride spaced apart by 5 min. This provided a 6-mm natural pupil for 1 hr.

TABLE 27–1
Profiles of Subjects

Subject	Age (year)	Amplitude of accommodation (D)	Sighting preference	Study participation
C.K.	27	10	Right	s[a]
T.C.	41	5	Right	s
T.K.	30	8	None	s
J.F.	50	2	Right	s
G.A.	39	4.5	Right	s
C.S.	43	4	None	s,a[b]
P.E.	39	6	Right	a
J.G.	52	2	Right	a

[a]s, binocular suppression study.
[b]a, accommodation study.

This dosage has been shown not to alter either the resting focus or the amplitude of the accommodative response (Baker *et al.*, 1983). A virtual 3-mm pupil was imaged in the natural pupil to provide the same depth of focus as used on the prior experiment. The target consisted of a maltese cross, which presents a broad range of spatial frequencies that did not change with target magnification.

The target was viewed binocularly through an anisometropic lens addition of 1.5 D worn before the left eye. After refractive errors were corrected and the anisometropic add was introduced, accommodation was stimulated dynamically with the SRI 3-D visual stimulators, which vary the stimulus to accommodation with a telecentric Badal-stimulus optometer (Fry, 1969). Dioptric changes were produced under computer control by a servocontrolled movable lens that adjusted the image distance of the accommodation stimulus before the Badal system.

Accommodation was stimulated with continuous sinusoidal and square-wave changes in dioptric vergence of 2.5 D at 0.2 Hz. The dioptric range of the stimulus was from the far point of the far eye (zero diopters) to 2.5 D. Accommodative responses to these stimuli were recorded from the right eye, irrespective of which eye viewed the target. The optometer was calibrated with a schematic eye having an adjustable axial length. Accommodation was stimulated monocularly for each eye and then binocularly. Optometer alignment was continuously adjusted to avoid eye position artifacts resulting from accommodative vergence of the right eye during monocular stimulation of the left eye. Subjects were instructed to stare at the center of the cross and to keep the target clear at all times. Accommodative responses of the right eye were recorded on a strip chart recorder and were analyzed in terms of their amplitude based on the calibration with the schematic eye. Characteristics of the three subjects who participated in this study are described in Table 27-1 (Study a). Ages ranged from 39 to 52 years. Ocular dominance and amplitude of accommodation were determined as described earlier. All subjects in both experiments had at least 40″ stereo acuity as measured with the Wirt ring test.

RESULTS

Binocular Depth of Focus

Three general patterns of response were seen in the measurement of depth of focus. These were alternating dominance (binocular summation at both near and far fixation distances), uniocular dominance (binocular summation at only one fixation distance), and partial dominance (incomplete binocular summation at either or both fixation distances). Depths of focus are presented in Table 27-2 and illustrated in three sets of histograms (Figs. 27-11 through 27-13). The left half of each figure illustrates the depth of focus measured monocularly for the right (far) eye and the left (near) eye. The right halves of the figures depict the binocular depth of focus measured during far and near viewing of fixation conditions (solid and dashed lines, respectively).

Individual dominance patterns and their corresponding sighting preferences and accommodative amplitudes are summarized in Table 27-3.

ALTERNATING DOMINANCE

Figure 27-11 illustrates results for two subjects who had complete binocular summation of their monocu-

TABLE 27–2
Depth of Focus in Centimeters (Diopters)

		Viewing conditions[b]			
Subject	Direction[a]	50 cm OD only	50 cm binocular	29 cm OS only	29 cm binocular
C.K.	P	24 (2.11)	20 (3.00)	16 (2.75)	15 (3.17)
	D	76(−0.68)	81(−0.77)	36(−0.72)	44(−1.25)
T.C.	P	22 (2.5)	19 (3.25)	14 (3.64)	14 (3.64)
	D	60(−0.33)	66(−0.60)	40(−1.00)	49(−1.46)
T.K.	P	28 (1.58)	14 (4.85)	15 (3.17)	15 (3.17)
	D	80(−0.75)	83(−0.8)	33(−0.47)	60(−1.83)
J.F.	P	31 (1.50)	30 (1.3)	16 (2.70)	15 (3.17)
	D	55(−0.25)	56(−0.25)	35(−0.64)	50(−1.50)
G.A.	P	31 (1.50)	30 (1.3)	13 (4.2)	13 (4.2)
	D	60(−0.33)	60(−0.33)	36(−0.72)	67(−2)
C.S.	P	25 (2.5)	15 (5.17)	16 (3.25)	15 (3.66)
	D	75(−0.15)	80(−0.25)	35(−0.14)	43(−0.68)

[a]P, proximal; D, distal.
[b]+2.00 D before OD, +3.5 D before OS except subject C.S. (+1.5 OD, +3.0 D OS). Values shown are relative to the 50cm or 29cm target position.

lar clear vision ranges when tested at either the near or far fixation distance. In these cases, each subject's depth of focus about a binocularly fixated target equaled the sum of the individual depths of focus for the two monocular measures.

The ability to achieve such summation seems to depend on the facility with which interocular suppression can be alternated. For those subjects whose depths of field summed under binocular conditions, independent of which eye was initially in focus, no

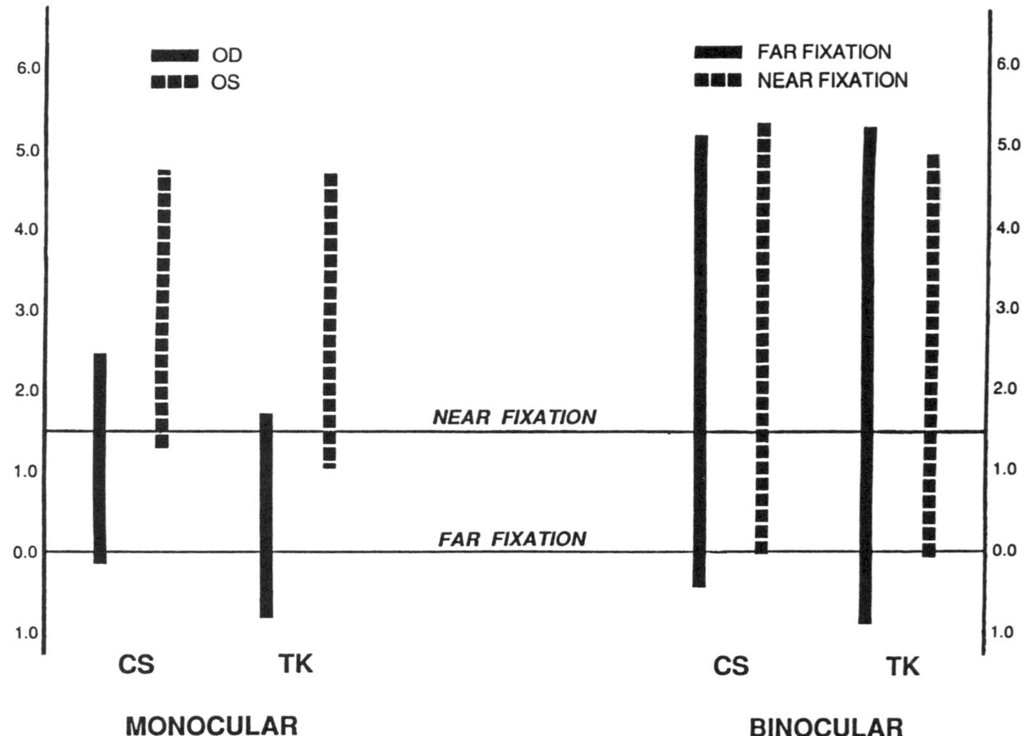

FIGURE 27-11. Monocular (left half) and binocular (right half) depths of focus for two subjects who showed an alternating dominance pattern of blur suppression. The vertical scale is in diopters.

TABLE 27–3
Performance Patterns

Suppression pattern	Subject	Sighting preference	Amplitude of accommodation (D)
Uniocular			
Right	J.F.	Right	2.0
Right	G.A.	Right	4.5
Partial			
Left	T.C.	Right	5.0
Right	C.K.	Right	10.0
Alternating	T.K.	None	4.0
	C.S.	None	8.0

clear sighting preference could be established (Table 27-3). These subjects alternated their monocular sighting preference from one eye to the other on successive sighting attempts.

UNIOCULAR DOMINANCE

Figure 27-12 illustrates results for two subjects who only manifested summation of the monocular depths of focus when tested at the near fixation distance. Uniocular dominance was manifested by these subjects when initial fixation under binocular anisometropic conditions was held in the clear zone for the dominant eyes. Initially the subject saw a clear movable target, which became blurred as the target moved out of the dominant eye's depth of focus.

Consider, for example, subject G.A. with the target fixated at 50 cm, which was conjugate to the retina of the right eye. The fixed target continued to be fixated as the movable target was brought closer to the subject. At 1.50 D relative to the starting point, the binocular image became blurred despite the fact that the movable target was within the clear zone of the left eye. This suggests an inability to switch ocular dominance from one eye to the other to maintain clarity.

The situation was then reversed, and initial fixation began at 29 cm, where the left eye's image was clear. Suppression of the blurred image was changed easily from the near eye to the far eye, and the image remained clear as it moved into the right (far) eye's depth of focus. This suggests that G.A. exhibited dominance of the right eye. A similar situation existed for subject J.F. Both subjects exhibited strong sighting preference for the right eye (Table 27-1).

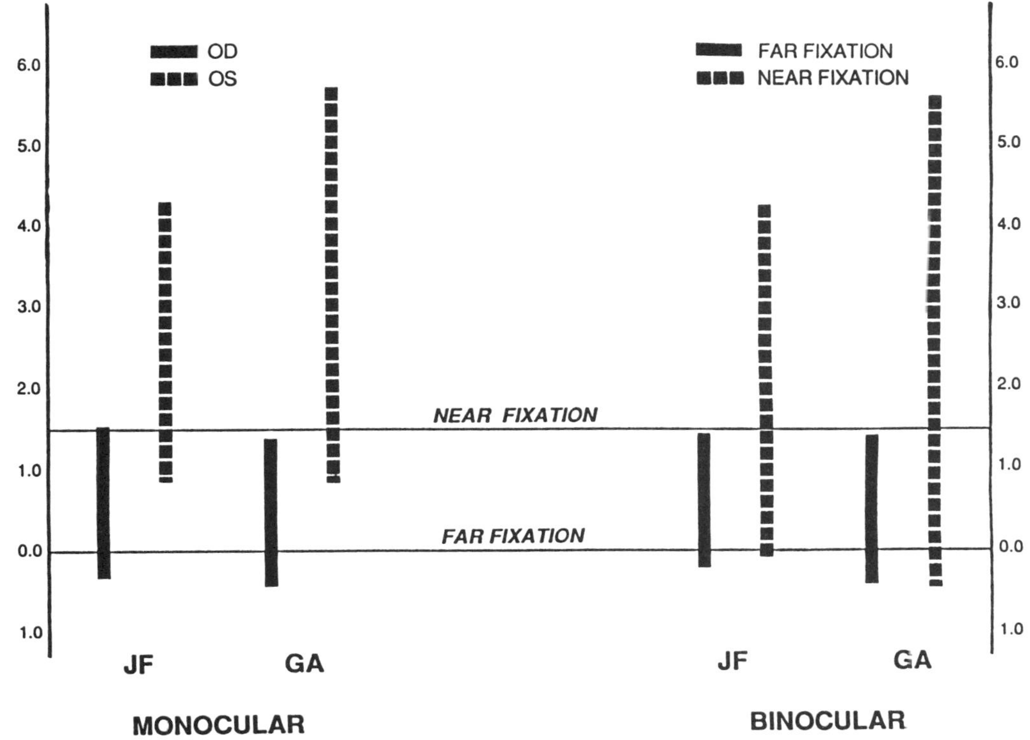

FIGURE 27-12. Monocular (left half) and binocular (right half) depths of focus for two subjects who showed a uniocular dominance pattern of blur suppression. The vertical scale is in diopters.

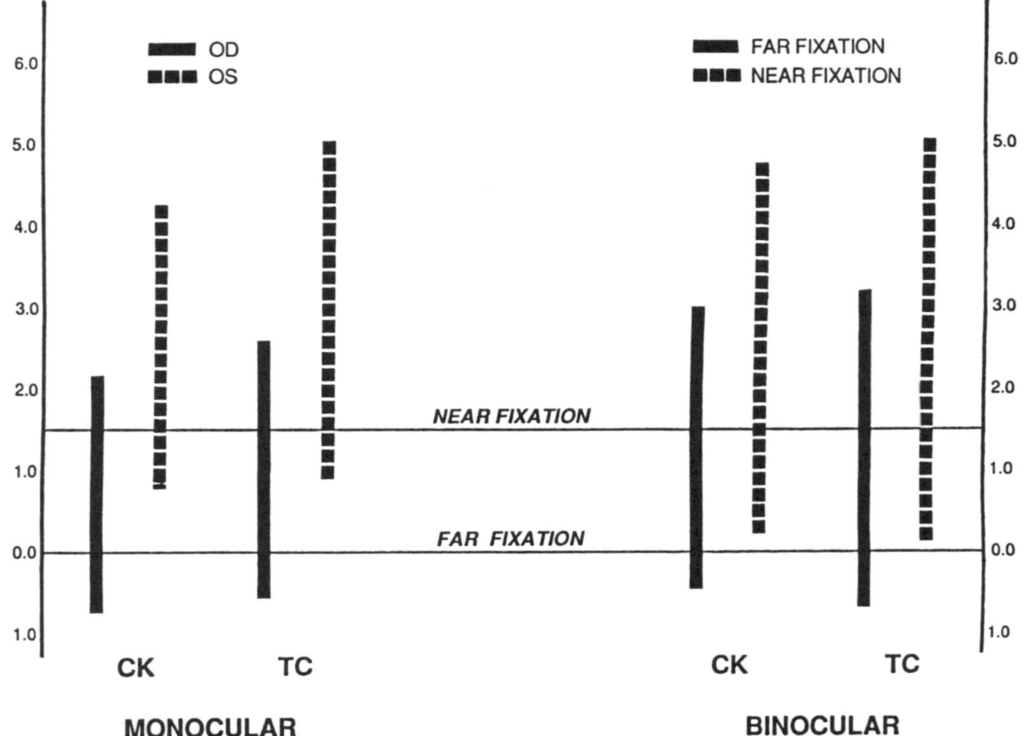

FIGURE 27-13. Monocular (left half) and binocular (right half) depths of focus for two subjects who showed a partial dominance pattern of blur suppression. The vertical scale is in diopters.

PARTIAL DOMINANCE

Figure 27-13 illustrates results for two subjects who showed partial binocular summation of the monocular depths of focus when tested at either the near or far distance. These two subjects, C.K. and T.C., despite their strong sighting preference, had depths of focus that showed partial summation. This suggests a limited ability to alternate interocular suppression. In general, these subjects showed some increase in the depth of focus in the binocular condition, but not as complete a summation as in the observed alternating dominance category.

Dynamic Accommodation

Figure 27-14 illustrates dynamic records of accommodative response to a 2.5-D accommodative stimulus presented as a square wave at 0.2 Hz to the right eye alone (A), the left eye alone (B), and both eyes (C) for subject P.E. This subject's greater amplitude of accommodation allows a clearer presentation of the response pattern characteristic of all three subjects. The right eye was corrected for optical infinity, and the left eye has a +1.50-D add over the far point correction. Accommodative responses were

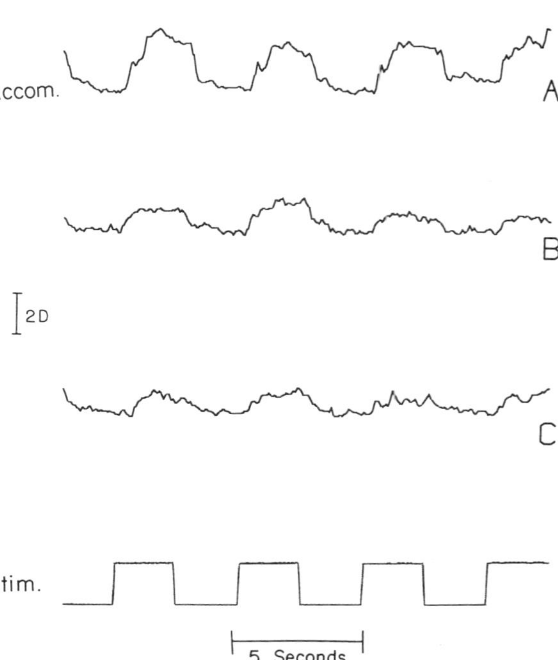

FIGURE 27-14. Typical prepresbyopic accommodative responses to a 2.5-D accommodative stimulus presented as a square wave to the right eye alone (A), left eye alone (B), and both eyes (C).

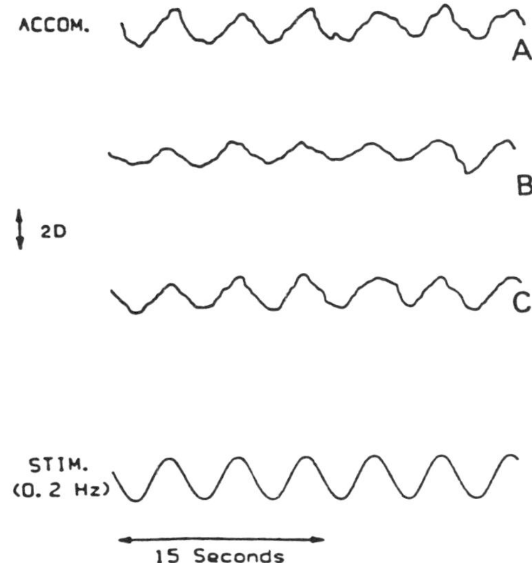

FIGURE 27-15. Typical prepresbyopic accommodative responses to a 2.5-D accommodative stimulus presented as a sine wave to the right eye alone (A), left eye alone (B), and both eyes (C).

observed when either eye was stimulated monocularly (Fig. 27-14A,B). However, little or no response occurred during binocular viewing conditions (Fig. 27-14C). All subjects reported no apparent blur or change in clarity during the binocular condition, which illustrates an interocular suppression of the out-of-focus image as the target stepped from the clear vision range of one eye to the other.

When continuous sinusoidal variations of defocus of 2.5 D were presented at the same temporal frequency (0.2 Hz), a different pattern of responses was observed. As illustrated in Fig. 27-15A,B, each eye tracked its monocular accommodative stimulus. During binocular viewing, accommodation continued in response to the continuously changing dioptric stimulus presented to one eye (usually the far eye) (Fig. 27-15C). Accordingly, as the target focus increased from zero diopters, the far eye stimulus evoked an accommodative response even when the dioptric stimulus came within range of the relaxed far point of the near eye wearing the 1.5-D lens. Instead of relaxing accommodation and allowing the near eye to receive a clear retinal image, accommodation remained under the control of the far eye throughout the 2.5-D stimulus range. Subject C.S., whose dominance shifted readily from one eye to the other in the first study and in sighting tests, was unable to shift dominance under the sinusoidal tracking conditions. The most presbyopic subject (J.G.), who required a full accommodative effort to maintain target clarity, also failed to relax accommodation as the target moved into the range where the near corrected eye could have provided a clear image with no accommodative effort.

DISCUSSION

Suppression of Anisometropic Blur

The success of the monovision correction is based on a process that suppresses blurred portions of the retinal image when there are interocular differences in image clarity. As demonstrated in Fig. 27-1, this suppression process is regional, since only blurred portions of the retinal image are suppressed by corresponding in-focus images perceived by the contralateral eye. The association of interocular suppression of blur with corresponding retinal points appears to account for the variation of suppression strength with target size shown in Fig. 27-3 through 27-8. As observed by Helmholtz (1962), the retinal image sizes of large targets that are in or out of focus are very similar, whereas the retinal image of a small target that is in focus is less than one tenth the size of its out-of-focus image. This size disparity for in- and out-of-focus targets appears to disrupt the interocular suppression of blur at high contrast levels such as encountered by motorists at night (McLendon *et al.*, 1968).

Reducing contrast by increasing background illumination reduces the perceived diameter of blur circles and allows the suppression of the blurred image. A small portion of this reduction of blur is caused by small changes in pupil size as background luminance is raised. For our two main subjects, pupil diameters were 4.0 mm and 6.0 mm when the background luminance was at its lowest level (0.5 cd/m^2), and they were 3.5 mm and 5 mm when the background luminance as at its highest level (620 cd/m^2). However, reductions in pupil size did not eliminate the blurred image seen monocularly by the ametropic eye under any of the background luminance levels. Clearly, the disappearance of the blurred image under binocular viewing conditions at high luminance levels was the result of a sensory inhibitory process, i.e., interocular suppression. It is also clear that a reduction of pupil size will facilitate the binocular suppression process by reducing the differential blur of the two ocular images. As seen in Figs. 27-3 and 27-4, interocular suppression is more effective with small than large amounts of anisometropic blur. Clearly, because pupillary constriction reduces blur, pupil size is an important factor in determining the prognosis for acceptance of a monovision correction.

The suppression process is not absolute. During suppression, some information from the blurred image continues to be utilized with the contralateral in-focus image so that retinal image disparities may be processed and stereoscopic depth be perceived (Larsen and Lachance, 1983; Lovasik and Szymkiw, 1985; Heath *et al.*, 1986). Previous studies of rivalry suppression also indicated that binocular rivalry can occur for conflicting portions of the stimulus while stereopsis and fusion are perceived simultaneously for corresponding components of the stimulus (Julesz and Miller, 1975). In addition, recent studies illustrate that stereoscopic depth can be perceived with binocular fusion stimuli that differ in spatial frequency by a factor of 2 (Schor *et al.*, 1984). Thus, even though monovision blur reduces the high-frequency content of one ocular image, the remaining medium- and low-frequency components can be fused with similar and higher-frequency detail that remains in the in-focus image of the contralateral eye. This common binocular spatial information is responsible for the minimal reduction of stereoacuity for monovision patients tested with high-contrast targets (McGill and Erickson, 1988).

The short-term adaptation experiment demonstrated a plasticity of the interocular suppression process. Prior to wearing the monovision correction, the strength of suppression was nearly equal for the two eyes. However, following adaptation to a +2.00-D monovision correction placed before the nonsighting eye for just 1 day, there was a strong aftereffect in which blur of the eye wearing the +2.00-D lens was more readily suppressed by the contralateral eye than it was in the preadapted state. Furthermore, in the postadapted state it was more difficult to suppress blur of the eye that was not adapted to the +2.00-D lens. The same aftereffect was produced in one subject (C.S.) by wearing the +2.00-D lens before the sighting eye (i.e., following adaptation it became easier to suppress blur of the nonsighting eye than of the sighting eye independently of which eye had adapted to the near prescription). This result is not surprising since during normal viewing conditions, both eyes have an opportunity to see clearly and suppress blur of the contralateral eye. The eye with the distance correction does this at remote viewing distances, and the eye with the near correction does this at near viewing conditions. Consequently, any imbalance of ocular dominance will manifest itself independently of which eye receives the near correction, since there is equal experience with near and far visual tasks.

DEPTH OF FOCUS

Our results demonstrate that the interocular suppression blur in monovision can be inhibited by the presence of strong sighting preference. On the other hand, the absence of a sighting preference appears to correlate with high flexibility in the interocular suppression of blur. The alternating dominance pattern exhibited by subjects T.K. and C.S. represents the ideal pattern for insuring constant suppression of blur throughout the monocular depths of focus. Such subjects would be expected to experience little difficulty maintaining clear binocular vision as gaze shifts from one distance to another. A continuum of increasing difficulty in achieving constant clear binocular vision would be expected for subjects with stronger sighting preference. Such subjects might experience intermittent episodes of blurred vision when the clear image is present only in the nonsighting eye. These results also suggest that, unlike binocularly balanced corrections, the successful monovision correction can provide clear vision within the far and near depths of focus simultaneously. That is, the binocular suppression of blur is regional for corresponding retinal areas.

ACCOMMODATIVE RESPONSE

An additional complication is present in young or nonpresbyopic monovision wearers because of the additional degree of freedom created by accommodation. One possible outcome of this viewing condition is unequal accommodation to compensate for the induced anisometropia. Several studies reviewed by Borish (1970) have failed to produce a consensus on the expected incidence or magnitude of this phenomenon. Based on the synchronous responses of the right eye regardless of whether the right (Fig. 27-14A, 27-15A) or left (Fig. 27-14B, 27-15B) eyes received the accommodative stimuli, we concluded that accommodation under these rapidly changing conditions was essentially consensual. For extended periods of viewing at one distance, however, unequal accommodation could occur in some patients.

Our results demonstrated a form of eye dominance for control of the consensual response of accommodation during anisometropic binocular viewing conditions. This dominance could defeat the intent of monovision correction under continuous tracking conditions (Fig. 27-15). Tonic aftereffects of accommodation would tend to bias accommodative control to the eye initially receiving the clear image. This aftereffect is larger in subjects with a small lag of accommodation (Schor *et al.,* 1986). However, it is uncommon for conditions that require accommodation to track smoothly over large stimulus ranges. More often, gaze is shifted abruptly from one distance to another (Fig. 27-14), much like the saccadic response to a square-wave stimulus. Under these conditions maintenance of clear vision is exchanged between the near and far eye. The eye controlling the steady state of static accommodation is the one whose far point is closest to the target distance.

Physiological Process Underlying the Interocular Suppression of Blur

Contrast differences between the two ocular images evoke a number of binocular inhibitory interactions such as binocular luster and Fechner's paradox. For example, a suprathreshold grating presented to one eye can be totally suppressed by another grating of slightly higher contrast presented to the contralateral eye if the two gratings are similar in orientation (within 10°) and spatial frequency (Abadi, 1967). Physiological correlates of binocular contrast suppression have been demonstrated in cat cortical neurons (Berardi *et al.,* 1986). Discharge rate of binocular cells in area 17 were recorded when a sinusoidal grating was presented to one eye while the other eye received a homogeneous field or a pattern of the same spatial frequency and orientation. Cell response to each stimulus was not substantially modified by the presence of the other stimulus unless the two eyes were presented with gratings of very different contrast. The response to the lower contrast was virtually suppressed. This mechanism of interocular suppression effectively reduces the contrast of the lower-contrast image seen by the two eyes, resulting in a further reduction of signal strength compared to background noise, with consequent elevation of stereothreshold. Suprathreshold disparities are large in comparison to the background noise such that binocular sensory fusion limits (Fig. 27-2) and suprathreshold stereoscopic depth matches (Ogle and Groch, 1956; Schor and Howarth, 1986) are unaffected by interocular differences in target contrast.

SUMMARY

Presbyopic contact lens patients with monocular corrections (monovision) see clearly at all distances by virtue of an interocular suppression of anisometropic blur that occurs regionally between corresponding retinal areas. This suppression fails to occur with small high-contrast targets viewed under low-luminance conditions. The effect of target size and contrast on interocular suppression blur was quantified by reducing contrast of a bright test spot, viewed binocularly while wearing various positive lenses monocularly, until the out-of-focus image was suppressed. The strength of interocular suppression was equivalent when the positive lens was before either eye. However, after subjects wore a positive lens over their nonsighting eye for 1 day, interocular suppression of blur became enhanced when the nonsighting eye was blurred, and it became reduced when the sighting eye was blurred. Successful monovision subjects suppressed blur at higher contrast levels than did unsuccessful subjects. This suppression mechanism increased the binocular depth of focus (DOF) in monovision wearers beyond the DOF measured monocularly. The expanded DOF disables accommodative response to step changes in focus, whereas accommodative responses to binocular sinusoidal variations in blur are controlled primarily by a dominant sighting eye. Single-unit studies indicate that the site of the interocular suppression of blur is in binocular cells of the primary visual cortex (Berardi *et al.,* 1986).

REFERENCES

Abadi, R., 1967, Induction masking—a study of some inhibitory interactions during dichoptic viewing, *Vision Res.* **16:**269–275.

Atchison, D. A., Smith, G., and Efron, N., 1979, The effect of pupil size on visual acuity in uncorrected and corrected myopia, *Am. J. Optom. Physiol. Opt.* **56:**315–323.

Baker, R., Brown, B., and Garner, L., 1983, Time course and variability of dark focus, *Invest. Ophthalmol. Vis. Sci.* **24:**1528–1531.

Berardi, N., Galk, L. L., Maffei, L., and Siliprandi, R., 1986, Binocular suppression in cortical neurones, *Exp. Brain Res.* **63:**581–584.

Borish, I. M., 1970, *Clinical Refraction,* 3rd ed., Professional Press, Chicago, pp. 173–174.

Brewster, D., 1856, *The Stereoscope, Its History, Theory and Constructing,* J. Murray, London.

Campbell, F., 1957, The depth of field of the human eye, *Optic Acta* **4:**157–164.

Campbell, F., and Gregory, M., 1960, Effect of size of pupil on visual acuity, *Nature* **187:**1121–1123.

Crawford, B. A., 1936, The dependence of pupil size upon external light stimulus under static and variable conditions, *Proc. R. Soc. Lond. [Biol.]* **121:**376–395.

Erickson, P., 1988, Potential range of clear vision in monovision, *J. Am. Opt. Assoc.* **59:**203–205.

Fahle, M., 1982, Binocular rivalry: Suppression depends on orientation and spatial frequency, *Vision Res.* **22:**787–800.

Fry, G. A., 1969, *Geometrical Optics,* Chilton, New York, p. 13.

Goodwin, R.T., and Romano, P. E., 1985, Stereoacuity degradation by experimental and real monocular and binocular amblyopia, *Invest. Ophthalmol. Vis. Sci.* **26:**917–923.

Heath, D. A., Hines, C., and Schivartz, F., 1986, Suppression behavior analyzed as a function of monovision addition power, *Am. J. Optom. Physiol. Opt.* **63:**198–201.

Helmholtz, H., 1962, *Treatise on Physiological Optics* (J. P. C. Southall, ed.), Dover, New York.

Julesz, B., and Miller, J., 1975, Independent spatial-frequency-tuned channels in binocular fusion and rivalry, *Perception* **4:**125–143.

Larsen, W. L., and Lachance, A., 1983, Stereoscopic acuity with induced refractive errors, *Am. J. Optom. Physiol. Opt.* **60:**509–513.

Levy, N. S., and Glick, E. B., 1974, Stereoscopic perception and snellen visual acuity, *Am. J. Ophthalmol.* **78:**722–724.

Lit, A., 1968, Presentation of experimental data, *J. Am. Optom. Assoc.* **39:**1098–1099.

Loshin, D. S., Loshin, M. S., and Comer, G., 1982, Binocular summation with monovision contact lens correction for presbyopia, *Int. Contact Lens Clin.* **9:**161–165.

Lovasik, J. V., and Szymkiw, M., 1985, Effects of aniseikonia, anisometropia, accommodation, retinal illuminance and pupil size on stereopsis, *Invest. Ophthalmol. Vis. Sci.* **26:**741–750.

McGill, E., and Erickson, P., 1988, Stereopsis in presbyopes wearing monovision and simultaneous vision bifocal contact lenses, *Am. J. Optom. Physiol. Opt.* **65:**619–626.

McLendon, J. H., Burcham, J. L., and Pheiffer, C. H., 1968, Presbyopic patterns and single vision contact lenses II, *South J. Optom.* **10:**7–12, 31, 36.

Ogle, K. N., and Groch, J., 1956, Stereopsis and unequal luminosities in the two eyes, *Am. Arch. Ophthalmol.* **54:**878–895.

Ogle, K. N., and Schwartz, J. T., 1959, Depth of focus of the human eye, *J. Opt. Soc. Am.* **49:**273–280.

Ong, J., and Burley, W. S., 1972, Effect of induced anisometropia an depth perception, *Am. J. Optom. Arch. Am. Acad. Optom.* **49:**333–335.

Peters, H. B., 1969, The influence of anisometropia an stereosensitivity, *Am. J. Optom. Arch. Am. Acad. Optom.* **46:**120–123.

Randle, R. J., 1970, Volitional control of visual accommodation, in: *Adaptation and Acclimatization in Aerospace Medicine* (H. J. Grunhofer, ed.), Garmisch-Partenkirchen.

Schor, C. M., and Erickson, P., 1988, Patterns of binocular suppression and accommodation in monovision, *Am. J. Optom. Physiol. Opt.* **65:**853–861.

Schor, C. M., and Heckmann, T., 1989, Interocular differences in contrast and spatial frequency: Effects on stereopsis and fusion, *Vision Res.* **29:**837–847.

Schor, C. M., and Howarth, P., 1986, Suprathreshold stereo-depth matches as a function of contrast and spatial frequency, *Perception* **15:**249–258.

Schor, C. M., Kotulak, J. C., and Tsuetaki, T., 1986, Adaption of tonic accommodation reduces accommodative lag and is masked in darkness, *Invest. Ophthalmol. Vision Sci.* **27:**820–827.

Schor, C. M., Landsman, L., and Erickson, P., 1987, Ocular dominance and the interocular suppression of blur in monovision, *Am. J. Optom. Physiol. Opt.* **64:**723–730.

Schor, C. M., Wood, I. C., and Ogawa, J., 1984, Spatial tuning of static and dynamic local stereopsis, *Vision Res.* **24:**573–578.

Tucker, J., and Charman, W. N., 1975, The depth of focus of the human eye for Snellen letters, *Am. J. Optom. Physiol. Opt.* **52:**3–21.

Woodhouse, J. M., 1975, The effect of pupil size on grating detection at various contrast levels, *Vision Res.* **15:**645–648.

III

Conclusions

LAWRENCE STARK

Part III has dealt with how we segregate or separate superimposed images. In the real world, we deal with overlapping complex images on our retina rather than working in the laboratory, where there are simple point targets against a black background. But how do the human eye and the human brain deal with this complexity of real extended images?

If one image is blurred, then directed attention to a particular plane enables the accommodation system to focus at this one image plane and so clarify that image plane.

As background, Professor Ciuffreda from the State University of New York College of Optometry discussed the depth of focus in blur in normals and in amblyopes. Professor Semmlow of Rutgers University described research using the stigmatoscopic method to ascertain and measure the plane conjugate to the retina and changes in accommodation with presbyopia. But what happens if the superimposed images are all in focus or both in focus, as in monovision? Professor Clifton Schor, from the School of Optometry of the University of California at Berkeley, presented some interesting material on summation of the depth of focus of the two monovision eyes in a number of subjects and of the rate of adaptation of this, although he was careful to point out that not all subjects could so summate their depth of focus or could adapt easily.

Supposing we go back to the original normal situation of having many superimposed images, and suppose for a moment they are all in clear focus, as might be possible, for example, with a multifocal progressive contact lens. We could then use certain important neural processes that deal with motion and stereo perception. Professor Michel Imbert from the University of Paris presented very exciting data from actual experiments, both neurophysiological experiments and behavioral experiments, on the development of depth vision. Professor Jeremy Wolfe, from the Massachusetts Institute of Technology, pointed out, in a very interesting summary of research that he and others have been carrying out over recent years, that the two apparently opposite processes of binocular fusion and binocular rivalry are both going on all the time and are interacting. Professor Christopher Tyler from the Smith–Kettlewell Eye Research Foundation in San Francisco discussed local, regional, and global processes that help to separate processes in motion and in depth. Professor Bridgeman, from the University of California in Santa Cruz, discussed the differences between sensory and motor maps of space and how these can be separated in experimental psychology studies.

In addition to these built-in neural processes, there are higher-level cognitive processes, and I contributed a chapter showing different eye movement scanpaths for different mental images of ambiguous figures, here there are no other processes except these higher cognitive processes that enable us to distinguish and form mental images of the ambiguous figures.

But finally, we may still have residual difficulties of great importance, and Dr. Haines from the NASA Ames Research Center in Moffet Field, California, showed some very interesting, carefully documented results using heads-up display, where the instrument panels of an airplane and (heaven forbid!) of an automobile, are presented on the windshield. He showed that pilots can be very confused because they are looking at but not seeing the distant image when they are looking at the instrument panel projected onto the windshield.

I feel that these contributors have started to understand the complexity of visual, neural, and cognitive processes that continually help our vision to deal with complex superimposed images.

LAWRENCE STARK • School of Optometry, University of California, Berkeley, California 94720.

Closing address by Mr. Bernard Maitenaz

I shall not reiterate by synthesizing the syntheses; indeed, these have shown us how much progress has been made. At the same time, we have seen that certain questions remain unanswered. The objective of the 4th Symposium has thus been attained: as with the preceding symposia, we aimed to determine the current state of knowledge and the main directions future research should adopt. The conclusions of the scientific congress have taught me that our understanding of the process which regulates sight and its evolution is beginning to improve, that much has yet to be done, but that work in this area abounds in exciting prospects, and is definitely advancing. The workshops have taught me that, apart from problems related to products and their development, there is a real concern for training, information and documentation - in short, for communication.

These findings are very important for us at ESSILOR. At the opening of the Congress on Monday, I briefly referred to our efforts in the research area, and quoted the figure of 30 million dollars. Let me call up another image: that of more than 300 or 350 people working on research and development problems in our company the whole year round. This population represents about a third of the congress assembled here today. We have to deal with problems of materials, processes, machines and procedures. To do so, we constitute teams whose disciplines range from mathematics to biochemistry, and include electronics, mechanics and of course optics. Three hundred people may be a lot, but they are not enough. Similarly, all disciplines do not suffice. That is why we need you, and that is why all the contacts made in the course of the Symposium are necessary and extremely useful. We need you at the initial phase of our work, for better knowledge of processes, of biological and physiological phenomena. We need you at the final phase, for product development and optimal adaptation to our customers. This explains our attributing numerous research contracts to outside groups, in France and elsewhere, to complement work carried out by our own teams. A number of these outside groups have reported on their activities in the course of this Symposium. More than twenty such studies are currently underway on various subjects.

While progress in terms of knowledge is the first priority of this Symposium, it is not the only one. A second aspect has obviously assumed considerable importance: this Symposium acts as a crossroads, a place for constructive meetings and fruitful exchanges. After these few days together, numerous participants have said it is now easier for them to understand the complementarity of the various disciplines grouped here, which act like successive and interpenetrating links to form a kind of chain whose object is improved vision. Schematically, five main categories of participants can be distinguished: scientists, ophthalmologists, opticians, teachers, and ESSILOR people. To a certain extent, one can add the media, whose role is information. I am sure that each of the five groups gathered here this week has learned something from the other four. That is how progress is made.

Particular thanks to organizers

As regards organizational aspects of the 4th Symposium, we deliberately opted for a dual system of organization: conferences on the one hand and workshops on the other. The workshops were organized by country, to allow the specificities of each to be taken into account. I believe this method was highly beneficial and most appreciated. On the other hand, organization along these lines obviously made it impossible to attend all events. I myself was the first victim of this arrangement, and regret not having been able to go to certain workshops or lectures. Similarly, I was not able to meet all the speakers as I had hoped to. Despite such difficulties, numerous contacts and exchanges took place in a very favorable atmosphere. The same constructive climate prevailed in meetings which rallied previously separate disciplines around subjects of common interest. This new approach appears most promising.

Many of you feel that organization of the Symposium has been excellent. I believe this was also due to each person's having been disciplined and, at the same time, indulgent regarding the small but inevitable imperfections which arise when caring for almost a thousand people. The congress has taken place in an environment we wished to be pleasant and far from the upheavals and problems of everyday life. I think we have succeeded in that respect. But if the congress and Symposium have been a success, it is thanks to your participation. This has been a studious event. Attendance at the lectures and workshops was remarkable—all the more so, given the temptations of the outside world. Success of the Symposium has also been the result of extremely thorough preparation, ensured notably by the Scientific Committee. Moreover, many of you prepared the Symposium by accepting to participate in preliminary surveys, studies and workshops, and to report on them here. The jury also played an important role by assuming the formidable task of selecting the best work presented. Personally and in the name of ESSILOR, I would like to thank all these persons.

Preparation of the Symposium was also begun more than a year ago by a team which, step by step, went about organizing the slightest details of our existence over the past week. In this respect, the experience and know-how we have steadily acquired since the 1st Symposium has been invaluable, for organizing both work and leisure activities. The gazelles from ESSILOR and the Agency which assisted them have done a remarkable job and deserve the warmest thanks. I am personally very proud of their work and contribution towards this event. I would like to thank our interpreters who have braved the most trying moments with warmth and a smile. I should also emphasize the role played by William Lenne, mastermind of the Symposium. I would like to announce that he has just been nominated to a key position at the Ministry of Research. Indeed, this is but due recognition of his merit and competence, and we are all the more happy for him in that he will remain very closely associated with ESSILOR's development. Personally, I would like to express my deepest thanks to William, for his work at ESSILOR and, in particular, in the context of this Symposium.

The 4th Symposium is now over. Each of us will leave with many issues to take up, buzzing with ideas and projects, and rich with memories. The question which naturally comes to mind is: will there be a 5th Symposium? If so, when and where? I do not yet have the answer to this question. There is obviously a substantial interest and a real demand. It is also true that our numbers should not increase any further, and should perhaps even be reduced. Here is a somewhat whimsical image to give you an idea of the difficulties involved: if tomorrow, on saying goodbye, we shook hands successively (rapidly - no longer than 10 seconds) rather than simultaneously, it would take us three hours. Let me follow up this theoretical calculation with another more practical example: I would like to have met all of you, to have welcomed you individually, to have had a word with each one of you. For me to have spent a mere 5 minutes with each participant would have required seven ten-hour days of solid conversation. I would have spent more time welcoming you than at the congress itself! This problem will indeed arise tomorrow when we part, so I would like to present my excuses to those I shall miss, as it will be impossible to say farewell to everyone.

We must now start thinking about solutions and possible ways of continuing our work. Of course, we shall inform you as soon as a direction has been defined, to allow for adequate preparation. But before then, we shall undoubtedly have the chance to meet up for other subjects or on other occasions, so may I conclude by simply saying "à bientôt". Thank you for your participation. I wish you all a pleasant journey and the best of luck for your future work. Thank you.

Closing speech by Mrs. N. Bellakhdar*

Gentlemen, thank you for allowing me the honor of addressing this closing session of the Fourth International Symposium on Presbyopia in our age-old town of Marrakesh, which has welcomed you with a particularly gentle climate for the season.

I would like to pay homage to Essilor for its continuous efforts to improve the comfort of presbyopes, who are becoming an increasingly demanding group. Indeed, such an evolution is natural, since this functional ametropia occurs at an age corresponding to the height of one's professional activity.

Over the past two days, I have personally been able to attend your work sessions and thus appreciate the excellent organization of the symposium and the remarkable quality of the various presentations. I would like to congratulate you on this.

I regret not having been able to welcome you here, but, in the name of the Moroccan Society of Ophthalmology, I wish all of you an excellent journey home. We hope that Morocco, or at least Marrakesh, will be a source of pleasant memories for you.

Thank you.

*Chairperson of the Moroccan Ophthalmology Society

Index